QUANTUM WELD DOTS

QUANTUM WELLS, WIRES AND DOTS
Theoretical and Computational Physics of Semiconductor Nanostructures

Third Edition

Paul Harrison
The University of Leeds, UK

A John Wiley and Sons, Ltd, Publication

Copyright ©2009 John Wiley & Sons Ltd, The Atrium, Southern Gate, Chichester,
West Sussex PO19 8SQ, England
Telephone (+44) 1243 779777

Email (for orders and customer service enquiries): cs-books@wiley.co.uk
Visit our Home Page on www.wileyeurope.com or www.wiley.com

Reprinted September 2010

Other Wiley Editorial Offices

John Wiley & Sons Inc., 111 River Street, Hoboken, NJ 07030, USA

Jossey-Bass, 989 Market Street, San Francisco, CA 94103-1741, USA

Wiley-VCH Verlag GmbH, Boschstr. 12, D-69469 Weinheim, Germany

John Wiley & Sons Australia Ltd, 42 McDougall Street, Milton, Queensland 4064, Australia

John Wiley & Sons (Asia) Pte Ltd, 2 Clementi Loop #02-01, Jin Xing Distripark, Singapore 129809

John Wiley & Sons Canada Ltd, 22 Worcester Road, Etobicoke, Ontario, Canada M9W 1L1

Wiley also publishes its books in a variety of electronic formats. Some content that appears in print may not be available in electronic books.

British Library Cataloguing in Publication Data

A catalogue record for this book is available from the British Library

ISBN-13   978-0-470-77098-6 (H/B)   978-0-470-77097-9 (P/B)

Typeset from LaTeX files provided by the author
Printed and bound in Great Britain by CPI Antony Rowe, Chippenham, Wiltshire.

To Claire, Hannah and Joe

CONTENTS

PREFACE

I have been working on theoretical and computational studies of the electronic, optical and magnetic properties of semiconductor heterostructures for well over a decade. During this time I have had to follow through various theoretical derivations from either books or research papers and frankly I've struggled time and time again. There never seems to be enough detail and what is commonly a couple of lines in a research paper can literally turn into many pages of precise mathematics.

With the advent of computers and their wide application to science, the emphasis of theoretical work has changed. Years ago, theoreticians aimed to produce short neat relationships, which summarised physical effects. The concentration dependency of the metal–insulator transition of Mott is an excellent example of a complex process summarised in a compact equation, which could then be evaluated by hand. The modern approach to the same problem would be to take a microscopic model of a semiconductor, calculate the bandstructure, and then vary the impurity concentration, thus deducing an empirical relationship from the numerics. The material system would then be changed and the process repeated. It would therefore appear that the old way was preferable, but in these times of materials modelling and device design, *quantitative results* are what we're paid for.

What this book is not

This book is unusual in that what it doesn't contain is as important as what it does contain, and furthermore, without the omissions this book would not exist. This book is not a book about condensed matter, nor is it a book about the physical properties of semiconductor heterostructures. It is not a book reporting the wealth of experimental measurements made on low–dimensional semiconductor systems, nor is it meant to be a general light reading book that you might cuddle up with in bed!

This book isn't even a review of all the methods that are available for calculating various properties. This book is merely an in-depth look at how quantities *can* be calculated. It is not meant to be the definitive guide; I'm sure there are better methods available than some of those presented here.

So all that remains is to say...

What this book is

This book is aimed at providing *all* of the essential information, both theoretical and computational, in order that the reader can, starting from essentially nothing, under-stand how the electronic, optical and transport properties of semiconductor heterostruc-tures are calculated. However, perhaps more importantly, starting from this low common denominator, this text is designed to lead the reader through a series of simple example theoretical and computational implementations, and slowly build from solid foundations, to a level where the reader can begin to initiate theoretical investigations or explanations of their own.

I believe that there are two aspects to theoretical work, with the first being to analyse and interpret experimental data, while the second is to advance new ideas. My hope is that this book will certainly facilitate the former and I believe that I will at least provide the knowledge and skills base from which quantified predictions can be developed from the beginnings of an idea.

I hope that this book will appeal to readers from outside the low dimensional semi-conductor community. Some of the examples developed will certainly be relevant to the semiconductor community at large, while the microscopic calculations presented could be of interest to other areas of condensed matter, such as carbon nanostructures, high–temperature superconductors, etc.

I have attempted here to write a book almost in the style of a mathematics course text. In such books they often describe briefly why differential equations or integration are important and then move on to show the standard techniques for solution, followed by examples and perhaps the application to real problems. Sometimes the books just state 'This is a binomial expansion and here's what you do'. In some ways this book follows both those routes. I expect that many readers will look at this present book having already a great deal of appreciation for their own particular problem. I would think that they have already quantified it in terms of, e.g. 'I must calculate the exciton binding energy'. Then they'll find that this book shows them exactly how to carry this out, and indeed provides the computer codes for them to achieve their aim quickly. I wouldn't expect a reader to pick this work up knowing nothing about solid state physics—it is not aimed at that particular person—and as I've stated already, there are many excellent books available which describe in detail the electronic, optical, transport, and other properties of semiconductors and semiconductor heterostructures. It is not my aim here to compete with these in any way; indeed I hope this present work will complement the earlier works.

Therefore this book was written to fill a need, namely collecting and documenting together derivations. It is a chance to set the mathematics in stone. By incorporating all of the steps in a derivation there is no possiblity of hiding away and 'glossing over' any point that is not fully understood. In many ways this means leaving oneself bare, for any mistakes or errors *will* be spotted, but this will give the opportunity for them to be corrected and hence allow the text to converge (perhaps through later editions) to a true record, which will be of use to future generations of students and researchers alike.

PAUL HARRISON

The University of Leeds

June 1999

The opportunity of correcting a dozen or so errors in the equations and one or two of the figures was taken for this, the first reprinting of this work. A list of these changes can be found on the book's web site.

PAUL HARRISON

The University of Leeds

April 2001

Perhaps the biggest change for this second edition has been brought about by the decision to drop the accompanying CD-ROM. This was done for a couple of reasons, the first one being that many of the people I met who had bought the book told me they only trusted their own codes and saw writing them as part of their own personal development. The second reason was that dropping the CD-ROM allowed the publisher to drop the price. With the introduction of a paperback version too, it is hoped that the book will be more accessible to students. To compensate for this the contents of the CD-ROM from the first edition will be published exactly as was on the book's website. Some of these codes have had updates already published so you'll need to consult with the relevant section, some of the codes will have been affected by the errata since discovered and some of the codes have basically been rewritten and improved, but you won't have access to them!

There is quite a bit of new material. In particular I've added sections on effects of magnetic fields on quantum wells, excited impurity levels, screening of the optical phonon interaction, acoustic and optical deformation potential scattering and spin-orbit coupling in the pseduopotential calculation. Perhaps more importantly, realising my own limitations I've commissioned contributions from some of my colleagues and this has resulted in new chapters on strained quantum wells and **k.p** theory. To keep the style of the book consistent I've edited these contributed works into my style.

PAUL HARRISON

The University of Leeds

March 2005

Again, in this third edition there is quite a bit of new material and again I've relied on the expertise of my colleagues. So, although I've expanded work on scattering rates into a new chapter on electron transport, it is my colleagues who have added the bulk of the new material with new chapters on quantum dots, optical waveguides and optical properties of quantum wells. So there are about 100 pages of new material. Again, I've done the final edit to make sure the book keeps its continuity of style and I hope it remains as readable as before. Even as I write these final words I realize there's still more that could be added, such as Monte Carlo simulations of electron transport, more advanced pseudopotential calculations of nanostructures, more on non-linear optics and perhaps even nanophotonics...but we'll leave that for another year!

PAUL HARRISON

The University of Leeds

June 2009

ACKNOWLEDGEMENTS

I am indebted to so many people with whom I've worked over the years that these acknowledgements have the potential for turning into a substantial work in themselves. Almost everybody with whom I've shared an office or a research project will look through this book and be able to say 'Ah yes, I helped him with that...'. I am truly grateful to everybody and it is such discussions that have in many ways motivated me to write this book.

In particular I would like to thank people such as John Davies, Winston Hagston and John Killingbeck who motivated me in my undergraduate days and spurred me on to my PhD work. At Newcastle I was lucky to be part of an excellent group with some truly great computational physicists. Perhaps the most important of these were Ian Morrison who in less than an hour really explained the bulk pseudopotential derivation to me and Jerry Hagon who put up with my endless computing questions. I must also include my tutors, Milan Jaros and David Herbert and, of course, my colleagues Andy Beavis and Richard Turton.

Without doubt the largest contribution to my knowledge base came during my formative years as a postdoctoral research assistant with Winston Hagston. Besides sparking my interest in quantum mechanics as an undergraduate and then luring me back to Hull, Winston showed the way in two of the major derivations that I document here. In particular, the first of the quantum confined impurity calculations I attempted, namely the spherically symmetric wave function, and together with Thomas Piorek, a substantial fraction of the electron–LO phonon scattering rate.

I am indebted to all of my research colleagues at Hull, particularly the theoreticians Tom Stirner, (the already mentioned) Thomas Piorek, Richard Roberts, Fei Long, and Jabar Fatah, all for humouring and implementing my ideas for heterostructure research.

Although I had already been working with electron–phonon scattering for a number of years and had implemented this on a computer, it was my colleague Paul Kinsler who showed me where the equations that I started with, actually came from. I believe this puts the phonon scattering rate work on a much surer footing and I am very grateful for his contribution.

I would also like to thank other colleagues at Leeds, in particular Bill Batty who on comparing the results of my implementation of the shooting technique for the solution of Schrödinger's equation, actually spotted an instance when a not insignificant error arose. This forced me to look again at the iterative equation for the case of a variable effective mass and to deduce the much more stable and accurate form presented in this work. Bill Batty also deserves a mention for proof reading of my manuscript along with Kate Donovan, Marco Califano and Byron Alderman.

For the second edition I am deeply indebted to my colleagues Zoran Ikonić and Vladimir Jovanović for contributing entirely new chapters. They are both helping to fill in gaps that this edition is certainly much better for. Some of my students have also pointed out some mistakes and opportunities for improvements in particular Jim Mc-Tavish, Nenad Vukmirović, Vladimir Jovanović, Ivana Savić and Craig Evans. Thanks guys the book is all the better for your hard work. Thanks also to all the people who contributed errata from the First Edition; these are all published on the book's website.

For this third edition I am very grateful to my colleagues Marco Califano, Craig Evans and Dragan Indjin, who have contributed entirely new chapters on their specialist areas of work. In turn, Dragan Indjin would like to thank Vitomir Milanović and Jelena Radovanović for their help. Again, I think the book is better for having contributions from specialists who can add more than I can on my own. They have all worked really hard to a tight deadline and I must thank them for putting up with me perpetually nagging them to finish their work. My interations with Alex Valavanis, Leon Lever and Will Freeman, in particular, have been very valuable. Thanks guys. Marie Barber and Zoran Ikonić deserve special mention for helping me in all aspects of my work and without them this new edition wouldn't exist.

Finally, I would like to express my gratitude to my employers, The University of Leeds and the School of Electronic and Electrical Engineering, for providing me in the first instance with a University Research Fellowship, which gave me an excellent platform on which to build a career in research. It was the flexibility of this scheme which allowed me the time to undertake the first edition. Since that time my employees have continued to support me and provide me with a position in which I can earn my living through research, teaching and administration, but still have enough time left over to work on the later editions.

P.H.

ABOUT THE AUTHOR(S)

Paul Harrison is currently working in the Institute of Microwaves and Photonics (IMP), which is a research institute within the School of Electronic and Electrical Engineering at the University of Leeds in the United Kingdom. He can always be found on the web, at the time of writing, at:

http://www.ee.leeds.ac.uk/homes/ph/

and always answers e-mail. Currently he can be reached at:

p.harrison@leeds.ac.uk or p.harrison@physics.org

Paul is working on a wide variety of projects, most of which centre around exploiting quantum mechanics for the creation of novel opto-electronic devices, largely, but not exclusively, in semiconductor *Quantum Wells, Wires and Dots*. Up-to-date information can be found on his web page. He is always looking for exceptionally well-qualified and motivated students to study for a PhD degree with him—if interested, please don't hesitate to contact him.

Zoran Ikonić was a Professor at the University of Belgrade and is now also a researcher in the IMP. His research interests and experience include the full width of semiconductor physics and optoelectronic devices, in particular, band structure calculations, strain-layered systems, carrier scattering theory, non-linear optics, as well as conventional and quantum mechanical methods for device optimization.

Vladimir Jovanović completed his PhD at the IMP on physical models of quantum well infrared photodetectors and quantum cascade lasers in GaN- and GaAs-based materials for near-, mid- and far-infrared (terahertz) applications.

Marco Califano is a Royal Society University Research Fellow based in the IMP at Leeds whose main interests focus on atomistic pseudopotential modelling of the electronic and optical properties of semiconductor nanostructures of different materials for applications in photovoltaics

Craig A. Evans completed his PhD on the optical and thermal properties of quantum cascade lasers in the School of Electronic and Electrical Engineering, University of Leeds in 2008. He then worked as a Postdoctoral Research Assistant in the IMP working in the field of rare-earth doped fibre lasers and integrated photonic device modelling and has now joined the staff of the school.

Dragan Indjin is an Academic Research Fellow in the IMP and has research interests in semiconductor nanostructures, non-linear optics, quantum computing and spintronics.

ABOUT THE BOOK

99.5% of this book was produced with 'open source' not-for-profit software. The text was prepared with LaTeX2ϵ using Wiley's own style (class) files. It was input by hand initially with the aid of the excellent 'VI Like Emacs' (vile) and then with the superb 'Vi IMproved' (vim/gvim). The schematic diagrams were prepared using 'xfig', the x-y plots with 'xmgr/xmgrace' and three-dimensional molecular models with 'RasMol'. 'BibView' was used to maintain a BIBTEX database in the earlier editions. Manuscript preparation has been under all three of the most popular operating systems by now!

Further information about the book, including errata and the software from the first edition is available on the book's web page, which is currently:

http://www.imp.leeds.ac.uk/qwwad/

INTRODUCTION

Since their discovery/invention by Esaki and Tsu in the 1970s, semiconductor quantum wells and superlattices have evolved from scientific curiosities to a means of probing the fundamentals of quantum mechanics, and more recently into wealth-creating semiconductor devices.

In this work, a brief resumé of quantum theory and solid state physics is given before launching into the main body of the book—the theoretical and computational framework of semiconductor heterostructures. The first chapter introduces the concepts of effective mass and envelope function approximations, which are two cornerstone theories, from a quite different perspective. Usually these two techniques are introduced from rigorous mathematics with some approximations—see, for example, the works by Bastard and Burt. The motivation behind this approach is to introduce these concepts from very simplistic intuitive, and at times, graphical arguments. The range of validity of such approximations is not being challenged; they are good theories and used many times in this book. This present approach is merely to reinforce these ideas in 'pictures rather than words'.

The aim of this book is to provide solid foundations in the theoretical methods necessary for calculating *some* of the basic electronic and optical properties of semiconductor quantum wells, wires and dots. Some background knowledge will be required; I often cite Ashcroft and Mermin [1] and Blakemore [2] for support material in solid state physics, and works such as Eisberg [3] and Weidner and Sells [4] for quantum theory. This present treatise should be considered to complement existing books in the field. I thoroughly recommend the books by Jaros [5], Davies and Long [6], Kelly [7], Turton [8], Ivchenko and Pikus [9], Shik [10], and Basu [11]. These texts will provide the literature reviews, and descriptive introductions, and also, in some cases, detailed theoretical treatments.

CHAPTER 1

SEMICONDUCTORS AND HETEROSTRUCTURES

1.1 THE MECHANICS OF WAVES

De Broglie (see reference [4]) stated that a particle of momentum p has an associated wave of wavelength λ given by the following

$$\lambda = \frac{h}{p} \tag{1.1}$$

Thus, an electron in a vacuum at a position $\mathbf{r}$ and away from the influence of any electromagnetic potentials, could be described by a *state function*, which is of the form of a wave, i.e.

$$\psi = e^{i(\mathbf{k} \bullet \mathbf{r} - \omega t)} \tag{1.2}$$

where t is the time, ω the angular frequency and the modulus of the wave vector is given by:

$$k = |\mathbf{k}| = \frac{2\pi}{\lambda} \tag{1.3}$$

The quantum mechanical momentum has been deduced to be a linear operator [12] acting upon the *wave function* ψ, with the momentum $\mathbf{p}$ arising as an eigenvalue, i.e.

$$-i\hbar\nabla\psi = \mathbf{p}\psi \tag{1.4}$$

1

Quantum Wells, Wires and Dots, Third Edition. P. Harrison
©2009 John Wiley & Sons, Ltd.

where

$$\nabla = \frac{\partial}{\partial x}\hat{\mathbf{i}} + \frac{\partial}{\partial y}\hat{\mathbf{j}} + \frac{\partial}{\partial z}\hat{\mathbf{k}} \tag{1.5}$$

which, when operating on the electron vacuum wave function in equation (1.2), would give the following:

$$-i\hbar\nabla e^{i(\mathbf{k}\cdot\mathbf{r}-\omega t)} = \mathbf{p}e^{i(\mathbf{k}\cdot\mathbf{r}-\omega t)} \tag{1.6}$$

and therefore

$$-i\hbar\left(\frac{\partial}{\partial x}\hat{\mathbf{i}} + \frac{\partial}{\partial y}\hat{\mathbf{j}} + \frac{\partial}{\partial z}\hat{\mathbf{k}}\right) e^{i(k_x x + k_y y + k_z z - \omega t)} = \mathbf{p}e^{i(\mathbf{k}\cdot\mathbf{r}-\omega t)} \tag{1.7}$$

$$\therefore -i\hbar\left(ik_x\hat{\mathbf{i}} + ik_y\hat{\mathbf{j}} + ik_z\hat{\mathbf{k}}\right) e^{i(k_x x + k_y y + k_z z - \omega t)} = \mathbf{p}e^{i(\mathbf{k}\cdot\mathbf{r}-\omega t)} \tag{1.8}$$

Thus the eigenvalue:

$$\mathbf{p} = \hbar\left(k_x\hat{\mathbf{i}} + k_y\hat{\mathbf{j}} + k_z\hat{\mathbf{k}}\right) = \hbar\mathbf{k} \tag{1.9}$$

which not surprisingly can be simply manipulated ($p = \hbar k = (h/2\pi)(2\pi/\lambda)$) to reproduce de Broglie's relationship in equation (1.1).

Following on from this, classical mechanics gives the kinetic energy of a particle of mass m as

$$T = \frac{1}{2}mv^2 = \frac{(mv)^2}{2m} = \frac{p^2}{2m} \tag{1.10}$$

Therefore it may be expected that the quantum mechanical analogy can also be represented by an eigenvalue equation with an operator:

$$\frac{1}{2m}\left(-i\hbar\nabla\right)^2\psi = T\psi \tag{1.11}$$

i.e.

$$-\frac{\hbar^2}{2m}\nabla^2\psi = T\psi \tag{1.12}$$

where T is the kinetic energy eigenvalue, and given the form of ∇ in equation (1.5) then:

$$\nabla^2 = \frac{\partial^2}{\partial x^2} + \frac{\partial^2}{\partial y^2} + \frac{\partial^2}{\partial z^2} \tag{1.13}$$

When acting upon the electron vacuum wave function, i.e.

$$-\frac{\hbar^2}{2m}\nabla^2 e^{i(\mathbf{k}\cdot\mathbf{r}-\omega t)} = T e^{i(\mathbf{k}\cdot\mathbf{r}-\omega t)} \tag{1.14}$$

then

$$-\frac{\hbar^2}{2m}\left(i^2 k_x^2 + i^2 k_y^2 + i^2 k_z^2\right) e^{i(\mathbf{k}\cdot\mathbf{r}-\omega t)} = T e^{i(\mathbf{k}\cdot\mathbf{r}-\omega t)} \tag{1.15}$$

Thus the kinetic energy eigenvalue is given by:

$$T = \frac{\hbar^2 k^2}{2m} \tag{1.16}$$

For an electron in a vacuum away from the influence of electromagnetic fields, then the total energy E is just the kinetic energy T. Thus the dispersion or energy versus

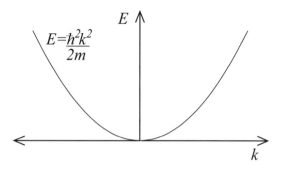

Figure 1.1   The energy versus wave vector (proportional to momentum) curve for an electron in a vacuum

momentum (which is proportional to the wave vector k) curves are parabolic, just as for classical free particles, as illustrated in Fig. 1.1.

The equation describing the total energy of a particle in this wave description is called the time-independent Schrödinger equation and, for this case with only a kinetic energy contribution, can be summarized as follows:

$$-\frac{\hbar^2}{2m}\nabla^2\psi = E\psi \tag{1.17}$$

A corresponding equation also exists that includes the time-dependency explicitly; this is obtained by operating on the wave function by the linear operator $i\hbar\partial/\partial t$, i.e.

$$i\hbar\frac{\partial}{\partial t}e^{i(\mathbf{k}\cdot\mathbf{r}-\omega t)} = i\hbar(-i\omega)e^{i(\mathbf{k}\cdot\mathbf{r}-\omega t)} \tag{1.18}$$

i.e.

$$i\hbar\frac{\partial}{\partial t}\psi = \hbar\omega\psi \tag{1.19}$$

Clearly, this eigenvalue $\hbar\omega$ is also the total energy but in a form usually associated with waves, e.g. a photon. These two operations on the wave function represent the two complimentary descriptions associated with *wave–particle duality*. Thus the second, i.e., time-dependent, Schrödinger equation is given by the following:

$$i\hbar\frac{\partial}{\partial t}\psi = E\psi \tag{1.20}$$

1.2   CRYSTAL STRUCTURE

The vast majority of the mainstream semiconductors have a face-centred cubic Bravais lattice, as illustrated in Fig 1.2. The lattice points are defined in terms of linear combinations of a set of *primitive lattice vectors,* one choice for which is:

$$\mathbf{a}_1 = \frac{A_0}{2}(\hat{\mathbf{j}}+\hat{\mathbf{k}}), \quad \mathbf{a}_2 = \frac{A_0}{2}(\hat{\mathbf{k}}+\hat{\mathbf{i}}), \quad \mathbf{a}_3 = \frac{A_0}{2}(\hat{\mathbf{i}}+\hat{\mathbf{j}}) \tag{1.21}$$

The *lattice vectors* then follow as the set of vectors:

$$\mathbf{R} = \alpha_1\mathbf{a}_1 + \alpha_2\mathbf{a}_2 + \alpha_3\mathbf{a}_3 \tag{1.22}$$

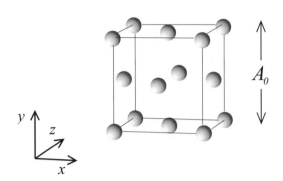

Figure 1.2   The face-centred cubic Bravais lattice

where α_1, α_2, and α_3 are integers.

The complete crystal structure is obtained by placing the atomic basis at each Bravais lattice point. For materials such as Si, Ge, GaAs, AlAs, InP, etc., this consists of two atoms, one at $(\frac{1}{8},\frac{1}{8},\frac{1}{8})$ and the second at $(-\frac{1}{8},-\frac{1}{8},-\frac{1}{8})$, in units of A_0.

Figure 1.3   The diamond (left) and zinc blende (right) crystal structures

For the group IV materials, such as Si and Ge, as the atoms within the basis are the same then the crystal structure is equivalent to diamond (see Fig. 1.3 (left)). For III–V and II–VI compound semiconductors such as GaAs, AlAs, InP, HgTe and CdTe, the cation sits on the $(-\frac{1}{8},-\frac{1}{8},-\frac{1}{8})$ site and the anion on $(+\frac{1}{8},+\frac{1}{8},+\frac{1}{8})$; this type of crystal is called the *zinc blende* structure, after ZnS, see Fig 1.3 (right). The only exception to this rule is GaN, and its important $In_xGa_{1-x}N$ alloys, which have risen to prominence in recent years due to their use in green and blue light emitting diodes and lasers (see for example [13]) these materials have the *wurtzite* structure (see reference [2] p. 47).

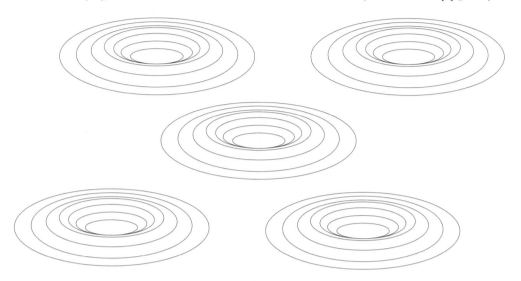

Figure 1.4 Schematic illustration of the ionic core component of the crystal potential across the {001} planes—a three-dimensional array of spherically symmetric potentials

From an electrostatics viewpoint, the crystal potential consists of a three-dimensional lattice of spherically symmetric ionic core potentials screened by the inner shell electrons (see Fig. 1.4), which are further surrounded by the covalent bond charge distributions that hold everything together.

1.3 THE EFFECTIVE MASS APPROXIMATION

Therefore, the crystal potential is complex; however using the principle of simplicity* imagine that it can be approximated by a constant! Then the Schrödinger equation derived for an electron in a vacuum would be applicable. Clearly though, a crystal isn't a vacuum so allow the introduction of an empirical fitting parameter called the effective mass, m^*. Thus the time-independent Schrödinger equation becomes:

$$-\frac{\hbar^2}{2m^*}\nabla^2\psi = E\psi \tag{1.23}$$

and the energy solutions follow as:

$$E = \frac{\hbar^2 k^2}{2m^*} \tag{1.24}$$

*Choose the simplest thing first; if it works use it, and if it doesn't, then try the next simplest!

This is known as the *effective mass approximation* and has been found to be very suitable for relatively low electron momenta as occur with low electric fields. Indeed, it is the most widely used parameterisation in semiconductor physics (see any good solid state physics book, e.g. [1, 2, 7]). Experimental measurements of the effective mass have revealed it to be anisotropic—as might be expected since the crystal potential along say the [001] axis is different than along the [111] axis. Adachi [14] collates reported values for GaAs and its alloys; the effective mass in other materials can be found in Landolt and Bornstein [15].

In GaAs, the reported effective mass is around 0.067 m_0, where m_0 is the rest mass of an electron. Fig. 1.5 plots the dispersion curve for this effective mass, in comparison with that of an electron in a vacuum.

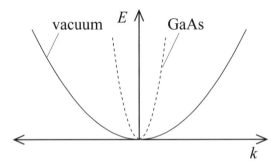

Figure 1.5   The energy versus wave vector (proportional to momentum) curves for an electron in GaAs compared to that in a vacuum

1.4   BAND THEORY

It has also been found from experiment that there are two distinct energy *bands* within semiconductors. The lower band is almost full of electrons and can conduct by the movement of the empty states. This band originates from the valence electron states which constitute the covalent bonds holding the atoms together in the crystal. In many ways, electric charge in a solid resembles a fluid, and the analogy for this band, labelled the *valence band* is that the empty states behave like bubbles within the fluid—hence their name *holes*.

In particular, the holes rise to the uppermost point of the valence band and just as it is possible to consider the release of carbon dioxide through the motion of beer in a glass, it is actually easier to study the motion of the bubble (the absence of beer), or in this case the motion of the hole.

In a semiconductor, the upper band is almost devoid of electrons. It represents excited electron states which are occupied by electrons promoted from localized covalent bonds into extended states in the body of the crystal. Such electrons are readily accelerated by an applied electric field and contribute to current flow. This band is therefore known as the *conduction band*.

Fig. 1.6 illustrates these two bands. Notice how the valence band is inverted—this is a reflection of the fact that the 'bubbles' rise to the top, i.e. their lowest energy states are at the top of the band. The energy difference between the two bands is known as the

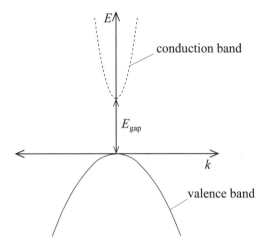

Figure 1.6 The energy versus wave vector curves for an electron in the conduction band and a hole in the valence band of GaAs

bandgap, labelled as E_{gap} on the figure. The particular curvatures used in both bands are indicative of those measured experimentally for GaAs, namely effective masses of around 0.067 m_0 for an electron in the conduction band, and 0.6 m_0 for a (heavy-)hole in the valence band. The convention is to put the zero of the energy at the top of the valence band. Note the extra qualifier 'heavy'. In fact, there is more than one valence band, and they are distinguished by their different effective masses. Chapter 15 will discuss band structure in more detail; this will be in the context of a microscopic model of the crystal potential which goes beyond the simple ideas introduced here.

1.5 HETEROJUNCTIONS

The effective mass approximation is for a bulk crystal, which means the crystal is so large with respect to the scale of an electron wave function that it is effectively infinite. Within the effective mass approximation, the Schrödinger equation has been found to be as follows:

$$-\frac{\hbar^2}{2m^*}\frac{\partial^2}{\partial z^2}\psi(z) = E\psi(z) \tag{1.25}$$

When two such materials are placed adjacent to each other to form a *heterojunction*, then this equation is valid within each, remembering of course that the effective mass could be a function of position. However the bandgaps of the materials can also be different (see Fig. 1.7).

The discontinuity in either the conduction or the valence band can be represented by a constant potential term. Thus the Schrödinger equation for any one of the bands, taking the effective mass to be the same in each material, would be generalized to:

$$-\frac{\hbar^2}{2m^*}\frac{\partial^2}{\partial z^2}\psi(z) + V(z)\psi(z) = E\psi(z) \tag{1.26}$$

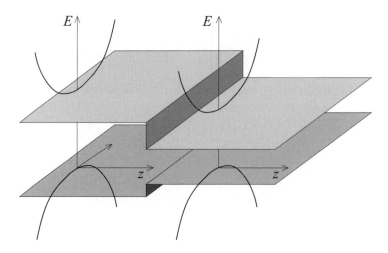

Figure 1.7 Two dissimilar semiconductors with different bandgaps joined to form a heterojunction; the curves represent the unrestricted motion parallel to the interface

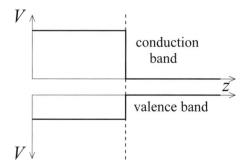

Figure 1.8 The one-dimensional potentials $V(z)$ in the conduction and valence band as might occur at a heterojunction (marked with a dashed line) between two dissimilar materials

In the above example, the one-dimensional potentials $V(z)$ representing the band discontinuities at the heterojunction would have the form shown in Fig. 1.8, noting that increasing hole energy in the valence band is measured downwards.

1.6 HETEROSTRUCTURES

Heterostructures are formed from multiple heterojunctions, and thus a myriad of possibilities exist. If a thin layer of a narrower-bandgap material 'A' say, is sandwiched between two layers of a wider-bandgap material 'B', as illustrated in Fig. 1.9 (left) then they form a double heterojunction. If layer 'A' is sufficiently thin for *quantum properties* to be exhibited, then such a band alignment is called a *single quantum well*.

If any charge carriers exist in the system, whether thermally produced intrinsic or extrinsic as the result of doping, they will attempt to lower their energies. Hence in this example, any electrons (solid circles) or holes (open circles) will collect in the quantum well (see Fig. 1.9). Additional semiconductor layers can be included in the heterostruc-

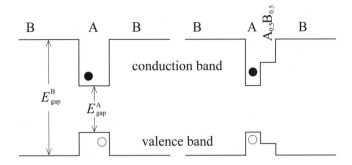

Figure 1.9   The one-dimensional potentials $V(z)$ in the conduction and valence bands for a typical single quantum well (left) and a stepped quantum well (right)

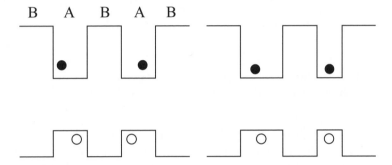

Figure 1.10   The one-dimensional potentials $V(z)$ in the conduction and valence band for typical *symmetric* (left) and *asymmetric* (right) double quantum wells

ture, for example a *stepped* or *asymmetric* quantum well can be formed by the inclusion of an alloy between materials A and B, as shown in Fig. 1.9 (right).

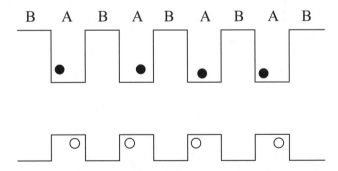

Figure 1.11   The one-dimensional potentials $V(z)$ in the conduction and valence band for a typical multiple quantum well or superlattice

Still more complex structures can be formed, such as symmetric or asymmetric *double quantum wells,* (see Fig. 1.10) and *multiple quantum wells* or *superlattices* (see Fig. 1.11). The difference between the latter is the extent of the interaction between the quantum wells; in particular, a multiple quantum well exhibits the properties of a collection of isolated single quantum wells, whereas in a superlattice the quantum wells do interact. The motivation behind introducing increasingly complicated structures is an attempt to tailor the electronic and optical properties of these materials for exploitation in devices. Perhaps the most complicated layer structure to date is the *chirped superlattice* active region of a mid-infrared laser [16].

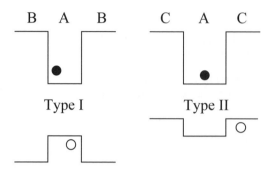

Figure 1.12   The one-dimensional potentials $V(z)$ in the conduction and valence bands for a typical Type-I single quantum well (left) compared to a Type-II system (right)

All of the structures illustrated so far have been examples of Type-I systems. In this type, the bandgap of one material is nestled entirely within that of the wider-bandgap material. The consequence of this is that any electrons or holes fall into quantum wells which are *within* the same layer of material. Thus both types of charge carrier are localized in the same region of space, which makes for efficient (fast) recombination. However other possibilities can exist, as illustrated in Fig. 1.12.

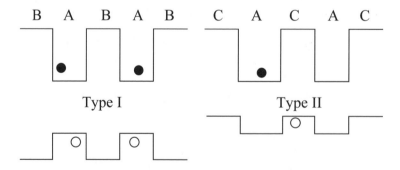

Figure 1.13   The one-dimensional potentials $V(z)$ in the conduction and valence bands for a typical Type-I superlattice (left) compared to a Type-II system (right)

In Type-II systems the bandgaps of the materials, say 'A' and 'C', are aligned such that the quantum wells formed in the conduction and valence bands are in different materials, as illustrated in Fig. 1.12 (right). This leads to the electrons and holes being confined in *different* layers of the semiconductor. The consequence of this is that the recombination times of electrons and holes are long.

1.7 THE ENVELOPE FUNCTION APPROXIMATION

Two important points have been argued:

1. The effective mass approximation *is* a valid description of bulk materials.

2. Heterojunctions between dissimiliar materials, both of which can be well represented by the effective mass approximation, can be described by a material potential which derives from the difference in the bandgaps.

The logical extension to point 2 is that the crystal potential of multiple heterojunctions can also be described in this manner, as illustrated extensively in the previous section.

Once this is accepted, then the electronic structure *can* be represented by the simple one-dimensional Schrödinger equation that has been aspired to:

$$-\frac{\hbar^2}{2m^*}\frac{\partial^2}{\partial z^2}\psi(z) + V(z)\psi(z) = E\psi(z) \tag{1.27}$$

The *envelope function approximation* is the name given to the mathematical justification for this series of arguments (see for example works by Bastard [17,18] and Burt [19,20]). The name derives from the deduction that physical properties can be derived from the slowly varying envelope function, identified here as $\psi(z)$, rather than the total wave function $\psi(z)u(z)$ where the latter is rapidly varying on the scale of the crystal lattice. The validity of the envelope function approximation is still an active area of research [20]. With the line of reasoning used here, it is clear that the envelope function approximation *can* be thought of as an approximation on the material and *not* the quantum mechanics.

Some thought is enough to appreciate that the envelope function approximation will have limitations, and that these will occur for very thin layers of material. The materials *are* made of a collection of a large number of atomic potentials, so when a layer becomes thin, these individual potentials will become significant and the global average of representing the crystal potential by a constant will breakdown, for example [21]. However, for the majority of examples this approach works well; this will be demonstrated in later chapters, and, in particular, a detailed comparison with an alternative approach which does account fully for the microscopic crystal potential will be made in Chapter 16.

1.8 THE RECIPROCAL LATTICE

For later discussions the concept of the *reciprocal lattice* needs to be developed. It has already been shown that considering electron wave functions as plane waves ($e^{i\mathbf{k}\cdot\mathbf{r}}$), as found in a vacuum, but with a correction factor called the effective mass, is a useful method of approximating the electronic bandstructure. In general, such a wave will not have the periodicity of the crystal lattice; however, for certain wave vectors it will. Such a set of wave vectors $\mathbf{G}$ are known as the *reciprocal lattice vectors* with the set of points mapped out by these primitives known as the *reciprocal lattice*.

If the set of vectors $\mathbf{G}$ *did* have the periodicity of the lattice, then this would imply that:

$$e^{i\mathbf{G}\cdot(\mathbf{r}+\mathbf{R})} = e^{i\mathbf{G}\cdot\mathbf{r}} \tag{1.28}$$

i.e. an electron with this wave vector $\mathbf{G}$ would have a wave function equal at all points in real space separated by a Bravais lattice vector $\mathbf{R}$. Therefore:

$$e^{i\mathbf{G}\bullet\mathbf{r}}e^{i\mathbf{G}\bullet\mathbf{R}} = e^{i\mathbf{G}\bullet\mathbf{r}} \tag{1.29}$$

which implies that

$$\mathbf{G}\bullet\mathbf{R} = 2\pi n, \qquad n \in \mathbb{Z} \tag{1.30}$$

Now learning from the form for the Bravais lattice vectors $\mathbf{R}$ given earlier in equation (1.22), it might be expected that the reciprocal lattice vectors $\mathbf{G}$ could be constructed in a similar manner from a set of three *primitive reciprocal lattice vectors*, i.e.

$$\mathbf{G} = \beta_1\mathbf{b}_1 + \beta_2\mathbf{b}_2 + \beta_3\mathbf{b}_3, \qquad \beta_1, \beta_2, \beta_3 \in \mathbb{Z} \tag{1.31}$$

With these choices then, the primitive reciprocal lattice vectors can be written as follows:

$$\mathbf{b}_1 = 2\pi\frac{\mathbf{a}_2 \times \mathbf{a}_3}{\mathbf{a}_1\bullet(\mathbf{a}_2 \times \mathbf{a}_3)} \tag{1.32}$$

$$\mathbf{b}_2 = 2\pi\frac{\mathbf{a}_3 \times \mathbf{a}_1}{\mathbf{a}_1\bullet(\mathbf{a}_2 \times \mathbf{a}_3)} \tag{1.33}$$

$$\mathbf{b}_3 = 2\pi\frac{\mathbf{a}_1 \times \mathbf{a}_2}{\mathbf{a}_1\bullet(\mathbf{a}_2 \times \mathbf{a}_3)} \tag{1.34}$$

It is possible to verify that these forms do satisfy equation (1.30):

$$\mathbf{G}\bullet\mathbf{R} = (\beta_1\mathbf{b}_1 + \beta_2\mathbf{b}_2 + \beta_3\mathbf{b}_3)\bullet(\alpha_1\mathbf{a}_1 + \alpha_2\mathbf{a}_2 + \alpha_3\mathbf{a}_3) \tag{1.35}$$

Now $\mathbf{b}_1$ is perpendicular to both $\mathbf{a}_2$ and $\mathbf{a}_3$, and so only the product of $\mathbf{b}_1$ with $\mathbf{a}_1$ is non-zero, and similarly for $\mathbf{b}_2$ and $\mathbf{b}_3$; hence:

$$\mathbf{G}\bullet\mathbf{R} = \beta_1\alpha_1\mathbf{b}_1\bullet\mathbf{a}_1 + \beta_2\alpha_2\mathbf{b}_2\bullet\mathbf{a}_2 + \beta_3\alpha_3\mathbf{b}_3\bullet\mathbf{a}_3 \tag{1.36}$$

and in fact, the products $\mathbf{b}_i\bullet\mathbf{a}_i = 2\pi$; therefore:

$$\mathbf{G}\bullet\mathbf{R} = 2\pi\left(\beta_1\alpha_1 + \beta_2\alpha_2 + \beta_3\alpha_3\right) \tag{1.37}$$

Clearly $\beta_1\alpha_1 + \beta_2\alpha_2 + \beta_3\alpha_3$ is an integer, and hence equation (1.30) is satisfied.

Using the face-centred cubic lattice vectors defined in equation (1.21), then:

$$\mathbf{a}_1\bullet(\mathbf{a}_2 \times \mathbf{a}_3) = \begin{vmatrix} a_{1x} & a_{1y} & a_{1z} \\ a_{2x} & a_{2y} & a_{2z} \\ a_{3x} & a_{3y} & a_{3z} \end{vmatrix} = \begin{vmatrix} 0 & \frac{A_0}{2} & \frac{A_0}{2} \\ \frac{A_0}{2} & 0 & \frac{A_0}{2} \\ \frac{A_0}{2} & \frac{A_0}{2} & 0 \end{vmatrix} \tag{1.38}$$

which gives:

$$\mathbf{a}_1\bullet(\mathbf{a}_2 \times \mathbf{a}_3) = 0 \times \begin{vmatrix} 0 & \frac{A_0}{2} \\ \frac{A_0}{2} & 0 \end{vmatrix} - \frac{A_0}{2}\begin{vmatrix} \frac{A_0}{2} & \frac{A_0}{2} \\ \frac{A_0}{2} & 0 \end{vmatrix} + \frac{A_0}{2}\begin{vmatrix} \frac{A_0}{2} & 0 \\ \frac{A_0}{2} & \frac{A_0}{2} \end{vmatrix} \tag{1.39}$$

$$\therefore \mathbf{a}_1\bullet(\mathbf{a}_2 \times \mathbf{a}_3) = 0 - \frac{A_0}{2}\left[0 - \left(\frac{A_0}{2}\right)^2\right] + \frac{A_0}{2}\left[\left(\frac{A_0}{2}\right)^2 - 0\right] \tag{1.40}$$

$$\therefore \mathbf{a}_1 \bullet (\mathbf{a}_2 \times \mathbf{a}_3) = 2 \left(\frac{A_0}{2} \right)^3 \tag{1.41}$$

Therefore, the first of the primitive reciprocal lattice vectors follows as:

$$\mathbf{b}_1 = 2\pi \left[2 \left(\frac{A_0}{2} \right)^3 \right]^{-1} \begin{vmatrix} \hat{\mathbf{i}} & \hat{\mathbf{j}} & \hat{\mathbf{k}} \\ \frac{A_0}{2} & 0 & \frac{A_0}{2} \\ \frac{A_0}{2} & \frac{A_0}{2} & 0 \end{vmatrix} \tag{1.42}$$

$$\therefore \mathbf{b}_1 = 2\pi \frac{1}{2} \left(\frac{2}{A_0} \right)^3$$

$$\times \left\{ \left[0 - \left(\frac{A_0}{2} \right)^2 \right] \hat{\mathbf{i}} - \left[0 - \left(\frac{A_0}{2} \right)^2 \right] \hat{\mathbf{j}} + \left[\left(\frac{A_0}{2} \right)^2 - 0 \right] \hat{\mathbf{k}} \right\} \tag{1.43}$$

$$\therefore \mathbf{b}_1 = \frac{2\pi}{A_0} \left(-\hat{\mathbf{i}} + \hat{\mathbf{j}} + \hat{\mathbf{k}} \right) \tag{1.44}$$

A similar calculation of the remaining primitive reciprocal lattice vectors $\mathbf{b}_2$ and $\mathbf{b}_3$ gives the complete set as follows:

$$\mathbf{b}_1 = \frac{2\pi}{A_0}(-\hat{\mathbf{i}} + \hat{\mathbf{j}} + \hat{\mathbf{k}}), \quad \mathbf{b}_2 = \frac{2\pi}{A_0}(\hat{\mathbf{i}} - \hat{\mathbf{j}} + \hat{\mathbf{k}}), \quad \mathbf{b}_3 = \frac{2\pi}{A_0}(\hat{\mathbf{i}} + \hat{\mathbf{j}} - \hat{\mathbf{k}}) \tag{1.45}$$

which are, of course, equivalent to the body-centred cubic Bravais lattice vectors (see reference [1], p. 68). Thus the reciprocal lattice constructed from the linear combinations:

$$\mathbf{G} = \beta_1 \mathbf{b}_1 + \beta_2 \mathbf{b}_2 + \beta_3 \mathbf{b}_3, \qquad \beta_1, \beta_2, \beta_3 \in \mathbb{Z} \tag{1.46}$$

is a body-centred cubic lattice with lattice constant $4\pi/A_0$. Taking the face-centred cubic primitve reciprocal lattice vectors in equation (1.45), then:

$$\mathbf{G} = \frac{2\pi}{A_0} \left[\beta_1(-\hat{\mathbf{i}} + \hat{\mathbf{j}} + \hat{\mathbf{k}}) + \beta_2(\hat{\mathbf{i}} - \hat{\mathbf{j}} + \hat{\mathbf{k}}) + \beta_3(\hat{\mathbf{i}} + \hat{\mathbf{j}} - \hat{\mathbf{k}}) \right] \tag{1.47}$$

$$\therefore \mathbf{G} = \frac{2\pi}{A_0} \left[(-\beta_1 + \beta_2 + \beta_3)\hat{\mathbf{i}} + (\beta_1 - \beta_2 + \beta_3)\hat{\mathbf{j}} + (\beta_1 + \beta_2 - \beta_3)\hat{\mathbf{k}} \right] \tag{1.48}$$

The specific reciprocal lattice vectors are therefore generated by taking different combinations of the integers β_1, β_2, and β_3. This is illustrated in Table 1.1.

It was shown by von Laue that, when waves in a periodic structure satisfied the following:

$$\mathbf{k} \bullet \hat{\mathbf{G}} = \frac{1}{2}|\mathbf{G}| \tag{1.49}$$

then diffraction would occur (see reference [1], p. 99). Thus the 'free' electron dispersion curves of earlier (Fig. 1.5), will be perturbed when the electron wave vector satisfies equation (1.49). Along the [001] direction, the smallest reciprocal lattice vector $\mathbf{G}$ is (0,0,2) (in units of $2\pi/A_0$). Substituting into equation (1.49) gives:

$$\mathbf{k} \bullet (0\hat{\mathbf{i}} + 0\hat{\mathbf{j}} + \hat{\mathbf{k}}) = \frac{1}{2} \times 2 \times \frac{2\pi}{A_0} \tag{1.50}$$

Table 1.1   Generation of the reciprocal lattice vectors for the face-centred cubic crystal by the systematic selection of the integer coefficients β_1, β_2, and β_3

β_1	β_2	β_3	$\mathbf{G}$ $(2\pi/A_0)$
0	0	0	$(0,0,0)$
1	0	0	$(\bar{1},1,1)$
0	1	0	$(1,\bar{1},1)$
0	0	1	$(1,1,\bar{1})$
-1	0	0	$(1,\bar{1},\bar{1})$
0	-1	0	$(\bar{1},1,\bar{1})$
0	0	-1	$(\bar{1},\bar{1},1)$
1	1	1	$(1,1,1)$
-1	-1	-1	$(\bar{1},\bar{1},\bar{1})$
1	1	0	$(0,0,2)$
1	0	1	$(0,2,0)$
0	1	1	$(2,0,0)$
-1	1	0	$(2,\bar{2},2)$
(etc.)			

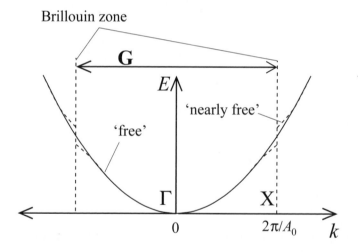

Figure 1.14   Comparison of the free and nearly free electron models

This then implies the electron will be diffracted when:

$$\mathbf{k} = \frac{2\pi}{A_0}\hat{\mathbf{k}} \tag{1.51}$$

Fig. 1.14 illustrates the effect that such diffraction would have on the 'free-electron' curves. At wave vectors which satisfy von Laue's condition, the energy bands are disturbed and an energy gap opens. Such an improvement on the parabolic dispersion curves of earlier, is known as the *nearly free electron model*.

The space between the lowest wave vector solutions to von Laue's condition is called the first *Brillouin zone*. Note that the reciprocal lattice vectors in any particular direction span the Brillouin zone. As mentioned above a face-centred cubic lattice has a body-centred cubic reciprocal lattice, and thus the Brillouin zone is therefore a three-dimensional solid, which happens to be a 'truncated octahedron' (see, for example reference [1], p. 89). High-symmetry points around the Brillouin zone are often labelled for ease of reference, with the most important of these, for this work, being the $k = 0$ point, referred to as 'Γ', and the $< 001 >$ zone edges, which are called the 'X' points.

CHAPTER 2

SOLUTIONS TO SCHRÖDINGER'S EQUATION

2.1 THE INFINITE WELL

The infinitely deep one-dimensional potential well is the simplest confinement potential to treat in quantum mechanics. Virtually every introductory level text on quantum mechanics considers this system, but nonetheless it *is* worth visiting again as some of the standard assumptions often glossed over, do have important consequences for one-dimensional confinement potentials in general.

The time-independent Schrödinger equation summarizes the wave mechanics analogy to Hamilton's formulation of classical mechanics [22], for *time-independent potentials*. In essence this states that the kinetic and potential energy components sum to the total energy; in wave mechanics, these quantities are the eigenvalues of linear operators, i.e.

$$\mathcal{T}\psi + \mathcal{V}\psi = E\psi \tag{2.1}$$

where the eigenfunction ψ describes the state of the system. Again in analogy with classical mechanics the kinetic energy operator for a particle of *constant* mass is given by the following:

$$\mathcal{T} = \frac{\mathcal{P}^2}{2m} \tag{2.2}$$

17

Quantum Wells, Wires and Dots, Third Edition. P. Harrison
©2009 John Wiley & Sons, Ltd.

where $\mathcal{P}$ is the usual quantum mechanical linear momentum operator:

$$\mathcal{P} = -i\hbar\nabla = -i\hbar\left(\frac{\partial}{\partial x}\mathbf{i} + \frac{\partial}{\partial y}\mathbf{j} + \frac{\partial}{\partial z}\mathbf{k}\right) \tag{2.3}$$

By using this form for the kinetic energy operator $\mathcal{T}$, the Schrödinger equation then becomes:

$$-\frac{\hbar^2}{2m}\left(\frac{\partial^2}{\partial x^2} + \frac{\partial^2}{\partial y^2} + \frac{\partial^2}{\partial z^2}\right)\psi + V(x,y,z)\psi = E\psi \tag{2.4}$$

where the function $V(x,y,z)$ represents the potential energy of the system as a function of the spatial coordinates. Restricting this to the one-dimensional potential of interest here, then the Schrödinger equation for a particle of mass m in a potential well aligned along the z-axis (as in Fig. 2.1) would be:

$$-\frac{\hbar^2}{2m}\frac{\partial^2}{\partial z^2}\psi(z) + V(z)\psi(z) = E\psi(z) \tag{2.5}$$

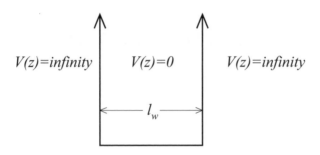

Figure 2.1   The one-dimensional infinite well confining potential

Outside of the well, $V(z) = \infty$, and hence the only possible solution is $\psi(z) = 0$, which in turn implies that all values of the energy E are allowed. Within the potential well, the Schrödinger equation simplifies to:

$$-\frac{\hbar^2}{2m}\frac{\partial^2}{\partial z^2}\psi(z) = E\psi(z) \tag{2.6}$$

which implies that the solution for ψ is a linear combination of the functions $f(z)$ which when differentiated twice give $-f(z)$. Hence try the solution:

$$\psi(z) = A\sin kz + B\cos kz \tag{2.7}$$

Substituting into equation (2.6) then gives:

$$\frac{\hbar^2 k^2}{2m}\left(A\sin kz + B\cos kz\right) = E\left(A\sin kz + B\cos kz\right) \tag{2.8}$$

$$\therefore \frac{\hbar^2 k^2}{2m} = E \tag{2.9}$$

Consideration of the boundary conditions will yield the, as yet unknown, constant k. With this aim, consider again the kinetic energy term for this system, i.e.

$$\mathcal{T} = -\frac{\hbar^2}{2m}\frac{\partial^2}{\partial z^2}\psi(z) \tag{2.10}$$

which can be rewritten as

$$T = -\frac{\hbar^2}{2m}\frac{\partial}{\partial z}\left(\frac{\partial}{\partial z}\psi(z)\right) \tag{2.11}$$

The mathematical form of this implies that, *as a minimum,* $\psi(z)$ must be continuous. If it is not, then the first derivative will contain poles that must be avoided if the system is to have finite values for the kinetic energy. Given that $\psi(z)$ has already been deduced as zero outside of the well, then $\psi(z)$ within the well must be zero at both edges too.

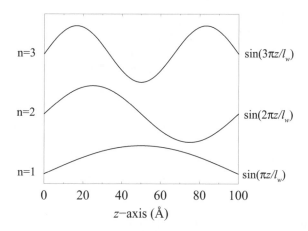

Figure 2.2 Solutions to the one-dimensional infinite well confining potential

If the origin is taken as the left-hand edge of the well as in Fig. 2.2, then $\psi(z)$ as defined in equation (2.8) can contain no cosine terms, i.e. $B = 0$, and hence $\psi(z) = A\sin kz$. In addition, for $\psi(0) = \psi(l_w) = 0$:

$$k = \frac{\pi n}{l_w} \tag{2.12}$$

where n is an integer, representing a series of solutions. Substituting into equation (2.9), then the energy of the confined states is given by:

$$E_n = \frac{\hbar^2\pi^2n^2}{2ml_w^2} \tag{2.13}$$

The only remaining unknown is the constant factor A, which is deduced by considering the normalization of the wave function; as $\psi^*(z)\psi(z)$ represents the probability of finding the particle at a point z, then as the particle must exist somewhere:

$$\int_0^{l_w}\psi^*(z)\psi(z)\ \mathrm{d}z = 1 \tag{2.14}$$

which gives $A = \sqrt{(2/l_w)}$, and therefore

$$\psi_n(z) = \sqrt{\frac{2}{l_w}}\sin\left(\frac{\pi n z}{l_w}\right) \tag{2.15}$$

Under the effective mass and envelope function approximations, the energy of an electron or hole in a hypothetical infinitely deep semiconductor quantum well can be calculated by using the effective mass m^* for the particle mass m of equation (2.13).

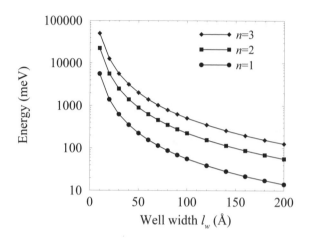

Figure 2.3   First three energy levels versus well width for an electron in a GaAs infinite potential well

Fig. 2.3 displays the results of calculations of the lowest three energy states of an electron in a GaAs well of width l_w surrounded by hypothetical infinite barriers (for these and all material parameters see Appendix A). All three states show the same monotonic behaviour, with the energy decreasing as the well width increases.

The sine function solutions derived for this system are completely standard and found extensively in the literature. Although it should be noted that the arguments developed for setting the boundary conditions, i.e. $\psi(z)$ continuous, also implied that the first derivative should be continuous too, use is never made of this second boundary condition. The limitations of solution imposed by this are avoided by saying that not only is the potential infinite outside the well, but in addition the Schrödinger equation is not defined in these regions—a slight contradiction with the deduction of the first boundary condition. This point, i.e. that there is still ambiguity in the choice of boundary conditions for commonly accepted solutions, will be revisited later in this chapter.

2.2   IN-PLANE DISPERSION

If the one-dimensional potential $V(z)$ is constructed from alternating thin layers of dissimilar semiconductors, then the particle, whether it be an electron or a hole, can move in the plane of the layers (see Fig 2.4).

In this case, all of the terms of the kinetic energy operator are required, and hence the Schrödinger equation would be as follows:

$$-\frac{\hbar^2}{2m}\left(\frac{\partial^2}{\partial x^2} + \frac{\partial^2}{\partial y^2} + \frac{\partial^2}{\partial z^2}\right)\psi + V(z)\psi = E\psi \qquad (2.16)$$

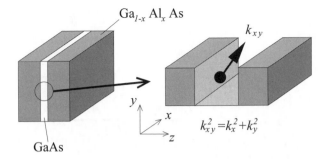

Figure 2.4   A GaAs/Ga$_{1-x}$Al$_x$As layered structure and the in-plane motion of a charge carrier

As the potential can be written as a sum of independent functions, i.e. $V = V(x) + V(y) + V(z)$, where it just happens in this case that $V(x) = V(y) = 0$, the eigenfunction of the system be written as a product:

$$\psi(x, y, z) = \psi_x(x)\psi_y(y)\psi_z(z) \tag{2.17}$$

Using this in the above Schrödinger equation, then:

$$-\frac{\hbar^2}{2m}\left(\frac{\partial^2\psi_x}{\partial x^2}\psi_y\psi_z + \frac{\partial^2\psi_y}{\partial y^2}\psi_x\psi_z + \frac{\partial^2\psi_z}{\partial z^2}\psi_x\psi_y\right) + V(z)\psi_x\psi_y\psi_z = E\psi_x\psi_y\psi_z \tag{2.18}$$

It is then possible to identify three distinct contributions to the total energy E, one from each of the perpendicular x-, y-, and z-axes, i.e. $E = E_x + E_y + E_z$. It is said that the motions 'de-couple' giving an equation of motion for each of the axes:

$$-\frac{\hbar^2}{2m}\frac{\partial^2\psi_x}{\partial x^2}\psi_y\psi_z = E_x\psi_x\psi_y\psi_z \tag{2.19}$$

$$-\frac{\hbar^2}{2m}\frac{\partial^2\psi_y}{\partial y^2}\psi_x\psi_z = E_y\psi_x\psi_y\psi_z \tag{2.20}$$

$$-\frac{\hbar^2}{2m}\frac{\partial^2\psi_z}{\partial z^2}\psi_x\psi_y + V(z)\psi_x\psi_y\psi_z = E_z\psi_x\psi_y\psi_z \tag{2.21}$$

Dividing throughout, then:

$$-\frac{\hbar^2}{2m}\frac{\partial^2\psi_x}{\partial x^2} = E_x\psi_x \tag{2.22}$$

$$-\frac{\hbar^2}{2m}\frac{\partial^2\psi_y}{\partial y^2} = E_y\psi_y \tag{2.23}$$

$$-\frac{\hbar^2}{2m}\frac{\partial^2\psi_z}{\partial z^2} + V(z)\psi_z = E_z\psi_z \tag{2.24}$$

The last component is identical to the one-dimensional Schrödinger equation for a confining potential $V(z)$ as discussed, for the particular case of an infinite well, in the last section. Consider the first and second components. Again, an eigenfunction f is sought which when differentiated twice returns $-f$; however, in this case it must be remembered that the solution will represent a moving particle. Thus the eigenfunction

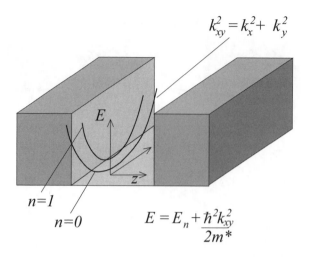

$$k_{xy}^2 = k_x^2 + k_y^2$$

$$E = E_n + \frac{\hbar^2 k_{xy}^2}{2m^*}$$

Figure 2.5   Schematic showing the in-plane $(k_{x,y})$ dispersion curves and the subband structure

must reflect a current flow and have complex components, so try the standard travelling wave, $\exp(ik_x x)$. Then:

$$-\frac{\hbar^2}{2m}\frac{\partial^2}{\partial x^2}\exp(ik_x x) = E_x \exp(ik_x x) \tag{2.25}$$

$$\therefore \frac{\hbar^2 k_x^2}{2m} = E_x \tag{2.26}$$

which is clearly just the kinetic energy of a wave travelling along the x-axis. A similar equation follows for the y-axis, and hence the in-plane motion of a particle in a one-dimensional confining potential, but of infinite extent in the x–y plane can be summarized as:

$$\psi_{x,y}(x,y) = \frac{1}{A}\exp[i(k_x x + k_y y)] \quad \text{and} \quad E_{x,y} = \frac{\hbar^2 |\mathbf{k}_{x,y}|^2}{2m} \tag{2.27}$$

Therefore, while solutions of Schrödinger's equation along the axis of the one-dimensional potential produce discrete states of energy $E_z = E_n$, in the plane of a semiconductor quantum well there is a continuous range of allowed energies, as illustrated in Fig. 2.5. In bulk materials, such domains are called 'energy bands', while in quantum well systems these energy domains associated with confined levels are referred to as 'subbands'. Therefore the effect of the one-dimensional confining potential is to remove a degree of freedom, thus restricting the momentum of the charge carrier from three-dimensions to two. It is for this reason that the states within quantum well systems are generally referred to as two-dimensional.

Later in this text, quantum wires and dots will be considered which further restrict the motion of carriers in two and three dimensions, respectively, thus giving rise to the terms one- and zero-dimensional states.

Summarizing then, within a semiconductor quantum well system the total energy of an electron or hole, of mass m^*, with in-plane momentum $k_{x,y}$, is equal to $E_z + E_{x,y}$, which is given by:

$$E = E_n + \frac{\hbar^2 |\mathbf{k}_{x,y}|^2}{2m^*} \tag{2.28}$$

2.3   DENSITY OF STATES

Therefore, the original confined states within the one-dimensional potential which could each hold two charge carriers of opposite spin, from the Pauli exclusion principle, broaden into subbands, thus allowing a continuous range of carrier momenta. In order to answer the question 'Given a particular number of electrons (or holes) within a subband, what is the distribution of their energy and momenta?', the first point that is required is a knowledge of the density of states, i.e. how many electrons can exist within a range of energies. In order to answer this point for the case of subbands in quantum wells, it is necessary first to understand this property in bulk crystals.

Following the idea behind Bloch's theorem (see reference [1] p.133) that an eigenstate within a *bulk* semiconductor, which can be written as $\psi = (1/\Omega) \exp(i\mathbf{k}_\bullet\mathbf{r})$, must display periodicity within the lattice, then if the unit cell is of side L:

$$\psi(x, y, z) = \psi(x + L, y + L, z + L) \tag{2.29}$$

$$\therefore \psi(x, y, z) = \frac{1}{\Omega} \exp\{i[k_x(x + L) + k_y(y + L) + k_z(z + L)]\} \tag{2.30}$$

$$\therefore \psi(x, y, z) = \frac{1}{\Omega} \exp[i(k_x x + k_y y + k_z z)] \exp[i(k_x L + k_y L + k_z L)] \tag{2.31}$$

Thus for the periodicity condition to be fulfilled, the second exponential term must be identical to 1, which implies that:

$$k_x = \frac{2\pi}{L}n_x \qquad\qquad k_y = \frac{2\pi}{L}n_y \qquad\qquad k_z = \frac{2\pi}{L}n_z \tag{2.32}$$

where n_x, n_y and n_z are integers. Each set of values of these three integers defines a distinct state, and hence the volume of **k**-space occupied by one state is $(2\pi/L)^3$. These states fill up with successively larger values of n_x, n_y and n_z, i.e. the lowest energy state has values (000), then permutations of (100), (110), etc., which gradually fill a sphere. At low temperatures, the sphere has a definite boundary between states that are all occupied followed by states that are unoccupied; the momentum of these states is called the *Fermi wave vector* and the equivalent energy is the *Fermi energy*. At higher temperatures, carriers near the edge of the sphere are often scattered to higher energy states, thus 'blurring' the boundary between occupied and unoccupied states. For a more detailed description see, for example, Ashcroft and Mermin [1].

Many of the interesting phenomena associated with semiconductors derive from the properties of electrons near the Fermi energy, as it is these electrons that are able to scatter into nearby states thus changing both their energy and momenta. In order to be able to progress with descriptions of, transport, for example (later in this book), it is necessary to be able to describe the density of available states.

The density of states is defined as the number of states per energy per unit volume of *real* space:

$$\rho(E) = \frac{\mathrm{d}N}{\mathrm{d}E} \tag{2.33}$$

In **k**-space, the total number of states N is equal to the volume of the sphere of radius **k**, divided by the volume occupied by one state and divided again by the volume of real space, i.e.

$$N = 2\,\frac{4\pi k^3}{3}\,\frac{1}{(2\pi/L)^3}\,\frac{1}{L^3} \tag{2.34}$$

$$\therefore N = 2 \, \frac{4\pi k^3}{3(2\pi)^3} \tag{2.35}$$

where the factor 2 has been introduced to allow for double occupancy of each state by the different carrier spins. Returning to the density of states, then:

$$\rho(E) = \frac{\mathrm{d}N}{\mathrm{d}E} = \frac{\mathrm{d}N}{\mathrm{d}k} \frac{\mathrm{d}k}{\mathrm{d}E} \tag{2.36}$$

Now equation (2.35) gives

$$\frac{\mathrm{d}N}{\mathrm{d}k} = 2 \, \frac{4\pi k^2}{(2\pi)^3} \tag{2.37}$$

In addition, the parabolic bands of effective mass theory give:

$$E = \frac{\hbar^2 k^2}{2m^*} \qquad \therefore k = \left(\frac{2m^* E}{\hbar^2} \right)^{\frac{1}{2}} \tag{2.38}$$

$$\therefore \frac{\mathrm{d}k}{\mathrm{d}E} = \left(\frac{2m^*}{\hbar^2} \right)^{\frac{1}{2}} \frac{E^{-\frac{1}{2}}}{2} \tag{2.39}$$

Which finally gives the density of states in bulk as:

$$\rho(E) = \frac{1}{2\pi^2} \left(\frac{2m^*}{\hbar^2} \right)^{\frac{3}{2}} E^{\frac{1}{2}} \tag{2.40}$$

Thus the density of states within a band, and around a minimum where the energy can be represented as a parabolic function of momentum, is continual and proportional to the square root of the energy.

The density of states in quantum well systems follows analogously; however, this time, as there are only two degrees of freedom, successive states represented by values of n_x and n_y fill a circle in k-space, as illustrated in Fig. 2.6. Such a situation has become known as a two-dimensional electron (or hole) gas (2DEG). Hence the total number of states per unit cross-sectional area is given by the spin degeneracy factor, multiplied by the area of the circle of radius k, divided by the area occupied by each state, i.e.

$$N^{\mathrm{2D}} = 2 \, \pi k^2 \, \frac{1}{(2\pi/L)^2} \frac{1}{L^2} \tag{2.41}$$

$$\therefore N^{\mathrm{2D}} = 2 \, \frac{\pi k^2}{(2\pi)^2} \tag{2.42}$$

$$\therefore \frac{\mathrm{d}N^{\mathrm{2D}}}{\mathrm{d}k} = \frac{k}{\pi} \tag{2.43}$$

In analogy to the bulk three-dimensional (3D) case, define:

$$\rho^{\mathrm{2D}}(E) = \frac{\mathrm{d}N^{\mathrm{2D}}}{\mathrm{d}E} = \frac{\mathrm{d}N^{\mathrm{2D}}}{\mathrm{d}k} \frac{\mathrm{d}k}{\mathrm{d}E} \tag{2.44}$$

As the in-plane dispersion curves are still described by parabolas, then reuse can be made of equation (2.39), as follows:

$$\rho^{\mathrm{2D}}(E) = \frac{k}{\pi} \left(\frac{2m^*}{\hbar^2} \right)^{\frac{1}{2}} \frac{E^{-\frac{1}{2}}}{2} \tag{2.45}$$

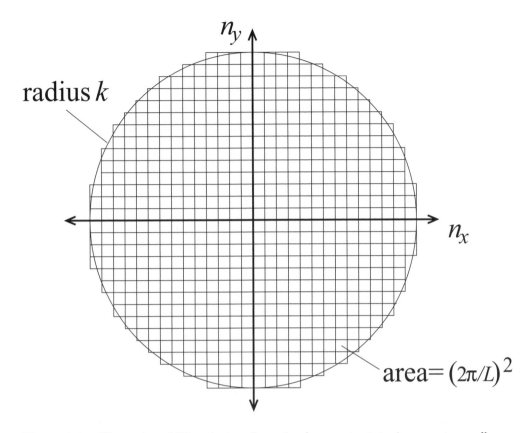

Figure 2.6   Illustration of filling the two-dimensional momenta states in a quantum well

By substituting for k in terms of the energy E, using equation (2.38) then finally the density of states for a single subband in a quantum well system is given by:

$$\rho^{2D}(E) = \frac{m^*}{\pi\hbar^2} \tag{2.46}$$

in agreement with Bastard [18] p.12.

If there are many (n) confined states within the quantum well system then the density of states ρ^{2D} at any particular energy is the sum over all subbands *below* that point, which can be written succinctly as:

$$\rho^{2D}(E) = \sum_{i=1}^{n} \frac{m^*}{\pi\hbar^2}\, \Theta(E - E_i) \tag{2.47}$$

where Θ is the unit step function. Fig. 2.7 gives an example of the two-dimensional density of states for a particular quantum well showing the first three confined levels. Note that the steps are of equal height and occur at the subband minima—which are not equally spaced.

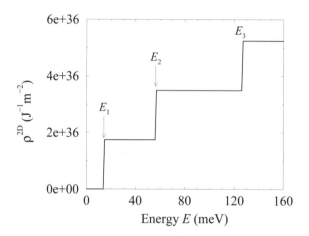

Figure 2.7 The density of states as a function of energy for a 200 Å GaAs quantum well surrounded by infinite barriers

2.4 SUBBAND POPULATIONS

The total number of carriers within a subband is given by the integral of the product of the probability of occupation of a state and the density of states. Given that the carriers are fermions, then clearly the probability of occupation of a state is given by Fermi–Dirac statistics; hence:

$$N = \int_{\text{subband}} f^{\text{FD}}(E)\, \rho(E)\ \mathrm{d}E \tag{2.48}$$

where the integral is over *all* of the energies of a given subband and, of course:

$$f^{\text{FD}}(E) = \frac{1}{\exp\left[(E - E_{\text{F}})/kT\right] + 1} \tag{2.49}$$

Note that E_{F} is not the Fermi energy in the traditional sense [1]; it is a 'quasi' Fermi energy which describes the carrier population *within* a subband. For systems left to reach equilibrium, the temperature T can be assumed to be the lattice temperature; however this is not always the case. In many quantum well devices which are subject to excitation by electrical or optical means, the 'electron temperature' can be quite different from the lattice temperature, and furthermore the subband population could be non-equilibrium and not able to be described by Fermi–Dirac statistics. For now, however, it is sufficient to discuss equilibrium electron populations and assume that the above equations are an adequate description.

Given a particular number of carriers within a quantum well, which can usually be deduced directly from the surrounding doping density, it is often desirable to be able to describe that distribution in terms of the quasi-Fermi energy E_{F}. With this aim, substitute the two-dimensional density of states appropriate to a single subband from

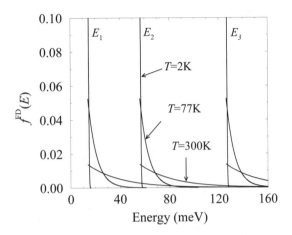

Figure 2.8 Effect of temperature on the distribution functions of the subband populations (all equal to 1×10^{10} cm^{-2}) of the infinite quantum well of Fig. 2.7

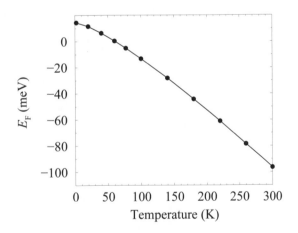

Figure 2.9 Effect of temperature on the quasi-Fermi energy describing the electron distribution of the ground state E_1

equation (2.46) into equation (2.48), then the carrier density, i.e. the number per unit area, is given by:

$$N = \int_{E_{\min}}^{E_{\max}} \frac{1}{\exp\left[(E - E_{\mathrm{F}})/kT\right] + 1} \frac{m^*}{\pi\hbar^2} \, \mathrm{d}E \qquad (2.50)$$

By putting:

$$E' = \frac{E - E_{\mathrm{F}}}{kT}, \quad \text{and then} \quad \mathrm{d}E' = \frac{\mathrm{d}E}{kT} \qquad (2.51)$$

equation (2.50) becomes:

$$N = \int_{(E_{\min}-E_{\mathrm{F}})/kT}^{(E_{\max}-E_{\mathrm{F}})/kT} \frac{kT}{\exp E' + 1} \frac{m^*}{\pi\hbar^2} \, dE' \tag{2.52}$$

which is a standard form (see, for example, Gradshteyn and Ryzhik [23] equation 2.313.2, p.112)

$$\int \frac{1}{1+e^x} \, dx = x - \ln(1 + e^x) \tag{2.53}$$

Hence:

$$N = \frac{m^* kT}{\pi\hbar^2} \left[E' - \ln\left(1 + e^{E'}\right) \right]_{(E_{\min}-E_{\mathrm{F}})/kT}^{(E_{\max}-E_{\mathrm{F}})/kT} \tag{2.54}$$

Evaluation then gives:

$$N = \frac{m^* kT}{\pi\hbar^2} \left\{ \left[\frac{E_{\max} - E_{\mathrm{F}}}{kT} - \ln\left(1 + e^{(E_{\max}-E_{\mathrm{F}})/kT}\right) \right] \right.$$
$$\left. - \left[\frac{E_{\min} - E_{\mathrm{F}}}{kT} - \ln\left(1 + e^{(E_{\min}-E_{\mathrm{F}})/kT}\right) \right] \right\} \tag{2.55}$$

The minimum of integration $E_{\min}$ is taken as the subband minima and the maximum $E_{\max}$ can either be taken as the top of the well, or even $E_{\mathrm{F}} + 10kT$, say, with the latter being much more stable at lower temperatures. Given a total carrier density N, the quasi-Fermi energy E_{F} is the only unknown in equation (2.55) and can be found with standard techniques. For an example of such a method, see Section 2.5.

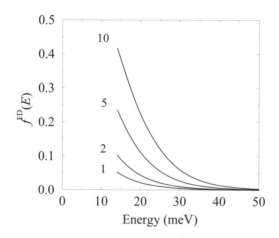

Figure 2.10   Effect of electron density, $N = 1, 2, 5, 10 \ (\times 10^{10})$ cm^{-2}, on the distribution function of the lowest subband of the infinite quantum well of Fig. 2.7

Fig. 2.8 gives an example of the distribution functions $f^{\mathrm{FD}}(E)$ for the first three confined levels within a 200 Å GaAs infinite quantum well. As the density of carriers, in this case electrons, have been taken as being equal and of value 1×10^{10}cm^{-2}, then the

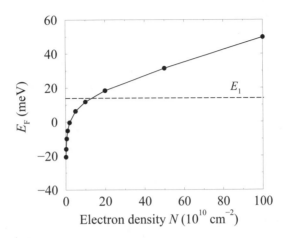

Figure 2.11   Effect of electron density on the quasi Fermi energy describing the distribution of the ground state E_1

distribution functions are all identical, but offset along the energy axis by the confinement energies. As mentioned above, at low temperatures the carriers tend to occupy the lowest available states, and hence the transition from states that are all occupied to those that are unoccupied is rapid—as illustrated by the 2 K data for all three subbands. As the temperature increases, the distributions broaden and a range of energies exist in which the states are partially filled, as can be seen by the 77 and 300 K data. Physically this broadening occurs due mainly to the increase in electron–phonon scattering as the phonon population increases with temperature (more of this in Chapter 10). Fig. 2.9 displays the Fermi energy E_F as a function of temperature T for the ground state of energy $E_1 = 14.031$ meV. At low temperatures, E_F is just above the confinement energy, since the electron density is fairly low $(1 \times 10^{10}$ cm$^{-2})$. As the temperature increases, E_F falls quite markedly and below the subband minima. If this seems counterintuitive, it must be remembered that E_F is a *quasi*-Fermi energy whose only physical meaning is to describe the population within a subband—it is not the true Fermi energy of the complete system.

Fig. 2.10 displays the distribution functions for a range of carrier densities, for this same ground state and at a lattice temperature of 77 K. Although not obvious from the mathematics, $f^{FD}(E)$ at any particular energy E appears to scale with N. The corresponding Fermi energy is illustrated in Fig. 2.11. Clearly, the Fermi energy starts below the subband minima at this mid-range temperature, as discussed above, and as expected increasing numbers of carriers in the subband increases the Fermi energy, i.e. the energy of the state whose probability of occupation is $1/2$.

2.5   FINITE WELL WITH CONSTANT MASS

While the infinitely deep confining potential has served well as a platform for developing the physics of two-dimensional systems, more relevant to alternating layers of dissimilar

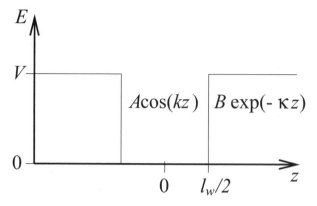

Figure 2.12   Solutions to the finite well potential

semiconductors is the finite quantum well model, which under both the effective mass and envelope function approximations looks like Fig. 2.12. In particular, a layer of GaAs 'sandwiched' between two thick layers of $Ga_{1-x}Al_xAs$ would form a type-I finite quantum well, where the conduction band has the appearance of Fig. 2.12, with the potential energy V representing the discontinuity in the conduction band edge between the materials.

Again taking the simplest starting case of a constant electron mass m^* throughout the dissimilar layers, and neglecting movement within the plane of the layers, then the standard Schrödinger equation can be written for each of the semiconductor layers as follows:

$$-\frac{\hbar^2}{2m^*}\frac{\partial^2}{\partial z^2}\psi(z) + V\psi(z) \;=\; E\psi(z), \qquad z \leq -\frac{l_w}{2} \tag{2.56}$$

$$-\frac{\hbar^2}{2m^*}\frac{\partial^2}{\partial z^2}\psi(z) \;=\; E\psi(z), \qquad -\frac{l_w}{2} \leq z \leq \frac{l_w}{2} \tag{2.57}$$

$$-\frac{\hbar^2}{2m^*}\frac{\partial^2}{\partial z^2}\psi(z) + V\psi(z) \;=\; E\psi(z), \qquad +\frac{l_w}{2} \leq z \tag{2.58}$$

Considering solutions to the Schrödinger equation for the central well region, then as in the infinite well case, the general solution will be a sum of sine and cosine terms. As the potential is symmetric, then the eigenstates will also have a definite symmetry, i.e. they will be either symmetric or antisymmetric. With the origin placed at the centre of the well, the symmetric (even parity) eigenstates will then be in cosine terms, while the antisymmetric (odd parity) states will be as sine waves.

For states confined to the well, the energy E must be less than the barrier height V, thus rearranging the Schrödinger equation for the right-hand barrier:

$$\frac{\hbar^2}{2m^*}\frac{\partial^2}{\partial z^2}\psi(z) = (V - E)\psi(z) \tag{2.59}$$

Therefore, a function f is sought which when differentiated twice gives $+f$. The exponential function fits this description, therefore consider a sum of growing $\exp(+\kappa z)$ and decaying $\exp(-\kappa z)$ exponentials. In the right-hand barrier, z is positive, and hence as z increases the growing exponential will increase too and without limit. The probability interpretation of the wave function requires that:

$$\int_{\text{all space}} \psi^*(z)\psi(z) \; dz = 1 \tag{2.60}$$

which further demands that:

$$\psi(z) \to 0 \quad \text{and} \quad \frac{\partial}{\partial z}\psi(z) \to 0, \qquad \text{as} \qquad z \to \pm\infty \tag{2.61}$$

These boundary conditions for states confined in wells will be used again and again and will be referred to as the *standard boundary conditions*. Using this result, the growing exponential components must be rejected and the solutions are for the even parity states, which would follow as:

$$\psi(z) = B\exp(\kappa z), \qquad z \le -\frac{l_w}{2} \tag{2.62}$$

$$\psi(z) = A\cos(kz), \qquad -\frac{l_w}{2} \le z \le \frac{l_w}{2} \tag{2.63}$$

$$\psi(z) = B\exp(-\kappa z), \qquad \frac{l_w}{2} \le z \tag{2.64}$$

Note for later that these wave functions are real, and that the eigenfunctions of this confined system carry no current and hence are referred to as stationary states. Using these trial forms of the wave function in their corresponding Schrödinger equations, gives the, as yet unknown constants:

$$k = \frac{\sqrt{2m^*E}}{\hbar}, \quad \text{and} \quad \kappa = \frac{\sqrt{2m^*(V-E)}}{\hbar} \tag{2.65}$$

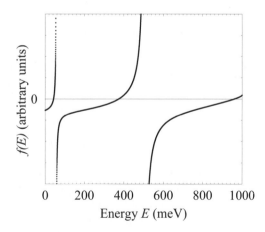

Figure 2.13 Illustration of $f(E)$ as a function of E for the even-parity solutions; where $l_w=100$ Å, $m^*=0.067\,m_0$ and $V=1000$ meV

In order to proceed, it is necessary to impose boundary conditions. Recalling the constant mass kinetic energy operator employed in equations (2.56)–(2.58), then in order to avoid infinite kinetic energies:

$$\text{both} \quad \psi(z) \quad \text{and} \quad \frac{\partial}{\partial z}\psi(z) \quad \text{must be continuous}$$

Consider the interface at $z = +l_w/2$; by equating ψ in the well and the barrier:

$$A\cos\left(\frac{kl_w}{2}\right) = B\exp\left(-\frac{\kappa l_w}{2}\right) \tag{2.66}$$

and equating the derivatives gives:

$$-kA\sin\left(\frac{kl_w}{2}\right) = -\kappa B\exp\left(-\frac{\kappa l_w}{2}\right) \tag{2.67}$$

Dividing equation (2.66) by equation (2.67) then gives:

$$-\frac{1}{k}\cot\left(\frac{kl_w}{2}\right) = -\frac{1}{\kappa} \tag{2.68}$$

$$\therefore k\tan\left(\frac{kl_w}{2}\right) - \kappa = 0 \tag{2.69}$$

Odd parity states would require the choice of wave function in the well region as a sine wave, and hence equation (2.63) would become $\psi = A\sin(kz)$; following through the same analysis as above gives the equation to be solved for the odd parity eigenenergies as:

$$k\cot\left(\frac{kl_w}{2}\right) + \kappa = 0 \tag{2.70}$$

Remembering that both k and κ are functions of the energy E, then equations (2.69) and (2.70) are also functions of E only. There are many ways of solving such single-variable equations, and for this particular case the literature often talks of 'graphical methods' [3, 7, 18]. While it is interesting to view the functional form of the equations, such methods are time consuming and inefficient and wouldn't be employed in the repetitive solution of many quantum wells. Computationally, it is much more effective just to treat equations (2.69) and (2.70) with standard techniques, such as Newton–Raphson iteration. In this technique, if $E^{(n)}$ is a first guess to the solution of $f(E) = 0$, then a better estimate is given by:

$$E^{(n+1)} = E^{(n)} - \frac{f(E^{(n)})}{f'(E^{(n)})} \tag{2.71}$$

The new estimate $E^{(n+1)}$ is then used to generate a second approximation to the solution $E^{(n+2)}$, and so on, until the successive estimates converge to a required accuracy.

In order to provide all of the required information to implement the solution, all that remains is to deduce $f'(E)$. For the even parity states:

$$f(E) = k\tan\left(\frac{kl_w}{2}\right) - \kappa \tag{2.72}$$

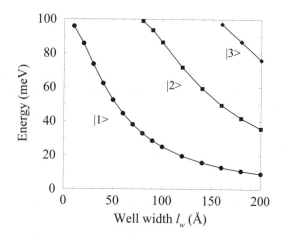

Figure 2.14   Energy levels in a GaAs single quantum well with constant effective mass $m^*=0.067m_0$ and $V=100$ meV

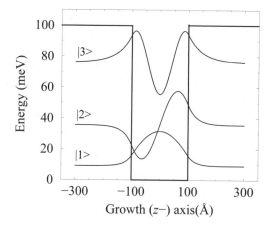

Figure 2.15   Eigenfunctions $\psi(z)$ for the first three energy levels of the 200 Å GaAs well of Fig. 2.14

and so therefore:

$$\frac{\mathrm{d}f}{\mathrm{d}E} = \frac{\mathrm{d}k}{\mathrm{d}E}\tan\left(\frac{kl_w}{2}\right) + k\sec^2\left(\frac{kl_w}{2}\right) \times \frac{l_w}{2}\frac{\mathrm{d}k}{\mathrm{d}E} - \frac{\mathrm{d}\kappa}{\mathrm{d}E} \qquad (2.73)$$

where

$$\frac{\mathrm{d}k}{\mathrm{d}E} = \frac{\sqrt{2m^*}}{2\sqrt{E}\hbar}; \qquad \frac{\mathrm{d}\kappa}{\mathrm{d}E} = -\frac{\sqrt{2m^*}}{2\sqrt{V-E}\hbar} \qquad (2.74)$$

For the odd-parity states:

$$f(E) = k \cot \left(\frac{k l_w}{2} \right) + \kappa \tag{2.75}$$

$$\therefore \frac{\mathrm{d}f}{\mathrm{d}E} = \frac{\mathrm{d}k}{\mathrm{d}E} \cot \left(\frac{k l_w}{2} \right) - k \csc^2 \left(\frac{k l_w}{2} \right) \times \frac{l_w}{2} \frac{\mathrm{d}k}{\mathrm{d}E} + \frac{\mathrm{d}\kappa}{\mathrm{d}E} \tag{2.76}$$

Fig. 2.13 illustrates the even parity $f(E)$. In practice, the finite number of discontinuities can lead to solutions being missed if the first guess to the solution $E^{(0)}$ is not close to the true solution. In order to circumvent this computational problem, $f(E)$ is calculated at discrete points along the E axis, separated by an energy thought to be smaller than the minimum separation between adjacent states (generally 1 meV), when $f(E)$ changes sign; then the Newton–Raphson is then implemented to obtain the solution accurately.

Figs. 2.14 and 2.15 summarise the application of the method to a GaAs single quantum well, surrounded by barrier of height 100 meV, with the same effective mass. Clearly, as the well width increases, then the energy levels all decrease, with the presence of excited states being also apparent at the larger well widths. The eigenstates are labelled according to their principle quantum number (energy order). The even and odd parities of the states within the well can be seen in Fig. 2.15.

2.6 EFFECTIVE MASS MISMATCH AT HETEROJUNCTIONS

Quantum wells are only fabricated by forming heterojunctions between *different* semiconductors. From an electronic viewpoint, the semiconductors are different because they have different band structures. The difference in perhaps the most fundamental property of a semiconductor, i.e. the band gap, (and its alignment) is accounted for by specifying a band offset, which has been labelled V. Of course, there are many other properties which are also different, such as the dielectric constant, the lattice constant and, what is considered the next most important quantity, the effective mass. It is generally accepted that the calculation of *static* energy levels within quantum wells should account for the variation in the effective mass across the heterojunction.

This problem has been continuously addressed in the literature [17, 24, 25] since the earliest work of Conley *et al.* [26] and BenDaniel and Duke [27], who derived the boundary conditions on solutions of the *envelope functions* as:

$$\text{both} \quad \psi(z) \quad \text{and} \quad \frac{1}{m^*} \frac{\partial}{\partial z} \psi(z) \quad \text{continuous} \tag{2.77}$$

by considering electron transport across a heterojunction. These boundary conditions have become known as the BenDaniel–Duke boundary conditions.

Therefore, applying this extension to the finite well of the previous section would require the Schrödinger equation to be specified in each region as follows:

$$-\frac{\hbar^2}{2m_b^*} \frac{\partial^2}{\partial z^2} \psi(z) + V \psi(z) = E \psi(z), \quad z \leq -\frac{l_w}{2} \tag{2.78}$$

$$-\frac{\hbar^2}{2m_w^*} \frac{\partial^2}{\partial z^2} \psi(z) = E \psi(z), \quad -\frac{l_w}{2} \leq z \leq \frac{l_w}{2} \tag{2.79}$$

$$-\frac{\hbar^2}{2m_b^*} \frac{\partial^2}{\partial z^2} \psi(z) + V \psi(z) = E \psi(z), \quad +\frac{l_w}{2} \leq z. \tag{2.80}$$

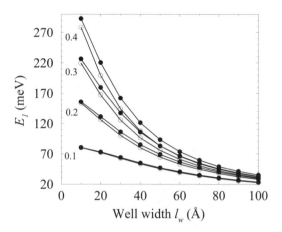

Figure 2.16   Electron ground state energy E_1 as a function of the width l_w of a GaAs well surrounded by $Ga_{1-x}Al_xAs$ barriers, calculated for both the constant mass-model (closed circles) and different barrier masses (open circles) and for a range of barrier alloy concentrations x ($= 0.1$, 0.2, 0.3, 0.4)

with the additional restraint of the matching conditions of equation (2.77).

The solutions follow as previously for the constant-mass case, in equations (2.62), (2.63) and (2.64), but now k and κ contain different effective masses:

$$k = \frac{\sqrt{2m_w^* E}}{\hbar} \qquad \kappa = \frac{\sqrt{2m_b^*(V - E)}}{\hbar} \tag{2.81}$$

The method of solution is almost identical: equating the envelope functions at the interface $z = +l_w/2$

$$A \cos\left(\frac{kl_w}{2}\right) = B \exp\left(-\frac{\kappa l_w}{2}\right) \tag{2.82}$$

and equating $1/m^*$ times the derivative gives

$$-\frac{kA}{m_w^*} \sin\left(\frac{kl_w}{2}\right) = -\frac{\kappa B}{m_b^*} \exp\left(-\frac{\kappa l_w}{2}\right) \tag{2.83}$$

Dividing equation (2.83) by equation (2.82) then:

$$f(E) = \frac{k}{m_w^*} \tan\left(\frac{kl_w}{2}\right) - \frac{\kappa}{m_b^*} = 0 \tag{2.84}$$

and similarly for the odd parity solutions, i.e.

$$f(E) = \frac{k}{m_w^*} \cot\left(\frac{kl_w}{2}\right) + \frac{\kappa}{m_b^*} = 0 \tag{2.85}$$

and obviously, equating the effective masses gives the original constant-mass equations. Again, for numerical solution, the derivatives are required, i.e. for even parity:

$$\frac{df}{dE} = \frac{1}{m_w^*} \frac{dk}{dE} \tan\left(\frac{kl_w}{2}\right) + \frac{k}{m_w^*} \sec^2\left(\frac{kl_w}{2}\right) \times \frac{l_w}{2} \frac{dk}{dE} - \frac{1}{m_b^*} \frac{d\kappa}{dE} \tag{2.86}$$

and for odd parity

$$\frac{\mathrm{d}f}{\mathrm{d}E} = \frac{1}{m_w^*}\frac{\mathrm{d}k}{\mathrm{d}E}\cot\left(\frac{kl_w}{2}\right) - \frac{k}{m_w^*}\csc^2\left(\frac{kl_w}{2}\right) \times \frac{l_w}{2}\frac{\mathrm{d}k}{\mathrm{d}E} + \frac{1}{m_b^*}\frac{\mathrm{d}\kappa}{\mathrm{d}E} \tag{2.87}$$

where

$$\frac{\mathrm{d}k}{\mathrm{d}E} = \frac{\sqrt{2m_w^*}}{2\sqrt{E}\hbar} \qquad \frac{\mathrm{d}\kappa}{\mathrm{d}E} = -\frac{\sqrt{2m_b^*}}{2\sqrt{V-E}\hbar} \tag{2.88}$$

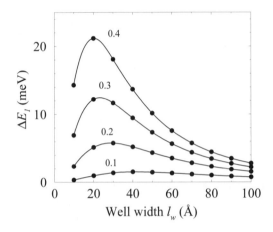

Figure 2.17   Energy difference $\Delta E_1 = E_1(m^*=\text{constant}) - E_1(m^*(z))$ for the structures shown in Fig. 2.16

Fig. 2.16 compares the electron ground state energy calculated with a constant GaAs mass, with the energy calculated with a material-dependent mass, for a single GaAs quantum well surrounded by $Ga_{1-x}Al_xAs$ barriers. All of the calculated ground state energies E_1 decrease with increasing well width and increase with increasing Al fraction in the barriers (barrier height $V \propto x$). The effective mass in $Ga_{1-x}Al_xAs$ is greater than in GaAs, hence the variable mass calculations give energies less than the constant-mass model for *all* systems considered here (see later).

Fig. 2.17 displays the calculated ground state energy difference between the two models, $\Delta E_1 = E_1(m^*=\text{constant}) - E_1(m^*(z))$. Clearly, and as would be expected, the larger the difference in the effective masses between the materials, the larger the difference in ground state energies.

2.7   THE INFINITE BARRIER HEIGHT AND MASS LIMITS

It is interesting to take theoretical models to certain limits as a means of verifying, or otherwise, their behaviour with what might be expected intuitively. Fig. 2.18 illustrates that, in the limit of large barrier heights V, the finite well model recovers the result of the infinite well model, which is what would be hoped for, thus increasing confidence in the derivation.

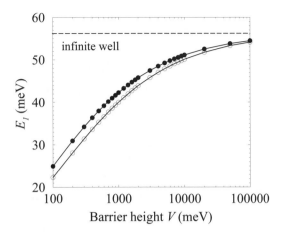

Figure 2.18 Electron ground state energy E_1 as a function of barrier height V, for a 100 Å GaAs finite well with constant mass (closed circles) and different barrier mass (fixed at mass in $Ga_{0.6}Al_{0.4}As$ (open circles))

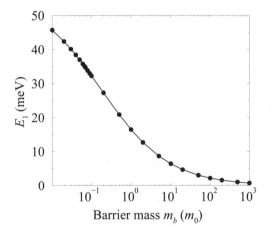

Figure 2.19 Electron ground state energy E_1 as a function of the mass in the barrier, for a 100 Å GaAs well with a barrier height fixed at that for $Ga_{0.6}Al_{0.4}As$

As has been found in the literature, Fig. 2.19 illustrates the results of allowing the barrier mass to increase without limit, while keeping all other parameters constant. The tendency for the ground state energy to tend towards zero and the unusual looking wave functions of Fig. 2.20 have been well documented [18] and are a direct consequence of the second boundary condition, ψ'/m^*.

It is worthwhile considering this limit still further. Fig. 2.21 reproduces the results of Fig. 2.19 but for a variety of well widths. Clearly, the ground state energy tends to

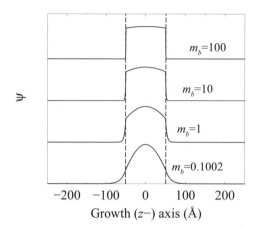

Figure 2.20   Electron ground state wave functions for several barrier masses, as given, for a 100 Å GaAs well with a barrier height fixed at that for $Ga_{0.6}Al_{0.4}As$

zero for all well widths:

$$\lim_{m_b \to \infty} E_1 = 0, \qquad \forall\ l_w \qquad\qquad (2.89)$$

which at first sight appears to violate Heisenberg's Uncertainty Principle, in that the ground state energy *can* be forced to zero for infinitesimal well widths, and thus the error in the measurement of position and momentum can be made arbitrarily small. This was an argument advanced by Hagston *et al.* [28]; however, direct evaluation of the variance in the position and momentum of these states will show that the uncertainty relationships are not violated. At the moment, the techniques for such a calculation have not been covered; hence such discussions will be returned to later in Section 3.15.

2.8   HERMITICITY AND THE KINETIC ENERGY OPERATOR

The changes in the boundary conditions at a heterojunction are only necessary under theoretical models, which parameterize physical quantities in terms of a variable mass. This is only encountered when applying the effective mass approximation to semiconductor heterostructures. Theories of semiconductor heterostructures do exist, which do not make this imposition (parameterisation)—within them all of the electron states, within both the conduction and valence band can be derived from the *constant* free electron mass. Therefore the kinetic energy operator remains:

$$T = -\frac{\hbar^2}{2m_0}\nabla^2 \qquad\qquad (2.90)$$

and hence the boundary conditions are set as continuity in both wave function and derivative—note *wave function*, and not *envelope function*. Such methods are the subject of later chapters, for example, Chapter 15.

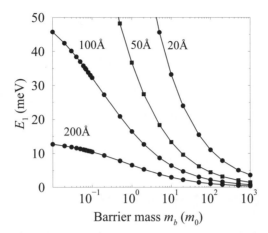

Figure 2.21   Electron ground state energy E_1 as a function of the mass in the barrier, for a variety of GaAs well widths, the barrier height is fixed at that for $Ga_{0.6}Al_{0.4}As$

However, within the auspices of this very successful model of semiconductor band structure, allowance *has* to be made for the additional complication of variable mass. As discussed in Section 2.6, this can be achieved with the original Schrödinger equation applied to each region, taken together with boundary conditions for matching solutions at the heterojunctions between the regions of different effective mass.

Fundamentally both of these conditions can be incorporated within one Hamiltonian, which is valid throughout all space at once. Its construction derives from the corner-stones of quantum mechanics—all physical observables can be represented by linear Hermitian operators [12]. With this aim, consider the classical (non-relativistic) kinetic energy:

$$T = \frac{p^2}{2m} \tag{2.91}$$

Quantum mechanically, the linear kinetic energy operator for a variable mass $m^*(z)$ say, is formed by replacing the classical quantities with the corresponding linear operators, giving:

$$\mathcal{T} = \frac{\mathcal{P}_z^2}{m^*(z)} \tag{2.92}$$

Inspection of equation (2.92) reveals that there is still ambiguity as to the *order* by which the operators will act upon the eigenvector. To resolve this, guidance is sought from a property of Hermitian operators, and that is they are equal to their own Hermitian conjugate (see Schiff [29] p.151), i.e.

$$\mathcal{T} = \mathcal{T}^\dagger \tag{2.93}$$

In addition, the Hermitian conjugate of a product is given by $(\mathcal{ABC})^\dagger = \mathcal{C}^\dagger \mathcal{B}^\dagger \mathcal{A}^\dagger$. Applying this to the effective mass kinetic energy operator then:

$$\mathcal{T} = \mathcal{P}_z \frac{1}{m^*(z)} \mathcal{P}_z \tag{2.94}$$

$$\therefore T = -\frac{\hbar^2}{2}\frac{\partial}{\partial z}\frac{1}{m^*(z)}\frac{\partial}{\partial z} \tag{2.95}$$

which would give a Schrödinger equation of the form:

$$-\frac{\hbar^2}{2}\frac{\partial}{\partial z}\frac{1}{m^*(z)}\frac{\partial}{\partial z}\psi(z) + V(z)\psi(z) = E\psi(z) \tag{2.96}$$

which in order to avoid differentiating discontinuous functions and producing infinities, clearly demands the boundary conditions of equation (2.77), i.e.,

$$\text{both} \quad \psi(z) \quad \text{and} \quad \frac{1}{m^*}\frac{\partial}{\partial z}\psi(z) \quad \text{are continuous}$$

2.9 ALTERNATIVE KINETIC ENERGY OPERATORS

The Hermiticity requirements above have been used to derive a kinetic energy operator and hence a single Hamiltonian which is valid throughout all space, and reproduces the accepted boundary conditions on the envelope functions of heterostructures. This form is, however, not unique. In particular, any combination of linear momentum operators and effective mass of the form:

$$T = m^{*\alpha}(z)P_z m^{*\beta}(z)P_z m^{*\alpha}(z), \quad \text{where} \quad 2\alpha + \beta = -1 \tag{2.97}$$

will also satisfy these requirements [30, 31].

The accepted choice is $\alpha = 0$, $\beta = -1$, but clearly there are infinitely many solutions. Morrow [32] and Hagston *et al.* [28] have independently forwarded $\alpha = -1/2$, $\beta=0$, giving:

$$T = \frac{1}{\sqrt{m^*(z)}}P_z P_z \frac{1}{\sqrt{m^*(z)}} \tag{2.98}$$

which can be written succinctly as:

$$T = Q^\dagger Q, \quad \text{where} \quad Q = P_z \frac{1}{\sqrt{m^*(z)}} \tag{2.99}$$

Transforming the wave function $\psi(z)$ into $\sqrt{m^*(z)}\phi(z)$ and substituting into the Schrödinger equation gives:

$$-\frac{\hbar^2}{2}\frac{1}{\sqrt{m^*(z)}}\frac{\partial^2}{\partial z^2}\frac{1}{\sqrt{m^*(z)}}\sqrt{m^*(z)}\phi(z) + V(z)\sqrt{m^*(z)}\phi(z) = E\sqrt{m^*(z)}\phi(z) \tag{2.100}$$

$$\therefore -\frac{\hbar^2}{2m^*(z)}\frac{\partial^2}{\partial z^2}\phi(z) + V(z)\phi(z) = E\phi(z) \tag{2.101}$$

which implies that $\phi(z)$ is an eigenfunction of this transformed Hamiltonian, but with the same eigenvalue E. This new entity, $\phi(z)$, satisfies the boundary conditions:

$$\text{both} \quad \phi(z) \quad \text{and} \quad \frac{\partial}{\partial z}\phi(z) \quad \text{are continuous}$$

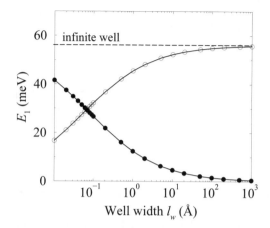

Figure 2.22 Effect of the two different kinetic energy operators on the ground state energy of a 100 Å GaAs well surrounded by $Ga_{0.8}Al_{0.2}As$ barriers: $\mathcal{T} = \frac{1}{m}\mathcal{P}^2$ (open circles); $\mathcal{T} = \mathcal{P}\frac{1}{m}\mathcal{P}$ (closed circles)

Therefore, discontinuous changes in the effective mass $m^*(z)$ at abrupt heterojunctions would give a discontinuity in the *wave function* ψ, a view that has been suggested elsewhere [33].

Fig. 2.22 compares the consequences of such an alternative kinetic energy operator on the ground state energy E_1 of a finite quantum well, with that of the accepted operator. In the limit of large barrier mass, the energy given by the new operator tends towards that of the infinitely deep quantum well. This appears intuitively satisfying, given that the wave function penetration into the barrier, $\exp(-\kappa z)$, has a similar dependence on the barrier mass m_b as on the height V, i.e.

$$\kappa = \frac{\sqrt{2m_b^*(V - E)}}{\hbar} \tag{2.102}$$

Thus mathematically speaking, increasing m_b would appear to produce the same effect as increasing V.

Galbraith and Duggan [25] have compared the results of similar calculations with experimental photoluminescence excitation measurements and have concluded that, for their single quantum well samples, the accepted form of $\mathcal{T} = \mathcal{P}\frac{1}{m}\mathcal{P}$ fits the data better. There is no doubt that this is not the last word on the form of the kinetic energy operator, and the associated boundary conditions, on the effective mass envelope functions which follow. However, for now it is best to conform to the commonly agreed standard and employ $\mathcal{T} = \mathcal{P}\frac{1}{m}\mathcal{P}$ throughout the remainder of this book.

2.10 EXTENSION TO MULTIPLE-WELL SYSTEMS

For a whole variety of reasons, semiconductor physicists and electronic device engineers need to design and fabricate heterostructures more complex than the single quantum well

[7]. The above techniques of solving Schrödinger's equation in each semiconductor layer separately and deducing the unknown coefficients by implementing boundary conditions can be applied to these multilayered systems. In particular, consider the asymmetric double quantum well of Fig. 2.23.

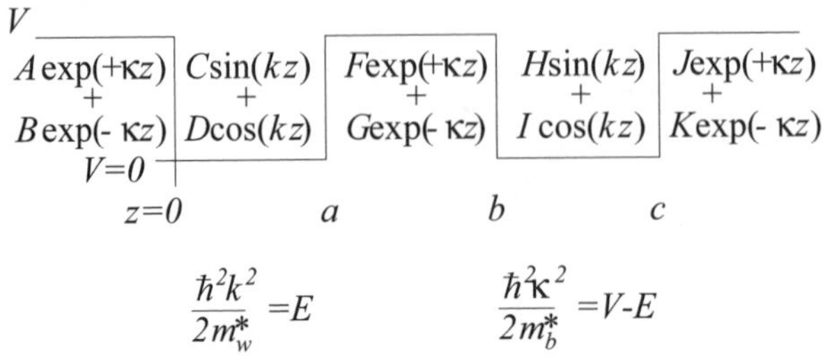

$$\frac{\hbar^2 k^2}{2m_w^*} = E \qquad\qquad \frac{\hbar^2 \kappa^2}{2m_b^*} = V\text{-}E$$

Figure 2.23 Solutions to Schrödinger's equation in a double quantum well

Choosing both wells to be of the same depth, and all of the barriers to be of the same height V, allows the simplification that k and κ are constant throughout the structure; the general solution to the Schrödinger equation in each region are as given in the figure. Proceeding as before, then matching envelope functions and the derivative divided by the mass at the interfaces, then considering $z = 0$ first:

$$A + B = D \tag{2.103}$$

$$\frac{1}{m_b^*}(\kappa A - \kappa B) = \frac{1}{m_w^*}kC \tag{2.104}$$

with $z = a$ then:

$$C \sin ka + D \cos ka = F \exp(+\kappa a) + G \exp(-\kappa a) \tag{2.105}$$

$$\frac{1}{m_w^*}(kC \cos ka - kD \sin ka) = \frac{1}{m_b^*}[\kappa F \exp(+\kappa a) - \kappa G \exp(-\kappa a)] \tag{2.106}$$

$z = b$:

$$F \exp(+\kappa b) + G \exp(-\kappa b) = H \sin kb + I \cos kb \tag{2.107}$$

$$\frac{1}{m_b^*}[\kappa F \exp(+\kappa b) - \kappa G \exp(-\kappa b)] = \frac{1}{m_w^*}(kH \cos kb - kI \sin kb) \tag{2.108}$$

$z = c$:

$$H \sin kc + I \cos kc = J \exp(+\kappa c) + K \exp(-\kappa c) \tag{2.109}$$

$$\frac{1}{m_w^*}(kH \cos kc - kI \sin kc) = \frac{1}{m_b^*}[\kappa J \exp(+\kappa c) - \kappa K \exp(-\kappa c)] \tag{2.110}$$

which can be rewritten more neatly in matrix form as:

$$\begin{pmatrix} 1 & 1 \\ \frac{1}{m_b^*}\kappa & -\frac{1}{m_b^*}\kappa \end{pmatrix} \begin{pmatrix} A \\ B \end{pmatrix} = \begin{pmatrix} 0 & 1 \\ \frac{1}{m_w^*}k & 0 \end{pmatrix} \begin{pmatrix} C \\ D \end{pmatrix} \tag{2.111}$$

$$\begin{pmatrix} \sin ka & \cos ka \\ \frac{1}{m_w^*}k \cos ka & -\frac{1}{m_w^*}k \sin ka \end{pmatrix} \begin{pmatrix} C \\ D \end{pmatrix}$$

$$= \begin{pmatrix} \exp(+\kappa a) & \exp(-\kappa a) \\ \frac{1}{m_b^*}\kappa\exp(+\kappa a) & -\frac{1}{m_b^*}\kappa\exp(-\kappa a) \end{pmatrix} \begin{pmatrix} F \\ G \end{pmatrix} \tag{2.112}$$

$$\begin{pmatrix} \exp(+\kappa b) & \exp(-\kappa b) \\ \frac{1}{m_b^*}\kappa\exp(+\kappa b) & -\frac{1}{m_b^*}\kappa\exp(-\kappa b) \end{pmatrix} \begin{pmatrix} F \\ G \end{pmatrix}$$

$$= \begin{pmatrix} \sin kb & \cos kb \\ \frac{1}{m_w^*}k\cos kb & -\frac{1}{m_w^*}k\sin kb \end{pmatrix} \begin{pmatrix} H \\ I \end{pmatrix} \tag{2.113}$$

$$\begin{pmatrix} \sin kc & \cos kc \\ \frac{1}{m_w^*}k\cos kc & -\frac{1}{m_w^*}k\sin kc \end{pmatrix} \begin{pmatrix} H \\ I \end{pmatrix}$$

$$= \begin{pmatrix} \exp(+\kappa c) & \exp(-\kappa c) \\ \frac{1}{m_b^*}\kappa\exp(+\kappa c) & -\frac{1}{m_b^*}\kappa\exp(-\kappa c) \end{pmatrix} \begin{pmatrix} J \\ K \end{pmatrix} \tag{2.114}$$

Labelling the 2×2 matrix for the left-hand side of the nth interface as $\mathbf{M}_{2n-1}$ and the corresponding matrix for the right-hand side of the interface as $\mathbf{M}_{2n}$, $n=1, 2, 3$, etc., then the above matrix equations would become:

$$\mathbf{M}_1 \begin{pmatrix} A \\ B \end{pmatrix} = \mathbf{M}_2 \begin{pmatrix} C \\ D \end{pmatrix} \tag{2.115}$$

$$\mathbf{M}_3 \begin{pmatrix} C \\ D \end{pmatrix} = \mathbf{M}_4 \begin{pmatrix} F \\ G \end{pmatrix} \tag{2.116}$$

$$\mathbf{M}_5 \begin{pmatrix} F \\ G \end{pmatrix} = \mathbf{M}_6 \begin{pmatrix} H \\ I \end{pmatrix} \tag{2.117}$$

$$\mathbf{M}_7 \begin{pmatrix} H \\ I \end{pmatrix} = \mathbf{M}_8 \begin{pmatrix} J \\ K \end{pmatrix} \tag{2.118}$$

Now equation (2.115) gives:

$$\begin{pmatrix} A \\ B \end{pmatrix} = \mathbf{M}_1^{-1}\mathbf{M}_2 \begin{pmatrix} C \\ D \end{pmatrix} \tag{2.119}$$

and equation (2.116) gives:

$$\begin{pmatrix} C \\ D \end{pmatrix} = \mathbf{M}_3^{-1}\mathbf{M}_4 \begin{pmatrix} F \\ G \end{pmatrix} \tag{2.120}$$

$$\therefore \begin{pmatrix} A \\ B \end{pmatrix} = \mathbf{M}_1^{-1}\mathbf{M}_2\mathbf{M}_3^{-1}\mathbf{M}_4 \begin{pmatrix} F \\ G \end{pmatrix} \tag{2.121}$$

and eventually:

$$\begin{pmatrix} A \\ B \end{pmatrix} = \mathbf{M}_1^{-1}\mathbf{M}_2\mathbf{M}_3^{-1}\mathbf{M}_4\mathbf{M}_5^{-1}\mathbf{M}_6\mathbf{M}_7^{-1}\mathbf{M}_8 \begin{pmatrix} J \\ K \end{pmatrix} \tag{2.122}$$

The product of the eight 2×2 matrices is still a 2×2 matrix; thus writing:

$$\begin{pmatrix} A \\ B \end{pmatrix} = \mathcal{M} \begin{pmatrix} J \\ K \end{pmatrix} \tag{2.123}$$

then

$$A = \mathcal{M}_{11}J + \mathcal{M}_{12}K \tag{2.124}$$

$$B = \mathcal{M}_{21}J + \mathcal{M}_{22}K \tag{2.125}$$

Again, the probability interpretation of the wave function implies that the wave function must tend towards zero into the outer barriers, i.e. the coefficients of the growing exponentials must be zero. *In this case*, with the origin at the 1st interface (see Fig. 2.23), then this implies that $B = 0$ and $J = 0$, and hence the second of the above equations would imply that $\mathcal{M}_{22}=0$. As all of the elements of $\mathcal{M}$ are functions of both k and κ, which are both in turn functions of the energy E, then an energy is sought which satisfies:

$$\mathcal{M}_{22}(E) = 0 \tag{2.126}$$

which can be found by standard numerical procedures, as discussed in Section 2.5. This approach and variations upon it are often referred to as the *transfer matrix technique*. Once the energy is known, the coefficients A to K follow simply and the envelope wave function can be deduced.

2.11   THE ASYMMETRIC SINGLE QUANTUM WELL

The above system has illustrated a method of solution for a system in which the confinement energy E was always less than the barrier height V. While this is often true, a class of quantum well structures exist in which there is more than one barrier height (or well depth); Fig. 2.24 illustrates one such system.

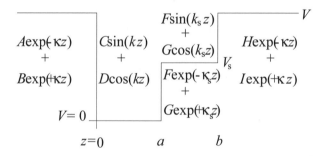

Figure 2.24   Solutions to Schrödinger's equation in a stepped asymmetric quantum well

A confined state could exist that has an energy below or above that of the step height V_s, hence the functional form of the solution to Schrödinger's equation in that region is dependent upon the energy E. This is illustrated schematically in Fig. 2.24. For states above the step, the wave function has the form of a normal well state, but with a k value different from that of the well region. For states of energy below the step potential V_s, the wave function resembles a barrier state. This causes difficulties computationally, as two transfer matrices have to be determined. A better approach is to write the wave function in *every* region (r) as a linear combination of travelling waves, i.e.

$$\psi_r = A' \exp\left(ik_r z\right) + B' \exp\left(-ik_r z\right) \tag{2.127}$$

where the coefficients A' and B' are now allowed to be complex and the wave vector k is as before:

$$k = \sqrt{\frac{2m^*}{\hbar^2}(E - V_r)} \tag{2.128}$$

Forming the transfer matrix from this standard form for the wave function and applying the boundary condition of decaying exponentials at both ends of the well structure, allows the method of solution as before. For now it is worthwhile confirming that ψ_r can take both the usual well- and barrier-state forms.

Consider regions where $E > V$:

$$\psi_r = A'(\cos k_r z + i \sin k_r z) + B'(\cos k_r z - i \sin k_r z) \tag{2.129}$$

$$\therefore \psi_r = (A' + B') \cos k_r z + i(A' - B') \sin k_r z \tag{2.130}$$

or by collecting real and imaginary components, then:

$$\psi_r = [\Re(A' + B') \cos k_r z - \Im(A' - B') \sin k_r z] \tag{2.131}$$
$$+ i [\Im(A' + B') \cos k_r z + \Re(A' - B') \sin k_r z] \tag{2.132}$$

Now eigenstates confined within potential wells are *stationary states*, which means that they have no time dependence and carry no current, i.e. their wave functions are real. Thus it would be expected that the solution would naturally yield such states, and in fact for the imaginary component of ψ_r to be zero, then:

$$\Im(A' + B') = 0 \qquad \text{and} \qquad \Re(A' - B') = 0 \tag{2.133}$$

which are simultaneously satisfied if:

$$B' = A'^* \tag{2.134}$$

This implies that in regions where the energy of the state of interest is greater than the potential, the solution will then give the coefficients as a complex conjugate pair in order to ensure a real wave function.

In regions where $E < V$, then:

$$k = i\sqrt{\frac{2m^*}{\hbar^2}(V_r - E)} = i\kappa \tag{2.135}$$

where κ is the usual decay constant as defined in equation (2.81). Then the general form for a barrier state is generated, i.e.

$$\psi_r = A' \exp(-\kappa z) + B' \exp(+\kappa z) \tag{2.136}$$

As mentioned above, application of the boundary conditions of $\psi \to 0$ as $z \to \pm\infty$, would force the choice that B' was zero and A' was real. In other regions where $E < V$, e.g. the step in Fig. 2.24 or the central barrier of Fig. 2.23, then both A' and B' should naturally arise as real.

2.12   ADDITION OF AN ELECTRIC FIELD

As illustrated in Fig. 2.25, the effect of an electric field along the growth (z-) axis is to add a linear potential, which for an electron of charge $-e$ in the conduction band will be written as $-eFz$. For small electric fields, the effect of this electric field on the confined

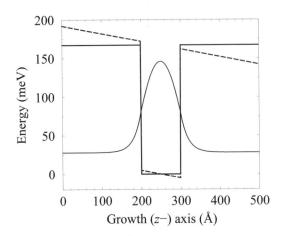

Figure 2.25   The ground state wave function of the zero-field potential (solid line) and the corresponding potential at a field of 10 kVcm^{-1} (dashed line)

energy levels within a quantum well can be approximated by first-order perturbation theory.

If the perturbing potential is written as V' then the change in the ground state energy level is given, for example, by Schiff [29], p.246:

$$\Delta E^{(1)} = \langle \psi_1 | V' | \psi_1 \rangle \tag{2.137}$$

with the electric field along the growth axis as the perturbation; then this translates to

$$\Delta E^{(1)} = \int_{-\infty}^{+\infty} \psi_1^*(z)(-eFz)\psi_1(z) \ dz \tag{2.138}$$

The ground state of the *symmetric* quantum well of Fig. 2.25 is an even function, and hence the integrand of equation (2.138) is odd. Therefore, evaluation of the integral gives zero, i.e. to first-order the addition of a *small* electric field has no effect on the *ground state* energy level. When applying the same logic to the first excited state, although ψ_2 is odd the integrand will still be an odd function, which again results in no change. There is currently much interest in the literature in breaking the symmetry of quantum wells in order to induce a first-order response in the ground state energy to an electric field [34–36].

Larger electric fields would require a more accurate evaluation of the perturbation as can be given by second-order perturbation theory—see Schiff [29], p.247:

$$\Delta E^{(2)} = \sum_{m=2}^{\infty} \frac{|\langle \psi_m | V' | \psi_1 \rangle|^2}{E_m - E_1} \tag{2.139}$$

which in this case translates to:

$$\Delta E^{(2)} = \sum_{m=2}^{\infty} \frac{|\int_{-\infty}^{+\infty} \psi_m^*(z)(-eFz)\psi_1(z) \ dz|^2}{E_m - E_1} \tag{2.140}$$

where the sum is over *all* excited states, including those with an energy $E_m > V$, i.e. the so called 'continuum' states. As the electric field F does not have a z-dependence, then clearly the change in the energy to second-order $\Delta E^{(2)}$ is proportional to F^2. In addition, a charged particle prefers to move to areas of lower potential, i.e. the electron within the quantum well of Fig. 2.25 moves to the right-hand side of the well, thus lowering its total energy, so indeed $\Delta E^{(2)} \propto -F^2$. This suppression of the confined energy level by an electric field is known as the 'quantum confined Stark effect' [37] and is commonly observed experimentally in heterostructures [35].

While such a perturbative treatise may be adequate for single quantum wells, its accuracy for complicated, perhaps multiple well systems and at high fields, is questionable. In order to improve upon this, a full (non-perturbative) solution to the heterostructure and field potentials is required. Within the envelope function and effective mass approximations, the Schrödinger equation within a particular material (i.e. region of constant mass) and with an applied electric field of strength F, is:

$$-\frac{\hbar^2}{2m^*}\frac{\partial^2 \psi}{\partial z^2} + [V(z) - eFz]\,\psi = E\psi \tag{2.141}$$

This new equation does not have the standard solution of linear combinations of trigonometric and exponential functions, and therefore a different approach is required. First, rearrange equation (2.141) to give:

$$\frac{\partial^2 \psi}{\partial z^2} - \frac{2m^*}{\hbar^2}\,[V(z) - eFz - E]\,\psi = 0 \tag{2.142}$$

Then by making the substitutions:

$$\alpha = \frac{2m^*}{\hbar^2}\,[V(z) - E] \tag{2.143}$$

and

$$\beta = \frac{2m^*}{\hbar^2}\,(-eF) \tag{2.144}$$

the Schrödinger equation becomes:

$$\frac{\partial^2 \psi}{\partial z^2} - (\alpha + \beta z)\,\psi = 0 \tag{2.145}$$

Consider the further substitution:

$$z' = \frac{\alpha + \beta z}{\gamma} \tag{2.146}$$

where γ is an, as yet, unknown constant. Then, since z' is a first-order linear function of z:

$$\frac{\partial^2 \psi}{\partial z^2} = \frac{\partial^2 \psi}{\partial z'^2} \times \left(\frac{\partial z'}{\partial z}\right)^2 \tag{2.147}$$

i.e. $$\frac{\partial^2 \psi}{\partial z^2} = \frac{\beta^2}{\gamma^2}\frac{\partial^2 \psi}{\partial z'^2} \tag{2.148}$$

Substituting into equation (2.145) then gives:

$$\frac{\beta^2}{\gamma^2}\frac{\partial^2 \psi}{\partial z'^2} - \gamma z'\psi = 0 \tag{2.149}$$

If γ^3 is set equal to β^2, then the full transformation is:

$$z' = \left(\frac{2m^*}{\hbar^2}\right)^{\frac{1}{3}} \left[\frac{V(z) - E}{(eF)^{\frac{2}{3}}} - (eF)^{\frac{1}{3}} z\right] \tag{2.150}$$

and the Schrödinger equation can be written as:

$$\frac{\partial^2 \psi}{\partial z'^2} - z'\psi = 0 \tag{2.151}$$

The reason for such a procedure is that equation (2.151) has a standard solution which is a linear combination of Airy functions (see Abramowitz and Stegun [38], p.446):

$$\psi(z') = A \operatorname{Ai}(z') + B \operatorname{Bi}(z') \tag{2.152}$$

The full solution proceeds by matching the wave functions at the heterojunctions according to the BenDaniel–Duke boundary conditions as before.

This is a standard procedure and many examples of its use can be found in the literature [39]. It is included along with the transfer matrix method of the previous sections for completeness. However, it can be appreciated that within these methods the study of a new quantum well system requires quite a large investment in developing new computational solutions. While a general computer code can be written to solve all semiconductor heterostructures, another one has to be written to solve all of these systems again with an electric field. The possibility of adding piezoelectric fields [40], which tilt the bands like an electric field, but with the additional possibility of being in the other direction, would require another computer code, and so on. While we continue in this vain for the remainder of this chapter, in the next chapter a numerical solution is introduced, which although occasionally has a few limitations, is still very powerful because of its *versatility*.

2.13   THE INFINITE SUPERLATTICE

Finite superlattices (or multiple quantum wells) can be treated with the transfer matrix technique, as described above for a double quantum well (a two period finite superlattice). Such approaches yield a discrete step of states due to the confinement along the growth axis of the semiconductor multilayer. However, when many identical quantum wells are stacked within the same semiconductor layer the electrons and holes see a periodic potential, which *can* appear to be infinite in exactly the same way as bulk crystals. When this occurs, the electron and hole wave functions are no longer localized but are of infinite extent and equally likely to be in any of the quantum wells. They are said to occupy 'Bloch states'.

As described in Section 2.11, and following Jaros [5], p.220, the envelope function within the well region can be written as:

$$\psi_w = A \exp\left(ik_w z\right) + B \exp\left(-ik_w z\right) \tag{2.153}$$

and in the barrier region as:

$$\psi_b = C \exp\left(ik_b z\right) + D \exp\left(-ik_b z\right) \tag{2.154}$$

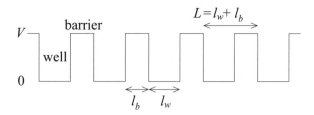

Figure 2.26   Band-edge potential profile of a superlattice of well and barrier widths, l_w and l_b respectively

where:

$$k_w = \sqrt{\frac{2m_w^*}{\hbar^2}E} \qquad\qquad k_b = \sqrt{\frac{2m_b^*}{\hbar^2}(E-V)} \qquad\qquad (2.155)$$

and it is assumed (for now) that $E > V$, so that k_b is real. Clearly, and as mentioned above, if the superlattice is infinite, then a particle is equally likely to be found in any well, which means that its wave function must be periodic with the lattice, i.e.

$$\psi(z) = \psi(z+L) \qquad\qquad (2.156)$$

For a travelling-wave state of the form, $\exp(ikz)$, then:

$$\psi(z+L) = \exp\left[ik(z+L)\right] = \exp(ikz)\exp(ikL) \qquad\qquad (2.157)$$

i.e. $\qquad \psi(z+L) = \psi(z)\exp(ikL) \qquad\qquad (2.158)$

By using this *periodic* form of the first BenDaniel–Duke boundary conditions (ψ continuous) at $z = L$, then:

$$\psi_w(L) = \psi_w(0)\exp(ikL) = \psi_b(L) \qquad\qquad (2.159)$$

Using the wave functions of equations (2.153) and (2.154), then:

$$(A+B)\exp(ikL) = C\exp(ik_bL) + D\exp(-ik_bL) \qquad\qquad (2.160)$$

By employing the same periodicity condition in the second of the BenDaniel–Duke boundary conditions (ψ/m), also at $z = L$, then:

$$\frac{ik_w}{m_w^*}(A-B)\exp(ikL) = \frac{ik_b}{m_b^*}\left[C\exp(ik_bL) - D\exp(-ik_bL)\right] \qquad\qquad (2.161)$$

Then using the BenDaniel–Duke boundary conditions at $z = l_w$:

$$A\exp(ik_wl_w) + B\exp(-ik_wl_w) = C\exp(ik_bl_w) + D\exp(-ik_bl_w) \qquad\qquad (2.162)$$

and

$$\frac{ik_w}{m_w^*}\left[A\exp(ik_wl_w) - B\exp(-ik_wl_w)\right] = \frac{ik_b}{m_b^*}\left[C\exp(ik_bl_w) - D\exp(-ik_bl_w)\right] \qquad (2.163)$$

Equations (2.160), (2.161), (2.162) and (2.163), can be rewritten as:

$$
\begin{pmatrix}
\exp\left(ikL\right) & \exp\left(ikL\right) & -\exp\left(ik_bL\right) & -\exp\left(-ik_bL\right) \\
\frac{k_w}{m_w^*}\exp\left(ikL\right) & -\frac{k_w}{m_w^*}\exp\left(ikL\right) & -\frac{k_b}{m_b^*}\exp\left(ik_bL\right) & \frac{k_b}{m_b^*}\exp\left(-ik_bL\right) \\
\exp\left(ik_wl_w\right) & \exp\left(-ik_wl_w\right) & -\exp\left(ik_bl_w\right) & -\exp\left(-ik_bl_w\right) \\
\frac{k_w}{m_w^*}\exp\left(ik_wl_w\right) & -\frac{k_w}{m_w^*}\exp\left(-ik_wl_w\right) & -\frac{k_b}{m_b^*}\exp\left(ik_bl_w\right) & \frac{k_b}{m_b^*}\exp\left(-ik_bl_w\right)
\end{pmatrix}
$$

$$
\times
\begin{pmatrix}
A \\
B \\
C \\
D
\end{pmatrix}
= 0
\tag{2.164}
$$

which implies that, for a solution other than the trivial $A = B = C = D = 0$, the determinant of the 4×4 must be zero, i.e.

$$
\begin{vmatrix}
\exp\left(ikL\right) & \exp\left(ikL\right) & -\exp\left(ik_bL\right) & -\exp\left(-ik_bL\right) \\
\frac{k_w}{m_w^*}\exp\left(ikL\right) & -\frac{k_w}{m_w^*}\exp\left(ikL\right) & -\frac{k_b}{m_b^*}\exp\left(ik_bL\right) & \frac{k_b}{m_b^*}\exp\left(-ik_bL\right) \\
\exp\left(ik_wl_w\right) & \exp\left(-ik_wl_w\right) & -\exp\left(ik_bl_w\right) & -\exp\left(-ik_bl_w\right) \\
\frac{k_w}{m_w^*}\exp\left(ik_wl_w\right) & -\frac{k_w}{m_w^*}\exp\left(-ik_wl_w\right) & -\frac{k_b}{m_b^*}\exp\left(ik_bl_w\right) & \frac{k_b}{m_b^*}\exp\left(-ik_bl_w\right)
\end{vmatrix}
$$

$$
= 0
\tag{2.165}
$$

Proceeding one step would give:

$$
\exp\left(ikL\right)
\begin{vmatrix}
-\frac{k_w}{m_w^*}\exp\left(ikL\right) & -\frac{k_b}{m_b^*}\exp\left(ik_bL\right) & \frac{k_b}{m_b^*}\exp\left(-ik_bL\right) \\
\exp\left(-ik_wl_w\right) & -\exp\left(ik_bl_w\right) & -\exp\left(-ik_bl_w\right) \\
-\frac{k_w}{m_w^*}\exp\left(-ik_wl_w\right) & -\frac{k_b}{m_b^*}\exp\left(ik_bl_w\right) & \frac{k_b}{m_b^*}\exp\left(-ik_bl_w\right)
\end{vmatrix}
$$

$$
-\exp\left(ikL\right)
\begin{vmatrix}
\frac{k_w}{m_w^*}\exp\left(ikL\right) & -\frac{k_b}{m_b^*}\exp\left(ik_bL\right) & \frac{k_b}{m_b^*}\exp\left(-ik_bL\right) \\
\exp\left(ik_wl_w\right) & -\exp\left(ik_bl_w\right) & -\exp\left(-ik_bl_w\right) \\
\frac{k_w}{m_w^*}\exp\left(ik_wl_w\right) & -\frac{k_b}{m_b^*}\exp\left(ik_bl_w\right) & \frac{k_b}{m_b^*}\exp\left(-ik_bl_w\right)
\end{vmatrix}
$$

$$
-\exp\left(ik_bL\right)
\begin{vmatrix}
\frac{k_w}{m_w^*}\exp\left(ikL\right) & -\frac{k_w}{m_w^*}\exp\left(ikL\right) & \frac{k_b}{m_b^*}\exp\left(-ik_bL\right) \\
\exp\left(ik_wl_w\right) & \exp\left(-ik_wl_w\right) & -\exp\left(-ik_bl_w\right) \\
\frac{k_w}{m_w^*}\exp\left(ik_wl_w\right) & -\frac{k_w}{m_w^*}\exp\left(-ik_wl_w\right) & \frac{k_b}{m_b^*}\exp\left(-ik_bl_w\right)
\end{vmatrix}
$$

$$
\exp\left(-ik_bL\right)
\begin{vmatrix}
\frac{k_w}{m_w^*}\exp\left(ikL\right) & -\frac{k_w}{m_w^*}\exp\left(ikL\right) & -\frac{k_b}{m_b^*}\exp\left(-ik_bL\right) \\
\exp\left(ik_wl_w\right) & \exp\left(-ik_wl_w\right) & -\exp\left(ik_bl_w\right) \\
\frac{k_w}{m_w^*}\exp\left(ik_wl_w\right) & -\frac{k_w}{m_w^*}\exp\left(-ik_wl_w\right) & -\frac{k_b}{m_b^*}\exp\left(ik_bl_w\right)
\end{vmatrix}
$$

which the reader can see becomes very tedious indeed. Fortunately use can be made of the constant mass case of Jaros [5], and generalized to account for $m_w \neq m_b$ as here, giving:

$$
\cos\left(k_wl_w\right)\cos\left(k_bl_b\right) - \sin\left(k_wl_w\right)\sin\left(k_bl_b\right)\left(\frac{m_b^2k_w^2 + m_w^2k_b^2}{2m_wm_bk_wk_b}\right) = \cos\left(kL\right)
\tag{2.166}
$$

If $E < V$, then following the arguments in Section 2.11, k_b becomes $i\kappa$, where:

$$
\kappa = \sqrt{\frac{2m_b^*}{\hbar^2}(V - E)}
\tag{2.167}
$$

Substituting for k_b into equation (2.166), and using the following identities:

$$\cos{(i\kappa l_b)} \equiv \cosh{(\kappa l_b)} \qquad \text{and} \qquad \sin{(i\kappa l_b)} \equiv i\sinh{(\kappa l_b)} \qquad (2.168)$$

then:

$$\cos{(k_w l_w)}\cosh{(\kappa l_b)} - \sin{(k_w l_w)}\sinh{(\kappa l_b)}\left(\frac{m_b^2 k_w^2 - m_w^2 \kappa^2}{2m_w m_b k_w \kappa}\right) = \cos{(kL)} \qquad (2.169)$$

The superlattice dispersion curves, i.e. the energy E of a particle as a function of its wave vector k, are obtained by solving equations (2.166) and (2.169). This is accomplished by using the same methods as in Section 2.5, i.e. equations (2.166) and (2.169) are rewritten in the form $f(E, k) = 0$, and solved for chosen values of k. Again, a Newton–Raphson iteration is efficient; however, in order to avoid having to deduce $f'(E, k)$, a finite difference expansion can be employed:

$$f'(E, k) \approx \frac{f(E + \delta E, k) - f(E - \delta E, k)}{2\delta E} \qquad (2.170)$$

Hence the Newton–Raphson iterative equation becomes:

$$E^{(n+1)} = E^{(n)} - \frac{2f(E^{(n)})\,\delta E}{f(E^{(n)} + \delta E) - f(E^{(n)} - \delta E)} \qquad (2.171)$$

where δE is smaller than adjacent solutions for E; typically 10^{-3} meV is a good choice. This convenient method of approximating derivatives with nearby values of the function will be returned to in the next chapter and dealt with there in much more detail.

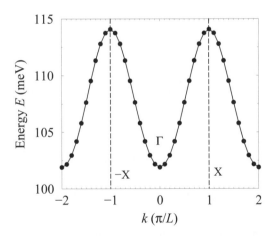

Figure 2.27   Lowest-energy solutions of the Kronig–Penney model of a 40 Å GaAs/40 Å Ga$_{0.6}$Al$_{0.4}$As superlattice, for a range of electron momenta along the growth (z-) axis

Fig. 2.27 shows the lowest-energy solutions of a 40 Å GaAs/40 Å Ga$_{0.6}$Al$_{0.4}$As superlattice, for a range of k values. It can be seen that the energy E

is a periodic function of k with period $2\pi/L$. The wave vector k was introduced earlier to describe the wave function as a travelling wave; clearly it can be identified as the momentum of a charge carrier along the growth (z-) axis of an *infinite* superlattice. Using Fig. 2.27, it can be seen that a definite minimum exists when the particle is at rest, i.e. $k=0$. As the carrier momentum increases, its energy increases too and reaches a maximum at $k = \pm\pi/L$. Thus the carrier within the superlattice occupies a continuous range of energies with a maxima and minima; this domain is the superlattice analogy to the energy bands of a crystal and is referred to as a *miniband*.

In addition the symmetry points of the $E–k$ curves are labelled as in bulk crystals, where $k=0$ is called Γ and $k = \pi/L$ is referred to as X. This domain in k-space has become known as the *superlattice Brillouin zone* and as in the bulk, is a convenient way to express the relationship between energy and momenta.

Just as single quantum wells can have more than one confined state, superlattices can have more than one miniband. Fig. 2.28 displays the two lowest energy solutions of the longer period 60 Å GaAs/60 Å Ga$_{0.6}$Al$_{0.4}$As superlattice; note on this scale that the lowest energy miniband looks almost flat. Clearly, the second miniband has the same periodicity as the first, although it should be noted that its minimum in energy occurs at the edge of the superlattice Brillouin zone and not at the centre. This effect is exploited in the recently developed 'inter-miniband laser' [41–43], which is a very promising high-power source of mid-infrared radiation. This device will be discussed in more detail later, i.e. in the chapter on electron scattering.

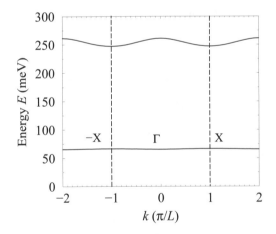

Figure 2.28   Two lowest-energy solutions of an 60 Å GaAs/60 Å Ga$_{0.6}$Al$_{0.4}$As superlattice

The structural dependence of the miniband is illustrated in Fig. 2.29 for a series of superlattices with equal well and barrier widths. The two solid curves display the energy as a function of l_w at Γ and X. It can be seen that these energies fall with increasing well width, as does the miniband width, defined as follows:

$$\text{miniband width} = |E(X) - E(\Gamma)| \tag{2.172}$$

where the modulus bars have been included to account for minibands such as the second, for which $E(X) < E(\Gamma)$. Interestingly, the top and bottom of the miniband straddle the energy of the corresponding single quantum well (given by the dashed line). As the wells become more isolated and further apart, the miniband narrows and the energies tend towards that of the single quantum well. This represents the cross-over from a superlattice to a multiple quantum well—a collection of non-interacting identical wells.

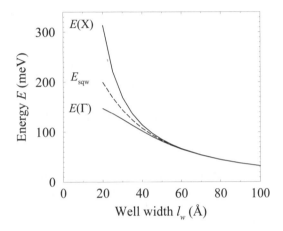

Figure 2.29 Energy at $k = 0$ (Γ) and $k = \pi/L$ (X), showing the lower and upper energy limits of the superlattice minibands, compared with the ground state energy of the single quantum well of the same width

As the miniband width changes with structure, the curvature of the E–k curve changes too. It is clear from Chapter 1 that the effective mass of a particle subject to an energy–momentum dependence of $E(\mathbf{k})$ can be derived from:

$$m^* = \frac{\hbar^2}{\partial^2 E/\partial \mathbf{k}^2} \qquad (2.173)$$

Therefore, when an electron or hole is moving along the growth axis of a superlattice, the energy–momentum relationship is such that they appear to have a different effective mass from the bulk. Fig. 2.30 plots this new effective mass at Γ for the range of structures in Fig. 2.29.

It can be seen from the figure that at very short periods the electron's miniband structure produces a new effective mass which is considerably smaller than the bulk value ($0.067\,m_0$). Thus electrons within this miniband will have a much greater response to an electric field than in bulk, just as from Newton's second law, '$F = ma$', a lower m requires a smaller force in order to produce the same acceleration. In strict semiconductor language, this is quantified in terms of the carrier mobility, which is given by:

$$\mu = \frac{e\tau_{\text{coll.}}}{m^*} \qquad (2.174)$$

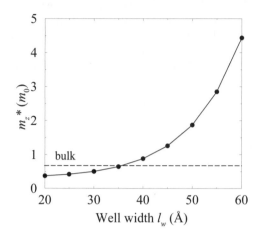

Figure 2.30   The new effective mass for an electron moving in the superlattice miniband

where $\tau_{\mathrm{coll.}}$ is the time between collisions, see [1], p.601, the reciprocal of which is usually referred to as the total scattering rate for all mechanisms, via phonons, ionised impurities, other carriers, etc. Clearly, a reduced effective mass within a miniband gives an increased mobility μ for transport within that miniband, *provided* that all other scattering terms remain the same.

   In this series of calculations, attention has been focussed on superlattices with a simple single quantum well unit cell, and indeed with equal well and barrier widths. If these restrictions are lifted, then there are an almost infinite number of possible superlattices that can be constructed, thus giving the opportunity to engineer, even design, the effective mass to suit the device.

2.14   THE SINGLE BARRIER

Hitherto, only semiconductor layered systems have been considered which form quantum wells, i.e. produce semiconductor layers which trap or confine the electrons along one axis. The 'opposite' of such structures also exist and are known as *barrier structures*. If a layer of a larger-band-gap material, e.g. $\mathrm{Ga}_{1-x}\mathrm{Al}_x\mathrm{As}$, is sandwiched between layers of a narrower-band-gap material, e.g. GaAs, then a potential barrier *can* result which repels carriers.

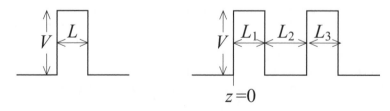

Figure 2.31   Single- and double-barrier structures

Fig. 2.31 illustrates such a single barrier structure. Electrons placed within this layered system simply collect in the GaAs outer (lower energy) regions and there are no quantum-confined energy states. These systems do, however, exhibit quantum behaviour when an electric field is applied perpendicular to the layers (along the growth (z-) axis). Electrons (or holes) arising from doping are accelerated and impinge upon the barrier and even when they have an energy E which is less than the potential energy height V of the barrier, they have a finite probability of passing through the barrier and appearing on the other side. This phenomenon is called *quantum mechanical tunnelling* or often just *tunnelling*. The bizarre nature of quantum mechanical tunnelling can be illustrated with a classical analogy, i.e. one would be pretty disturbed, if when kicking a football against a brick wall, it went straight through! Yet this is exactly what happens with *some* electrons and holes when they meet a potential barrier within a crystal. Tunnelling is an everyday phenomenon which occurs in a range of semiconductor devices, some of which appear in consumer electronics products, see, for example, Sze [44].

One way of quantifying the proportion of electrons that tunnel through a barrier is in terms of the *transmission coefficient* which is defined as the probability that any single electron impinging on a barrier structure will tunnel and contribute to the current flow through the barrier. Ferry has produced a comprehensive analysis of the transmission coefficient for a single-barrier structure (see [45], p.60). Suffice here to quote Ferry's result in ([45], equation 3.12), for a constant effective mass across the structure, i.e. the transmission coefficient at an energy E for a barrier of width L and height V is given by:

$$T(E) = \frac{1}{1 + \left(\frac{k^2+\kappa^2}{2k\kappa}\right)^2 \sinh^2\left(\kappa L\right)}, \qquad \text{for} \qquad E < V \qquad (2.175)$$

where as usual:

$$k = \frac{\sqrt{2m^*E}}{\hbar} \quad \text{and} \quad \kappa = \frac{\sqrt{2m^*(V-E)}}{\hbar} \qquad (2.176)$$

For values of the carrier energy E greater than the barrier height V, $\kappa \to ik'$ (as in Section 2.13), and hence:

$$T(E) = \frac{1}{1 + \left(\frac{k^2-k'^2}{2kk'}\right)^2 \sin^2\left(k'L\right)} \qquad \text{for} \qquad E > V \qquad (2.177)$$

where

$$k = \frac{\sqrt{2m^*E}}{\hbar} \quad \text{and} \quad k' = \frac{\sqrt{2m^*(E-V)}}{\hbar} \qquad (2.178)$$

The mathematics show, that for $E > V$, the transmission coefficient would be expected to oscillate, with resonances when the sine term is zero. These occur when:

$$k'L = n\pi \qquad (2.179)$$

which is implied when:

$$E = \frac{(n\hbar\pi)^2}{2m^*L^2} + V \qquad (2.180)$$

The squared dependence implies that the resonances when $T = 1$ occur at larger and larger intervals in E, which can be clearly seen in Figs. 2.32 and 2.33.

Fig. 2.32 displays the transmission coefficient as a function of the energy E and for a range of barrier widths L. For this range of calculations, the barrier height V was fixed at

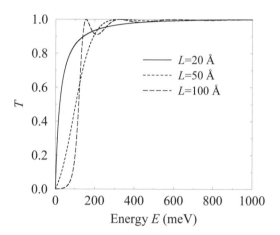

Figure 2.32   Transmission coefficient as a function of the energy through a single barrier for different barrier widths

100 meV, and below this energy the thinner the barrier, then the higher the probability of tunnelling. For $E > V$, the situation is more complex due to the oscillatory nature of T. The trend, however, as highlighted by the curves, is that the thicker the barrier, the closer the first resonance ($T = 1$) is to the top of the barrier. This can be understood from equation (2.180):

$$\lim_{L \to \infty} E^{n=1}_{\text{resonance}} = \lim_{L \to \infty} \frac{(\hbar \pi)^2}{2m^* L^2} + V = V \tag{2.181}$$

Conversely, for a fixed L and a variable V, the first resonance occurs at the same point above the barrier height; this is clearly illustrated in Fig. 2.33.

2.15 THE DOUBLE BARRIER

If two barriers are placed a reasonably small distance apart (in the same crystal, perhaps a few nm) then the system is known as a *double barrier* (see Fig. 2.31), and has quite different transmission properties to the single barrier. Datta [46], p.33, has deduced the transmission $T(E)$ dependence for the restricted case of symmetric barriers, while Ferry [45], p.66, has considered asymmetric barriers. In this formalism, allowance will be made for differing barrier widths as well as the discontinuous change in the effective mass between well and barrier materials. Thus, with the aim of deducing the new $T(E)$, consider the solutions to Schrödinger's equation within each region for $E < V$:

$$\psi(z) = A \exp(ikz) + B \exp(-ikz), \qquad z < I_1 \tag{2.182}$$
$$\psi(z) = C \exp(\kappa z) + D \exp(-\kappa z), \quad I_1 < z < I_2 \tag{2.183}$$
$$\psi(z) = F \exp(ikz) + G \exp(-ikz), \quad I_2 < z < I_3 \tag{2.184}$$
$$\psi(z) = H \exp(\kappa z) + J \exp(-\kappa z), \quad I_3 < z < I_4 \tag{2.185}$$
$$\psi(z) = K \exp(ikz) + L \exp(-ikz), \quad I_4 < z \tag{2.186}$$

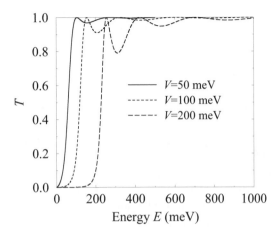

Figure 2.33   Transmission coefficient as a function of the energy through a single 100 Å barrier for different barrier heights

where k and κ have their usual forms as given in equation (2.176) and the positions of the interfaces have been labelled I_1, I_2, I_3, and I_4, respectively. Using the standard BenDaniel–Duke boundary conditions at each interface gives the following $z = I_1 = 0$:

$$A + B = C + D \tag{2.187}$$

$$\frac{1}{m_w}(ikA - ikB) = \frac{1}{m_b}(\kappa C - \kappa D) \tag{2.188}$$

$z = I_2 = L_1$:

$$C \exp(\kappa I_2) + D \exp(-\kappa I_2) = F \exp(ikI_2) + G \exp(-ikI_2) \tag{2.189}$$

$$\frac{1}{m_b}[\kappa C \exp(\kappa I_2) - \kappa D \exp(-\kappa I_2)] = \frac{1}{m_w}[ikF \exp(ikI_2) - ikG \exp(-ikI_2)] \tag{2.190}$$

$z = I_3 = L_1 + L_2$:

$$F \exp(ikI_3) + G \exp(-ikI_3) = H \exp(\kappa I_3) + J \exp(-\kappa I_3) \tag{2.191}$$

$$\frac{1}{m_w}[ikF \exp(ikI_3) - ikG \exp(-ikI_3)] = \frac{1}{m_b}[\kappa H \exp(\kappa I_3) - \kappa J \exp(-\kappa I_3)] \tag{2.192}$$

$z = I_4 = L_1 + L_2 + L_3$:

$$H \exp(\kappa I_4) + J \exp(-\kappa I_4) = K \exp(ikI_4) + L \exp(-ikI_4) \tag{2.193}$$

$$\frac{1}{m_b}[\kappa H \exp(\kappa I_4) - \kappa J \exp(-\kappa I_4)] = \frac{1}{m_w}[ikK \exp(ikI_4) - ikL \exp(-ikI_4)] \tag{2.194}$$

The method of solution is the transfer matrix technique as before, writing the above equations in matrix form:

$$\mathbf{M}_1 \begin{pmatrix} A \\ B \end{pmatrix} = \mathbf{M}_2 \begin{pmatrix} C \\ D \end{pmatrix} \tag{2.195}$$

$$\mathbf{M}_3 \begin{pmatrix} C \\ D \end{pmatrix} = \mathbf{M}_4 \begin{pmatrix} F \\ G \end{pmatrix} \tag{2.196}$$

$$\mathbf{M}_5 \begin{pmatrix} F \\ G \end{pmatrix} = \mathbf{M}_6 \begin{pmatrix} H \\ J \end{pmatrix} \tag{2.197}$$

$$\mathbf{M}_7 \begin{pmatrix} H \\ J \end{pmatrix} = \mathbf{M}_8 \begin{pmatrix} K \\ L \end{pmatrix} \tag{2.198}$$

Then, as before, the coefficients of the outer regions can be linked by forming the transfer matrix, i.e.

$$\begin{pmatrix} A \\ B \end{pmatrix} = \mathbf{M}_1^{-1}\mathbf{M}_2\mathbf{M}_3^{-1}\mathbf{M}_4\mathbf{M}_5^{-1}\mathbf{M}_6\mathbf{M}_7^{-1}\mathbf{M}_8 \begin{pmatrix} K \\ L \end{pmatrix} \tag{2.199}$$

Clearly, this 2×2 matrix equation still has four unknowns and can't be solved—it is at this point that additional boundary conditions have to be imposed from physical intuition. Whereas before, the standard boundary conditions, i.e. $\psi(z) \to 0$ as $z \to \pm\infty$, were used to solve for the confined states within quantum wells, in these barrier structures these are not appropriate since the travelling waves in the outer layers can have infinite extent. The standard procedure is to assume, quite correctly, that all of the charge carriers approach the double barrier from the same side, as would occur when as part of a biased device, as illustrated schematically in Fig. 2.34. Furthermore, if it is assumed that there are no further heterojunctions to the right of the structure, then no further reflections can occur and the wave function beyond the structure can only have a travelling wave component moving to the right, i.e. the coefficient L must be zero.

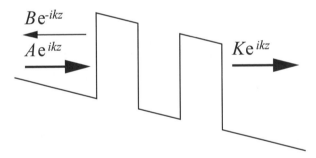

Figure 2.34 The wave function after imposition of the boundary conditions

Therefore, if the 2×2 matrix of equation (2.199) is written as $\mathcal{M}$, then:

$$\begin{pmatrix} A \\ B \end{pmatrix} = \mathcal{M} \begin{pmatrix} K \\ 0 \end{pmatrix} \tag{2.200}$$

$$\therefore A = \mathcal{M}_{11}K \tag{2.201}$$

and the ratio of transmitted to incident current, i.e. the transmission coefficient, is simply:

$$T(E) = \frac{K^*K}{A^*A} = \frac{1}{\mathcal{M}_{11}^*\mathcal{M}_{11}} \tag{2.202}$$

Fig. 2.35 gives an example of the form of $T(E)$ for barriers of height 100 meV and width 100 Å as a function of the distance L_2 between them. The effective masses

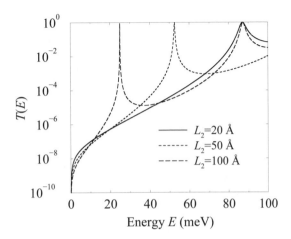

Figure 2.35   Transmission coefficient as a function of the energy through a double barrier of width 100 Å and height 100 meV, separated by a distance L_2

were both taken to be equal to the bulk Γ valley electron mass of 0.067 m_0. It can be seen from the figure that the curves contain Dirac δ-functions at certain energies E below the potential barrier height V. This is quite unlike the single barrier case. At these *resonance* energies, the double-barrier system appears transparent and has a transmission coefficient of 1. The wave functions of these states are localized between the barriers and are often referred to as *quasi-bound states* since they resemble the bound states of quantum well structures. However, they are not stationary states in that electrons or holes in such states will eventually scatter into the lower energy states outside of the barriers.

The effect of an increasing barrier height V is shown in Fig. 2.36. It can be seen that, away from a resonance, an increasing barrier height leads, as would be expected, to a decrease in the transmission coefficient T. The classical explanation for this would be, 'it is harder for the electrons to tunnel through higher barriers'. The *resonance energies* increase with increasing barrier height due to confinement effects, and the appearance of the second resonance at higher energies is a reflection on the existence of a second quasi bound state.

While the transmission coefficient represents a very important parameterisation of the properties of a double-barrier structure, it itself is not a measurable quantity. In fact, the properties of such two-terminal electronic devices are generally inferred (or summarized) from their current–voltage characteristics (I–V curve).

When an electric field is placed across such a double barrier structure, any charge carriers present in the semiconductor, intrinsic or extrinsic, constitute a current which approaches the left-hand barrier. These charge carriers have a distribution of energy and momenta, often a Fermi-Dirac distribution. Those carriers that are of the same energy as the resonance are able to pass right through the double barrier without hindrance—a phenomenum which has become known as *resonant tunnelling*. As the applied electric field (applied voltage) is increased, the number of carriers with the resonance energy

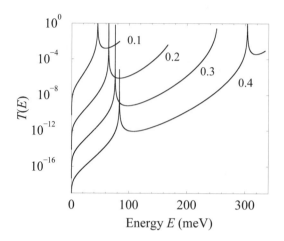

Figure 2.36   Transmission coefficient as a function of the energy through a 100 Å $Ga_{1-x}Al_xAs$ /50 Å GaAs/100 Å $Ga_{1-x}Al_xAs$ double barrier, for x=0.1, 0.2, 0.3, and 0.4

increases and peaks as the Fermi level of the semiconductor to the left of the first barrier is brought into alignment. Therefore, the current gradually increases. At higher fields, the current falls away and a period of negative differential resistance ensues [47, 48]. Such a current–voltage characteristic has been exploited in high frequency circuits, thus bringing the *resonant tunnelling diode* to prominence as a very useful electronic device [49].

There are various models of the current–voltage properties of different levels of complexity, the simplest of which would probably be to return to the idea that the current at any particular field would be equal to the number of carriers that tunnelled through the structure. This in turn would be the probability of a particular carrier tunnelling, multiplied by the number of carriers at that energy, i.e.

$$I \propto \int_{\text{band}} T(E) f^{\text{FD}}(E) \rho^{\text{3D}}(E) \ \mathrm{d}E \tag{2.203}$$

As the carriers approaching the barrier structure are in a bulk band, then the integral is over their energies, and the Fermi–Dirac distribution function and density of states have the bulk (3D) forms.

Fig. 2.37 outlines this model. The electric field dependence is introduced via f^{FD} and ρ^{3D}. As the electric field is increased, the bottom of the bulk band is increased in energy relative to the centre of the double-barrier structure by an amount $\Delta E = eF(L_1 + L_2/2)$. Hence, the reference energy (the band minimum) will increase, and by using equation (2.40), the density of states at some energy E, measured from the band minimum at the centre of the well, would become:

$$\rho(E) = \frac{1}{2\pi^2} \left(\frac{2m^*}{\hbar^2} \right)^{\frac{3}{2}} (E - \Delta E)^{\frac{1}{2}} \ \Theta(E - \Delta E) \tag{2.204}$$

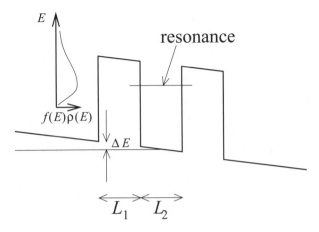

Figure 2.37   Simple model of current flow through a double barrier structure

where the unit step function ensures that the energy of the carriers impinging on the left-hand barrier E is greater than ΔE. At the bottom of the band, $E = \Delta E$, and $\rho(E) = 0$. In addition, the Fermi energy of the bulk carriers increases by the same amount, i.e. from equation (2.49):

$$f^{\mathrm{FD}}(E) = \frac{1}{\exp\left\{[E - (E_{\mathrm{F}} + \Delta E)]/kT\right\} + 1} \tag{2.205}$$

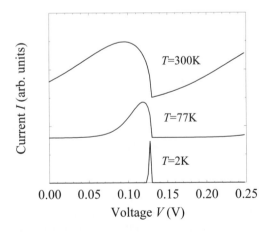

Figure 2.38   Current–voltage curve for the 100 Å $\mathrm{Ga_{0.8}Al_{0.2}As}$/50 Å GaAs/100 Å $\mathrm{Ga_{0.8}Al_{0.2}As}$ of earlier, obtained at different temperatures

Fig. 2.38 illustrates the results obtained by using this model. The current is obtained in arbitrary units and the voltage V has been defined simply as the potential difference across the structure at a particular field, i.e. $V = F(L_1 + L_2 + L_3)$. The I–V curves have been plotted for several temperatures for the $x = 0.2$ structure of Fig. 2.36. The single resonance within this system shows itself as a single peak in the current. Clearly, the current peak broadens as the temperature increases, with this being a direct result of the broadening of the carrier distribution (the number of carriers in a given energy range), given by the product $f^{FD}(E)\rho(E)$ appearing in the integral for the current, and illustrated schematically in Fig. 2.37. At low temperatures, the carriers occupy a small region of energy space around the band minima. As the field is increased, current only begins to flow when this narrow distribution is brought into line with the resonance energy, and hence there is a narrow current peak. As the temperature increases, the carrier distribution broadens, therefore there is a greater range of applied voltages that give some degree of alignment of the carriers with the resonance energy. The peak occurs when the peak of the distribution is aligned with the resonance energy. At voltages above this, the number of carriers available for tunnelling decreases, and hence the current also decreases.

The non-zero current at zero field for $T = 300$ K is a consequence of this simplistic model. It arises because even at zero field, there is a finite number of carriers in the broadened distribution which are aligned with the resonance energy and are therefore able to tunnel.

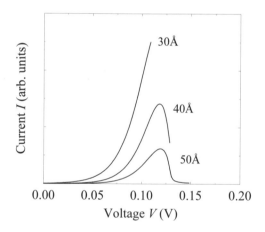

Figure 2.39   Current–voltage curves obtained at 77 K for a variety of barrier widths the structure of Fig. 2.38

Fig. 2.39 displays the results of the same simple model as a function of barrier width for the same structure of Fig. 2.38, but this time at the fixed temperature of 77 K. Again, as noted before, thinner barriers allow the electrons to tunnel more easily and hence give a higher current. An interesting physical point can be inferred from this data. The incomplete nature of the peaks in the 40 and 30 Å data is an indication that a significant fraction of the electrons are passing over the top of the barriers. While the current–voltage model advanced in this section can account for such a situation, the transmission coefficient versus energy data is terminated at the top of the barrier, and so only the fraction of electrons below this are included in the integration for the current in equation (2.203). It is left as an exercise for the reader to generalise the transmission coefficient theory to account for the case when $E > V$, and therefore improve the current–voltage model. This can be achieved by using a similar approach to that of Section 2.13.

The above discussion has been a simple introduction to the modelling of I–V curves for barrier structures, but nonetheless it shows some of the features of real devices. For a much more complete and in depth study see, for example, Mizuta and Tanoue [49].

2.16   EXTENSION TO INCLUDE ELECTRIC FIELD

An obvious improvement to the above model would be to account for the changes in the transmission coefficient as a function of the applied electric field. By using the substitution as before (equation (2.150)), i.e.

$$z' = \left(\frac{2m^*}{\hbar^2}\right)^{\frac{1}{3}} \left[\frac{V(z) - E}{(eF)^{\frac{2}{3}}} - (eF)^{\frac{1}{3}} z\right] \tag{2.206}$$

the solution in each region can then be written as a linear combination of Airy functions, just as for the general electric field case of equation (2.152), i.e.

$$\psi(z) = A\text{Ai}(z') + B\text{Bi}(z'), \qquad z < I_1 \tag{2.207}$$
$$\psi(z) = C\text{Ai}(z') + D\text{Bi}(z'), \qquad I_1 < z < I_2 \tag{2.208}$$
$$\psi(z) = F\text{Ai}(z') + G\text{Bi}(z'), \qquad I_2 < z < I_3 \tag{2.209}$$
$$\psi(z) = H\text{Ai}(z') + J\text{Bi}(z'), \qquad I_3 < z < I_4 \tag{2.210}$$
$$\psi(z) = K\text{Ai}(z') + L\text{Bi}(z'), \qquad I_4 < z \tag{2.211}$$

while Airy functions *can* be difficult to work with numerically, an immediate advantage over the zero-field case is that this solution is valid for both $E < V$ and $E > V$, and hence generalisation to this form produces two benefits. The method of solution is analogous to the zero-field case, in that application of the BenDaniel–Duke boundary conditions yields two equations for each interface, which in this case gives eight equations. The unknown coefficients, A and B, are linked to K and L as before by forming the transfer matrix, and are solved by imposition of a boundary condition, which is again a travelling wave in the direction of $+z$ to the right of the barrier structure. It is left to the interested reader to follow through such a derivation. A very general implementation for multiple barrier structures has been reported in the literature by Vatannia and Gildenblat [39].

2.17   MAGNETIC FIELDS AND LANDAU QUANTISATION

If a magnetic field is applied externally to a *non-magnetic* semiconductor heterostructure then the constant effective mass Hamiltonian (familiar from equation (2.5)):

$$\mathcal{H} = -\frac{\hbar^2}{2m^*}\frac{\partial^2}{\partial z^2} + V(z) \tag{2.212}$$

which can be written:

$$\mathcal{H} = \frac{\mathcal{P}^2}{2m^*} + V(z) \tag{2.213}$$

becomes [50–53]:

$$\mathcal{H} = \frac{1}{2m^*}\left(\mathcal{P} + e\mathbf{A}\right)^2 \mp \frac{1}{2}g^*(z)\mu_B B + V(z) \tag{2.214}$$

where the kinetic energy operator becomes modified by the magnetic field vector potential $\mathbf{A}$, and the second term produces a splitting, known as the 'gyromagnetic spin splitting' between the spin-up ($-$ sign) and spin-down ($+$ sign) electrons. g^* is known as the 'Landé factor' and is really a function of z as it depends on the material; however, it is generally assumed to be constant and approximately 2 for conduction band electrons, μ_B is the Bohr magneton and B is the magnitude of the magnetic flux density which is assumed aligned along the growth (z-) axis.

Although the heterostructure potential $V(z)$ remains one-dimensional, the vector potential means that the wave functions are not necessarily one-dimensional so the Schrödinger equation must be written:

$$\left[\frac{1}{2m^*}\left(\mathcal{P} + e\mathbf{A}\right)^2 \mp \frac{1}{2}g^*(z)\mu_B B + V(z)\right]\psi(x,y,z) = E\psi(x,y,z) \tag{2.215}$$

The magnetic field produces a parabolic potential along one of the in-plane axes, the x-axis say, leaving the particle free to move (with a wave vector k_y, say) along the other axis. The standard approach is to employ the Landau gauge $\mathbf{A} = Bx\mathbf{e}_y$, then following the notation of Savić et al. [54] the wave function can be written in the separable form $\psi(x,y,z) = \psi_x(x)\psi_y(y)\psi_z(z)$, i.e.

$$\psi(x,y,z) = \frac{1}{\sqrt{L_y}}\psi_j(x - X_{k_y})e^{ik_y y}\psi(z) \tag{2.216}$$

where L_y is a normalization constant (the length of the structure along the y-axis, $X_{k_y} = -k_y l_B^2$, where $l_B = \sqrt{\hbar/eB}$ is the Landau length, j is an index over the $\psi_j(x - X_{k_y})$ harmonic oscillator solutions of the parabolic potential and $\psi(z)$ is the usual one-dimensional envelope function of the heterostructure potential without the magnetic field. The harmonic oscillator solutions will be generated numerically in Section 3.5, however they can also be expressed analytically as:

$$\psi_j(x - X_{k_y}) = \frac{1}{\pi^{\frac{1}{4}}\sqrt{2^j j! l_B}}\exp\left(-\frac{(x - X_{k_y})^2}{2l_B^2}\right)H_j\left(\frac{x - X_{k_y}}{l_B}\right) \tag{2.217}$$

where H_j is the jth Hermite polynomial, see Liboff [55] page 200 for a detailed description of the analytical solutions of the harmonic oscillator.

Taking g^* as a constant equal to 2 and the Bohr magneton as 9.274×10^{-24} JT^{-1} then it can be seen that even in a relatively high magnetic field of 10 T, the difference in energy between the spin-up and spin-down electrons is:

$$g^*\mu_B B \approx 2 \times 9.274 \times 10^{-24}\mathrm{JT}^{-1} \times 10\ \mathrm{T} \approx 1\ \mathrm{meV} \tag{2.218}$$

which on the scales of typical quantum well systems and typical carrier densities is relatively small and can generally be ignored. Thus the total energy therefore can be

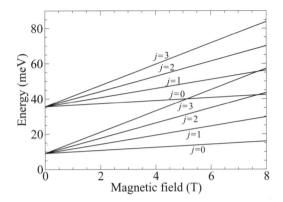

Figure 2.40 The magnetic field induced Landau levels of a 200 Å GaAs quantum well surrounded by $Ga_{1-x}Al_xAs$ barriers of height 100 meV (same structure as Fig. 2.15), with constant effective mass of $0.067m_0$. Note only the first 4 Landau levels associated with each quantum well subband are shown for clarity.

considered to be composed of just two components: the usual energy eigenvalue E_n associated with the nth state of the quantum well without a magnetic field and the harmonic oscillator type energy associated with the in-plane cyclotron motion $(j+\frac{1}{2})\hbar\omega_c$, i.e.

$$E_{n,j} = E_n + \left(j + \frac{1}{2}\right)\hbar\omega_c \qquad (2.219)$$

where the 'cyclotron frequency' is the usual one from bulk semiconductors [1,2,56]:

$$\omega_c = \frac{eB}{m^*} \qquad (2.220)$$

Fig. 2.40 shows the Landau level splitting induced by the application of an external magnetic field on a 200 Å GaAs quantum well surrounded by $Ga_{1-x}Al_xAs$ barriers, assuming a constant electron effective mass of $0.067m_0$. The figure illustrates the lowest four Landau levels for each of the two lowest energy confined states within the quantum well. A more accurate calculation could include the variation in effective mass of the carrier between the various layers; this would be done by calculating the weighted mean or expectation value of the effective mass according to the probability of finding the carrier in each of the semiconductor layers.

2.18 IN SUMMARY

The consideration of simple layered semiconductor heterostructures with analytical forms for the solutions to Schrödinger's equation has allowed exploration and discovery of the properties of two-dimensional systems. Such models are invaluable and allow a whole range of physical observables of experimental layer structures and electronic devices to be explained. However, implementing these methods computationally *can* be tedious, in that a computer program, written to calculate the energy levels of a single quantum well, has to be rewritten for a double quantum well. In addition, a different program would be needed to solve a triangular well, and furthermore some potential profiles, which can be fabricated, such as diffused quantum wells, have no analytical solution at all. While such continuously varying structures can be expressed with flat step potentials for each monolayer [57], the treatment of an electric field in this manner is questionable.

In the next chapter, a simple, but very general *numerical* solution to Schrödinger's equation will be derived, which will overcome all of these difficulties.

CHAPTER 3

NUMERICAL SOLUTIONS

3.1 SHOOTING METHOD

As a starting point, consider a general, but simple (constant-mass) form for the time-independent Schrödinger equation, the analytical solutions of which have been extensively studied in the previous chapter:

$$-\frac{\hbar^2}{2m^*}\frac{\partial^2}{\partial z^2}\psi(z) + V(z)\psi(z) = E\psi(z) \tag{3.1}$$

where the one-dimensional potential $V(z)$ will remain undefined and again $\psi(z)$ is the wave function representing the particle of interest *while under the effective mass and envelope function approximations*. This, of course, can be written as follows:

$$-\frac{\hbar^2}{2m^*}\frac{\partial^2}{\partial z^2}\psi(z) + [V(z) - E]\,\psi(z) = 0 \tag{3.2}$$

The problem now is to find a numerical method for the solution of both the energy eigenvalues E and the eigenfunctions $\psi(z)$ for *any* $V(z)$.

With this aim, consider expanding the second-order derivative in terms of finite differences. For example, as in Fig. 3.1, the first derivative of any function is defined as:

$$\lim_{\Delta z \to 0}\frac{\Delta f}{\Delta z} = \frac{\mathrm{d}f}{\mathrm{d}z} \tag{3.3}$$

Quantum Wells, Wires and Dots, Third Edition. P. Harrison
©2009 John Wiley & Sons, Ltd.

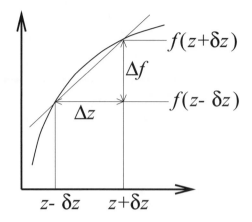

Figure 3.1   The first derivative of a function

It suits the purpose here to retain the approximate form, i.e.

$$\frac{\mathrm{d}f}{\mathrm{d}z} \approx \frac{\Delta f}{\Delta z} = \frac{f(z + \delta z) - f(z - \delta z)}{2\delta z} \qquad (3.4)$$

Hence, the second derivative follows as:

$$\frac{\mathrm{d}^2 f}{\mathrm{d}z^2} \approx \frac{\left.\frac{\mathrm{d}f}{\mathrm{d}z}\right|_{z+\delta z} - \left.\frac{\mathrm{d}f}{\mathrm{d}z}\right|_{z-\delta z}}{2\delta z} \qquad (3.5)$$

By using the finite difference forms in equation (3.4) for the first derivatives, then:

$$\frac{\mathrm{d}^2 f}{\mathrm{d}z^2} \approx \frac{\left[\frac{f(z+2\delta z)-f(z)}{2\delta z}\right] - \left[\frac{f(z)-f(z-2\delta z)}{2\delta z}\right]}{2\delta z} \qquad (3.6)$$

$$\therefore \frac{\mathrm{d}^2 f}{\mathrm{d}z^2} \approx \frac{f(z + 2\delta z) - 2f(z) + f(z - 2\delta z)}{(2\delta z)^2} \qquad (3.7)$$

As δz is an, as yet, undefined small step length along the z-axis, and as it only appears in equation (3.7) with the factor 2, then this finite difference representation of the second derivative can be simplified slightly by substituting δz for $2\delta z$, i.e.

$$\frac{\mathrm{d}^2 f}{\mathrm{d}z^2} \approx \frac{f(z + \delta z) - 2f(z) + f(z - \delta z)}{(\delta z)^2} \qquad (3.8)$$

Using this form for the second derivative in the original Schrödinger equation and taking the step length δz as sufficiently small that the approximation is good, i.e. drop the '$\approx$' in favour of '$=$', then:

$$-\frac{\hbar^2}{2m^*}\left[\frac{\psi(z + \delta z) - 2\psi(z) + \psi(z - \delta z)}{(\delta z)^2}\right] + [V(z) - E]\,\psi(z) = 0 \qquad (3.9)$$

$$\therefore \psi(z + \delta z) - 2\psi(z) + \psi(z - \delta z) = \frac{2m^*}{\hbar^2}(\delta z)^2\,[V(z) - E]\,\psi(z) \qquad (3.10)$$

which can finally be written as:

$$\psi(z + \delta z) = \left[\frac{2m^*}{\hbar^2} (\delta z)^2 \left(V(z) - E \right) + 2 \right] \psi(z) - \psi(z - \delta z) \tag{3.11}$$

Equation (3.11) implies that if the wave function is known at the two points $(z - \delta z)$ and z, then the value of the wave function at $(z + \delta z)$ can be calculated for any energy E. This iterative equation forms the basis of a standard method of solving differential equations numerically, and is known as the *shooting method* [58].

Using two known values of the wave function $\psi(z - \delta z)$ and $\psi(z)$, a third value, i.e. $\psi(z + \delta z)$, can be predicted. Using this new point $\psi(z + \delta z)$, together with $\psi(z)$ and by making the transformation $z + \delta z \rightarrow z$, a fourth point, $\psi(z + 2\delta z)$, can be calculated, and so on. Hence the complete wave function can be deduced for any particular energy. The solutions for stationary states have wave functions which satisfy the standard boundary conditions, i.e.

$$\psi(z) \rightarrow 0 \quad \text{and} \quad \frac{\partial}{\partial z} \psi(z) \rightarrow 0, \quad \text{as} \quad z \rightarrow \pm\infty \tag{3.12}$$

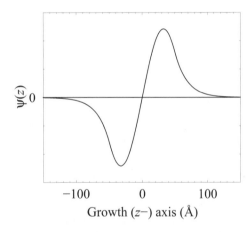

Figure 3.2   An odd-parity wave function—the first excited state of a symmetric quantum well

The first two values of the wave function necessary to start off the procedure *can* be deduced by using simple symmetry arguments, namely that if the potential $V(z)$ is symmetric, then the eigenstates must be either symmetric (even parity) or anti-symmetric (odd parity). If the state of interest has odd parity, e.g. the first excited state of a symmetric quantum well, as illustrated in Fig. 3.2, then the wave function at the centre of the well (call this the origin $z = 0$ for now) must be zero. Correspondingly, a small displacement along the growth (z-) direction must yield a finite value for the wave function. The actual magnitude is not relevant since the energy eigenvalues of the linear Schrödinger equation are unchanged if the wave function is scaled by any given number. Hence in this case, the following starting conditions could be chosen:

$$\psi(0) = 0; \qquad \psi(\delta z) = 1 \tag{3.13}$$

Given this, it remains to find the eigenvalue energy E. As stated above, the value of E corresponding to a stationary state, or more specifically in this case a confined state within a well potential, is that value which produces a wave function conforming to the standard boundary conditions. As E is an unknown in equation (3.11), then ψ is really a function of both position z *and* energy E, since given the starting conditions a wave function can be generated for any E—although it will not always tend to zero at infinity and be a stationary state. Thus the wave function should be written as $\psi(z, E)$, in which case solutions are sought to the equation:

$$\psi(\infty, E) = 0 \tag{3.14}$$

which can be found by using standard techniques such as the Newton–Raphson iteration (see, for example, Section 2.5).

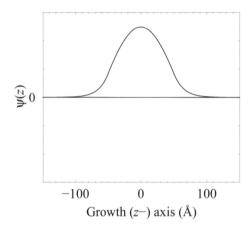

Figure 3.3   An even-parity wave function—the first excited state of a symmetric quantum well

If the eigenstate of interest has even parity, e.g. the ground state of a symmetric quantum well, as shown in Fig. 3.3, then new starting conditions must be deduced. In particular, as the value of the wave function at the origin is non-zero, then $\psi(0) = 1$ can be chosen. In addition, as the wave function is symmetric then $\psi(-\delta z) = \psi(+\delta z)$; substituting both of these expressions into equation (3.11) then gives:

$$\psi(+\delta z) = \left\{ \frac{2m^*}{\hbar^2} (\delta z)^2 \left[V(0) - E \right] + 2 \right\} \times 1 - \psi(+\delta z) \tag{3.15}$$

$$\therefore \psi(+\delta z) = \frac{1}{2} \left\{ \frac{2m^*}{\hbar^2} (\delta z)^2 \left[V(0) - E \right] + 2 \right\} \tag{3.16}$$

3.2   GENERALIZED INITIAL CONDITIONS

While such initial (starting) conditions for symmetric potentials are very useful, they are restrictive in that there are many systems which do not have symmetric potentials, with perhaps the most obvious being the quantum well with an electric field applied (see Section 2.12). With the aim of deducing more general starting conditions for the iterative equation (equation 3.11): Note that at the moment the potentials of interest are all confining potentials, and hence as stated above many times, all wave functions satisfy the standard boundary conditions:

$$\psi(z) \to 0 \quad \text{and} \quad \frac{\partial}{\partial z}\psi(z) \to 0, \quad \text{as} \quad z \to \pm\infty \tag{3.17}$$

In addition, learning from the analytical solutions of the previous chapter, the wave function decays exponentially into the end barriers (see Fig. 3.4). The decay constant κ

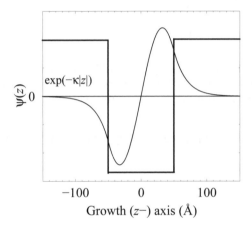

Figure 3.4 Exponential decay of the wave function into the end barrier

is also known as follows:

$$\kappa = \frac{\sqrt{2m^*(V - E)}}{\hbar} \tag{3.18}$$

Hence: upon choosing the first value of the wave function $\psi(z - \delta z) = 1$, then

$$\psi(z - \delta z) = \psi(z) \exp\left(-\kappa|\delta z|\right) \tag{3.19}$$

Therefore, given the starting value $\psi(z - \delta z)$, the next value $\psi(z)$ can be calculated, thus implementing the boundary condition of exponential growth.

In practice, however, these boundary conditions *can* be unreliable, not because of the starting conditions they impose, but because of the boundary conditions that must be sought at the other end of the potential structure. The mathematics imply that energies must be sought for which ψ tends to zero; however, the original choice of $\psi = 1$ at the start of the potential structure immediately implies an asymmetry. Generally, this asymmetry is very small, but however, not always. In order to correct this asymmetry, the, at first sight bizarre, starting conditions are chosen as:

$$\psi(z - \delta z) = 0 \quad \text{and} \quad \psi(z) = 1 \tag{3.20}$$

Such starting conditions for the shooting method solution of the Schrödinger equation are important because of their generality. They are applicable to all potential profiles, whether symmetric, or not, whether the outer barrier is flat, or not, and whether the eigenstate of interest is symmetric, anti-symmetric, or without definite parity.

These conditions can be partially justified mathematically. As multiplying an eigenstate (wave function ψ) by a constant doesn't affect the eigenvalue (the energy E), then if the first wave function point, $\psi(z - \delta z)$, was taken as a small, but finite value, say $\delta\psi$, then the second starting point, $\psi(z)$, could be given any value larger than $\delta\psi$, say $N\delta\psi$, where N is a large number, while still giving exponential growth and without changing the energy eigenvalue. The third value of the wave function simply follows from the shooting equation (equation 3.11) as follows:

$$\psi(z + \delta z) = \left\{ \frac{2m^*}{\hbar^2} (\delta z)^2 \left[V(z) - E \right] + 2 \right\} N\delta\psi - \delta\psi \tag{3.21}$$

$$\therefore \psi(z + \delta z) = \left(\left\{ \frac{2m^*}{\hbar^2} (\delta z)^2 \left[V(z) - E \right] + 2 \right\} N - 1 \right) \delta\psi \tag{3.22}$$

These new starting conditions of 0 and 1 merely represent the limit of large N.

While the above only offers partial justification, full vindication will be given by the results of the convergence tests in the following sections.

3.3   PRACTICAL IMPLEMENTATION OF THE SHOOTING METHOD

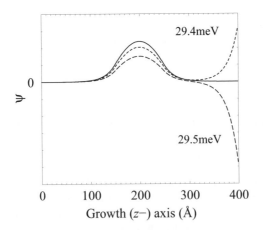

Figure 3.5   Numerically obtained wave functions above and below the true solution at $E = 29.43$ meV, for an electron in a 150 Å $Ga_{0.8}Al_{0.2}As/$ 100 Å GaAs/ 150 Å $Ga_{0.8}Al_{0.2}As$ single quantum well

Fig. 3.5 illustrates the practical implementation of the shooting method. The energy is varied systematically until the wave function switches from diverging to $+\infty$ to $-\infty$; clearly, an energy exists between these values for which the wave function will tend

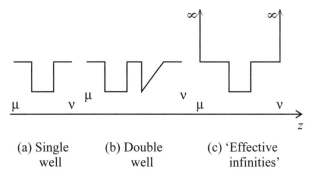

(a) Single   (b) Double   (c) 'Effective
    well   well   infinities'

Figure 3.6   Schematic illustration of the extent of the potential $V(z)$, where $\mu < z < \nu$

smoothly to zero. A Newton–Raphson iteration is then begun to converge on the true solution, in this case 29.43 meV.

In practice, the wave function iteration is begun a finite distance to the left of the first well in the potential and is halted a finite distance to the right of the last well (see Fig. 3.6). The potential, which can be merely a list of numbers specifying $V(z)$ at regular intervals along the z-axis, defines the step length, the extent of the wave functions, and the effective infinities. The latter are the points by which the wave function is considered to have converged towards zero at both limits of the potential profile. Seeking solutions based on this criterion is, in effect, equivalent to applying infinite potentials a sufficient distance into the outer barriers of our finite structure, as shown in Fig. 3.6(c). The positions of these effective infinities, 'μ' and 'ν', which are the lower and upper limits of the z-domain, should be chosen to be of sufficient extent so as not to affect the eigenvalues, defining this mathematically:

$$\lim_{\mu \to -\infty} E = E_n, \quad \text{and} \lim_{\nu \to +\infty} E = E_n \tag{3.23}$$

where E_n are the set of true eigenenergies.

In Fig. 3.7 the left- and right-hand barriers are varied in thickness to illustrate the effect on the solution for the energy produced by this numerical method, for the single quantum well of above. Clearly, the energy converges to a constant, which is the true stationary state—in this case, the ground state E_1. Thus if the energy is the only motivation for the calculation, any barrier width beyond, say 150 Å, will suffice. Although note this can vary depending on the barrier height, well width and carrier effective mass.

If the wave function is also desired, e.g. to be used as an input for calculating another property, such as the exciton binding energy or an electron–phonon scattering rate, then greater care has to be taken. While very large outer barrier layers can be chosen to ensure convergence of the eigenenergy, this can have a detrimental effect on the wave function, as can be seen in Fig. 3.8. At too narrow a barrier width, e.g. the 60 Å case, although the energy may be returned to within 1 meV, the wave function doesn't satisfy the second of the standard boundary conditions, i.e. $\psi' = 0$. With a 200 Å barrier, which corresponds to just after the energy has converged, the wave function is as expected and satisfies both the standard boundary conditions—this represents the optimal barrier width. However, at very large barrier widths, such as 600 Å as illustrated, the wave function is beginning to diverge and as stated above, although the energy obtained from such a solution is reliable, the wave function cannot be used to calculate other properties.

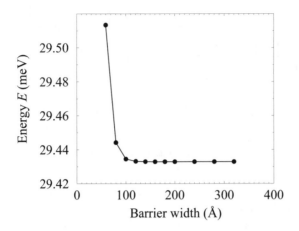

Figure 3.7   Effect of barrier width (effective infinities, μ and ν) on energy

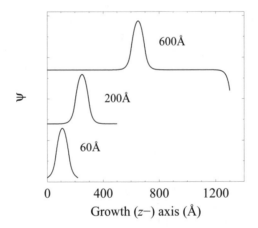

Figure 3.8   Effect of barrier width (effective infinities, μ and ν) on the solution to the wave function

The energy E can always be obtained to far higher accuracy than would be required for comparison of say, electron confinement energies with photoluminescence data. Computer codes typically assign 16-figure accuracy to a double precision number, and the shooting method can be used to determine E to 12 of these, for example. However, even at this high accuracy the wave function at the end of an iteration, i.e. the value of $\psi(\infty, E) = \psi(\nu, E)$, cannot be made equal to zero for large barrier widths, simply because E cannot be stored to enough significant figures. If 32-figure accuracy could be used in the computing, then the wave function could be returned realistically for larger barrier widths.

The wave functions obtained from this numerical method are not normalised, i.e. they do not satisfy:

$$\int_{\text{all space}} \psi^*(z)\psi(z) \ dz = 1 \tag{3.24}$$

This can easily be achieved with the following transformation:

$$\psi(z) \rightarrow \frac{\psi(z)}{\sqrt{\int_{\text{all space}} \psi^*(z)\psi(z) \ dz}} \tag{3.25}$$

This numerical solution to a still simplistic Schrödinger equation is of use as it enables comparison with a number of curved potentials that have analytical solutions, such as the parabolic and the Pöschl–Teller potentials.

3.4   HETEROJUNCTION BOUNDARY CONDITIONS

In the previous chapter it was shown that the conditions for matching solutions at interfaces between dissimilar materials were fixed as soon as the Hamiltonian was decided. This is also reflected here within the shooting method. Consider again the shooting equation for the constant-mass Hamiltonian ($\frac{1}{m^*}P_z P_z$) (equation (3.11)):

$$\psi(z + \delta z) = \left\{ \frac{2m^*}{\hbar^2}(\delta z)^2 \left[V(z) - E \right] + 2 \right\} \psi(z) - \psi(z - \delta z) \tag{3.26}$$

Figure 3.9   Calculating a new wave function point across a discontinuity in the potential

Consider also that the two known points of the wave function, i.e. $\psi(z - \delta z)$ and $\psi(z)$, lie within a quantum well with $V(z) = 0$, but the new point to be calculated lies across a heterojunction in a barrier (as in Fig. 3.9) then:

$$\psi(z + \delta z) = \left[\frac{2m^*}{\hbar^2}(\delta z)^2 \left(-E \right) + 2 \right] \psi(z) - \psi(z - \delta z) \tag{3.27}$$

Then in the limit of decreasing step length δz, i.e. eliminating terms in $(\delta z)^2$:

$$\psi(z + \delta z) = 2\psi(z) - \psi(z - \delta z) \tag{3.28}$$

$$\therefore \psi(z + \delta z) - \psi(z) = \psi(z) - \psi(z - \delta z) \tag{3.29}$$

which implies continuity of the *derivative* across a heterojunction, as expected for this constant effective mass Hamiltonian.

3.5 THE PARABOLIC POTENTIAL WELL

The parabolic potential well is a good testing ground for the accuracy of the numerical method introduced, in that it allows progression from the standard step potentials that have been dealt with over and over again, to a curved profile, but still with analytical solutions to compare with.

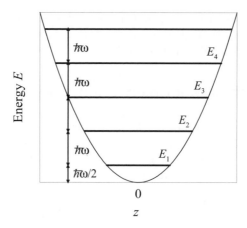

Figure 3.10 The quantum mechanical harmonic oscillator

The parabolic potential well is a direct analogy to the harmonic oscillator of classical mechanics in that the potential is, of course, proportional to the square of the displacement, i.e. following Eisberg [3]:

$$V(z) = C\frac{z^2}{2} \tag{3.30}$$

where C is a constant. If a particle of mass m is displaced from the equilibrium position by an amount Δz, then the restoring force is proportional to $-\partial V/\partial z = -Cz$, thus causing oscillations of amplitude Δz and angular frequency:

$$\omega = \sqrt{\frac{C}{m}} \tag{3.31}$$

The behaviour of a quantum particle, such as an electron, within such a potential is quite different. The Schrödinger equation for the constant-mass case would then become:

$$-\frac{\hbar^2}{2m^*}\frac{\partial^2}{\partial z^2}\psi + C\frac{z^2}{2}\psi = E\psi \tag{3.32}$$

Substituting ω for C, then:

$$-\frac{\hbar^2}{2m^*}\frac{\partial^2}{\partial z^2}\psi + m^*\omega^2\frac{z^2}{2}\psi = E\psi \tag{3.33}$$

Table 3.1   Convergence test of the shooting method solution for a parabolic quantum well; the energy eigenvalues E_n for the lowest 10 confined states—note that the value for E_1 implies $\hbar\omega = 871.876$ meV

n	E_n (meV)	$E_n/(2E_1)$
1	435.938	0.500
2	1307.809	1.500
3	2179.671	2.500
4	3051.522	3.500
5	3923.341	4.500
6	4794.990	5.500
7	5665.755	6.498
8	6532.518	7.492
9	7382.747	8.468
10	8156.309	9.355

The method of solution of this eigenvalue problem has been covered many times in standard texts, for example [3, 45] or alternatively [12]. It is suffice here to quote the result, i.e. the energy levels are quantised and given by:

$$E_{n+1} = \left(n + \frac{1}{2}\right)\hbar\omega \tag{3.34}$$

where the subscript $(n+1)$ has been introduced in order for the notation to agree with that developed so far, i.e. the ground state being labelled '1'.

Fig. 3.10 illustrates the solution; the ground state has an energy $\hbar\omega/2$ even at the absolute zero of temperature, thus coining the term the *zero point energy*. Above this the equal-spaced energy steps form a ladder, which has prompted suggestions for exploitation in non-linear optics [59].

Focusing on parabolic quantum wells within semiconductor multilayers, then it has to be acknowledged, that as only finite potentials are available through the standard band offsets, the potential profile will resemble that shown in Fig. 3.11, i.e. a *finite parabolic quantum well*, rather than that without limit as shown in Fig. 3.10.

Taking the central well to be of width a and the outer barriers to have widths b, the parabolic quantum well can then be specified by the variation in the alloy component x along the z-axis, say in $Ga_{1-x}Al_xAs$. In particular, if x is allowed to vary from a minimum value x_{min} to a maximum x_{max}, then the well profile is given by the following quadratic relationship:

$$x(z) = x_{min} + \frac{[z - (b - a/2)]^2\,(x_{max} - x_{min})}{(a/2)^2} \tag{3.35}$$

Table 3.1 displays the energies of the lowest ten levels within a $Ga_{1-x}Al_xAs$ parabolic quantum well. The layer thicknesses have been taken as $a = b = 100$ Å, and the effective mass has been taken as a constant, i.e. $0.067\,m_0$; however, the maximum alloy concentration, x_{max}, has been taken artificially as high as '10', in order to produce a large number of confined levels. As $E_1 = \hbar\omega/2$, then the third column of data displays the 'half-integer' $(n + \frac{1}{2})$. Clearly, the agreement with the analytical theory is exact for the lower states, although as the levels approach the top of the barrier, they experience the finiteness of the potential, and hence some discrepancy with the infinite parabola solution arises.

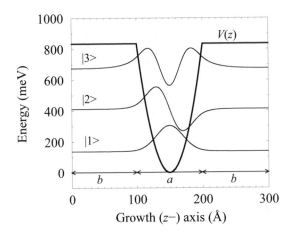

Figure 3.11   The three confined eigenstates of a finite parabolic quantum well

The energy step $\hbar\omega$ between the levels is a useful quantity for describing the properties of the well, even though ω doesn't represent the angular frequency of oscillation. With this in mind, consider the conversion of the alloy concentration of equation (3.35) into a potential profile (see Appendix A); the conduction band profile of $Ga_{1-x}Al_xAs$ would then become:

$$V(z) = \Delta V_{CB}\, x(z) \left(E_g^{AlAs} - E_g^{GaAs} \right) \tag{3.36}$$

Recalling that $\partial V/\partial z = Cz$, then:

$$C = \frac{1}{z}\frac{\partial V}{\partial z} = \frac{1}{z}\Delta V_{CB} \left(E_g^{AlAs} - E_g^{GaAs} \right) \frac{\partial}{\partial z} x \tag{3.37}$$

Ignoring the origin shift, $b - a/2$:

$$C = \frac{1}{z}\Delta V_{CB} \left(E_g^{AlAs} - E_g^{GaAs} \right) \frac{8z \left(x_{max} - x_{min} \right)}{a^2} \tag{3.38}$$

$$\therefore C = \Delta V_{CB} \left(E_g^{AlAs} - E_g^{GaAs} \right) \frac{8 \left(x_{max} - x_{min} \right)}{a^2} \tag{3.39}$$

Then, ω follows from equation (3.31) as:

$$\omega = \sqrt{8\Delta V_{CB} \left(E_g^{AlAs} - E_g^{GaAs} \right) \left(x_{max} - x_{min} \right)}\ \frac{1}{a\sqrt{m}} \tag{3.40}$$

By using the values corresponding to the data in Table 3.1, i.e. $V_{CB} = 0.67$, $E_g^{AlAs} - E_g^{GaAs} = 1247$ meV, $x_{max} = 10$, $x_{min} = 0$, $a = 100$ Å and $m = 0.067\, m_0$ then $\hbar\omega =$

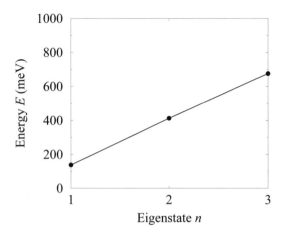

Figure 3.12   Eigenenergies for a GaAs/AlAs parabolic quantum well

871.879 meV, which is in excellent agreement with the numerical result as displayed in the table.

Figs. 3.11 and 3.12 display the wave functions and energies, respectively, of the three confined states of the more realistic parabolic quantum well, with $x_{max}=1$. The wave functions resemble those of the single quantum well and the energy levels are to all intents and purposes equally spaced.

In summary, it has been demonstrated that the simple shooting-method solution of the Schrödinger equation, taken together with the stated starting conditions, can produce energy levels in exact agreement with the analytical values for the curved potential of the parabolic quantum well. In addition, equation (3.40), summarizes the behaviour of parabolic wells within semiconductor multilayers. The energy level spacing is inversely proportional to both the well width and the square root of the mass.

3.6   THE PÖSCHL–TELLER POTENTIAL HOLE

The modified Pöschl–Teller potential hole [60], an example of which is shown in Fig. 3.13, is important in that it resembles the profile of a diffused quantum well, but has the advantage of analytical solutions. Therefore as is currently the aim, it will serve well for validation of the shooting method *and* the choice of starting conditions described in Section 3.1.

The potential is given by the following [61]:

$$V(z) = -\frac{\hbar^2}{2m^*}\alpha^2\frac{\lambda(\lambda-1)}{\cosh^2\alpha z} \tag{3.41}$$

where α is known as the width parameter and λ as the depth parameter (see Fig. 3.14). Clearly, values of $0 < \lambda < 1$ give potential barriers, $\lambda = 1$ gives a flat band and the area of interest here, namely potential wells, are given by values of $\lambda > 1$.

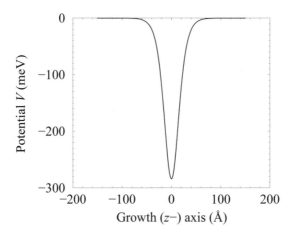

Figure 3.13   Pöschl–Teller potential hole; $\alpha = 0.05$ Å^{-1} and $\lambda = 2.0$

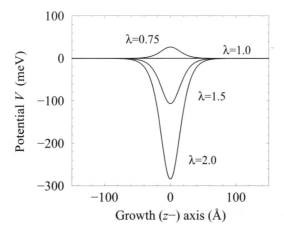

Figure 3.14   Pöschl–Teller potentials as a function of the depth parameter λ for a fixed width parameter α of 0.05 Å^{-1}

The eigenvalues of the resulting Schrödinger equation are given by [60]:

$$E_n = -\frac{\hbar^2 \alpha^2}{2m^*}(\lambda - 1 - n)^2 \tag{3.42}$$

Table 3.2 compares the analytical solutions given by equation (3.42) for the two lowest energy confined states of the Pöschl–Teller potential holes of Fig. 3.14, with that obtained by the numerical shooting method. The agreement is very good for all values of λ and for both the ground state, of energy E_1, and the first excited state, of energy E_2. However, for the largest value of λ used, which does represent a very deep well, of a magnitude

Table 3.2   Comparison of the numerical solution with the analytical solution of the Pöschl–Teller potential with a step length δz of 1 Å

λ	Analytical solution		Numerical solution	
	E_1 (meV)	E_2 (meV)	E_1 (meV)	E_2 (meV)
1.5	−35.541154	—	−35.442724	—
2.0	−142.164614	—	−142.178093	—
5.0	−2274.633827	−1279.481528	−2274.940451	−1280.518108
10.0	−11515.333749	−9098.535308	−11517.014128	−9105.712331

not found in semiconductor heterostructures, the discrepancy between the solutions has increased to 2 and 7 meV for the ground and first excited states respectively. This discrepancy can be removed by increasing the accuracy of the numerical solution.

3.7   CONVERGENCE TESTS

This is an appropriate point to perform the final convergence test, i.e. the final test of numerical accuracy of the shooting method solution of Schrödinger's equation, as derived in Section 3.1. As discovered in the previous section, while the agreement between the analytical solution of the Pöschl–Teller potential and the numerical solution is generally very good, at times, for certain parameter values, discrepancies can occur. This is often a reflection of the choice of the computational values, for example, and probably most importantly, the resolution in the definition of the potential profile, i.e. the step length δz. The user has complete control over this value, and hence an improved comparison with analytical solutions (should they exist) can be achieved, but as always, at the inconvenience of increased computational time.

Table 3.3   Comparison of the numerical solution with the analytical solution of the Pöschl–Teller potential as a function of the number (N) of points per Å

N (Å^{-1})	Analytical solution		Numerical solution	
	E_1 (meV)	E_2 (meV)	E_1 (meV)	E_2 (meV)
1	−2274.633827	−1279.481528	−2274.940451	−1280.518108
2	−2274.633827	−1279.481528	−2274.710431	−1279.740343
5	−2274.633827	−1279.481528	−2274.646081	−1279.522923
10	−2274.633827	−1279.481528	−2274.636891	−1279.491876
20	−2274.633827	−1279.481528	−2274.634593	−1279.484115
50	−2274.633827	−1279.481528	−2274.633950	−1279.481942
100	−2274.633827	−1279.481528	−2274.633858	−1279.481631

Table 3.3 compares the analytical solutions corresponding to the $\lambda = 5$ potential hole of Table 3.2 with the corresponding numerical solution as a function of the number of

potential points $V(z)$ per Å along the growth axis. The step length δz in Å is just $1/N$. It can be seen that as the number of points increases the discrepancy between the data decreases (obviously the analytical solutions are not a function of N). At the upper limit of the data in the table, the solutions agree to 7 or 8 significant figures. This is well beyond the accuracy of spectroscopic data, which in the highest quality of semiconductor heterostructure samples, might yield transition energies to 0.1 meV, e.g. [62], a precision which is reached for $N = 1$.

In summary, for the majority of applications, a step length δz of 1 Å is adequate.

3.8 EXTENSION TO VARIABLE EFFECTIVE MASS

Hitherto, the numerical solution has focused on the constant-mass Schrödinger equation, and this has served the purposes of development and application well. However in the real world, just as in the analytical solutions before, the Schrödinger equation of real interest accounts for the possibility of the effective mass varying between dissimilar semiconductor layers.

With the aim of generalising the numerical solution for this situation, consider the Schrödinger equation (equation (2.96)), i.e.

$$-\frac{\hbar^2}{2}\frac{\partial}{\partial z}\frac{1}{m^*(z)}\frac{\partial}{\partial z}\psi(z) + V(z)\psi(z) = E\psi(z) \tag{3.43}$$

which can be written:

$$\frac{\partial}{\partial z}\frac{1}{m^*(z)}\frac{\partial}{\partial z}\psi(z) = \frac{2}{\hbar^2}\left[V(z) - E\right]\psi(z) \tag{3.44}$$

The variable mass kinetic energy operator can be expanded directly to give:

$$\frac{\partial}{\partial z}\left[\frac{1}{m^*(z)}\right]\frac{\partial}{\partial z}\psi(z) + \frac{1}{m^*(z)}\frac{\partial^2}{\partial z^2}\psi(z) = \frac{2}{\hbar^2}\left[V(z) - E\right]\psi(z) \tag{3.45}$$

and then:

$$-\frac{1}{[m^*(z)]^2}\frac{\partial}{\partial z}m^*(z)\frac{\partial}{\partial z}\psi(z) + \frac{1}{m^*(z)}\frac{\partial^2}{\partial z^2}\psi(z) = \frac{2}{\hbar^2}\left[V(z) - E\right]\psi(z) \tag{3.46}$$

However, shooting equations derived from this point by expanding the derivatives in terms of finite differences have lead to significant computational inaccuracies in systems with a large discontinuous change in the effective mass $m^*(z)$, as occurs, e.g. in GaAs/AlAs quantum wells. The source of the inaccuracy is thought to arise from the δ-function nature of $\partial m^*(z)/\partial z$.

A more robust scheme can be derived by expanding the left-hand derivative, $\partial/\partial z$, in equation (3.44), first giving the following:

$$\frac{\frac{1}{m^*(z+\delta z)}\left.\frac{\partial\psi(z)}{\partial z}\right|_{z+\delta z} - \frac{1}{m^*(z-\delta z)}\left.\frac{\partial\psi(z)}{\partial z}\right|_{z-\delta z}}{2\delta z} = \frac{2}{\hbar^2}\left[V(z) - E\right]\psi(z) \tag{3.47}$$

$$\therefore \frac{1}{m^*(z+\delta z)}\left.\frac{\partial\psi}{\partial z}\right|_{z+\delta z} - \frac{1}{m^*(z-\delta z)}\left.\frac{\partial\psi}{\partial z}\right|_{z-\delta z} = \frac{2(2\delta z)}{\hbar^2}\left[V(z) - E\right]\psi(z) \tag{3.48}$$

Recalling the centred finite difference expansion for the first derivative, i.e.

$$\frac{\partial f}{\partial z}\bigg|_z = \frac{f(z+\delta z) - f(z-\delta z)}{2\delta z} \tag{3.49}$$

Then:

Table 3.4 Comparison of the numerical solution with the analytical solution for a single quantum well with differing effective masses in the well and barrier

Well width (Å)	Analytical solution		Numerical solution	
	E_1 (meV)	E_2 (meV)	E_1 (meV)	E_2 (meV)
20	126.227914	—	126.204335	—
40	80.111376	—	80.087722	—
60	53.276432	166.522007	53.260451	167.766634
80	37.619825	137.330295	37.609351	137.308742
100	27.884814	106.557890	27.877769	106.535858
120	21.463972	83.606781	21.459068	83.589149
160	13.820474	54.647134	13.817861	54.636614
200	9.629394	38.282383	9.627854	38.275914

$$\frac{1}{m^*(z+\delta z)}\left[\frac{\psi(z+2\delta z) - \psi(z)}{2\delta z}\right] - \frac{1}{m^*(z-\delta z)}\left[\frac{\psi(z) - \psi(z-2\delta z)}{2\delta z}\right]$$

$$= \frac{2(2\delta z)}{\hbar^2}[V(z) - E]\psi(z) \tag{3.50}$$

$$\therefore \left[\frac{\psi(z+2\delta z) - \psi(z)}{m^*(z+\delta z)}\right] - \left[\frac{\psi(z) - \psi(z-2\delta z)}{m^*(z-\delta z)}\right]$$

$$= \frac{2(2\delta z)^2}{\hbar^2}[V(z) - E]\psi(z) \tag{3.51}$$

By gathering terms in $\psi(z)$ on the right-hand side, then:

$$\frac{\psi(z+2\delta z)}{m^*(z+\delta z)} + \frac{\psi(z-2\delta z)}{m^*(z-\delta z)}$$

$$= \left\{\frac{2(2\delta z)^2}{\hbar^2}[V(z) - E] + \frac{1}{m^*(z+\delta z)} + \frac{1}{m^*(z-\delta z)}\right\}\psi(z) \tag{3.52}$$

Making the transformation, $2\delta z \to \delta z$, then gives:

$$\frac{\psi(z+\delta z)}{m^*(z+\delta z/2)} = \left\{\frac{2(\delta z)^2}{\hbar^2}[V(z) - E] + \frac{1}{m^*(z+\delta z/2)} + \frac{1}{m^*(z-\delta z/2)}\right\}\psi(z)$$

$$- \frac{\psi(z-\delta z)}{m^*(z-\delta z/2)} \tag{3.53}$$

which is the variable effective-mass shooting equation, and is solved according to the boundary conditions (as in Section 3.1). The effective mass m^* can be found at the intermediate points, $z \pm \delta z/2$, by taking the mean of the two neighbouring points at z and $z \pm \delta z$. Clearly, equation 3.53 collapses back to the original form in equation (3.11) when m^* is constant.

Table 3.5   Comparison of the numerical solution with the analytical solution for a single GaAs quantum well surrounded by $Ga_{0.25}Al_{0.75}As$ barriers, with differing effective masses in the well and barrier

Well width (Å)	Analytical solution		Numerical solution	
	E_1 (meV)	E_2 (meV)	E_1 (meV)	E_2 (meV)
20	270.764969	—	270.357809	—
40	129.450671	512.393113	129.247261	512.047324
60	75.774672	307.023049	75.669905	306.736857
80	49.774624	201.176843	49.715086	200.986873
100	35.204788	141.851288	35.168092	141.724429
120	26.217761	105.411306	26.193670	105.323933
160	16.158835	64.812071	16.146909	64.766509
200	10.949827	43.869233	10.943102	43.842871

Table 3.4 compares the ground state and first excited state energy levels, E_1 and E_2, respectively, calculated with this extended shooting equation, with the analytical solution from Section 2.6, for a GaAs quantum well surrounded by $Ga_{0.8}Al_{0.2}As$ barriers. In this series of calculations, the step length δz was taken as 1 Å and it can be seen from the data in the table that the agreement is very good for both the ground state energy E_1 and the first excited state energy E_2 across the range of well widths. The discrepancy between the solutions of the two methods is largest for the excited state of the narrower wells, at which point it is of the order of 1 meV. For the wider wells, the discrepancy reduces to less than 0.1 meV. Such accuracies are entirely acceptable when modelling, e.g. experimental spectroscopic data.

As this shooting equation for a variable effective mass will be used widely, it is worthwhile performing a few more convergence tests in order to increase confidence in its applicability. In particular, Table 3.5 repeats the calculations of the previous table but with a much higher barrier Al concentration. The effect of this is twofold, i.e. there is an increased difference in the potential between the well and barrier, but more importantly for this present section, there is an increased difference in the effective masses. The discrepancies between the analytical solution and the numerical solution are of a similar order as before, for this step length δz of 1 Å.

However, as before, the agreement between the numerical solution and that obtained from the analytical form can be improved by increasing the computational accuracy of the shooting method, i.e. increasing the number of points per Å (decreasing δz). This is highlighted in Table 3.6, which compares the solutions from the two methods for a decreasing step length δz.

Table 3.6   Comparison of the numerical solution with the analytical solution for a 20 Å single GaAs quantum well surrounded by $Ga_{0.25}Al_{0.75}As$ barriers, as a function of the number (N) of points per Å in the mesh—note the step length $\delta z = 1/N$ (Å)

N (Å^{-1})	Analytical solution E_1 (meV)	Numerical solution E_1 (meV)
2	270.764969	270.663106
4	270.764969	270.739499
6	270.764969	270.753649
8	270.764969	270.758601
10	270.764969	270.760894
12	270.764969	270.762139

When using the form for the variable effective mass shooting equation in equation (3.53), the BenDaniel–Duke boundary conditions are 'hard-wired' in. Thus it is unnecessary to repeat the analysis described in Section 3.4 in order to recover them.

3.9   THE DOUBLE QUANTUM WELL

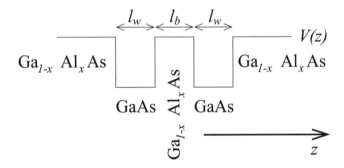

Figure 3.15   The band profile of symmetric GaAs/Ga$_{1-x}$Al$_x$As double quantum well

The convergence tests of the numerical solution to Schrödinger's equation have provided the confidence to apply it to systems for which the analytical solutions have not (in this book!) been developed or do not exist.

The simplest example, of the former would be the symmetric double quantum well of Fig. 3.15. The potential function $V(z)$ required for the numerical solution is simply the (in this case) conduction band edge as given in the figure.

Fig. 3.16 displays the results of calculations of the lowest two energy states as a function of the central barrier width for a double quantum well with the Al barrier concentration $x = 0.2$ and a fixed well width $l_w = 60$ Å. When the wells are separated by a large distance, the interaction between the eigenstates localized within each well is very small and the wells behave as two independent single quantum wells. However,

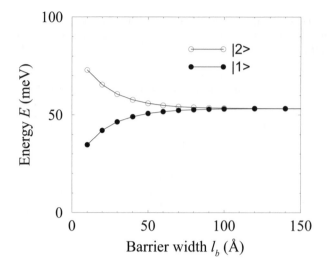

Figure 3.16   The confinement energies of the lowest two states of a symmetric double quantum well as a function of the central barrier width

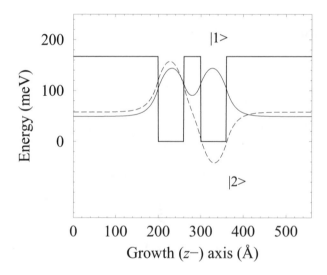

Figure 3.17   The wave functions of the lowest two energy levels of the symmetric double quantum well with a central barrier width of 40 Å

as illustrated in Fig. 3.16, as the central barrier is decreased, the energy levels interact, with one being forced to higher energies and the other to lower energies. This is directly analogous to the formation of a pair of bonding and anti-bonding orbitals when two hydrogen atoms are brought together to form a hydrogen molecule. In that case, as here, the wave function of the lower (bonding) state is a sum of the wave functions of the separate atoms (wells), and the higher (anti-bonding) state is a difference in wave functions, i.e.

$$\Psi_{\text{bonding}} = \frac{1}{\sqrt{2}} \left(\psi_1 + \psi_2 \right); \quad \Psi_{\text{anti-bonding}} = \frac{1}{\sqrt{2}} \left(\psi_1 - \psi_2 \right) \tag{3.54}$$

In the case of hydrogen, both electrons move into the lower energy orbital, thus reducing the total energy of the two-hydrogen-atom system to form a chemically bound hydrogen molecule. In the case of the double quantum well here, a similar situation occurs with the electron spins aligning in an 'anti-parallel' arrangement in order to satisfy the Pauli exclusion principle.

Fig. 3.17 displays the wave functions of the double quantum well with a central barrier width of 40 Å. Clearly, they form a symmetric and anti-symmetric pair, with the former being of lower energy, as discussed for the hydrogen molecule as above.

3.10   MULTIPLE QUANTUM WELLS AND FINITE SUPERLATTICES

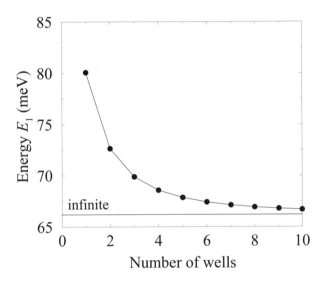

Figure 3.18   The ground state energy of a N-well heterostructure

Real layered structures can only contain a finite number of quantum wells, so infinite superlattices don't actually exist. Thus, the Kronig–Penney model derived in the previous chapter has to be used carefully; i.e. it will give the energy and dispersion relationships of carriers *near the centre of a many-period quantum well system*.

One way of discovering exactly how many quantum wells are required before a finite structure resembles an infinite would be to look at the ground state energy as a function of the number of periods. Fig. 3.18 does exactly this for N repeats of a 40 Å GaAs/40 Å $Ga_{0.8}Al_{0.2}As$ unit cell capped with 200 Å $Ga_{0.8}Al_{0.2}As$ barriers. By the time there are ten wells, the ground state energy is within 1 meV of that given by the Kronig–Penney model of an infinite structure.

While ten wells are sufficient from an energy perspective for this particular finite system to resemble the infinite case, the picture given by the wave function is quite

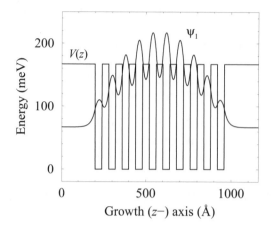

Figure 3.19   The ground state wave function of a finite superlattice

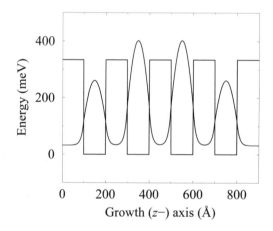

Figure 3.20   The ground state wave function of a multiple quantum well

different, as illustrated in Fig. 3.19. Clearly, this system is a superlattice in that the wave function 'associated' with each well has a significant overlap with that of the adjacent well. In addition, the finiteness of the structure leads to the feature that the electron is not equally likely to be in any of the wells.   Compare this with the wave function for the heterostructure based on a 100 Å GaAs/100 Å $Ga_{0.6}Al_{0.4}As$ unit cell, as in Fig. 3.20, which is clearly a multiple quantum well in that the wave function reaches zero between the wells. It should be noted, however, that the effect of finiteness is *still* apparent in that the probability of the electron being in any of the quantum wells is not equal.

3.11 ADDITION OF ELECTRIC FIELD

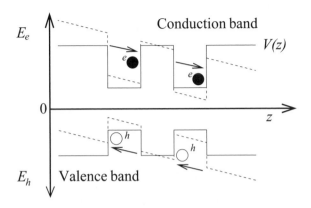

Figure 3.21   Schematic illustration of the tilting of the conduction and valence bands under the influence of an electric field

Unlike the previous analytical forms for the solution of Schrödinger's equation (described in Chapter 2), the numerical solution which includes the effect of an electric field on a heterostructure is simplicity itself. The potential energy is simply added to the potential term within the appropriate shooting equation (equations (3.11) or (3.53)), i.e.

$$V(z) \rightarrow V(z) + qF(z - z_0) \tag{3.55}$$

where for an electron, $q = -e$ and for a hole, $q = +e$. The position z_0 represents the origin of the field, often chosen to be the centre of the well. Fig. 3.21 shows the effect of an electric field on both the conduction- and valence-band potentials; note that increasing hole energies are measured downwards. Thus, any electrons in the double well are pulled to the right-hand side, while holes are pulled to the left, thus producing space charge or a polarization of the carriers.

3.12 QUANTUM CONFINED STARK EFFECT

Section 2.12 demonstrated through perturbation theory the well-known phenomenon of the quantum-confined Stark effect. Fig. 3.22 illustrates this energy level suppression as a function of field for the ground state of a single quantum well.

In order to validate the perturbation theory approach of earlier, the curve represents a parabolic fit to the calculated data which are displayed as points on the figure. It was found that the ground state energy over the range of fields investigated, could be represented very well by the parabola:

$$E_1(F) = E_1(0) - 0.00036F^2 \tag{3.56}$$

where $E_1(0)$ refers to the ground state energy (53.310 meV) at zero field, and the electric field strength F is in units of kVcm^{-1}.

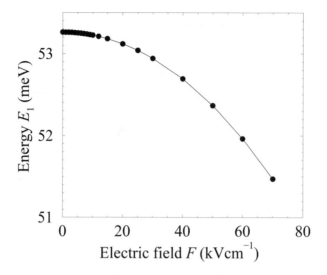

Figure 3.22   The effect of an electric field on the ground state energy E_1 of a 200 Å $Ga_{0.8}Al_{0.2}As/60$ Å GaAs/200 Å $Ga_{0.8}Al_{0.2}As$ single quantum well

3.13   FIELD–INDUCED ANTI-CROSSINGS

As a further example of electric-field-induced phenomena and the versatility of the numerical solution, consider a double quantum well, as shown in Fig. 3.15 but with the left-hand well wider than the right. This asymmetry breaks the degeneracy of the two confined states and results in the lower energy state being more localized in the wider left-hand well and the higher energy state being associated with the narrower right-hand well.

On the addition of an electric field, the left-hand side of the structure increases in potential, while the right hand side decreases, and thus the energy levels are brought closer together. A priori, it is not clear what happens in such a situation. However, the system can be solved numerically for all field points in order to reveal the particulars of the quantum mechanics. Fig. 3.23 displays the results of such a calculation.

The two energy levels *are* brought closer together, and as the field increases their separation ΔE $(= E_2 - E_1)$, which is initially decreasing linearly, reaches a minimum value and then increases again. Within the linear sections, ΔE is merely equal to the difference in field potentials between the centres of the wells, in this case:

$$\Delta E = eF\Delta z = eF(30 \text{ Å} + 30 \text{ Å} + 25 \text{ Å}) \tag{3.57}$$

At the minimum separation, the wave functions reveal a change in character—the lowest energy state switches from being localized in the left-hand well to the right-hand well, and vice versa for the higher energy level. This phenomenon is referred to as an *anti-crossing*.

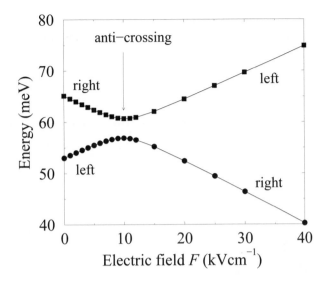

Figure 3.23   The two lowest energy confined states of a 200 Å $Ga_{0.8}Al_{0.2}As$/60 Å GaAs/60 Å $Ga_{0.8}Al_{0.2}As$/50 Å GaAs/200 Å $Ga_{0.8}Al_{0.2}As$ double quantum well as a function of the applied electric field

3.14 SYMMETRY AND SELECTION RULES

Overlap integrals can be a very quick way of understanding or explaining complex inter-actions and are often instrumental in determining selection rules for carrier scattering events. The simplest overlap integral is defined as follows:

$$O = \langle \psi_n | \psi_{n'} \rangle = \int_{-\infty}^{+\infty} \psi_n^* \psi_{n'} \ dz \qquad (3.58)$$

If ψ_n and $\psi_{n'}$ are eigenstates of the same Hamiltonian, whether they be wave or envelope functions, then they are orthonormal, i.e.

$$\langle \psi_n | \psi_{n'} \rangle = \delta_{nn'} \qquad (3.59)$$

Hence, any scattering event which involves simply O will follow this selection rule. As will be encountered later in Chapter 10, the relative strength of an *interband* electron–hole recombination (both excitonic and free-carrier) is proportional to O, even though one of the eigenstates corresponds to a conduction-band electron and the other to a valence-band hole; under the envelope function approximation, they are eigenstates of different Hamiltonians. For symmetric potentials, such as a single quantum well, the states have exact parity, and thus O is non-zero, so corresponding to an allowed transition, for $\Delta n=0$, 2, etc.

Electron–electron scattering will also be discussed in Chapter 10; this is a scattering event between two initial states, resulting in two new states. The rate therefore contains an overlap between all four wave functions, as follows:

$$\text{rate} \propto \int_{-\infty}^{+\infty} \psi_i^* \psi_j^* \ \psi_f \psi_g \ dz \qquad (3.60)$$

Again, for electrons in symmetric potentials, the wave functions have definite parity, and thus for a two subband system, the $|2\rangle|2\rangle \rightarrow |1\rangle|1\rangle$ transition is allowed, while the $|2\rangle|2\rangle \rightarrow |2\rangle|1\rangle$ event is not.

Occasionally it is worthwhile to define another form for the overlap integral, i.e. the overlap integral of the moduli:

$$O_{\mathrm{mod}} = \langle |\psi_n| \mid |\psi_{n'}| \rangle = \int_{-\infty}^{+\infty} |\psi_n^*| \, |\psi_{n'}| \; \mathrm{d}z \tag{3.61}$$

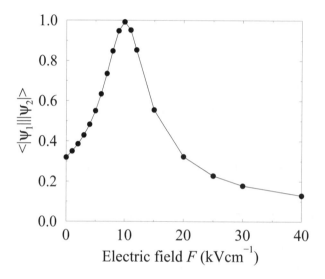

Figure 3.24   The overlap integral of the modulus of the wave functions undergoing the anti-crossing in Fig. 3.23

For example, Fig. 3.24 displays the overlap integral of the wave function moduli for the double quantum well of the previous section. As discussed there, the eigenstates undergo an anti-crossing at a field of 10 kVcm^{-1}. Initially, the wave functions are localized in separate wells, and then as the field is increased the energy levels are brought into alignment and the wave functions 'spread out' over both wells. As both states are simultaneously eigenstates of the same Hamiltonian, O remains zero; however, the nature of the anti-crossing is reflected in the peak of O_{mod}.

3.15   THE HEISENBERG UNCERTAINTY PRINCIPLE

This is an opportune moment to digress slightly and use the techniques now available to explore a fundamental building block of quantum mechanics, namely Heisenberg's Uncertainty Principle. In particular, the aspect of interest in this work will be 'Do the eigenstates of one-dimensional finite potentials always obey the principle and how close to the limit can they be taken?'

Mathematically, the Uncertainty Principle is given by the following [63]:

$$\Delta x_i \Delta p_j \gtrsim \frac{\hbar}{2} \delta_{ij} \tag{3.62}$$

where the δ function implies that for an uncertainty in the determination of the particle's position along the growth (z-) axis, there is no physical limit to the accuracy of any determination in the particle's in-plane momentum.

Restricting interest to the direction of confinement, then the uncertainty relationship simplifies to:

$$\Delta z \Delta p_z \gtrsim \frac{\hbar}{2} \tag{3.63}$$

where

$$(\Delta z)^2 = \langle z^2 \rangle - \langle z \rangle^2 \quad \text{and} \quad (\Delta p_z)^2 = \langle p_z^2 \rangle - \langle p_z \rangle^2 \tag{3.64}$$

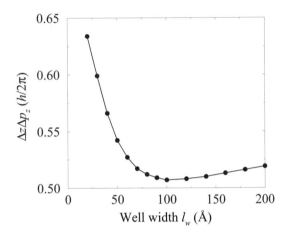

Figure 3.25 The product of the uncertainties in the position and momentum for an electron in a GaAs single quantum well surrounded by barriers of 100 meV height and with constant mass, as a function of the well width

In the previous chapter, Section 2.7 discussed the relevance of Heisenberg's Uncertainty relationship with respect to solutions to Schrödinger's equation in the limit of infinite barrier mass. At that point, the techniques had not been developed—the analytical forms for the wave functions do not lend themselves well to the calculation of Δz and Δp_z. Now, however, numerical solutions are at hand and the calculation can proceed.

The individual components contributing to Δz and Δp_z are expectation values of the state of interest, calculated in the usual way, i.e.

$$\langle z \rangle = \int_{-\infty}^{+\infty} \psi^*(z) \, z \, \psi(z) \, dz \tag{3.65}$$

$$\langle z^2 \rangle = \int_{-\infty}^{+\infty} \psi^*(z) \, z^2 \, \psi(z) \, dz \tag{3.66}$$

$$\langle p_z \rangle = -i\hbar \int_{-\infty}^{+\infty} \psi^*(z) \, \frac{\partial}{\partial z} \psi(z) \, dz \tag{3.67}$$

$$\langle p_z^2 \rangle = -\hbar^2 \int_{-\infty}^{+\infty} \psi^*(z) \frac{\partial^2}{\partial z^2} \psi(z) \ dz \tag{3.68}$$

Whereas analytical calculations of $\langle p_z \rangle$ and $\langle p_z^2 \rangle$ are tedious, with the numerical formalism and the familiarity built up with finite difference expansions, both quantities follow simply as:

$$\langle p_z \rangle = -i\hbar \int_{-\infty}^{+\infty} \psi^*(z) \frac{\psi(z + \delta z) - \psi(z - \delta z)}{2\delta z} \ dz \tag{3.69}$$

and

$$\langle p_z^2 \rangle = -\hbar^2 \int_{-\infty}^{+\infty} \psi^*(z) \frac{\psi(z + \delta z) - 2\psi(z) + \psi(z - \delta z)}{(\delta z)^2} \ dz \tag{3.70}$$

Fig. 3.25 plots $\Delta z \Delta p_z$ for a typical single quantum well, as a function of the well width. The form of the curve is very interesting in that it is a non-monotonic function with a minimum at around $l_w = 100$ Å. This can be understood intuitively in terms of the spatial coordinate—as the well width is reduced from large values, Δz, which might be considered as the well width, decreases. However, at the same time the confinement energy increases and at narrow well widths lies just below the top of the well. At this point, the wave function 'spills over' the barriers, thus leading to an increase in Δz.

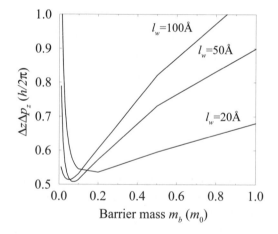

Figure 3.26 The product of the uncertainties in the position and momentum for an electron in a GaAs single quantum well surrounded by $Ga_{0.6}Al_{0.4}As$ barriers, as a function of barrier mass and for various well widths

Interestingly, the minimum in $\Delta z \Delta p_z$ is just above the $\hbar/2$ limit given by Heisenberg's Uncertainty Principle. This is perhaps even more dramatically illustrated by the curves shown in Fig. 3.26, which plots $\Delta z \Delta p_z$ versus the barrier mass for a variety of well widths. These data correspond to the single quantum well shown in Fig. 2.21, which at the point of introduction was thought to imply a violation of Heisenberg's Uncertainty Principle, but now, however, the calculations reveal this not to be the case.

The non-monotonic behaviour as a function of well width is again apparent from the ordering of the curves for any particular mass. In addition, a non-monotonic behaviour as a function of the electron mass in the barrier can be seen. The minimum value of $\Delta z \Delta p_z$ on Fig. 3.26 occurs for $l_w = 50$ Å, as would be expected from earlier. This occurs for a barrier mass which is very close to the well mass of $0.067\, m_0$ and is equal to 0.508 $\hbar$, which is very close, but still larger than $\hbar/2$.

3.16   EXTENSION TO INCLUDE BAND NON-PARABOLICITY

For semiconductor heterostructures with relatively low barrier heights and low carrier densities, the electrons cluster around the subband minima and their energy is in turn reasonably close to the bulk conduction band minima, i.e. within a couple of hundred meV, compared to a bandgap of the order of 1.5 eV. In this region, the band minima, both the bulk conduction band and the in-plane subband, can be described by a parabolic E–$\mathbf{k}$ curve, i.e. in the usual form:

$$E = \frac{\hbar^2 |\mathbf{k}|^2}{2m^*} \tag{3.71}$$

where $\mathbf{k}$ can be read as a three-dimensional vector for the bulk, or as a two-dimensional in-plane vector $|\mathbf{k}_{x,y}|$ for a subband.

However, in situations where the electrons are forced up to higher energies, by either large barrier heights and narrow wells, or in the case of very high carrier densities, equation (3.71) becomes more approximate. This is even more important for holes in the valence band. The approximation can be improved upon by adding addition terms into the polynomial expansion for the energy. For example, no matter how complex the band structure along a particular direction, it clearly has inversion symmetry and hence can always be represented by an expansion in even powers of k, if sufficient terms are included, i.e. the energy E can always be described by:

$$E = a_0 k^0 + a_2 k^2 + a_4 k^4 + a_6 k^6 + a_8 k^8 + \ldots = \sum_{i=0}^{\infty} a_{2i} k^{2i} \tag{3.72}$$

Usually when discussing single band models, as has been focused on entirely so far, the energy origin is set at the bottom of the band of interest, and thus $a_0 = 0$. In addition, truncating the series at k^2, gives $a_2 = \hbar^2/(2m^*)$.

The next best approximation is to include terms in k^4, hence accounting for band non-parabolicity, as displayed in Fig. 3.27, with:

$$E = a_2 k^2 + a_4 k^4 = \frac{\hbar^2}{2m^*} \left(k^2 + \beta k^4 \right) \tag{3.73}$$

where

$$\beta = a_2 \frac{2m^*}{\hbar^2} \tag{3.74}$$

Remembering the basic definition of effective mass, i.e.

$$m^* = \hbar^2 \left(\frac{\partial^2 E}{\partial k^2} \right)^{-1} \tag{3.75}$$

then clearly m^* remains a function of k, unlike the case of parabolic bands where it is a constant. As it is a function of k, then it is also a function of energy, and indeed

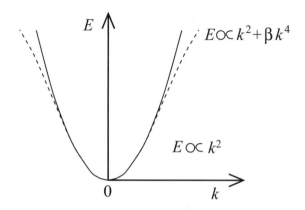

Figure 3.27   Non-parabolicity of the bulk and subband structure

band non-parabolicity is accounted for by allowing the effective mass to have an energy dependence [64]:

$$m^*(E) = m^*(0)[1 + \alpha(E - V)] \tag{3.76}$$

where V is the barrier height and the parameter α is given by [65, 66]:

$$\alpha = \left[1 - \frac{m^*(0)}{m_0}\right]^2 / E_g \tag{3.77}$$

where E_g is the semiconductor bandgap. With this new form of the effective mass, the Schrödinger equation then becomes:

$$-\frac{\hbar^2}{2} \frac{\partial}{\partial z} \frac{1}{m^*(z, E)} \frac{\partial}{\partial z} \psi(z) + V(z)\psi(z) = E\psi(z) \tag{3.78}$$

which can be solved by using the iterative shooting equation (equation (3.53)) as before, but with the additional feature of adjusting the effective mass for each energy E.

Fig. 3.28 displays the effect of the inclusion of non-parabolicity on the ground state energy of a single GaAs quantum well surrounded by $Ga_{1-x}Al_xAs$ barriers. This figure plots ΔE, i.e. the energy calculated including non-parabolicity minus the energy which does not include non-parabolicity. Clearly, non-parabolicity increases the energy calculated, and the increase is (generally) larger for an increasing barrier height and a decreasing well width. The latter two points reflect the expected conclusions formed earlier, i.e. that as the electron (or hole) is forced up the E–$\mathbf{k}$ curve, the parabolic description of the band becomes more approximate.

The barrier material, $Ga_{1-x}Al_xAs$, becomes indirect as x increases beyond 0.42, thus producing an extra complication. Many real layered structures are grown with concentrations below this value, and therefore the majority of examples in this present book do conform to this criterion. Given this, the *maximum* effect of band non-parabolicity is largely given by the bottom curve of Fig. 3.28, i.e. 2 meV. The chapters that follow are centred around the development of techniques and methods and the illustration of generic physics which is applicable across material systems, so band non-parabolicity will not be employed in future examples of calculations based on the GaAs/$Ga_{1-x}Al_xAs$ material system.

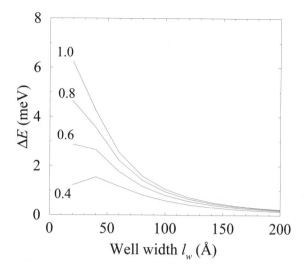

Figure 3.28   The difference in ground state energy of an electron at the subband minima in a GaAs single quantum well, with and without non-parabolicity, $\Delta E = E_1(\text{with}) - E_1(\text{without})$, for a range of $\text{Ga}_{1-x}\text{Al}_x\text{As}$ barrier concentrations x

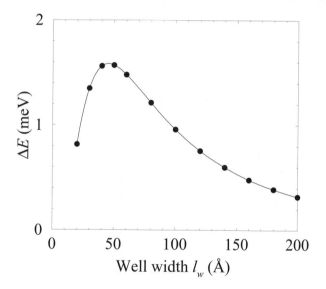

Figure 3.29   The difference in ground state energy of an electron in a single $\text{In}_{0.53}\text{Ga}_{0.47}\text{As}/\text{In}_{0.52}\text{Al}_{0.48}\text{As}$ quantum well, with and without non-parabolicity, $\Delta E = E_1(\text{with}) - E_1(\text{without})$

One material in which this effect is thought to be more important is the quaternary $\text{In}_{1-x-y}\text{Al}_x\text{Ga}_y\text{As}$ system, which has become an important material system for inter-subband lasers (more of this later in Chapter 10) because of its large conduction band offset. In particular, the $\text{In}_{0.52}\text{Al}_{0.48}\text{As}$ and $\text{In}_{0.53}\text{Ga}_{0.47}\text{As}$ alloys are lattice-matched to the readily available InP substrates (see [15] p. 196).

In fact, Fig. 3.29 demonstrates that the effect is relatively small, but nevertheless significant, at narrower well widths. Indeed, energy level spacings in intersubband lasers are designed in relation to phonon energies which are typically a few 10s of meV, and hence it is debatable whether a 2 meV effect should be included. The effect is larger when the $In_{0.53}Ga_{0.47}As$ wells are surrounded by strained AlAs barriers, so providing even more offset, and this material system has been the subject of several recent papers [64, 67, 68].

3.17   POISSON'S EQUATION

All of the techniques are now in place to be able to solve the Schrödinger equation for any heterostructure for which the band-edge potential profile defining the structure is known. However, all of the theoretical methods and examples described so far have concentrated solely on solving systems for a single charge carrier. In many devices such models would be inadequate as large numbers of charge carriers, e.g. electrons, can be present in the conduction band. In order to decide whether or not typical carrier densities would give rise to a *significant* additional potential on top of the usual band-edge potential terms (which will be labelled specifically as V_{CB} or V_{VB}), it then becomes necessary to solve the electrostatics describing the system.

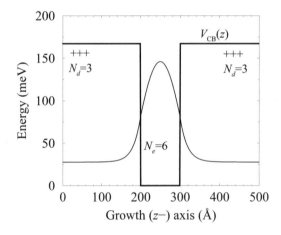

Figure 3.30   A modulation-doped single quantum well

When considering the case of an *n*-type material, then (although obvious) it is worth stating that the number of 'free' electrons in the conduction band is equal to the number of positively charged ionized donors in the heterostructure. In the example shown in Fig. 3.30 this equates to six donors becoming ionized and supplying six electrons into the quantum well—thus the system maintains charge neutrality. Fig. 3.30 is a *modulation-doped* system in that the doping is located in a position where the free carriers it produces will become spatially separated from the ions.

The additional potential term $V_\rho(z)$ arising from this, or any other charge distribution ρ, can be expressed by using Poisson's equation:

$$\nabla^2 V_\rho = -\frac{\rho}{\epsilon} \tag{3.79}$$

where ϵ is the permittivity of the material, i.e. $\epsilon = \epsilon_r \epsilon_0$. The solution is generally obtained via the electric field strength $\mathbf{E}$; recalling that:

$$\mathbf{E} = -\nabla V \tag{3.80}$$

the potential then would follow in the usual way [69]:

$$V_\rho(\mathbf{r}) = -\int_{-\infty}^{\mathbf{r}} \mathbf{E} \cdot \mathbf{dr} \tag{3.81}$$

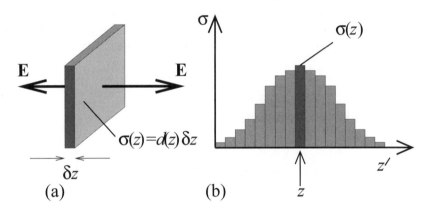

Figure 3.31   Electric field strength from an infinite plane of charge of volume density $d(z)$ and thickness δz

Given that the potential profiles, $V_{\mathrm{CB}}(z)$ for example, are one-dimensional, then they will also produce a one-dimensional charge distribution. In addition, remembering that the quantum wells are assumed infinite in the x–y plane then any charge density $\rho(z)$ can be thought of as an infinite plane, i.e. a *sheet*, with areal charge density $\sigma(z)$ and thickness δz, as shown in Fig. 3.31(a). Such an infinite plane of charge produces an electric field perpendicular to it, and with a strength:

$$\mathbf{E} = \frac{\sigma}{2\epsilon} \tag{3.82}$$

Note that, as the sheet is infinite in the plane, then the field strength is constant *for all distances from the plane*. The total electric field strength due to many of these planes of charge, as shown in Fig. 3.31(b), is then just the sum of the individual contributions as follows:

$$\mathbf{E}(z) = \sum_{z'=-\infty}^{\infty} \frac{\sigma(z')}{2\epsilon} \mathrm{sign}(z - z') \tag{3.83}$$

where the function sign is defined as

$$\mathrm{sign}(z) = +1, \quad z \geq 0; \qquad \mathrm{sign}(z) = -1, \quad z < 0 \tag{3.84}$$

and has been introduced to account for the vector nature of $\mathbf{E}$, i.e. if a single sheet of charge is at a position z', then for $z > z'$, $\mathbf{E}(z) = +\sigma/(2\epsilon)$, whereas for $z < z'$, $\mathbf{E}(z) = -\sigma/(2\epsilon)$. Note further that it is only the charge neutrality, i.e. there are as many ionized donors (or acceptors) in the system as there are electrons (or holes), or expressed mathematically:

$$\sum_{z=-\infty}^{+\infty} \sigma(z) = 0 \tag{3.85}$$

which ensures that the electric field, and hence the potential, go to zero at large distances from the charge distribution. For the case of a doped semiconductor, there would be two contributions to the charge density $\sigma(z)$, where the first would be the ionized impurities and the second the free charge carriers themselves. While the former would be known from the doping density in each semiconductor layer, as defined at growth time, the latter would be calculated from the probability distributions of the carriers in the heterostructure. Thus if $d(z)$ defines the volume density of the dopants at position z, where the planes are separated by the usual step length δz, then the total number of carriers, per unit cross-sectional area, introduced into the heterostructure is given by:

$$N = \int_{-\infty}^{+\infty} d(z) \ \mathrm{d}z \tag{3.86}$$

The net charge density in any of the planes follows as:

$$\sigma(z) = q\left[N\psi^*(z)\psi(z) - d(z)\right]\delta z \tag{3.87}$$

where q is the charge on the extrinsic carriers. The step length δz selects the proportion of the carriers that are within that 'slab' and converts the volume density of dopant, $d(z)$ into an areal density.

If the charge carriers are distributed over more than one subband, then the contribution to the charge density $\sigma(z)$ would have to be summed over the relevant subbands, i.e.

$$\sigma(z) = q\left(\sum_{i=1}^{n} N_i\psi_i^*(z)\psi_i(z) - d(z)\right)\delta z \tag{3.88}$$

where $\sum_{i=1}^{n} N_i = N$.

Fig. 3.32 shows the areal charge density along the growth axis for a 100 Å GaAs well, n-type doped to $2\times10^{18}\mathrm{cm}^{-3}$, surrounded by 200 Å undoped $Ga_{0.8}Al_{0.2}As$ barriers. The ionized donors yield a constant contribution to σ within the well of $d(z)\delta z = 2\times10^{24}\mathrm{m}^{-3}\times1\text{Å} = +2\times10^{14}\mathrm{m}^{-2}$, in each of the 1 Å thick slabs. Hence, the total number N of electrons in the quantum well is $100\times2\times10^{14}\mathrm{m}^{-2} = 2\times10^{12}\mathrm{cm}^{-2}$. By assuming that the electrons introduced by such doping all occupy the ground state of the quantum well, then the curve on top of the ionized impurity background clearly resembles $-\psi^*\psi$, as expected from the mathematics. The discontinuities in σ occur at the edges of the doping profiles and *are* of magnitude $2\times10^{14}\mathrm{m}^{-2}$, again as expected.

There are a number of points to note about Fig. 3.33, which plots the electric field strength $\mathbf{E}$ due to the charge distribution (as defined in equation (3.83)) along the growth axis of the heterostructure. First, the field does reach zero at either end of the structure, which implies charge neutrality. In addition, the zero field point at the centre of the structure reflects the symmetry of the charge distribution. The electric field strength

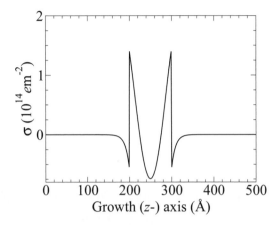

Figure 3.32   Areal charge density σ for a 100 Å GaAs well, n-type doped to $2\times10^{18}\,\mathrm{cm}^{-3}$, surrounded by undoped Ga$_{0.8}$Al$_{0.2}$As barriers

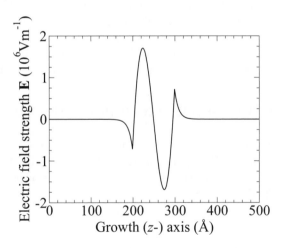

Figure 3.33   The electric field strength **E** due to the charge distribution shown in Fig. 3.32

itself is not an observable, merely an intermediate quantity which *can* be useful to plot from time to time; the quantity which is significant is, of course, the potential due to this charge distribution. Fig. 3.34 plots the potential as calculated from equation (3.81), as usual defining the origin, in this case for the potential, at the effective infinity μ at the left-hand edge of the barrier–well–barrier structure.

Again, the symmetry of the original heterostructure and doping profiles are reflected in the symmetric potential. The potential is positive at the centre of the well since the

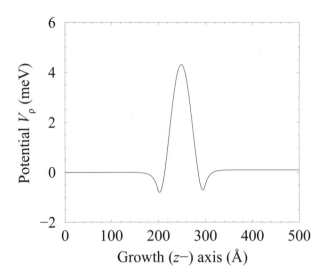

Figure 3.34 The potential due to the ionized donor/electron charge distribution

system under consideration consists of electrons in the conduction band, so any test charge used to probe the potential is also an electron which would be repelled by the existing charge. The carrier density in this single quantum well is reasonably high at $2 \times 10^{12} \mathrm{cm}^{-2}$, and this produces a potential of up to 4 meV; while this is small compared to the conduction band offset, which is usually of the order of 100 or 200 meV or more, it could still have a measurable effect on the energy eigenvalues of the quantum well.

3.18 SELF-CONSISTENT SCHRÖDINGER–POISSON SOLUTION

The energy eigenvalues are calculated by considering the introduction of a further test electron into the system and incorporating the potential due to the carrier density already present into the standard Schrödinger equation, i.e. the potential term $V(z)$ in equations (2.96) or (3.43) becomes:

$$V(z) \rightarrow V_{\mathrm{CB}}(z) + V_{\rho}(z) \tag{3.89}$$

where V_{CB} represents the band edge potential at zero doping and the potential due to the non-zero number of carriers, i.e. the charge density ρ, is represented by the function V_{ρ}.

The numerical shooting method, described in detail earlier in this chapter, can be used without alteration to solve for this new potential, which will thus yield new energies and wave functions. The latter is an important point since the potential due to the charge distribution is itself dependent on the wave functions. Therefore, it is necessary to form a closed loop solving Schrödinger's equation, calculating the potential due to the resulting charge distribution, adding it to the original band-edge potential, solving Schrödinger's equation again, and so on—a process which is illustrated schematically in Fig. 3.35.

The process is repeated until the energy eigenvalues converge; at this point the wave functions are simultaneously solutions to both Schrödinger's and Poisson's equations—

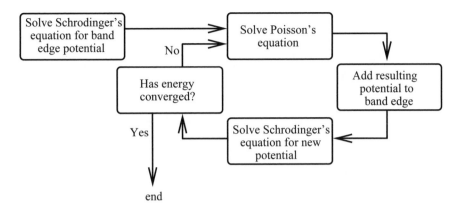

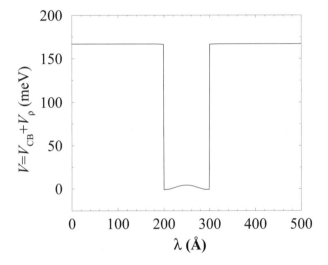

Figure 3.35   Block diagram illustrating the process of self-consistent iteration

the solutions are described as *self-consistent*, rather like Hartree's approach to solving many electron atoms (see e.g. reference [3] p. 396).

Figure 3.36   The sum of the band-edge potential V_{CB} and Poisson's potential V_ρ for the single quantum well of Fig. 3.32

Fig. 3.36 shows the result of adding the potential due to the charge distribution V_ρ, as displayed in Fig. 3.34, to the original band-edge potential V_{CB} for the single quantum well of the previous section. The perturbation, even at this relatively high carrier density of 2×10^{12}cm^{-2}, is rather small compared with the barrier height, for instance. Nonetheless, it is important to calculate the effect of this perturbation on the electron energy levels by continuing with the iterative process and looking for convergence of the resulting energy solutions. This process is illustrated in Fig. 3.37.

The first iterative loop produces the majority of the change in the energy level from the single carrier system to the doped system. Subsequent iterations produce only minor refinements to the energy level, which has certainly converged by the eighth loop.

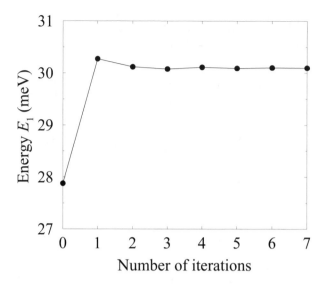

Figure 3.37   The ground state energy of the single quantum well of Fig. 3.32 as a function of the number of iterative steps

Altogether the ground state energy has changed by about 2 meV when accounting for this finite charge density.

3.19   COMPUTATIONAL IMPLEMENTATION

As this work is about how to actually implement a computational method to provide quantitative predictions, *in addition to* deriving the theoretical equations in the first place, it is important to be aware that the self-consistent Schrödinger–Poisson solution presented here, can at times diverge.

The doping density in the last series of examples in the previous two sections was chosen such that it produced a noticeable perturbation to the total potential profile, as illustrated in Fig. 3.36. In fact, this total carrier density of 2×10^{12}cm^{-2} is fairly high *when contained within just one quantum well*. The majority of calculations later in this text, on active regions of unipolar lasers for example, will have carrier densities much lower than this. However, in order to illustrate the intricacies of the computational method, consider now an even higher doping density, i.e. double that of the previous.

The first energy calculated in the process is that of the ground state for an undoped system, which is obviously the same as before. However, the higher doping density leads to different Poisson potentials, and hence subsequent iterations *do* yield different energy levels. Although just one iterative step is sufficient to yield a result to within 1 meV of the self-consistent solution, the calculations with the same step length (δz=1 Å) as before, diverge as the iteration proceeds. This is a manifestation of the computational implementation—the smallest asymmetry in the potential profile is exaggerated with each iterative loop. Fig. 3.38 also illustrates the solution, which is merely to increase the computational accuracy by way of reducing the step length δz in the original dis-

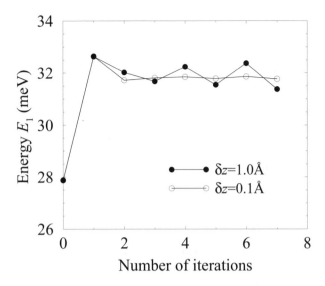

Figure 3.38   The ground state energy of the single quantum well of Fig. 3.32, but doped to $4\times10^{18}\text{cm}^{-3}$ in this case, thus giving a total carrier density of $4\times10^{12}\text{cm}^{-2}$

cretization of the potential. Asymmetric potentials, such as step quantum wells, or quantum wells with electric fields applied, are naturally more resistant to such effects.

3.20   MODULATION DOPING

Although mention has been made of quantum well systems in which doping in the barriers leads to a spatial separations of the ions and charge carriers, i.e. which collect in a quantum well, quantitative calculations presented thus far have not considered these modulation-doped systems. Fig. 3.39 shows the band-edge potential, V_{CB}, and the self-consistent potential, $V_{CB} + V_\rho$, for a system of the type illustrated in Fig. 3.30, i.e. an undoped single quantum well surrounded by doped barriers; with the full layer definition thus being: 100 Å $\text{Ga}_{0.8}\text{Al}_{0.2}\text{As}$ doped n-type to $2\times10^{17}\text{cm}^{-3}$; 100 Å GaAs undoped; 100 Å $\text{Ga}_{0.8}\text{Al}_{0.2}\text{As}$ doped n-type to $2\times10^{17}\text{cm}^{-3}$.

The electrons introduced into the system are physically separated from the ionized donors, so therefore instead of an ion/charge carrier plasma, the mobile charge in this case is often referred to as a *two-dimensional electron gas*. The physical separation leads to a reduction in the ionized impurity scattering and hence increased electron mobilities for in-plane (x–y) transport, a feature which is exploited in *High-Electron-Mobility Transistors* (HEMTs).

3.21   THE HIGH-ELECTRON-MOBILITY TRANSISTOR

The high-electron-mobility transistor (HEMT), the heterostructure field-effect transistor (HFET), the modulated-doped field-effect transistor (MODFET), or even the two-dimensional electron-gas field-effect transistor (TEGFET), are all names which refer to

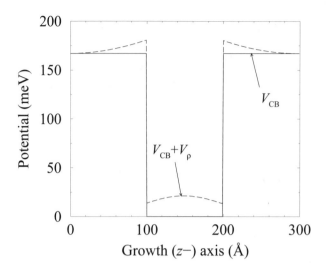

Figure 3.39   The band-edge potential and the self-consistent potential of a modulation-doped single quantum well

a transistor designed to exploit the high in-plane $(x-y)$ mobility, which arises when a (usually single) heterojunction is modulation-doped. In essence, a high-bandgap material such as $Ga_{1-x}Al_xAs$ is doped n-type, and upon ionization of the donors the electrons move to the lower energy levels available in the narrower bandgap material, such as GaAs—this situation is very similar to the single quantum well of the previous section, but with just one barrier layer in this case.

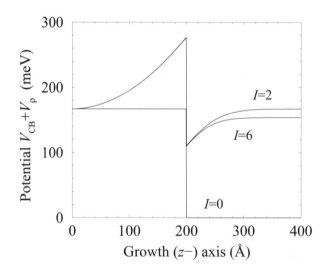

Figure 3.40   The evolution of the self-consistent potential $V_{CB} + V_\rho$, with the number of iterations I, for a single modulation doped heterojunction

Fig. 3.40 shows the evolution of the self-consistent potential, $V_{CB} + V_\rho$, with each iterative step. After the first solution of Schrödinger's equation, the potential V consists entirely of the band-edge potential V_{CB} and clearly defines the single heterojunction. However, as the iteration proceeds the charge carriers are drawn towards the interface and form a triangular well potential—a process often referred to as *band bending*. This is vividly illustrated by Fig. 3.41, which shows the corresponding evolution of the electron wave functions.

This section has merely 'scratched the surface' of transistor modelling, and has been used to illustrate the applicability of quantum and electrostatic theory to everyday electronic devices, as well as the more esoteric semiconductor structures concentrated on in this work. For more information on heterojunction-based transistors see Kelly [7], Chapter 16, or more specialized works, such as [70–73].

3.22   BAND FILLING

One way of measuring the energy levels of quantum well systems is via the use of spectroscopy, which involves illuminating the semiconductor with light from a laser. If the photon energy is greater than the bandgap then, in the bulk material, electrons are excited from the valence-band states to the conduction-band, which depending in more detail on the excitation energy, can produce pairs of free-electrons and free-holes, or excitons (see Chapter 6). In a quantum well system, the photons must have additional energy in order to overcome the electron- and hole- ground state confinement energies as well.

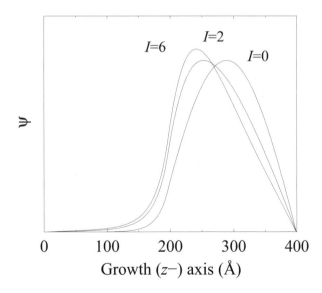

Figure 3.41   The evolution of the self-consistent wave functions, with the number of iterations I, for a single modulation doped heterojunction

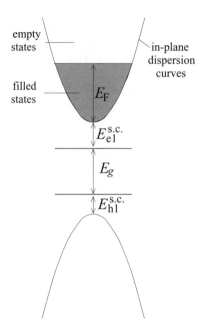

Figure 3.42 Schematic representation of the increase in band edge due to band filling

In a doped system, e.g. n-type, many electrons already exist in the conduction band, and thus spectroscopic measurements will observe the self-consistent solutions, $E_{\text{el}}^{\text{s.c.}}$ and $E_{\text{h1}}^{\text{s.c.}}$ for the confinement energies.

However, in addition there is another effect which may or may not be large enough to observe. Thus far, it has been assumed that there are an unlimited number of states available to electrons (or holes) when they are introduced as extrinsic carriers; note however (see Chapter 2) that the density of states has been derived for two-dimensional systems. Thus excitation can only occur when the photon energy is sufficiently large to excite an electron into an *empty* state, see Fig. 3.42. At low temperatures, the Fermi edge defines a sharp boundary between the filled states $f^{\text{FD}}(E) = 1$ and the empty states $f^{\text{FD}}(E) = 0$, and hence the *minimum* excitation energy must overcome this additional component, i.e.

$$E = E_g + E_{\text{el}}^{\text{s.c.}} + E_{\text{h1}}^{\text{s.c.}} + E_{\text{F}} \qquad (3.90)$$

This phenomenon of an apparently increased 'band edge' to absorption is known as *band filling*. This can also occur in undoped systems under very high excitation intensities which produce large numbers of both electrons and holes, which in turn can fill all of the available states and thus increase the minimum required photon energy for absorption. It can be accounted for under the present scheme by simply calculating the quasi-Fermi energy, E_{F}, as described in Section 2.4.

CHAPTER 4

DIFFUSION

4.1 INTRODUCTION

Any substance will attempt to diffuse from an area where it is present in high concentrations to an area of low concentration. For example, obvious though it seems, if the curve $x(z)$ in Fig. 4.1 represented the concentration of water in a trough then the water would fall very rapidly from the region of high concentration to the region of low! Crude though it may seem, this is an example of diffusion. It occurs also for gaseous systems, e.g. smoke gradually disperses in an enclosed room; however, the point of interest in this present work is diffusion of material species in solids.

In the context of semiconductors, and in particular semiconductor heterostructures, it is clear that diffusion of *material species* could be important as their very nature derives from discontinuous changes in materials. Fig. 4.1 could therefore represent a dopant, in either a bulk semiconductor, i.e. a *homojunction*, or an alloy component, e.g. Al at a $GaAs/Ga_{1-x}Al_xAs$ *heterojunction*.

Diffusion at such boundaries is a strong possibility, particularly during the elevated temperatures often used during growth, or the lower but more prolonged heating that may occur during normal device operation. Any movement of material, e.g. Al at a $GaAs/Ga_{1-x}Al_xAs$ heterojunction will 'blur' the interface, i.e. the change from one material to the other will occur via a range of intermediatary alloys. Such a process is known as *interface mixing* and is represented schematically in Fig. 4.2. The change

Quantum Wells, Wires and Dots, Third Edition. P. Harrison
©2009 John Wiley & Sons, Ltd.

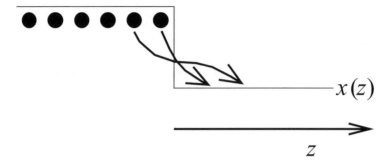

Figure 4.1   Simple illustration of diffusion; diffusant moves from areas of high concentration to low

in profile of the junction will inevitably alter the electronic properties of the system, which will, in turn, affect the device characteristics and therefore the operating lifetime. The motivation behind modelling diffusion is to understand such time dependency of the properties of quantum well systems, in order to be able to control or prevent it, design a device in which its impact will be minimised, or predict the lifetime of a device. Besides these detrimental effects, controlled diffusion is sometimes initiated deliberately in order to tailor the optical and electronic properties of quantum wells systems [59]. For a comprehensive treatment of utilizing quantum well interdiffusion for photonics, see the review by Li [74].

Diffusion was first put on a quantitative basis by Fick in 1855 [75] who derived two laws. The first of these stated that the steady-state flux through a plane is proportional to, but in the opposite direction to, the concentration gradient, i.e.

$$\text{flux} = -\mathcal{D}\nabla c \tag{4.1}$$

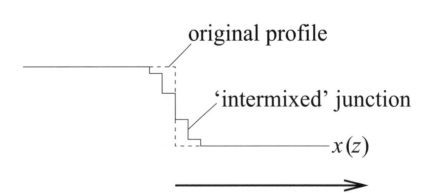

Figure 4.2   An intermixed heterojunction; note that the total amount of diffusant represented by the area under the $x(z)$ curve, is constant

while the second showed that the time dependency of the concentration is given by the following:

$$\frac{\partial c}{\partial t} = -\nabla \cdot \text{flux} \tag{4.2}$$

$$\therefore \frac{\partial c}{\partial t} = \nabla \cdot (\mathcal{D}\nabla c) \tag{4.3}$$

which in one dimension would become:

$$\frac{\partial c}{\partial t} = \frac{\partial}{\partial z}\mathcal{D}\frac{\partial c}{\partial z} \tag{4.4}$$

For a simple introduction to these laws, see p.66 of the text by Callister [76].

Many different particular solutions of Fick's first and second laws can be found, with perhaps the most common one being the error function solution for diffusion, described by a constant $\mathcal{D}$, at the interface between two semi-infinite slabs. Such a model is commonly employed to model diffusion at a semiconductor heterojunction (see, e.g. [59, 77–80]).

Crank [81] and Shewmon [82] offer several solutions based upon numerical methods and involving complex initial diffusant profiles and variable diffusion coefficients. However, as with Tuck [83], each situation is considered separately and a new numerical solution is sought. It would be advantageous if a general, probably numerical, method could be developed which was applicable to all of these situations and also new ones.

Such a method would attempt to model diffusion processes, such as those mentioned above, in terms of an average *diffusion coefficient* which may be a constant or a simple function of the material properties. The mechanisms by which the diffusion proceeds, e.g. vacancy assisted [84], are not of interest. Atomistic models of diffusion *do* exist and are generally centred around a Monte Carlo simulation of the movement of individual atoms within the crystal (see, e.g. [85]).

For the purpose of this present work, solutions are demonstrated within semiconductor heterostructures constructed of an alloy $A_{1-x}B_xC$, where the system can be most generally represented by the z-dependence of the alloy component, i.e. $x = x(z)$. The methods developed are equally applicable to a dopant distribution whose profile may be $c(z)$, as labelled above.

4.2   THEORY

As mentioned above, the *most general* one-dimensional diffusion equation for a diffusant distribution represented by $x(z)$ is given by Fick's second law for non-steady state diffusion [81]:

$$\frac{\partial x}{\partial t} = \frac{\partial}{\partial z}\left(\mathcal{D}\frac{\partial x}{\partial z}\right) \tag{4.5}$$

where t is the time and the diffusion coefficient $\mathcal{D}$ could have temporal t, spatial z and concentration x dependencies, i.e. $\mathcal{D} = \mathcal{D}(x, z, t)$. Given this, then the derivative with respect to z operates on both factors, resulting in the following:

$$\frac{\partial x}{\partial t} = \frac{\partial \mathcal{D}}{\partial z}\frac{\partial x}{\partial z} + \mathcal{D}\frac{\partial^2 x}{\partial z^2} \tag{4.6}$$

which is a second-order $(\partial^2/\partial z^2)$ non-linear $((\partial/\partial z)^2)$ differential equation. In any given problem it is likely that two of the following unknowns will be known (!):

- The initial diffusant profile, $x(z, t = 0)$;

- The final diffusant profile, $x(z, t)$;

- The diffusion coefficient, $\mathcal{D} = \mathcal{D}(x, z, t)$.

The problem will be to deduce the third unknown. This could manifest itself in several ways:

a. given the initial diffusant profile and the diffusion coefficient, predict the diffusant profile a certain time into the future;

b. given the initial and final diffusant profiles, calculate the diffusion coefficient;

c. given the final diffusant profile and the diffusion coefficient, calculate the initial diffusant profile.

Knowing the versatility achieved by the *numerical* shooting method solution to Schröd-inger's equation as discussed in the previous chapter, then clearly a numerical solution would again be favourable. Learning from the benefits of expanding the derivatives in the Schrödinger equation with finite differences, this would then appear to offer a promising way forward.

Recall the finite difference approximations to first and second derivatives, i.e.

$$\frac{\partial f}{\partial z} \approx \frac{f(z + \delta z) - f(z - \delta z)}{2\delta z} \tag{4.7}$$

and

$$\frac{\partial^2 f}{\partial z^2} \approx \frac{f(z + \delta z) - 2f(z) + f(z - \delta z)}{(\delta z)^2} \tag{4.8}$$

Then equation (4.6) can be expanded to give:

$$\frac{x(z, t + \delta t) - x(z, t)}{\delta t} = \left[\frac{\mathcal{D}(x, z + \delta z, t) - \mathcal{D}(x, z - \delta z, t)}{2\delta z} \right]$$

$$\times \left[\frac{x(z + \delta z, t) - x(z - \delta z, t)}{2\delta z} \right]$$

$$+ \mathcal{D}(x, z, t) \left[\frac{x(z + \delta z, t) - 2x(z, t) + x(z - \delta z, t)}{(\delta z)^2} \right] \tag{4.9}$$

Notice that the derivative with respect to time has been written as:

$$\frac{x(z, t + \delta t) - x(z, t)}{\delta t}, \quad \text{rather than} \quad \frac{x(z, t + \delta t) - x(z, t - \delta t)}{2\delta t} \tag{4.10}$$

as would be expected from the expansion in equation (4.7). In fact, in this case where this is the only time derivative, the two are equivalent.

Assuming the most common class of problem, as highlighted above, as point (a), namely that the function $x(z, t)$ is known when $t = 0$, i.e. it is simply the initial profile of the diffusant, and the diffusion coefficient $\mathcal{D}$ is fully prescribed, then it is apparent from equation (4.9) that the concentration x at any point z can be calculated a short time interval δt into the future, provided that the concentration x is known at small spatial steps δz either side of z. This approach to the solution of the differential equation is known as a *numerical simulation*. It is not a mathematical solution, but rather a computational scheme which has been derived to mirror the physical process.

It has already been mentioned that the diffusion coefficient $\mathcal{D}$ could be a function of x, z, and t, with the form of $\mathcal{D}$ being used to define the class of diffusion problem, e.g.

(i). $\mathcal{D} = D_0$, a constant, for simple diffusion problems.

(ii). $\mathcal{D} = D(x)$, a function of the concentration as encountered in non-linear diffusion problems [86]. Note, that as $x = x(z)$, then $\mathcal{D}$ is intrinsically a function of position too.

(iii). $\mathcal{D} = D(z)$, a function of position only, as could occur in ion implantation problems [87]. Here, the diffusion coefficient could be linearly dependent on the concentration of vacancies for example, where the latter itself is depth dependent.

(iv). $\mathcal{D} = D(t)$, a function of time, as could occur during the annealing of radiation damage. For example, ion implantation can produce vacancies which aid diffusion [88, 89]. During an anneal, the vacancy concentration decreases as the lattice is repaired, which in turn alters the diffusion coefficient.

4.3 BOUNDARY CONDITIONS

Thus, given the initial diffusant profile and a fully prescribed diffusion coefficient, everything is in place for predicting the profile of the diffusant at any time in the future, except for the conditions at the ends of the system. These cannot be calculated with the iterative equation (equation (4.9)), as the equation requires points that lie outside the z-domain.

For diffusion from an infinite source, it may be appropriate to fix the diffusant concentration x at the two end points, e.g. $x(z = 0, t) = x(z = 0, t = 0)$. Alternatively, the concentrations x at the limits of the z-domain could be set equal to the adjacent points, which can be deduced from equation (4.9). Physically, this defines the semiconductor structure as a closed system, with the total amount of diffusant remaining the same. It is these latter 'closed system' boundary conditions which will be employed exclusively in the following examples.

4.4 CONVERGENCE TESTS

Fig. 4.3 shows the result of allowing the diffusant profile in Fig. 4.1 to evolve to equilibrium, using the closed-system boundary conditions as described above. Clearly the 'closed' nature of the system can be seen—the total amount of diffusant remains the same, and ultimately as would be expected for the 'water step', the concentration reaches a constant value. In the case of water, this could be looked upon as minimizing the potential energy.

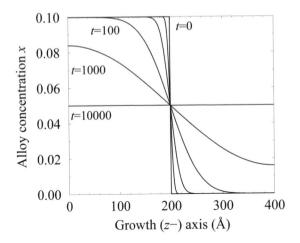

Figure 4.3   Time evolution of adjacent finite slabs of $Ga_{0.9}Al_{0.1}As/GaAs$ with a constant diffusion coefficient of $\mathcal{D} = D_0 = 10$ $Å^2s^{-1}$

If the diffusion process can be described by a constant diffusion coefficient, $\mathcal{D} = D_0$, then the general diffusion equation, equation (4.5), and its equivalent computation form, (equation (4.9)) simplifies to the following:

$$\frac{\partial x}{\partial t} = D_0 \frac{\partial^2 x}{\partial z^2} \tag{4.11}$$

which, as mentioned above, has error function solutions for the case of diffusion at the interface ($z = 0$) of a semi-infinite slab of concentration x_0 [81], i.e.

$$x(z) = \frac{x_0}{2} \text{erfc} \frac{z}{2\sqrt{D_0 t}} \tag{4.12}$$

where erfc is the complementary error function see reference [38], p.295. This technique can be imposed on multiple heterojunctions by linearly superposing solutions [77]. Fig. 4.4 compares just such an error function solution* with the numerical solution for a 200 Å single GaAs quantum well surrounded by 200 Å $Ga_{0.9}Al_{0.1}As$ barriers after 100 s of diffusion described by a constant coefficient D_0=10 $Å^2s^{-1}$. Clearly, the numerical method advocated here exactly reproduces the analytical solution.

As with any numerical method, convergence tests must be performed. In the present case, these are necessary in order to determine that the results are independent of the intervals δz and δt. For the characteristic dimensions of interest in the following problems, namely nanostructures ($\sim$100—1000 Å) and macroscopic annealing times ($\sim$100—1000 s), $\delta z = 1$ Å and $\delta t = 0.01$ s were found to be satisfactory.

From a computational viewpoint, the technique is useful since, in general, it is anticipated that the numerical complexity scales linearly with the physical complexity. In

---

*Thanks are expressed here to T. Stirner for providing the error function data

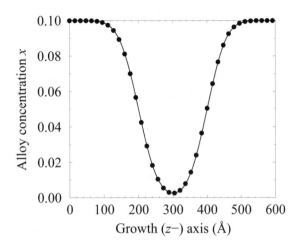

Figure 4.4   Comparison between the numerical solution (solid symbols) and the standard error function solution (continuous line) for the concentration profile $x(z)$ of a 200 Å $Ga_{0.9}Al_{0.1}As/200$ Å GaAs /200 Å $Ga_{0.9}Al_{0.1}As$ single quantum well, after 100 s of diffusion described by a constant diffusion coefficient $D_0 = 10$ Å^{2}s^{-1}

actuality, it has been found that, for the series of problems discussed below, the number of spatial points required for the calculation is independent of the variability in the initial concentration profile. Increasingly larger microstructures, i.e. superlattices with more and more periods, do require more points (i.e. δz remains constant), but the computational effort required is proportional to the number of points. For a given structure, but with an increasingly complex dependence of the diffusion coefficient, e.g. $\mathcal{D}$ having a stronger functional dependence on the concentration x, then smaller time intervals δt are required. Again, however, halving the time interval merely doubles the necessary computational effort.

Having therefore established a numerical method for simulating diffusion and after validating it by comparison to the accepted analytical form for a special case, it is now an appropriate time to demonstrate the solution's versatility by visiting, in turn, the examples (i)–(iv) listed in Section 4.2.

4.5   CONSTANT DIFFUSION COEFFICIENTS

The case of a constant diffusion coefficient has already been touched upon in the last section, as its analytical solutions represent an important test for the numerical solution forwarded here. Such a numerical solution can be used to illustrate two specific points for this situation.

In particular, the error function solutions given by equation (4.12) imply *universality*, in that they are dependent upon the value of the product $D_0 t$ only, and not the individual values themselves. Therefore, a system with a high diffusion coefficient can give the same diffused profile after a short time as a system with a low diffusion coefficient that is allowed to diffuse for longer periods.

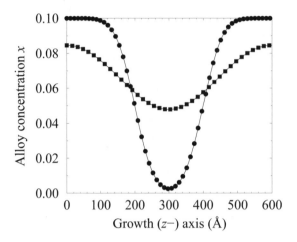

Figure 4.5   Diffusion profiles for a 200 Å $Ga_{0.9}Al_{0.1}As$/200 Å GaAs /200 Å $Ga_{0.9}Al_{0.1}As$ single quantum well, with constant diffusion coefficient D_0, and $D_0t=1000$ $Å^2s^{-1}$ (continuous line plus solid circles) and $D_0t=10000$ $Å^2s^{-1}$ (continuous line plus solid squares)

The continuous lines in Fig. 4.5 display the results of diffusing a single quantum well for 100 (bottom) and 1000 s (top curve) with $D_0 = 10$ $Å^2s^{-1}$. In comparison with this, the solid symbols are the results of diffusing the same initial system for just 10 s but with $D_0=100$ (circles) and 1000 $Å^2s^{-1}$ (squares). Clearly, the numerical solution reproduces the universality as well—a point which is not at all obvious from the numerical form of the diffusion equation in equation (4.9).

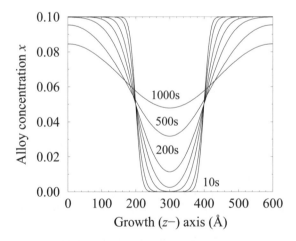

Figure 4.6   Diffusion profiles for the single quantum well of Fig. 4.5 with constant diffusion coefficient $D_0 = 10$ $Å^2s^{-1}$, and diffusing times $t = 10, 20\ 50, 100, 200, 500,$ and 1000 s

Fig. 4.6 displays a succession of diffusant profiles for the same single quantum well as before, again with $D_0 = 10 \text{ Å}^2\text{s}^{-1}$, where the evolution of the profile at longer diffusing times is clear. Therefore given a semiconductor heterostructure and an annealing time, it is apparent that the diffused profile can be predicted, *provided* that the diffusion coefficient characterising the process is known from other experiments. The direct determination of the diffused profile on such short length scales is difficult (although not impossible) from traditional techniques used for bulk analysis, such as *Secondary Ion Mass Spectroscopy* (SIMS) (see reference [7], Section 2.4). However, alternative spectroscopic techniques can be used to infer the extent of diffusion by its effect on other observables. In particular, as mentioned in Chapter 3, photoluminescence and photoluminescence excitation can be used to measure the excitation energy of electrons across the band gap. As the electron and hole confinement energies, E_e and E_h, are dependent upon the band-edge profiles of the conduction- and valence-band edges, respectively, the excitation energy ($E = E_g + E_e + E_h$, ignoring the exciton binding energy), is then also dependent upon the amount of diffusion.

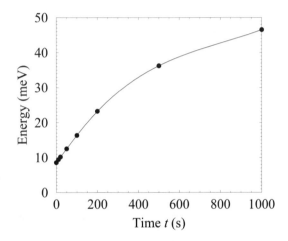

Figure 4.7   The corresponding ground state electron energies for the diffused profiles given in Fig. 4.6

The curve in Fig. 4.7 illustrates the change in the electron energy corresponding to the diffusion process of Fig. 4.6. The magnitude of the variation is substantial and easily detectable by experiment [78–80, 90–92]. For a theoretical interpretation of such experiments, *including* the change in exciton binding energy, see [77].

4.6   CONCENTRATION DEPENDENT DIFFUSION COEFFICIENT

There are many examples for which the diffusion coefficient of a species is a function of the concentration of that species itself [83, 93, 94].

The work by Tuck is perhaps the most interesting as it makes a detailed comparison with experiment. Tuck considered the diffusant of a dopant, Zn in bulk GaAs, a well-studied system in which the 'substitutional-interstitial' mechanism is thought to dominate [95]. In this system the dopant (Zn) sits at two different positions within the crystal lattice, i.e. as the name implies both substitutionally and at interstitial sites. Atoms on the substitutional sites, which are in the majority, diffuse very slowly; however, the minority of interstitial atoms diffuse so readily that they dominate the evolution of the dopant profile with time.

The diffusion coefficient describing the process is concentration (c) dependent, and is given mathematically as:

$$\mathcal{D} = kc^n \tag{4.13}$$

hence the diffusion equation becomes:

$$\frac{\partial c}{\partial t} = \frac{\partial}{\partial z}\left(kc^n \frac{\partial c}{\partial z}\right) \tag{4.14}$$

The constant k was deduced from experiment, and the index n, for this system, was found to be 2. Tuck considered several situations and implemented solutions which agreed very well with experiment. In this work, the numerical simulation can be implemented directly.

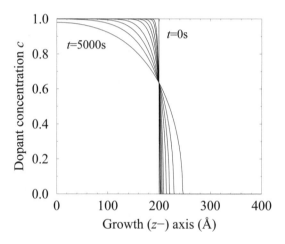

Figure 4.8   Concentration dependent ($\mathcal{D} = kc^2$) diffusion

Fig. 4.8 presents examples of the concentration profiles that such a concentration-dependent diffusion coefficient yields. The initial concentration c was just taken to be unity and k was taken as 1 $\text{Å}^2\text{s}^{-1}$. Note that, unlike the case of the constant diffusion coefficient, the curves do not all pass through the midpoint of the initial profile height, as shown in Fig. 4.3. Most importantly, the curves are of the same form as those given in Tuck [83] and hence agree with experiment.

4.7   DEPTH DEPENDENT DIFFUSION COEFFICIENT

As mentioned above, diffusion is becoming widely used as a post-growth method for fine tuning the structural, and hence electronic and optical properties of semiconductor heterostructures. Ion implantation (see Kelly [7], Section 3.4 for an introduction) is one method of stimulating diffusion; it is controllable, reproducible and its spatial resolution can be used to pattern a semiconductor wafer. Ion implantation can also be used as a way of introducing new species into a crystal, e.g. optically active rare earth ions into a host semiconductor [96]. However, the focus of interest here is in ion implantation as a means of enhancing the diffusion of the material species already present [89, 90, 92]. The implantation itself produces lattice damage, displacing atoms from their equilibrium positions and thus creating interstitial atoms and vacancies. As in the previous example, the interstitial atoms can diffuse very readily and in addition the presence of vacancies provides a route for the diffusion of substitutional species.

The diffusion coefficient describes the average speed by which atoms diffuse, and clearly the more vacancies and interstitials in a given region of material, then the faster diffusion will proceed. Hence, in order to attempt to quantify the diffusion coefficient, it is necessary to have a knowledge of the amount of lattice damage that an implantation produces; this can be calculated with a Monte Carlo simulation of the atoms impinging on the crystal planes. Such computer simulations, known as T.R.I.M. codes, are well understood and documented (see, for example [97]). An example of the output from such a simulation is given in Fig. 4.9.

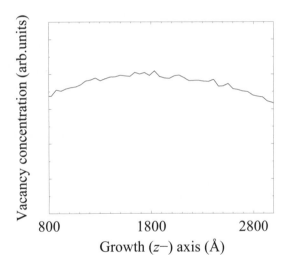

Figure 4.9   Typical depth dependency of vacancy concentration given by T.R.I.M. code data for 200 keV Ar^{2+} ions impinging on a [001] CdTe surface

The effect of 'vacancy-assisted' diffusion from such a profile has been calculated for a $CdTe/Cd_{1-x}Mn_xTe$ multiple quantum well [87, 98]. In this work, the depth dependence will be exaggerated merely to produce a more interesting variation in the diffusion. The depth dependence of the vacancy concentration could actually be described *empirically* by a Gaussian distribution of the form:

$$\rho^{\text{vacancy}} = \rho_0^{\text{vacancy}} \exp\left[-\frac{(z-z_0)^2}{2\sigma^2}\right] \quad (4.15)$$

where z_0 is the depth of the maximum, e.g. $z_0 \sim 1800$ Å in Fig. 4.9, and, of course, the standard deviation σ is related to the width of the distribution.

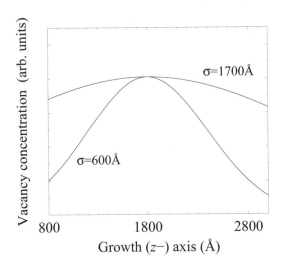

Growth ($z-$) axis (Å)

Figure 4.10   Gaussian fit to the T.R.I.M. code data of Fig. 4.9 together with an exaggerated depth dependence

Fig. 4.10 displays the Gaussian fit to the T.R.I.M. data of Fig. 4.9, described by the parameter $\sigma = 1700$ Å, together with the modified data for the purpose of this example, given by $\sigma = 600$ Å.

In real systems where intensive investigations for a particular semiconductor multi-layer with a particular ion implantation dosage are carried out, it would be necessary to relate the diffusion coefficient to the absolute vacancy concentration; however, in these present demonstrator examples of the numerical solution to the diffusion equation, it suffices to say let $\mathcal{D}$ be proportional to ρ^{vacancy}, and furthermore, choose the constant of proportionality to be 10 Å^{2}s^{-1}, i.e.

$$\mathcal{D} = 10 \exp\left[-\frac{(z-z_0)^2}{2\sigma^2}\right] \text{Å}^2\text{s}^{-1} \tag{4.16}$$

Given this, Fig. 4.11 displays both the initial and final alloy concentration profile after 200 s of diffusion described by equation (4.16) for a generic 150 Å AC/50 Å $A_{1-x}B_xC$ superlattice/multiple quantum well. As expected, the central wells have diffused considerably more than those near the edges where the diffusion coefficient is lower. The exaggerated z-dependence of $\mathcal{D}$ has fulfilled its goal in this illustration of producing a much clearer depth dependence than previously published for a realistic system (see, for example [98]).

Photoluminescence measurements on diffused systems such as those shown in Fig. 4.11 would exhibit a broadened emission line, as the photogenerated carriers in the central wells would have a different energy to those in the outer wells. This phenomenon of line broadening in a diffused superlattice has been observed by Elman *et al.* [90] and subsequently modelled theoretically [98].

Alternatively, the structure, including the depth dependence of the diffusion, can be mapped directly by using *double-crystal X-ray diffraction* (DCXRD) [99], a technique

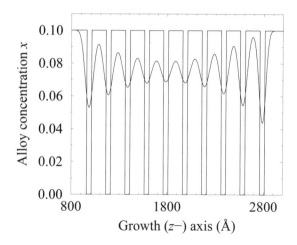

Figure 4.11 Alloy concentration profile x after 200 s of diffusion described by the depth dependent coefficient of equation (4.16)

which has showed itself to be very valuable in monitoring the progress of interface intermixing [89].

4.8 TIME DEPENDENT DIFFUSION COEFFICIENT

Vacancy-enhanced diffusion will continue for as long as the vacancies are present, but one way to control their lifetime and 'freeze' the diffusant profile is to anneal out the radiation damage. The annealing process thermally activates the interstitials back into vacancies and so restores order to the crystal. This subsequent annealing process *could* be described by a time-dependent diffusion coefficient, perhaps of the form:

$$\mathcal{D} = \mathcal{D}(z) \exp\left(-\frac{t}{\tau}\right) \tag{4.17}$$

thus presenting the opportunity to complete the examples of the functional dependencies listed in Section 4.2. $\mathcal{D}(z)$ is the initial (time-independent) depth-dependent diffusion coefficient due to the vacancy distribution. Using the form in equation (4.16), the diffusion coefficient would then be fully specified by:

$$\mathcal{D} = 10 \exp\left[-\frac{(z - z_0)^2}{2\sigma^2}\right] \exp\left(-\frac{t}{\tau}\right) Å^2 s^{-1} \tag{4.18}$$

Making use of the ion-implantation-enhanced diffusion profile of Fig. 4.11 as a starting point, Fig. 4.12 shows the results of simulating annealing out the lattice damage with a time-dependent diffusion coefficient of the form shown in equation (4.18). In this case, the decay time τ of the vacancy concentration was taken to be 100 s, and so the curves represent the points at which the vacancy concentration is a fraction 1, $e^{-\frac{1}{2}}$, e^{-1}, and e^{-2} of its original value. Clearly, the curves are converging to a point, which represents the region at which the ion-implantation-enhanced diffusion has been frozen.

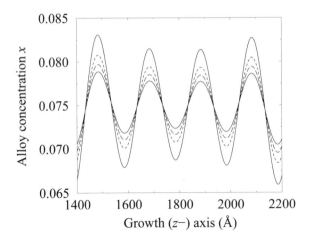

Figure 4.12   Annealing out radiation damage—an example of time-dependent diffusion, with annealing times of 0, 50, 100, and 200 s giving the concentration profiles of *decreasing* amplitude

4.9 δ-DOPED QUANTUM WELLS

In the previous sections, examples have been given of simulating diffusion for all of the various forms of diffusion coefficient that can exist. It is clear that the computational method can be extended to include combinations of all three dependencies, and indeed in the final example the diffusion coefficient had both depth and time dependency. It now serves a purpose to follow through an example of a diffusion problem that is of direct relevance to semiconductor heterostructures, and to calculate the subsequent effects on an observable, in this case, the quantum-confinement energies.

Contemporary epitaxial growth techniques, such as molecular beam epitaxy (MBE) and chemical beam epitaxy (CBE), etc., allow for the possibility of growing very thin layers of semiconductor material. Another possibility, is the potential for these techniques to lay down very thin layers of dopant atoms. Fig. 4.13(a) represents bulk doping of a layer, as used in HEMTs and the majority of semiconductor heterostructure devices. Fig. 4.13(b) represents the dopant profile for a single δ-layer in a quantum well, although of course, it could be in a barrier or at an interface. Such thin layers of dopant are called *delta-layers*, as the dopant profile resembles the Dirac δ-function, and the whole process is known as δ-doping; for a comprehensive treatise of this technique see Schubert [100].

One of the main problems affecting δ-doped layers is diffusion of the dopant which can often occur during the high temperatures employed during growth. A δ-layer designed to extend over just one monolayer can in fact spread over several layers. This would affect the electronic and optical properties of the device, in particular the electron- or hole-confinement energies, and scattering from the distribution of ionised impurities would be different. In this section, concentration will be focused on the first of these points, as the techniques to model such a system have already been covered in this work.

Fig. 4.14 displays the dopant profile of a single-monolayer-thick δ-layer when subject to diffusion described by a constant coefficient $D_0{=}1$ Å^{2}s^{-1}. Clearly, the total amount of dopant remains constant, and hence as the profile broadens, the height of the peak decreases. If the dopant were p-type, e.g. beryllium (Be) in GaAs, then at elevated

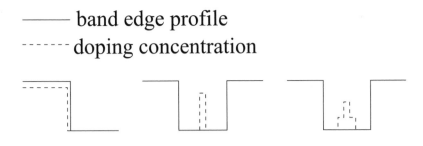

a) Bulk doping b) δ-doping   c) Diffusion

Figure 4.13   Doping and δ-doping

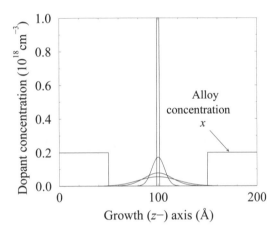

Figure 4.14   Dopant profile after 0, 20, 100 and 200 s (in order of decreasing maximum height) of diffusion described by a constant diffusion coefficient of 1 Å²s⁻¹, together with the alloy concentration x of the $Ga_{1-x}Al_xAs$ /GaAs/$Ga_{1-x}Al_xAs$ single quantum well

temperatures the acceptors would be ionised and donate holes which would become localised in the quantum well. By using the methods outlined in Chapter 3, the Coulomb potential due to this negative acceptor/positive hole distribution can be calculated and incorporated in the self-consistent solution of Schrödinger's and Poisson's equations.

Assuming that all of the holes occupy the lowest heavy-hole subband in the quantum well, then Fig. 4.15 shows the self-consistent potential V_ρ due to the charge distribution for various diffusion times. Clearly, as the δ-doped layer broadens, then the depth of the potential decreases. Interestingly, the potential seems to tend towards zero; this is because the diffused acceptor distribution begins to resemble the self-consistent hole wave functions thus leading to an almost complete cancellation of the positive and negative potentials.

Fig. 4.16 displays the self-consistent heavy-hole energy as a function of the diffusion time. At this relatively low doping density, the effect of diffusion on the heavy-hole

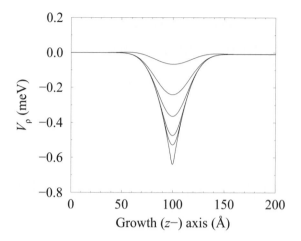

Figure 4.15 The self-consistent potential V_ρ due to the acceptor/hole distribution in the δ-doped quantum well of Fig. 4.14, for diffusion times, in order of decreasing depth, of 0, 10, 20, 50, 100, and 200 s

energy is small; nonetheless, this example serves the purpose of demonstrating the applicability of the diffusion equation solution to systems of interest for devices, while the inclusion of the self-consistent Schrödinger's and Poisson's solution, shows the power of combining such techniques together. Note that, as a constant diffusion coefficient was employed, then all of the curves are universal in $D_0 t$.

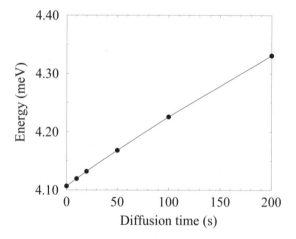

Figure 4.16 The self-consistent heavy-hole energy as a function of the diffusion time

4.10   EXTENSION TO HIGHER DIMENSIONS

The one-dimensional diffusion equation can be generalized to higher dimensions [81], i.e.

$$\frac{\partial x}{\partial t} = \nabla_\bullet (\mathcal{D} \nabla x) \tag{4.19}$$

This can again be expanded in terms of finite differences to give an iterative equation in analogy with equation (4.9). This time, however, the concentration at a point for a time interval δt into the future, is dependent on the diffusion coefficient $\mathcal{D}$ and the concentration x, which may both be functions of all of the spatial coordinates. Therefore, an iteration needs to be performed for each spatial dimension. Following this, an important check must be made that the resulting new concentration is independent of the order in which the iterations were performed.

CHAPTER 5

IMPURITIES

5.1 DONORS AND ACCEPTORS IN BULK MATERIAL

The ability to introduce impurities directly into the lattice was the most important technological advance in the development of semiconductors as electronic materials. This allows their electronic properties to be tailored to suit the engineer's needs. Atoms that are introduced with an additional electron more than that required to form the chemical bonds with neighbouring atoms are easily ionized, thus donating electrons to the crystal. Such atoms are known as donors. Alternatively, impurity atoms can be incorporated into the lattice, which are an electron short of that needed to form the chemical bonds. These atoms are able to accept electron from nearby bonds, thus generating an empty state in the valence band. Atoms of this type are known as acceptors. This *hole* is then mobile within the lattice and can contribute to the conductivity. Doping different regions of the same lattice with donors and acceptors leads to a *p-n* junction, which was the basis of the first transistor—the bipolar.

The interest here is not with the transport properties of doped semiconductors and their exploitation as the electronic materials of commercial devices such as diodes and transistors, as this has been covered very successfully by many other authors, see for example, Sze [101]. The interest here *is* in the properties of impurity states within semiconductor heterostructures. These states have a bearing on the electronic and op-

Quantum Wells, Wires and Dots, Third Edition. P. Harrison
©2009 John Wiley & Sons, Ltd.

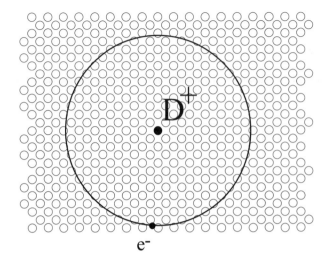

Figure 5.1   Schematic representation of a donor in a bulk semiconductor

tical properties of these materials, and again it is the evaluation of the fundamentals governing their behaviour that is the primary concern.

Fig. 5.1 shows a schematic representation of a neutral (occupied) donor in a bulk semiconductor. The regular lattice represents a {100} plane. An example of a donor in silicon is the group V element phosphorus. In a compound semiconductor such as gallium arsenide or cadmium telluride, the dopant would occupy a cation site. Thus the atoms of Fig. 5.1 would be gallium or cadmium, respectively. In GaAs, a typical donor is silicon, where the latter is forced to sit on the Ga site by growth under an As overpressure. The group IV element Si has one electron more than the group III element Ga and it donates this to the lattice. Note that a Si atom occupying an As site would be an acceptor.

It can be seen from the figure that the unionized donor resembles the electron–proton pair constituting a hydrogen atom; the physics of this is well understood and was initially described with a semi-classical model by Bohr [4] and reinforced later with full quantum mechanical descriptions [3, 29].

The binding energy of this neutral donor–electron pair is given by the following:

$$E_{\text{D}^0} = -\frac{m_e^* e^4}{32\pi^2 \hbar^2 \epsilon_r^2 \epsilon_0^2} \tag{5.1}$$

and the Bohr radius follows as:

$$\lambda = \frac{4\pi \epsilon_r \epsilon_0 \hbar^2}{m_e^* e^2} \tag{5.2}$$

In this case the mass of the donor atom is much larger than the electron mass and hence the simpler 'infinite nucleus mass' variant can be employed. Later, in the chapter dealing with excitons (Chapter 6), this hydrogenic model will again be employed; however, on that occasion the hole mass is of the same order of magnitude as the electron mass and therefore cannot be ignored. Note that the only material parameters needed are the electron effective mass and the permittivity of the host material, and thus the neutral donor binding energy should be independent of the impurity atom. Experimental data reproduced by Sze [44] substantiate this result, for the majority of cases in GaAs.

Taking typical values for GaAs, e.g. the Γ valley electron effective mass $m_e^* = 0.067$ m_0 and the static dielectric constant $\epsilon_r = 13.18$ [14], then:

$$E_{D^0} = -5.3 \text{ meV} \qquad \text{and} \qquad \lambda = 103 \text{ Å}$$

which is very close to the measured value of 5.8 meV [44]. Correspondingly for CdTe, $m_e^* = 0.096 m_0$ and $\epsilon_r = 10.6$, and therefore:

$$E_{D^0} = -11.7 \text{ meV} \qquad \text{and} \qquad \lambda = 58 \text{ Å}$$

The Bohr radii produced by these simple calculations represent an important result. The lattice constant of GaAs is 5.65 Å, and hence 103 Å represents around 18 unit cells along any radius from the donor to the electron orbit. Or, alternatively, each face-centred-cubic unit cell contains four lattice points, or eight atoms, and has a volume of $A_0^3 = (5.65 \text{ Å})^3$, and hence the volume occupied by one atom is $A_0^3/8$. Thus the number of atoms within the spherical electron orbit of radius λ is given by:

$$\text{number of atoms} = \frac{4\pi\lambda^3/3}{A_0^3/8} \tag{5.3}$$

which for GaAs gives, 203 000 atoms. This seems like quite a large number and hence provides justification for using the bulk value of the permittivity ϵ_r. For systems with a much smaller Bohr radius, which might contain substantially fewer atoms, it would be apparent that the electromagnetic properties of the crystal are quite different from the bulk and care must be taken in choosing the value for the permittivity. A recent work [102], has shown by careful comparison of exciton binding calculations with detailed experimental work, that the permittivity required to produce agreement lies between the static, ϵ_s, and infinite, ϵ_∞, frequency values. The binding energies themselves are small relative to the bandgap and, obviously, negative, which implies that they lie just below the conduction-band edge.

At low temperatures, there are few lattice vibrations (phonons) and hence the electrons remain bound to the donors. However, as the temperature is increased, the number of phonons within the lattice increases and the donors can become ionized, thus liberating electrons into the conduction band of the crystal. The occupancy (proportion of ionized donors) can be represented by statistics [1, 2], but this is not of particular concern here as most experiments are performed at liquid helium temperatures where all of the donors can be considered occupied, or at room temperature where they can be considered ionized.

At first sight, the direct analogy of acceptors in p-type material would appear to be described by the above equations, but with the hole mass replacing the electron mass and the binding energy now referring to a hole bound to a negatively charged acceptor. The hydrogenic model for a hole bound to an acceptor would give:

$$E_{A^0} = -\frac{m_h^* e^4}{32\pi^2 \hbar^2 \epsilon_r^2 \epsilon_0^2} \qquad \text{and} \qquad \lambda = \frac{4\pi\epsilon_r\epsilon_0\hbar^2}{m_h^* e^2}$$

Taking the typical values for GaAs of $m_{hh} = 0.62 \ m_0$ and $\epsilon_r = 13.18$ again, then:

$$E_{A^0} = -49.0 \text{ meV} \qquad \text{and} \qquad \lambda = 11 \text{ Å}$$

Unlike the donor case, this doesn't agree well with experimentally measured values [44]. In fact, in practice the acceptor state is much more complex for a number of reasons. First, often the valence band at the Γ minimum consists of two degenerate states (the light- and heavy-holes), and thus it is not clear which effective mass should be employed, or indeed whether the hole is a mixture of both light- and heavy-hole states. Although, however, this degeneracy is usually broken within a quantum well, due to the differing effect of the confining potential on the effective masses. Secondly, the much larger heavy-hole mass has the effect of producing a much smaller Bohr radius λ, i.e. the hole orbits much more closely to the central Coulombic potential, as can be seen from the calculated radius above. Consequently the approximation that the bulk relative permittivity ϵ describes the electromagnetic response of the lattice is questionable, although a technique for accounting for this problem will be introduced later, in Section 5.6. The problem of point defects will be revisited much later in Chapter 15 and dealt with by a microscopic model, which will take these effects into account. For now concentration will be focused on donors.

5.2 BINDING ENERGY IN A HETEROSTRUCTURE

When a donor is placed within a quantum well structure, the situation is considerably more complex than in the bulk, due to two additional degrees of freedom. First, the binding energy depends upon the confining potential due to the quantum well structure. In its simplest form this would be the well width, schematically represented in Fig. 5.2. Beyond this, however, lies the possibility of a more complex heterostructure. The donor binding energy will be different, e.g. in a double quantum well, a superlattice and a diffused quantum well; in fact, it is just as sensitive to structure as a lone electron is.

Secondly, the donor binding energy and wave function are also a function of the donor position *within* the heterostructure. The binding energy is, of course, different for a donor at the centre of a quantum well than it is for a donor at the edge of a well, as illustrated in Fig. 5.3.

Any theoretical study of the properties of neutral donors in a heterostructure necessitates solving the standard Schrödinger equation for the particular structure and with the inclusion of the additional Coulombic term representing the donor potential. Within the envelope function and effective mass approximations, the Hamiltonian for an electron confined in a heterostructure and in the presence of a shallow donor is merely the standard Hamiltonian of earlier, plus an additional term due to the Coulombic interaction as follows:

$$\mathcal{H} = -\frac{\hbar^2}{2m^*}\nabla^2 + V(z) - \frac{e^2}{4\pi\epsilon r'} \tag{5.4}$$

In the interests of generality, the potential $V(z)$ describing the conduction-band-edge potential of the heterostructure will remain unspecified. The displacement between the electron and the donor is given by:

$$r'^2 = x^2 + y^2 + (z - r_d)^2 = x^2 + y^2 + z'^2 \tag{5.5}$$

where r_d is the position of the donor along the growth (z) direction; note the origin of the x–y plane has been defined on the donor atom for convenience. Note also, that for now the effective mass m^* has been assumed to be constant, which leads to simplifications in

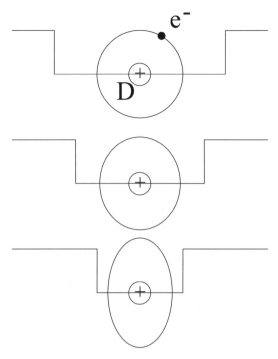

Figure 5.2   Schematic representation of the effect of quantum well width on the donor wave function

the analysis. The increased localization of the electron within the well ensures that this approximation has little effect on the final numerical values. The extension to include the effects of a varying effective mass will be discussed later.

Methods of solution of the Schrödinger equation have centred around two basic approaches. The first of these involves expanding the electron wave function as a linear combination of Gaussian functions [103, 104]. While this technique has been successful in calculating the properties of donors in simple quantum well structures, the generalization to more complex structures, including graded gap materials and systems where piezo electric fields are present, is non-trivial.

The other category of approach is based on the *variational principle*. In this method, a trial wave function is chosen whose functional form may contain one or more unknown parameters. These parameters are varied systematically and the expectation value of the energy calculated for each set. The variational principle [3, 29] states that *the lowest energy obtained is the closest approximation to the true state of the system.*

The success of variational approaches centres around the general choice of the trial wave function. A common choice is a product of two terms [105], i.e.

$$\Psi = \psi(z)e^{-\frac{r'}{\lambda}} \tag{5.6}$$

where r' is the electron–donor separation and λ is a variational parameter; thus the second factor is a simple hydrogenic wave function. The function $\psi(z)$ is the uncorrelated eigenfunction, i.e. the straightforward wave function calculated in Chapters 2 and 3, of the electron in the quantum well *without* the donor [106, 107].

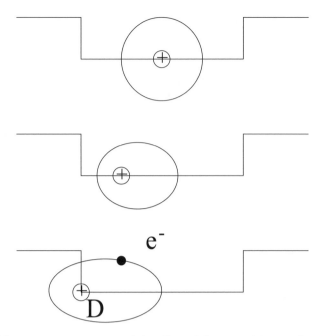

Figure 5.3   Schematic representation of the effect of donor position on the wave function

The latter restriction can be removed and a much more general choice of the donor wave function Ψ can be made. The motivation being generality, the problem is recast in a form which is suitable for numerical solution as described in Chapter 3, thus making it applicable to any quantum well structure, be it a double quantum well, a diffused quantum well, a graded gap quantum well, etc. With this aim, the trial wave function to be employed here is influenced by the above, but for now will be written as follows:

$$\Psi = \chi(z)\xi(x, y, z - r_d) \tag{5.7}$$

The one-dimensional envelope function $\chi(z)$ is yet to be determined and the function $\xi(x, y, z - r_d)$ is expected to be a hydrogenic type factor which is a function of the spatial coordinates and one or more variational parameters. It can be seen that forming the Schrödinger equation with the Hamiltonian of equation (5.4) and the trial wave function Ψ of equation (5.7) will lead to a term $\nabla^2\Psi$; it is worthwhile to derive this now. Consider:

$$\frac{\partial^2 \Psi}{\partial x^2} = \frac{\partial}{\partial x}\frac{\partial}{\partial x}[\chi(z)\xi(x, y, z - r_d)] \tag{5.8}$$

$$\therefore \frac{\partial^2 \Psi}{\partial x^2} = \frac{\partial}{\partial x}\left[\chi(z)\frac{\partial\xi}{\partial x}\right] = \chi\frac{\partial^2\xi}{\partial x^2} \tag{5.9}$$

and similarly for y. Consider now:

$$\frac{\partial^2 \Psi}{\partial z^2} = \frac{\partial}{\partial z}\left(\frac{\partial\chi}{\partial z}\xi + \chi\frac{\partial\xi}{\partial z}\right) \tag{5.10}$$

$$\therefore \frac{\partial^2 \Psi}{\partial z^2} = \frac{\partial^2\chi}{\partial z^2}\xi + 2\frac{\partial\chi}{\partial z}\frac{\partial\xi}{\partial z} + \chi\frac{\partial^2\xi}{\partial z^2} \tag{5.11}$$

Hence

$$\nabla^2 \Psi = \left(\nabla_z^2 \chi\right)\xi + 2\nabla_z \chi \nabla_z \xi + \chi \nabla^2 \xi \tag{5.12}$$

where the subscript z implies derivatives with respect to z only. Then, forming the Schrödinger equation with the Hamiltonian of equation (5.4) gives the following:

$$-\frac{\hbar^2}{2m^*}\left\{[\nabla_z^2 \chi(z)]\xi + 2\nabla_z \chi(z)\nabla_z \xi + \chi(z)\nabla^2 \xi\right\} - \frac{e^2}{4\pi\epsilon r'}\chi(z)\xi + V(z)\chi(z)\xi$$

$$= E\chi(z)\xi \tag{5.13}$$

where E is the total energy of the system. The donor binding energy E_{D^0} is equal to the difference between this and the standard confinement energy of the electron in the heterostructure without the donor present, i.e. using the notation of Chapter 2, the ground state energy for the electron in the well without the donor would be E_1, and hence:

$$E_{D^0} = E - E_1 \tag{5.14}$$

In order to proceed further, it is necessary to be more specific about the form of the hydrogenic factor ξ. In a free hydrogen atom the ground state wave function would be spherically symmetric and given by:

$$e^{-\frac{r}{\lambda}}, \qquad \text{where} \qquad r^2 = x^2 + y^2 + z^2 \tag{5.15}$$

where λ is known as the Bohr radius. Indeed, this is the form also used for donors in bulk materials. However, in heterostructures there is a loss of translational symmetry along the growth (z-) axis and hence it cannot be assumed that the spatial coordinate is spherically symmetric. For now, the hydrogenic term can be taken as:

$$\xi(x, y, z - r_d) = e^{-\frac{r''}{\lambda}} \tag{5.16}$$

where the spatial coordinate r'' is, of course, a function of x, y and $(z - r_d)$. Again, λ is referred to as the Bohr radius, but now it will be employed as a variational parameter in order to minimize the total energy of the system. Substituting for ξ into equation (5.13) gives:

$$-\frac{\hbar^2}{2m^*}\left\{[\nabla_z^2 \chi(z)]e^{-\frac{r''}{\lambda}} + 2\nabla_z \chi(z)\nabla_z e^{-\frac{r''}{\lambda}} + \chi(z)\nabla^2 e^{-\frac{r''}{\lambda}}\right\}$$

$$-\frac{e^2}{4\pi\epsilon r'}\chi(z)e^{-\frac{r''}{\lambda}} + V(z)\chi(z)e^{-\frac{r''}{\lambda}} = E\chi(z)e^{-\frac{r''}{\lambda}} \tag{5.17}$$

Multiplying by $e^{-\frac{r''}{\lambda}}$ and integrating over the x–y plane, leads to an equation of the form:

$$-\frac{\hbar^2}{2m^*}\left[\nabla_z^2 \chi(z)I_1 + 2\nabla_z \chi(z)I_2 + \chi(z)I_3\right]$$

$$-\frac{e^2}{4\pi\epsilon}\chi(z)I_4 + V(z)\chi(z)I_1 = E\chi(z)I_1 \tag{5.18}$$

where the integrals I_j, $(j = 1, 2, 3, 4)$ are defined as follows:

$$I_1 = \int_0^\infty \int_0^\infty e^{-\frac{2r''}{\lambda}} \, dx \, dy \tag{5.19}$$

$$I_2 = \int_0^\infty \int_0^\infty e^{-\frac{r''}{\lambda}} \nabla_z e^{-\frac{r''}{\lambda}} \; dx \; dy \tag{5.20}$$

$$I_3 = \int_0^\infty \int_0^\infty e^{-\frac{r''}{\lambda}} \nabla^2 e^{-\frac{r''}{\lambda}} \; dx \; dy \tag{5.21}$$

$$I_4 = \int_0^\infty \int_0^\infty \frac{e^{-\frac{2r''}{\lambda}}}{r'} \; dx \; dy \tag{5.22}$$

Returning to the Schrödinger equation (equation (5.18)) then:

$$\nabla_z^2 \chi(z) I_1 + 2\nabla_z \chi(z) I_2 + \chi(z) \left\{ I_3 + \frac{2m^*}{\hbar^2} \frac{e^2}{4\pi\epsilon} I_4 - \frac{2m^*}{\hbar^2} [V(z) - E] I_1 \right\} = 0 \tag{5.23}$$

Remembering that the integrals I_1, I_2, I_3, and I_4 are just real numbers, then equation (5.23) is just a linear second-order differential equation, very much like the one met in Chapter 3. The energy E can be solved *for any choice of* λ by expanding the derivatives in finite differences and forming an iterative shooting algorithm as before.

First though for simplicity sake, it is better to simplify equation (5.23), and rewrite it as:

$$\alpha \nabla_z^2 \chi(z) + \beta \nabla_z \chi(z) + \gamma \chi(z) = 0 \tag{5.24}$$

Then, by using the finite difference expansions:

$$\nabla_z \chi(z) = \frac{\chi(z + \delta z) - \chi(z - \delta z)}{2\delta z} \tag{5.25}$$

and

$$\nabla_z^2 \chi(z) = \frac{\chi(z + \delta z) - 2\chi(z) + \chi(z - \delta z)}{(\delta z)^2} \tag{5.26}$$

equation (5.24) becomes:

$$\frac{\chi(z + \delta z) - 2\chi(z) + \chi(z - \delta z)}{(\delta z)^2} + \frac{\beta}{\alpha} \left[\frac{\chi(z + \delta z) - \chi(z - \delta z)}{2\delta z} \right] + \frac{\gamma}{\alpha} \chi(z) = 0 \tag{5.27}$$

Finally, gathering the terms in $\chi(z + \delta z)$ on the left-hand side gives:

$$\left(1 + \frac{\beta}{2\alpha} \delta z \right) \chi(z + \delta z) = \left(-1 + \frac{\beta}{2\alpha} \delta z \right) \chi(z - \delta z) + \left[2 - (\delta z)^2 \frac{\gamma}{\alpha} \right] \chi(z) \tag{5.28}$$

which is an iterative shooting equation and can be solved subject to the standard boundary conditions, $\chi(z) \to \infty$ as $z \to \pm\infty$. In practice, this involves choosing exponential-growth starting conditions deep inside the quantum barrier and then progressively calculating $\chi(z)$ at points along the z-axis across the structure. The energy E is varied until the function $\chi(z)$ satisfies the standard boundary condition, i.e. it tends to zero at the other end of the structure. For more details, see again Chapter 3.

The variational aspect of the calculation arises from the value of the unknown constant λ. This is varied systematically with the aim of minimising the total energy. The system coordinates are therefore represented by the minimum energy and the corresponding value of the Bohr radius λ.

The derivation thus far is independent of the choice of the form of the hydrogenic exponential term, $\exp(-r''/\lambda)$, and this summarizes the versatility of this approach. In the following sections, the various choices that can be made for this 'orbital' term will be investigated. In addition, the power of this method is represented by the generality of the conduction-band-edge potential $V(z)$. No assumptions have been made relating to the form of this function and hence the method is applicable to all forms of $V(z)$, i.e. the formalism can be applied to any heterostructure.

5.3   TWO-DIMENSIONAL TRIAL WAVE FUNCTION

The simplest choice that can be made for the hydrogenic factor of the trial wave function for an electron confined in a quantum well structure in the presence of a positively charged donor, is one that is dependent only upon an in-plane coordinate, say, for example, as follows:

$$r'' = \sqrt{x^2 + y^2} = r_\perp \tag{5.29}$$

Therefore:

$$\xi = \exp\left(-\frac{\sqrt{x^2 + y^2}}{\lambda}\right) = \exp\left(-\frac{r_\perp}{\lambda}\right) \tag{5.30}$$

This choice replaces the spherical symmetry perhaps expected, a priori, with circular symmetry and is often referred to as the two-dimensional (2D) form. These are an unfortunate choice of words as, in fact, ξ has cylindrical symmetry, and hence is infinitely extended along the z-axis. This is another example of the word 'dimension' being interchanged with 'coordinate'.

Given this specific form for the spatial coordinate r'', then it is possible to calculate the integrals I_j explicitly, and hence all of the terms in the Schrödinger equation, (equation (5.23)) and the corresponding iterative shooting equation (5.28) will be known.

Consider first:

$$I_1 = \int_0^\infty \int_0^\infty e^{-\frac{2r_\perp}{\lambda}} \, dx \, dy \tag{5.31}$$

In this case, and indeed all that follows, it makes sense to exploit the circular symmetry by transforming the integral from the Cartesian coordinates, x and y, to plane polar coordinates $r_\perp$ and θ, i.e.

$$I_1 = \int_0^{2\pi} \int_0^\infty e^{-\frac{2r_\perp}{\lambda}} \, r_\perp dr_\perp \, d\theta \tag{5.32}$$

As the integrand, and all that follow, have no angular dependences then the θ integral is trivial, i.e.

$$I_1 = 2\pi \int_0^\infty e^{-\frac{2r_\perp}{\lambda}} \, r_\perp dr_\perp \tag{5.33}$$

This can now be solved with integration by parts, remembering the standard formula:

$$\int_a^b u \frac{dv}{dx} \, dx = [uv]_a^b - \int_a^b v \frac{du}{dx} \, dx \tag{5.34}$$

and that the independent variable is $r_\perp$, i.e. $x = r_\perp$. By choosing:

$$u = r_\perp \qquad \text{and} \qquad \frac{dv}{dr_\perp} = e^{-\frac{2r_\perp}{\lambda}} \tag{5.35}$$

then:

$$\frac{du}{dr_\perp} = 1 \qquad \text{and} \qquad v = -\frac{\lambda}{2}e^{-\frac{2r_\perp}{\lambda}} \tag{5.36}$$

Substituting into equation (5.34) gives:

$$I_1 = 2\pi\left\{\left[-\frac{\lambda r_\perp}{2}e^{-\frac{2r_\perp}{\lambda}}\right]_0^\infty - \int_0^\infty -\frac{\lambda}{2}e^{-\frac{2r_\perp}{\lambda}}\,dr_\perp\right\} \tag{5.37}$$

$$\therefore I_1 = 2\pi\left[\left(-\frac{\lambda r_\perp}{2}-\frac{\lambda^2}{4}\right)e^{-\frac{2r_\perp}{\lambda}}\right]_0^\infty \tag{5.38}$$

$$\therefore I_1 = 2\pi\frac{\lambda^2}{4} \tag{5.39}$$

Next, consider I_2 as defined in equation (5.20)

$$I_2 = \int_0^\infty\int_0^\infty e^{-\frac{r_\perp}{\lambda}}\nabla_z e^{-\frac{r_\perp}{\lambda}}\,dx\,dy \tag{5.40}$$

Immediately it can be seen, that with this particular choice of $r'' = r_\perp$, the derivative with respect to z is zero, i.e.

$$\nabla_z e^{-\frac{r_\perp}{\lambda}} = 0 \tag{5.41}$$

$$\therefore I_2 = 0 \tag{5.42}$$

Now, consider the third integral, as specified in equation (5.21):

$$I_3 = \int_0^\infty\int_0^\infty e^{-\frac{r_\perp}{\lambda}}\nabla^2 e^{-\frac{r_\perp}{\lambda}}\,dx\,dy \tag{5.43}$$

First it is necessary to calculate the terms arising from the ∇^2, so consider:

$$\frac{\partial}{\partial x}e^{-\frac{r_\perp}{\lambda}} = \frac{\partial}{\partial r_\perp}e^{-\frac{r_\perp}{\lambda}}\frac{\partial r_\perp}{\partial x} = -\frac{1}{\lambda}e^{-\frac{r_\perp}{\lambda}}\frac{\partial}{\partial x}\sqrt{x^2+y^2} = -\frac{x}{\lambda r_\perp}e^{-\frac{r_\perp}{\lambda}} \tag{5.44}$$

and then:

$$\frac{\partial^2}{\partial x^2}e^{-\frac{r_\perp}{\lambda}} = \frac{\partial}{\partial x}\left(-\frac{x}{\lambda r_\perp}e^{-\frac{r_\perp}{\lambda}}\right) \tag{5.45}$$

$$\therefore \frac{\partial^2}{\partial x^2}e^{-\frac{r_\perp}{\lambda}} = -\frac{1}{\lambda r_\perp}e^{-\frac{r_\perp}{\lambda}} - \frac{x}{\lambda}e^{-\frac{r_\perp}{\lambda}}\frac{\partial}{\partial x}\left(\frac{1}{r_\perp}\right) - \frac{x}{\lambda r_\perp}\frac{\partial}{\partial x}e^{-\frac{r_\perp}{\lambda}} \tag{5.46}$$

which gives:

$$\frac{\partial^2}{\partial x^2}e^{-\frac{r_\perp}{\lambda}} = \left(-\frac{1}{\lambda r_\perp}+\frac{x^2}{\lambda r_\perp^3}+\frac{x^2}{\lambda^2 r_\perp^2}\right)e^{-\frac{r_\perp}{\lambda}} \tag{5.47}$$

and similarly for the y-direction. Again there are no terms arising from differentiation with respect to z. Adding the two contributing terms together gives:

$$\nabla^2 e^{-\frac{r_\perp}{\lambda}} = \left(\frac{\partial^2}{\partial x^2}+\frac{\partial^2}{\partial y^2}\right)e^{-\frac{r_\perp}{\lambda}} = \left(-\frac{1}{\lambda r_\perp}+\frac{1}{\lambda^2}\right)e^{-\frac{r_\perp}{\lambda}} \tag{5.48}$$

Hence the integral I_3, as in equation (5.43), with the usual transformation to plane

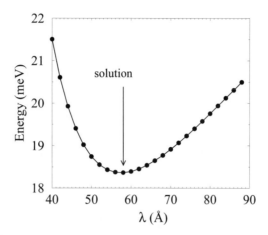

Figure 5.4   Total energy E of the electron as a function of the variational parameter λ, as an illustration of the variational principle

polar coordinates becomes:

$$I_3 = 2\pi \int_0^\infty e^{-\frac{r_\perp}{\lambda}} \left(-\frac{1}{\lambda r_\perp} + \frac{1}{\lambda^2} \right) e^{-\frac{r_\perp}{\lambda}} \, r_\perp dr_\perp \tag{5.49}$$

By using equations (5.33) and (5.39), obtain:

$$I_3 = 2\pi \left(\int_0^\infty -\frac{1}{\lambda r_\perp} e^{-\frac{2r_\perp}{\lambda}} \, r_\perp dr_\perp + \frac{1}{\lambda^2} I_1 \right) \tag{5.50}$$

$$\therefore I_3 = 2\pi \left\{ \left[-\frac{1}{\lambda} e^{-\frac{2r_\perp}{\lambda}} \times -\frac{\lambda}{2} \right]_0^\infty + \frac{1}{\lambda^2} I_1 \right\} \tag{5.51}$$

which gives:

$$I_3 = 2\pi \left(-\frac{1}{4} \right) \tag{5.52}$$

Finally, for the two-dimensional case, consider the only remaining integral:

$$I_4 = 2\pi \int_0^\infty \frac{e^{-\frac{2r_\perp}{\lambda}}}{r'} \, r_\perp dr_\perp \tag{5.53}$$

Consider, say for example, the substitution $r'^2 = r_\perp^2 + (z - r_d)^2 = r_\perp^2 + z'^2$. In this case the limits of integration change as follows. When $r_\perp = 0$, $r' = |z'|$ and when $r_\perp = \infty$, $r' = \infty$, hence:

$$I_4 = 2\pi \int_{|z'|}^\infty \frac{e^{-\frac{2\sqrt{r'^2 - z'^2}}{\lambda}}}{r'} \, r' dr' \tag{5.54}$$

Note that the lower limit for r' is $|z'|$ and not just z'—this is an important point! This change of limit will be used repeatedly in this analysis. The modulus bars are present as r' is an absolute distance; it is not a vector quantity, and is therefore always positive.

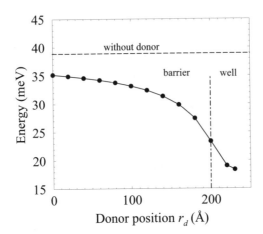

Figure 5.5 Total energy E of the electron in a 60 Å CdTe quantum well surrounded by $Cd_{0.9}Mn_{0.1}Te$ barriers in the presence of a donor at position r_d

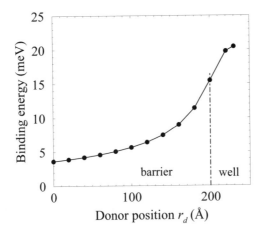

Figure 5.6 Neutral donor binding energy E_{D^0} in the 60 Å CdTe quantum well as before, as a function of the donor position r_d

Now this integration cannot be performed analytically and it is always prudent to avoid evaluating definite integrals numerically where the upper limit is infinity. After all exactly what value of infinity would one use? The form for I_4 can be manipulated further however, into a more manageable integral between 0 and 1, which can then be evaluated accurately and with confidence. The mathematical trick* introduced here will

*Credit to Winston Hagston for this

be used again and again in this present chapter and also later in Chapter 6.

Put $\qquad r' = |z'| \cosh\theta \qquad$ then $\qquad dr' = |z'| \sinh\theta \; d\theta$

and the limits become zero and infinity, which for now seems to be contrary to what was intended. Hence:

$$I_4 = 2\pi \int_0^\infty \exp\left(-\frac{2\sqrt{z'^2 \cosh^2\theta - z'^2}}{\lambda}\right) \; |z'| \sinh\theta \; d\theta \qquad (5.55)$$

$$\therefore I_4 = 2\pi \int_0^\infty \exp\left(-\frac{2|z'| \sinh\theta}{\lambda}\right) \; |z'| \sinh\theta \; d\theta \qquad (5.56)$$

Using the substitution $w = e^{-\theta}$, then when $\theta = 0$, $w = 1$, and when $\theta = \infty$, $w = 0$. Also, remembering:

$$\sinh\theta = \frac{1}{2}\left(e^\theta - e^{-\theta}\right), \qquad \text{then} \qquad \sinh\theta = \frac{1}{2}\left(\frac{1}{w} - w\right) \qquad (5.57)$$

equation (5.56) therefore becomes:

$$I_4 = 2\pi \int_1^0 \exp\left[-\frac{2|z'|\frac{1}{2}\left(\frac{1}{w} - w\right)}{\lambda}\right] \; |z'|\frac{1}{2}\left(\frac{1}{w} - w\right) \left(-\frac{dw}{w}\right) \qquad (5.58)$$

which gives, finally:

$$I_4 = 2\pi \int_0^1 \exp\left[-\frac{|z'|\left(\frac{1}{w} - w\right)}{\lambda}\right] \; |z'|\frac{1 - w^2}{2w^2} \; dw \qquad (5.59)$$

Fig. 5.4 demonstrates the principle behind the variational calculation, using the above analysis, by employing a 60 Å CdTe well surrounded by 200 Å $Cd_{0.9}Mn_{0.1}Te$ barriers for illustration. The graph plots the total energy E as a function of the variational parameter λ, for an electron in the presence of a donor at the centre of the well. The system assumes the lowest energy state possible, and hence the electron moves to a Bohr orbit of radius $\lambda = 58$ Å, with energy $E = 18.366$ meV.

Fig. 5.5 shows the results of calculations for a range of donor positions across the same 60 Å CdTe quantum well. The $Cd_{1-x}Mn_xTe$ system has been chosen to illustrate the method merely because the binding energy is larger, as calculated earlier, so it seems a bit more interesting than the more common $Ga_{1-x}Al_xAs$ system. In fact, $Cd_{1-x}Mn_xTe$ is quite similar to $Ga_{1-x}Al_xAs$; i.e. it is a direct material, with a similar bandgap, and CdTe forms type I quantum wells with a valence band offset similar to that of $GaAs/Ga_{1-x}Al_xAs$. None of the results obtained are particularly features of the choice of material, and re-emphasizing the aim of this present book, it is the theory and computational methods and the generic deductions that are important, and *not the absolute values* of the calculations. As mentioned already, all of the necessary tools are provided for the reader to quite quickly repeat the calculations for the material system of interest to them. Therefore, returning to Fig. 5.5, as probably expected the electron energy E is lowest (and therefore the donor binding energy E_{D^0} highest) for a donor at the centre of the well. It would be expected *a priori* that the electron energy would

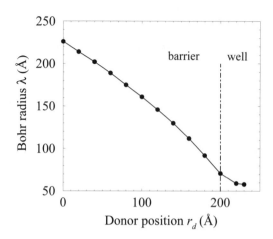

Figure 5.7 Bohr radius λ as a function of the donor position r_d across the 60 Å CdTe quantum well of above

return to the value without the donor present, as the donor is moved further and further away from the well, i.e.

$$\lim_{r_d \to \infty} E = E_1, \quad \text{or} \quad \lim_{r_d \to \infty} E_{D^0} = 0 \qquad (5.60)$$

This second point is highlighted by the plot of the magnitude of the binding energy, again as a function of donor position, in Fig. 5.6. For completeness, the corresponding plot of the variational parameter λ is displayed in Fig. 5.7. Clearly, there is a correlation between λ and E_{D^0}, with the larger E_{D^0} giving the smaller λ; physically this implies that the binding energy increases as the radius of the electron orbit is decreased.

In order to produce the data plotted in Fig. 5.8, the electron ground state energies E_1 were calculated without the donor present, and with the donor present E, for a donor fixed at the centre of the well, as a function of well width. This figure illustrates two points that are both important convergence tests for this analysis and for the analysis given in Chapter 3. First, as the well width increases, then the electron ground state energy without the donor present, i.e. E_1, decreases monotonically and tends towards zero. Secondly, and perhaps more importantly, and relevant to this section, as the well width increases and the contribution to the electron energy due to quantum confinement decreases, the total energy for the electron tends towards the binding energy of the neutral donor in bulk (11.7 meV, as calculated earlier).

This latter result is perhaps represented more clearly in Fig. 5.9. The binding energy is a non-monotonic function of the well width and peaks at a relatively small width value. This is a very similar result to that of the more complex case of exciton binding energies which will appear again in Chapter 6; in the latter case, such non-monotonic behaviour has been observed experimentally. The non-monotonic behaviour in both cases is merely due to the probability of the electron being within the well. For very narrow wells, the increasing confinement energy, as seen in Fig. 5.8, pushes the electron closer to the top of the well, eventually forcing it to 'spill over the top', thus leading to decreases in the

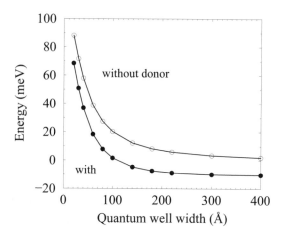

Figure 5.8   Energy of an electron in a quantum well, both with and without a donor at the centre of the well, as a function of the well width

binding energy. Returning to the earlier point, as the well width increases then the binding energy of the donor tends towards its bulk value, as previously hoped.

These results show that the theoretical approach is sound and, when implemented computationally, the model assumes simple limits according to sensible physical thinking. Therefore, confidence can be placed in this method. However, this form for the wave function would probably never be used in real calculations. The reason being, that while it appears to be the simplest choice of trial wave function, and although it does appear to follow the expected limits, a three-dimensional trial wave function, as discussed in the next section, supersedes it. Calculations will show that a spherically symmetric wave function gives lower variational energies, which are therefore considered to be better approximations, and in addition, the computational implementation of the mathematics is more efficient than in this, the two-dimensional case.

5.4   THREE-DIMENSIONAL TRIAL WAVE FUNCTION

Driven by knowledge of bulk material and the spherical symmetry of the hydrogenic wave function, the next obvious choice of the spatial coordinate r'' is given by:

$$r'' = \sqrt{x^2 + y^2 + (z - r_d)^2} \qquad (5.61)$$

which is just the electron–donor separation, defined earlier as r'. Again, the investment in the general formalism developed in Section 5.2 pays dividends in that the full requirements for the calculation merely require an evaluation of the integrals I_j, $(j = 1, 2, 3, 4)$ but with the new form for r''.

Consider first I_1 as originally defined in equation (5.19). Using the new form for $r'' = r'$ and switching to plane polar coordinates gives:

$$I_1 = \int_0^\infty \int_0^\infty e^{-\frac{2r''}{\lambda}} \; \mathrm{d}x \, \mathrm{d}y = \int_0^\infty e^{-\frac{2r'}{\lambda}} \; 2\pi \, r_\perp \; \mathrm{d}r_\perp \qquad (5.62)$$

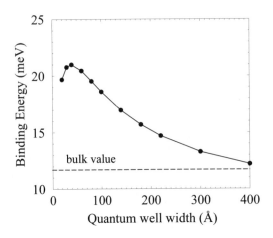

Figure 5.9   The neutral donor binding energy E_{D^0} for a donor at the centre of a well, as a function of the well width

Now:

$$r'^2 = r_\perp^2 + z'^2 \tag{5.63}$$

and therefore $r'\,dr' = r_\perp dr_\perp$. Preparing for the substitution r' for $r_\perp$, then when $r_\perp = 0$, $r' = |z'|$, and when $r_\perp = \infty$, $r' = \infty$, and hence:

$$I_1 = \int_{|z'|}^{\infty} e^{-\frac{2r'}{\lambda}} \, 2\pi \, r' \, dr' \tag{5.64}$$

which can be integrated by parts in the same manner as in Section 5.3, with the following choices:

$$u = r' \qquad \text{and} \qquad \frac{dv}{dr'} = e^{-\frac{2r'}{\lambda}} \tag{5.65}$$

which gives:

$$\frac{du}{dr'} = 1 \qquad \text{and} \qquad v = -\frac{\lambda}{2} e^{-\frac{2r'}{\lambda}} \tag{5.66}$$

Thus, applying the above to equation (5.64) gives:

$$I_1 = 2\pi \left\{ \left[-\frac{\lambda r'}{2} e^{-\frac{2r'}{\lambda}} \right]_{|z'|}^{\infty} - \int_{|z'|}^{\infty} -\frac{\lambda}{2} e^{-\frac{2r'}{\lambda}} \, dr' \right\} \tag{5.67}$$

$$\therefore I_1 = 2\pi \left[-\frac{\lambda r'}{2} e^{-\frac{2r'}{\lambda}} - \frac{\lambda^2}{4} e^{-\frac{2r'}{\lambda}} \right]_{|z'|}^{\infty} \tag{5.68}$$

which after evaluation gives:

$$I_1 = 2\pi \left(\frac{\lambda |z'|}{2} + \frac{\lambda^2}{4} \right) e^{-\frac{2|z'|}{\lambda}} \tag{5.69}$$

Turning attention to the second integral, and using the particular expression for r'' and plane polar coordinates, then:

$$I_2 = \int_0^\infty \int_0^\infty e^{-\frac{r''}{\lambda}} \nabla_z e^{-\frac{r''}{\lambda}} \; dx \; dy = \int_0^\infty e^{-\frac{r'}{\lambda}} \nabla_z e^{-\frac{r'}{\lambda}} \; 2\pi \; r_\perp dr_\perp \tag{5.70}$$

$$\therefore I_2 = 2\pi \int_0^\infty e^{-\frac{r'}{\lambda}} \frac{\partial}{\partial r'} e^{-\frac{r'}{\lambda}} \frac{\partial r'}{\partial z'} \; r_\perp dr_\perp \tag{5.71}$$

Now:

$$r' = \sqrt{x^2 + y^2 + (z - r_d)^2} = \sqrt{x^2 + y^2 + z'^2} \tag{5.72}$$

and therefore:

$$\frac{\partial r'}{\partial z'} = \frac{1}{2\sqrt{x^2 + y^2 + z'^2}} \times 2z' = \frac{z'}{r'} \tag{5.73}$$

Note here the z' and *not* $|z'|$. Using this result in equation (5.71) then:

$$I_2 = 2\pi \int_0^\infty e^{-\frac{r'}{\lambda}} \left(-\frac{z'}{r'\lambda} \right) e^{-\frac{r'}{\lambda}} \; r_\perp dr_\perp \tag{5.74}$$

but $r'^2 = r_\perp^2 + z'^2$, and therefore $r'dr' = r_\perp dr_\perp$. Using this substitution and changing the limits of integration, then obtain:

$$I_2 = 2\pi \int_{|z'|}^\infty \left(-\frac{z'}{r'\lambda} \right) e^{-\frac{2r'}{\lambda}} \; r'dr' \tag{5.75}$$

$$\therefore I_2 = 2\pi \left[\frac{z'}{2} e^{-\frac{2r'}{\lambda}} \right]_{|z'|}^\infty \tag{5.76}$$

and finally:

$$I_2 = 2\pi \left(-\frac{z'}{2} e^{-\frac{2|z'|}{\lambda}} \right) \tag{5.77}$$

Writing I_3 in plane polar coordinates with the three-dimensional form for r'', then:

$$I_3 = 2\pi \int_0^\infty e^{-\frac{r'}{\lambda}} \nabla^2 e^{-\frac{r'}{\lambda}} \; r_\perp \; dr_\perp \tag{5.78}$$

In order to proceed, it is necessary to evaluate the differential. Consider:

$$\nabla_x e^{-\frac{r'}{\lambda}} = \frac{\partial}{\partial r'} e^{-\frac{r'}{\lambda}} \frac{\partial r'}{\partial x} \tag{5.79}$$

which, because of the isotropy of the exponential term in this case, yields, in the same manner as above in equation (5.73):

$$\nabla_x e^{-\frac{r'}{\lambda}} = -\frac{x}{r'\lambda} e^{-\frac{r'}{\lambda}} \tag{5.80}$$

Differentiating again:

$$\nabla_x^2 e^{-\frac{r'}{\lambda}} = -\frac{1}{r'\lambda} e^{-\frac{r'}{\lambda}} - \frac{x}{\lambda} \times -\frac{1}{r'^2} \frac{\partial r'}{\partial x} e^{-\frac{r'}{\lambda}} - \frac{x}{r'\lambda} \times -\frac{1}{\lambda} e^{-\frac{r'}{\lambda}} \frac{\partial r'}{\partial x} \tag{5.81}$$

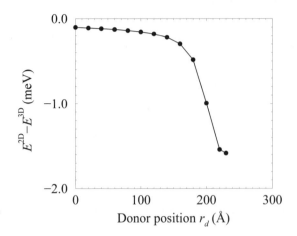

Figure 5.10   Difference in total energy for the two- and three-dimensional cases, as a function of donor position r_d across the 60 Å CdTe well

$$\therefore \nabla_x^2 e^{-\frac{r'}{\lambda}} = \left(-\frac{1}{r'\lambda} + \frac{x^2}{r'^3\lambda} + \frac{x^2}{r'^2\lambda^2}\right) e^{-\frac{r'}{\lambda}} \tag{5.82}$$

and similarly for $\nabla_y^2 e^{-\frac{r'}{\lambda}}$. The same is also true for $\nabla_z^2 e^{-\frac{r'}{\lambda}}$ and can be followed through by noting, however, that:

$$\nabla_z^2 e^{-\frac{r'}{\lambda}} = \nabla_{z'}^2 e^{-\frac{r'}{\lambda}} \times \left(\frac{\partial z'}{\partial z}\right)^2, \quad \text{where} \quad \frac{\partial z'}{\partial z} = 1 \tag{5.83}$$

Gathering all of the terms together, obtain:

$$\nabla^2 e^{-\frac{r'}{\lambda}} = \left(\nabla_x^2 + \nabla_y^2 + \nabla_z^2\right) e^{-\frac{r'}{\lambda}} = \left(-\frac{2}{r'\lambda} + \frac{1}{\lambda^2}\right) e^{-\frac{r'}{\lambda}} \tag{5.84}$$

Using this form in equation (5.78) then:

$$I_3 = 2\pi \int_0^\infty e^{-\frac{r'}{\lambda}} \left(-\frac{2}{r'\lambda} + \frac{1}{\lambda^2}\right) e^{-\frac{r'}{\lambda}} r_\perp \, dr_\perp \tag{5.85}$$

which, using equation (5.64) gives the following:

$$I_3 = \frac{I_1}{\lambda^2} + 2\pi \int_0^\infty \left(-\frac{2}{r'\lambda}\right) e^{-\frac{2r'}{\lambda}} r_\perp \, dr_\perp \tag{5.86}$$

Again, substituting r' for $r_\perp$, then gives:

$$I_3 = \frac{I_1}{\lambda^2} + 2\pi \left(-\frac{2}{\lambda}\right) \int_{|z'|}^\infty \frac{e^{-\frac{2r'}{\lambda}}}{r'} r' \, dr' \tag{5.87}$$

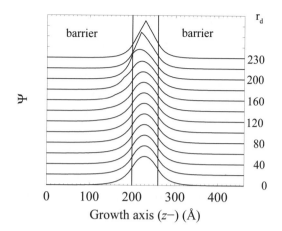

Figure 5.11   The electron wave function Ψ, as in equation (5.7), for the donor positions given on the right-hand axis

$$\therefore I_3 = \frac{I_1}{\lambda^2} + 2\pi \left(-\frac{2}{\lambda}\right) \left[-\frac{\lambda}{2} e^{-\frac{2r'}{\lambda}}\right]_{|z'|}^{\infty} \tag{5.88}$$

Using the final form for I_1, as in equation (5.69), then:

$$I_3 = 2\pi \left(\frac{|z'|}{2\lambda} - \frac{3}{4}\right) e^{-\frac{2|z'|}{\lambda}} \tag{5.89}$$

Finally, for the three-dimensional case, I_4 becomes:

$$I_4 = 2\pi \int_0^{\infty} \frac{e^{-\frac{2r'}{\lambda}}}{r'} \, r_\perp \, dr_\perp \tag{5.90}$$

which, on changing the variable to r', becomes trivial, i.e.

$$\therefore I_4 = 2\pi \int_{|z'|}^{\infty} \frac{e^{-\frac{2r'}{\lambda}}}{r'} \, r' \, dr' \tag{5.91}$$

and therefore:

$$I_4 = 2\pi \left(\frac{\lambda}{2} e^{-\frac{2|z'|}{\lambda}}\right) \tag{5.92}$$

Looking upon these results from a computational viewpoint, it can then be seen that this 3D trial wave function has an immediate advantage over the previous, seemingly simpler, 2D case, namely that all of the integrals I_1, I_2, I_3 and I_4 have analytical expressions. Indeed, evaluation of the integrals and the minimization of the energy is computationally much less demanding than previously.

Fig. 5.10 shows the change in the total energy E of the electron between the previous 2D trial wave function and the more complex 3D case presented in this section. It is

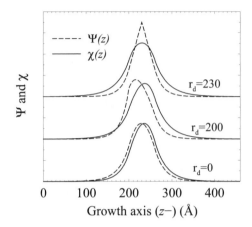

Figure 5.12   Comparison between the total electron wave function $\Psi(z)$ and the numerically determined envelope $\chi(z)$ for three different donor positions

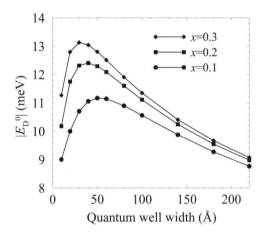

Figure 5.13   Magnitude of the donor binding energy E_{D^0} in GaAs/Ga$_{1-x}$Al$_x$As as a function of quantum well width, for a variety of barrier alloys x

clear that the energy E is lower for *all* donor positions across this (a typical) quantum well. The difference between the two trial wave functions is smallest when the separation between the donor and the electron is larger, i.e. when the donor is deep in the barrier ($r_d = 0$ Å) and the electron is, as always, centred in the well ($z = 230$ Å). Thus it might be concluded that the 2D wave function is a reasonable approximation when the donor is in the barrier, although as the graph shows, for donors in the well, considerably lower energies can be obtained by using a spherical hydrogenic term. Recalling the variational principle, then the lower energies obtained imply that the 3D approximation to the wave function is a more accurate representation than the 2D case. When coupled together with the computational advantage, as mentioned above, then the argument in favour of the 3D trial wave function is clear.

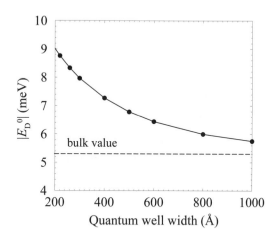

Figure 5.14   Magnitude of the donor binding energy E_{D^0} for donors at the centre of GaAs wells of large width, surrounded by $Ga_{0.9}Al_{0.1}As$ barriers

Fig. 5.11 displays the total wave function, $\Psi = \chi(z)e^{-\frac{r'}{\lambda}}$, for the range of donor positions across the quantum well. It can be seen that the wave function Ψ resembles the one-particle wave function ψ for an electron without a donor present, for donors in the barrier, i.e. $r_d \lesssim 160$ Å. As the donor approaches the electron wave function, i.e. nears the barrier-well interface at $z = 200$ Å, then the electron is drawn distinctly to the left towards the donor. The influence of the hydrogenic factor $e^{-\frac{r'}{\lambda}}$ can be seen for the donor positions $r_d = 220$ and 230 Å within the well.

In the 2D case, the total wave function was given by:

$$\Psi(z) = \chi(z)e^{-\frac{\sqrt{x^2+y^2}}{\lambda}} \tag{5.93}$$

and hence the z-dependence is merely $\Psi(z) = \chi(z)$. Furthermore, it was found that the numerically determined envelope $\chi(z)$ was a very close approximation to the electron wave function $\psi(z)$ without the donor present. In this case, as is clear from Fig. 5.11, $\Psi(z) \neq \chi(z)$; this is illustrated more clearly in Fig. 5.12.

Moving on to the $GaAs/Ga_{1-x}Al_xAs$ material system and employing the bulk values of $m^* = 0.067m_0$ and $\epsilon = 13.18$, Fig. 5.13 shows the effect of well width on the neutral donor binding energy, for donors at the centre of the well, for a variety of barrier compositions. As would be expected from earlier results, E_{D^0} peaks at a narrow well width and then tails off towards the bulk value. This important limit is explored further in Fig. 5.14; it is clear from this figure that the convergence is very close, and this helps give justification to the methods developed. The variation in the Bohr radius is displayed in Fig. 5.15, and it too converges well to a value of 104 Å at very large well widths. This compares admirably with the value deduced from the simple bulk hydrogenic model of

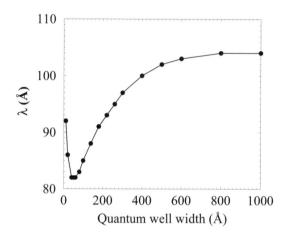

Figure 5.15   The Bohr radius λ as a function of well width for donors at the centres of GaAs wells surrounded by $Ga_{0.9}Al_{0.1}As$ barriers

103 Å, calculated at the beginning of the chapter. The very small difference could arise from the finite λ increment employed of 1 Å.

5.5   VARIABLE-SYMMETRY TRIAL WAVE FUNCTION

In the two previous sections, simple high-symmetry trial wave functions have been chosen to illustrate the method. Indeed, often they provide approximate, but quick, numerical results that may generally suffice. However, a little thought highlights their deficiencies. The aim of this work, like the solution of many physical systems with the variational principle, is to choose a trial wave function which is applicable no matter what the system parameters may be. Which in this case would mean, no matter what the well width, barrier height and donor position are.

The spherical 3D trial wave function was chosen as this is the form found in bulk materials, i.e. systems which do not have a confining potential. Hence it might be expected that this trial wave function is most appropriate for systems which most closely resemble bulk systems i.e. those with only small confinement, such as wide wells and/or low barrier heights. For example, for a donor in the centre of a wide well, the envelope $\chi(z)$ would look like $\cos kz$, which would be slowly varying, and hence the total wave function Ψ would resemble the bulk. Conversely, a large confinement potential might be expected to give a more 2D-like wave function.

It is clear, therefore that, while the two trial wave functions considered so far are useful, a more general form can be taken for the hydrogenic factor, a variable symmetry term, which can move between the 2D and 3D forms should it prove energetically favourable to do so. Equation (5.94) summarizes the most general choice for the hydrogenic factor of the trial wave function:

$$r'' = \sqrt{x^2 + y^2 + \zeta^2(z - r_d)^2} \tag{5.94}$$

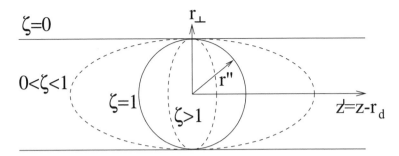

Figure 5.16   Schematic illustration of the variable symmetry relative motion term in the donor trial wave function

where ζ is a second variational parameter, which like λ will be adjusted systematically in order to minimize the total energy E of the system. Hence:

$$\xi = \exp\left(-\frac{\sqrt{x^2 + y^2 + \zeta^2 z'^2}}{\lambda}\right) \tag{5.95}$$

The effect of the parameter ζ on the symmetry of ξ is illustrated in Fig. 5.16.

Again, the investment made earlier, in structuring the problem in simple units, means that all that needs to be done in order to specify all of the quantities in the differential equation (5.23) exactly, is to evaluate the integrals I_j with the new form for r''. Consider I_1 in the usual plane polar coordinates, i.e.

$$I_1 = 2\pi \int_0^\infty e^{-\frac{2r''}{\lambda}} \, r_\perp \, dr_\perp \tag{5.96}$$

Now $r''^2 = r_\perp^2 + \zeta^2 z'^2$, and therefore $r'' dr'' = r_\perp dr_\perp$, considering the limits, i.e.

$$\text{when} \quad r_\perp = 0, \quad r'' = \zeta|z'|$$

$$\text{and when} \quad r_\perp = \infty, \quad r'' = \infty$$

Therefore:

$$I_1 = 2\pi \int_{\zeta|z'|}^\infty e^{-\frac{2r''}{\lambda}} \, r'' \, dr' \tag{5.97}$$

As previously, this can be integrated by parts. Choosing:

$$u = r'', \quad \text{and} \quad \frac{dv}{dr''} = e^{-\frac{2r''}{\lambda}}$$

then:

$$\frac{du}{dr''} = 1, \quad \text{and} \quad v = -\frac{\lambda}{2} e^{-\frac{2r''}{\lambda}}$$

Therefore:

$$I_1 = 2\pi \left\{ \left[-\frac{r'' \lambda}{2} e^{-\frac{2r''}{\lambda}} \right]_{\zeta|z'|}^\infty - \int_{\zeta|z'|}^\infty \left(-\frac{\lambda}{2} \right) e^{-\frac{2r''}{\lambda}} \, dr'' \right\} \tag{5.98}$$

and so:

$$I_1 = 2\pi \left[\left(-\frac{r''\lambda}{2} - \frac{\lambda^2}{4} \right) e^{-\frac{2r''}{\lambda}} \right]_{\zeta|z'|}^{\infty} \tag{5.99}$$

leading finally to:

$$I_1 = 2\pi \left(\frac{\zeta|z'|\lambda}{2} + \frac{\lambda^2}{4} \right) e^{-\frac{2\zeta|z'|}{\lambda}} \tag{5.100}$$

Secondly:

$$I_2 = 2\pi \int_0^\infty e^{-\frac{r''}{\lambda}} \nabla_z e^{-\frac{r''}{\lambda}} \, r_\perp \, dr_\perp \tag{5.101}$$

In order to proceed, the differential must be evaluated, i.e.

$$\nabla_z e^{-\frac{r''}{\lambda}} = \frac{\partial}{\partial z} e^{-\frac{r''}{\lambda}} = \frac{\partial}{\partial z''} e^{-\frac{r''}{\lambda}} \frac{\partial z''}{\partial z}, \quad \text{where} \quad z'' = \zeta z' \tag{5.102}$$

With this final substitution, i.e. the use of z'', then this form resembles the 3D version from earlier, and use can be made of some of the results. For example, using equation (5.80), then:

$$\nabla_z e^{-\frac{r''}{\lambda}} = -\frac{z''}{r''\lambda} e^{-\frac{r''}{\lambda}} \frac{\partial z''}{\partial z} \quad \left(\text{note that} \quad \frac{\partial z''}{\partial z} = \zeta \right) \tag{5.103}$$

Therefore, equation (5.101) becomes:

$$I_2 = 2\pi \int_0^\infty e^{-\frac{r''}{\lambda}} \left(-\frac{\zeta z''}{r''\lambda} \right) e^{-\frac{r''}{\lambda}} \, r_\perp \, dr_\perp \tag{5.104}$$

Returning to z' and substituting r'' for $r_\perp$, then obtain:

$$I_2 = 2\pi \int_{\zeta|z'|}^\infty \left(-\frac{\zeta^2 z'}{r''\lambda} \right) e^{-\frac{2r''}{\lambda}} \, r'' \, dr'' \tag{5.105}$$

As the values r'' cancel, the integration then becomes trivial, and indeed:

$$I_2 = 2\pi \left(-\frac{\zeta^2 z'}{2} \right) e^{-\frac{2\zeta|z'|}{\lambda}} \tag{5.106}$$

In order to evaluate I_3, it is again necessary to deduce $\nabla^2 e^{-\frac{r''}{\lambda}}$ first. In a similar manner to equation (5.102), then:

$$\nabla_z^2 e^{-\frac{r''}{\lambda}} = \nabla_{z''}^2 e^{-\frac{r''}{\lambda}} \left(\frac{\partial z''}{\partial z} \right)^2 = \zeta^2 \nabla_{z''}^2 e^{-\frac{r''}{\lambda}} \tag{5.107}$$

Considering z'' as the direct analogy of z' for the 3D case, then use can be made of the earlier result in equation (5.82), hence:

$$\nabla_z^2 e^{-\frac{r''}{\lambda}} = \zeta^2 \left(-\frac{1}{r''\lambda} + \frac{z''^2}{r''^3\lambda} + \frac{z''^2}{r''^2\lambda^2} \right) e^{-\frac{r''}{\lambda}} \tag{5.108}$$

Using the equivalent forms for $\nabla_x^2 e^{-\frac{r''}{\lambda}}$ and $\nabla_y^2 e^{-\frac{r''}{\lambda}}$, also given in equation (5.82) and summing gives:

$$\nabla^2 e^{-\frac{r''}{\lambda}} = \left[\frac{-1 - \zeta^2}{r''\lambda} + \frac{1}{\lambda^2} + \frac{(\zeta^2 - 1)z''^2}{r''^3\lambda} + \frac{(\zeta^2 - 1)z''^2}{r''^2\lambda^2} \right] e^{-\frac{r''}{\lambda}} \tag{5.109}$$

Noting that $z'' = \zeta z'$, then:

$$\nabla^2 e^{-\frac{r''}{\lambda}} = \left[\frac{-1 - \zeta^2}{r''\lambda} + \frac{1}{\lambda^2} + \frac{(\zeta^4 - \zeta^2)z'^2}{r''^3\lambda} + \frac{(\zeta^4 - \zeta^2)z'^2}{r''^2\lambda^2} \right] e^{-\frac{r''}{\lambda}} \tag{5.110}$$

As a check on the analysis so far, putting $\zeta = 0$ and $r'' = r_\perp$ does indeed yield the equivalent equation for the 2D case (equation (5.48)), while putting $\zeta = 1$ and $r'' = r'$ gives the 3D case expressed in equation (5.84).

Therefore, everything is in place for the evaluation of the third integral, which now becomes:

$$I_3 = 2\pi \int_0^\infty \left[\frac{-1 - \zeta^2}{r''\lambda} + \frac{1}{\lambda^2} + \frac{(\zeta^4 - \zeta^2)z'^2}{r''^3\lambda} + \frac{(\zeta^4 - \zeta^2)z'^2}{r''^2\lambda^2} \right] e^{-\frac{2r''}{\lambda}} r_\perp \, dr_\perp \tag{5.111}$$

Unfortunately each of the four terms needs to be handled separately, so writing equation (5.111) as:

$$I_3 = 2\pi \left(I_{31} + I_{32} + I_{33} + I_{34} \right) \tag{5.112}$$

Then the first integral is trivial, while the second can be solved by parts, thus giving:

$$I_{31} = \frac{-1 - \zeta^2}{2} e^{-\frac{2\zeta|z'|}{\lambda}}, \quad \text{and} \quad I_{32} = \left(\frac{\zeta|z'|}{2\lambda} + \frac{1}{4} \right) e^{-\frac{2\zeta|z'|}{\lambda}} \tag{5.113}$$

The third in this short series cannot be solved analytically; however, it can again be manipulated into an integral between 0 and 1, and hence quickly and accurately evaluated. It is a worthwhile exercise to follow this one through; by changing the variable of integration from $r_\perp$ to r'', then obtain:

$$I_{33} = \int_{\zeta|z'|}^\infty \left[\frac{(\zeta^4 - \zeta^2)z'^2}{r''^3\lambda} \right] e^{-\frac{2r''}{\lambda}} r'' \, dr'' \tag{5.114}$$

By putting $r'' = \zeta|z'| \cosh\theta$, then $dr'' = \zeta|z'| \sinh\theta \, d\theta$, and it follows that:

$$I_{33} = \int_0^\infty \frac{(\zeta^4 - \zeta^2)z'^2}{\lambda\zeta|z'|\cosh^2\theta} e^{-\frac{2\zeta|z'|\cosh\theta}{\lambda}} \sinh\theta \, d\theta \tag{5.115}$$

Again, making the substitution $w = e^{-\theta}$, and rewriting the sinh and cosh terms as in equation (5.57), yields the final form:

$$I_{33} = \int_0^1 \frac{2(\zeta^3 - \zeta)|z'|}{\lambda} \exp\left[-\frac{\zeta|z'|(\frac{1}{w} + w)}{\lambda} \right] \frac{1 - w^2}{(1 + w^2)^2} \, dw \tag{5.116}$$

Similarly

$$I_{34} = \int_0^1 \frac{2(\zeta^4 - \zeta^2)z'^2}{\lambda^2} \exp\left(-\frac{\zeta|z'|(\frac{1}{w} + w)}{\lambda} \right) \frac{1 - w^2}{w(1 + w^2)} \, dw \tag{5.117}$$

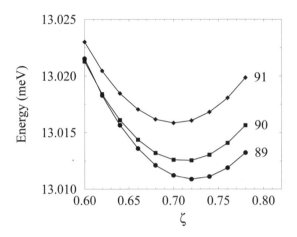

Figure 5.17   Total energy E of an electron bound to a donor as a function of the variable symmetry parameter ζ for $\lambda = 89$, 90 and 91 Å, thus illustrating the convergence of the two parameter variational calculation

At last, finally I_4, it follows similarly, with the variable of integration $r_\perp$ needing to be switched in order to match the denominator r', thus allowing manipulation to give:

$$I_4 = 2\pi \int_0^1 \exp\left[-\frac{2|z'|\sqrt{\left(\frac{1-w^2}{2w}\right)^2 + \zeta^2}}{\lambda}\right] |z'|\frac{1-w^2}{2w^2} \, \mathrm{d}w \qquad (5.118)$$

It should be noted that putting $\zeta=1$ in the above integrals, I_1, I_2, and I_3, reassuringly, gives the same result as those found for the 3D case. Unfortunately, the manipulation of I_4 with $\zeta=1$ is much more difficult and lies beyond the scope of this present work. However, numerical convergence tests have shown that putting $\zeta=1$ does yield the 3D energies, while putting $\zeta=0$ gives the 2D energies, although in both cases it is necessary to use a fine mesh, i.e. a small value of δz in equation (5.28).

Fig. 5.17 gives an example of the output from this two-parameter variational calculation. For the record, this calculation is for a donor at the centre of a 100 Å GaAs well surrounded by $Ga_{0.9}Al_{0.1}As$ barriers. For the sake of clarity, only the Bohr radii corresponding to the minima (at $\lambda = 89$ Å) and above are shown. It can be seen from this figure that the change in energy due to the ζ variation is very small, and certainly beyond the scope of experimental measurement. Nonetheless, the graph shows that the wave function will assume a value of $\zeta \approx 0.72$. This will have implications for systems where the extent and nature of the wave function are important, e.g. electron scattering from neutral donors, donor-bound excitons, and the electrical excitation of impurities.

Table 5.1 compares the energies obtained with all three trial wave functions for donor positions across a 60 Å CdTe well, surrounded by 200 Å $Cd_{0.9}Mn_{0.1}Te$ barriers (hence the centre of the well is at 230 Å). Note, as mentioned above, that the trial wave functions which contain integrals that require numerical evaluation, i.e. the 2D and the variable symmetry case which contain the integrals between 0 and 1 with respect

Table 5.1   Total energies for an electron bound to a donor as a function of donor position in a 60 Å CdTe quantum well surrounded by $Cd_{0.9}Mn_{0.1}Te$ barriers, for the three different trial wave functions (data first appeared in [108], data reproduced by permission of Academic Press)

r_d (Å)	$E(\zeta = 0)$ (meV)	$E(\zeta = 1)$ (meV)	$E(\zeta)$ (meV)
150	30.72	30.48	30.44
160	29.89	29.60	29.56
170	28.83	28.48	28.44
180	27.45	27.01	26.97
190	25.62	25.05	25.02
200	23.22	22.50	22.47
210	20.49	19.71	19.67
220	18.30	17.58	17.53
230	17.47	16.80	16.74

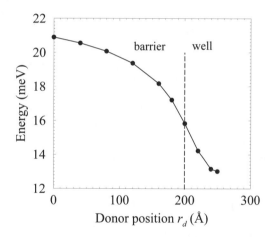

Figure 5.18   Energy E as a function of donor position across a 100 Å GaAs well surrounded by $Ga_{0.9}Al_{0.1}As$ barriers

to w, demand around eight wave function points per Å for complete convergence. The data displayed earlier in Fig. 5.10 as a comparison between the 2D and the 3D case, was obtained quickly with the standard mesh of 1 point per Å, i.e. $\delta z = 1$ Å. This was acceptable at that point in order to illustrate merely that the 3D case gave lower variational energies than the 2D case. However, for the purposes of demonstrating that the variable symmetry case gives lower energies still, more accurate calculations were necessary, and indeed for Table 5.1, δz in the iterative equation (equation (5.28)) was required to be 0.1 Å. The disadvantage to moving to such accuracy is, of course, the increased computational effort, which is roughly inversely proportional to δz; hence an

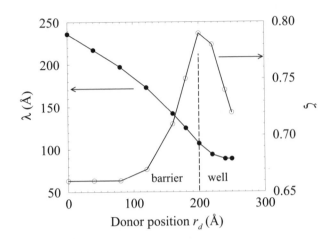

Figure 5.19   Bohr radius λ and symmetry parameter ζ as a function of donor position

increase by a factor of 10 in the number of wave function points leads to a similar increase in computational time. Table 5.1 clearly illustrates that the variable symmetry trial function gives the lowest energies, for all donor positions, although in terms of energy the 3D wave function is a very good approximation also, with energies differing by less than 0.1 meV. In comparison with the more accurate evaluation of the energy for the 2D case, this differs by up to 0.7 meV.

Figs. 5.18 and 5.19 illustrate the results of detailed calculations for the GaAs/ $Ga_{1-x}Al_xAs$ system. The energies in Fig. 5.18 are similar in form to those obtained earlier, with the interest here lying with the symmetry of the wave function, as given in Fig. 5.19. Most obviously, it is clear that the values of ζ imply that the donor wave function will assume symmetries that are neither 2D or 3D, thus justifying the investment in the variable symmetry analysis. The value of ζ is non-monotonic as a function of donor position, with a local minimum for donors at the centre of the well and a peak for donors near the well–barrier interface; this is the exact same behaviour as reported by Roberts *et al.* [108] for the CdTe-Cd$_{1-x}$Mn$_x$Te system and hence appears to be a generic result. Typically, it is found that $\zeta \sim 0.7$, and hence the variable symmetry trial wave function looks like the following:

$$\Psi = \chi(z) \exp \left(\frac{\sqrt{x^2 + y^2 + z'^2/2}}{\lambda} \right) \tag{5.119}$$

i.e. the coefficient of z'^2 is half that of x and y. Therefore, the hydrogenic factor is midway between the 2D and 3D case, with a very definite prolate spheroid shape (refer again to Fig. 5.16).

In conclusion, regarding the various advantages and disadvantages of the three trial wave functions, clearly the last, namely the variable symmetry case, gives the lowest variational energies and hence is the most accurate physical representation. Without such an investment in complexity and computational time, it would not have been apparent that donors assume this mixed 2D and 3D symmetry, i.e. wave functions resembling prolate spheroids. However, the computational demands are *substantial* and it has been shown that donor energies can be obtained within 0.1 meV of the lowest possible value, by using a simple 3D (spherical) trial wave function. This wave function has the advantage that all of the relevant integrations can be performed analytically, which leads to a quick evaluation of the total energy E.

5.6 INCLUSION OF A CENTRAL CELL CORRECTION

As mentioned earlier, the use of the bulk dielectric constant has to be employed with a note of caution. In the majority of cases its use is probably fine; however, it is worthwhile noting that there is a scheme for accounting for a change in the permittivity of the material as the electron approaches the donor. This occurs as the degree of electronic shielding is reduced and is known as the *central cell correction* .

This extension takes the form of an additional factor in the Coulombic potential term, which, for the purpose of illustration, is taken to be as follows:

$$-\frac{e^2}{4\pi\epsilon r'} \longrightarrow -\frac{e^2}{4\pi\epsilon r'}\left(1 + \Xi e^{-\frac{r'}{\Lambda}}\right) \tag{5.120}$$

where Λ is a parameter describing the extent of the correction; the latter can be thought of as a screening length. As $r' \longrightarrow 0$, it is expected that:

$$-\frac{e^2}{4\pi\epsilon r'}\left(1 + \Xi e^{-\frac{r'}{\Lambda}}\right) \longrightarrow -\frac{e^2}{4\pi\epsilon_0 r'} \tag{5.121}$$

where ϵ_0 is the permittivity of free space. Taking this limit gives:

$$\frac{1}{\epsilon}\left(1 + \Xi\right) = \frac{1}{\epsilon_0} \tag{5.122}$$

which when rearranged gives:

$$\Xi = \frac{\epsilon}{\epsilon_0} - 1 \tag{5.123}$$

The effect of these changes on the analysis carried out above occurs solely in the integral I_4, which for the most general case would become:

$$I_4' = 2\pi \int_0^\infty \left(1 + \Xi e^{-\frac{r'}{\Lambda}}\right) \frac{e^{-\frac{2r''}{\lambda}}}{r'} \, r_\perp \, dr_\perp \tag{5.124}$$

If the particular 3D case were chosen, then this gives:

$$I_4' = I_4 + 2\pi\Xi\frac{\Lambda\lambda}{\lambda + 2\Lambda}e^{-\frac{|z'|}{\Lambda}}e^{-\frac{|z'|}{\lambda}} \tag{5.125}$$

Recent work has demonstrated the effect of such a central cell correction for this 3D case [109]. While the physical reasoning behind the addition of such a correction seems sound, here, as always, caution needs to be taken in choosing a value for Λ. Cynically, Λ could be looked upon as merely another parameter (although it is *not* a variational parameter), that can be varied in order to produce agreement between experiment and theory.

The evaluation of I_4' for the most general, variable symmetry case will be left as an exercise for the reader!

5.7 SPECIAL CONSIDERATIONS FOR ACCEPTORS

Hitherto, concentration has been focused on donor energies in heterostructures, although, as discussed earlier in this chapter, *in principle* all of the analysis above is also applicable to the calculation of acceptor levels. Agreement between the hydrogenic model of the acceptor in bulk and experiment needs to account for both the degeneracy of the valence band and the reduced screening of the negatively charged acceptor ion potential. In heterostructures, however, the confinement potential lifts the degeneracy of light- and heavy-holes, so the only additional consideration that needs to be made relates to the permittivity.

The procedure would be to deduce empirically the central cell correction parameter Λ, by comparing the calculated acceptor binding energy E_{A^0} predicted by the above analysis and adjusting Λ to give agreement with experiment, for any well-characterized quantum well system. For a particular material system and acceptor atom, the screening length Λ *should be* constant. Therefore it can be used in further predictive calculations for different well widths, barrier heights, dopant positions, etc.

Acceptor levels within quantum wells have been calculated successfully, as discussed by Bastard [18]. Masselink *et al.* [110] have achieved good agreement with the experimental measurements of Miller *et al.* [111] for the binding energy of carbon acceptors in GaAs-Ga$_{1-x}$Al$_x$As quantum wells.

5.8 EFFECTIVE MASS AND DIELECTRIC MISMATCH

A great deal of attention has been paid in the literature to the role of effective mass mismatch at interfaces in semiconductor heterostructures (see Bastard [18]) and to the effect that this may have on donor energies. In addition, Fraizzoli *et al.* [112] have studied in detail the role of a dielectric constant mismatch at interfaces between dissimilar materials and its effect on shallow donor impurity levels in GaAs-Ga$_{1-x}$Al$_x$As quantum well structures. Such considerations can be readily incorporated in the present approach. At first sight it might be thought that this can be achieved by making allowance for the fact that both m^* and ϵ, which appear in the original Schrödinger equation, are a function of z. In as far as constructing the Schrödinger equation, this is true for the latter. However, as discussed in detail in Chapter 3, the change to a variable effective mass necessitates substantial alterations to the quantum mechanical kinetic energy operator. Again, as before, the first term in equation (5.4) would need to be substituted with:

$$-\frac{\hbar^2}{2m^*}\frac{\partial^2}{\partial z^2} \longrightarrow -\frac{\hbar^2}{2}\frac{\partial}{\partial z}\frac{1}{m^*(z)}\frac{\partial}{\partial z} \tag{5.126}$$

In principle, the analysis could be followed through again, with re-evaluation of the integrals affected by this substitution.

The generalization to include the material dependency of the permittivity ϵ is simpler to deal with. It can be achieved by absorbing this function into the integral I_4 which arises from the Coulombic term, i.e. with the appropriate change in the coefficient, I_4 would become:

$$I_4' = 2\pi \int_0^\infty \frac{e^{-\frac{2r''}{\lambda}}}{\epsilon(z)r'} \, r_\perp \, dr_\perp \tag{5.127}$$

as ϵ depends on the electron position and not simply the electron–donor separation as usual. It is clear, however, that evaluation of this integral is non-trivial. To the author's knowledge, neither of these generalizations to the theoretical approach in this chapter have been implemented.

While some authors deem it necessary to include such dependencies in calculations [103], it must be remembered that often the interest lies in systems of weak alloys such as $Ga_{1-x}Al_xAs$ and $Cd_{1-x}Mn_xTe$ where the effective mass change between the well and barrier is relatively small. Nonetheless, the mechanisms by which the theory presented here can be extended have been mapped out and here lies an opportunity for the interested reader to explore such systems further.

5.9 BAND NON-PARABOLICITY

Small well widths and large potential barriers could require the inclusion of the non-parabolicity of the conduction band [106]. Ekenberg [113] described the inclusion of non-parabolicity on the subband structure of quantum wells. This method can account accurately for a variety of physical phenomena, but is analytically complicated. Simpler procedures have been proposed by various authors for the more complex problem of a donor in a quantum well. For example, Chaudhuri and Bajaj [106] have used the following simple replacement:

$$m^* \longrightarrow m^*(E) = a_0 + a_1 E + a_2 E^2 + a_3 E^3 + \cdots + a_n E^n \qquad (5.128)$$

where a_n represents a series of constants. Given the values of these constants, the effects of non-parabolicity can be incorporated into equation (5.17) simply by making the effective mass a function of the energy E. This extension is certainly much more straightforward than the two described in the previous section.

In relation to the calculations below, it should be noted that Chaudhuri and Bajaj [106] showed that, even with relatively large potential barriers, band non-parabolicity was only significant for wells narrower than half the Bohr radius of the neutral donor, which, for the case of $Ga_{1-x}Al_xAs$, would be <50 Å, and for $Cd_{1-x}Mn_xTe$, would be <35 Å.

5.10 EXCITED STATES

Just as there are many solutions to the Schrödinger equation for an isolated hydrogen atom, there are also many more solutions representing excited energy states of a donor in a heterostructure. Recalling the hydrogen-atom solutions [4]:

$$\psi_{1s} = e^{-\frac{r}{\lambda}}; \quad \psi_{2s} = \left(2 - \frac{r}{\lambda}\right)e^{-\frac{r}{2\lambda}}; \quad \psi_{2p_x} = xe^{-\frac{r}{2\lambda}} \qquad (5.129)$$

and similarly for $2p_y$ and $2p_z$. The corresponding eigenenergy involves only the ground state and the principle quantum number, i.e. $E = E_1/n^2$, so very simply $E_2 = E_1/4$, $E_3 = E_1/9$, etc. In bulk semiconductors, the neutral donor does also exhibit these states (see [2], p. 314).

The situation is more complex in semiconductor heterostructures as the one-dimensional potential due to the layer structure breaks the symmetry of the spherical potential, and hence the wave function is more complex than for the hydrogen atom. Much detail has been given as to solving the corresponding ground state under these conditions. In an analogy with the hydrogen atom the first excited state of the donor might be written as follows:

$$\Psi_{2s} = \chi(z)\left(1 - \frac{\alpha r''}{\lambda_{2s}}\right) e^{-\frac{r''}{\lambda_{2s}}} \tag{5.130}$$

where λ_{2s} has been labelled specifically as it cannot be assumed, a priori, that $\lambda_{2s} = 2\lambda_{1s}$. The constant α has been introduced, and is determined by ensuring orthogonality between the ground state and this the first excited state, i.e.

$$\langle\Psi_{1s}|\Psi_{2s}\rangle = \int_0^\infty \Psi_{1s}^* \Psi_{2s} \ d\tau = 2\pi \int_{-\infty}^\infty \int_0^\infty \Psi_{1s}^* \Psi_{2s} \ dr_\perp \ dz = 0 \tag{5.131}$$

Studies do exist in the literature of the excited states of donors in heterostructures [114], but to the author's present knowledge the extension utilizing the general form for $\Psi = \chi(z)\xi(x, y, z - r_d)$ has not yet been attempted. A simple alternative approach which yields reasonable results is presented later in this chapter.

5.11 APPLICATION TO SPIN–FLIP RAMAN SPECTROSCOPY IN DILUTED MAGNETIC SEMICONDUCTORS

5.11.1 Diluted magnetic semiconductors

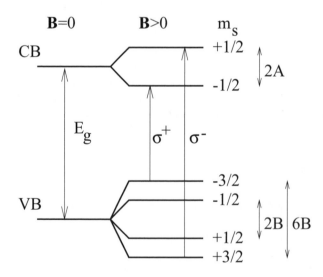

Figure 5.20   Zeeman splitting of conduction (CB) and valence (VB) bands of a diluted magnetic semiconductor within an external magnetic field **B**

Diluted magnetic semiconductorsdiluted magnetic semiconductors (DMSs) [115, 116] are important because of the strong exchange interaction between the hybridized sp^3–d orbitals of the magnetic ions and charge carriers. This manifests itself most clearly in the giant Zeeman splittings observed in both the conduction and valence bands when the material is placed in an external magnetic field. At low magnetic ion concentrations, the materials generally exhibit 'frustrated' paramagnetism, with the number of spin singlet states being reduced by nearest-neighbour anti-ferromagnetic spin-pairing [117]. At low magnetic fields (< 8 T) these spins remain locked and cannot contribute to the paramagnetism of the material; however, experiments carried out under very high magnetic fields have been able to break these spin-doublets. The magnetic ion itself, usually Mn^{2+}, sits substitutionally on a cation site and can generally be incorporated to high concentrations. The most common DMS is $Cd_{1-x}Mn_xTe$, while others include $Zn_{1-x}Mn_xS$, and more recently $Ga_{1-x}Mn_xAs$. Fig. 5.20 shows the Zeeman effect in a DMS material, with the vertical arrows linking the heavy-hole states ($\pm 3/2$) and the electron states, thus illustrating the allowed interband transitions under circularly polarized light, i.e. $|hh_{+3/2}\rangle \rightarrow |e_{+1/2}\rangle$ is the σ^- transition, and $|hh_{-3/2}\rangle \rightarrow |e_{-1/2}\rangle$ is the σ^+ transition.

The magnitudes of the conduction and valence band splittings are given in terms of the variables A and B as follows:

$$A = -\frac{1}{6}N_0\alpha x \langle S_z \rangle ; \qquad B = -\frac{1}{6}N_0\beta x \langle S_z \rangle \qquad (5.132)$$

where $N_0\alpha$ and $N_0\beta$ are constants (220 and 880 meV, respectively, in $Cd_{1-x}Mn_xTe$). The expectation value of the magnetic ion spin along the z-axis $\langle S_z \rangle$ is given by:

$$\langle S_z \rangle = S_0(x)B_J(\mathbf{B}, T_{\text{eff}}) \qquad (5.133)$$

where $S_0(x)$ is the effective spin of the magnetic ions and B_J is a Brillouin function describing the response of the spins in a magnetic fields $\mathbf{B}$. The effective spin S_0 accounts for the proportion of magnetic ions which are spin-paired with a nearest neighbour and cannot respond to the alignment induced by the magnetic field. Alternatively the effective spin can be considered as the concentration of spin-singlet states, i.e.

$$xS_0(x) = \frac{5}{2}\overline{x} \qquad (5.134)$$

where the spin of the Mn^{2+} ions are $5/2$. Recent theoretical studies [117] have calculated $\overline{x}$ in agreement with experiment [118] and shown that for moderate fields (≈ 8 T), i.e. when the splittings have saturated, but before the nearest neighbour spin pairings are broken, the maximum value of $x\langle S_z(x)\rangle$ occurs at a manganese concentration $x \approx 0.15$ and is equal to 0.105. Hence, in $Cd_{1-x}Mn_xTe$ the maximum splitting in the conduction band is ≈ 23 meV, and for the heavy-holes in the valence band it is ≈ 92 meV. The paramagnetic behaviour falls off with increasing temperature.

When applying a magnetic field to a semiconductor heterostructure, the direction of the field becomes important. For fields parallel to the growth axis (z-) (the Faraday configuration), the splittings are still well represented by Fig. 5.20. However, application of the magnetic field along the plane of the wells, i.e. the Voigt configuration, leads to mixing of the light- and heavy-hole valence states, thus producing a much more complex band structure. However, this lies beyond the scope of this short introduction.

Thus far, the $CdTe/Cd_{1-x}Mn_xTe$ system has been employed to illustrate donor binding energy calculations; now, however, specific use will be made of these magnetic properties in an application of the binding-energy calculations, namely spin–flip Raman spectroscopy.

5.11.2   Spin-flip Raman spectroscopy

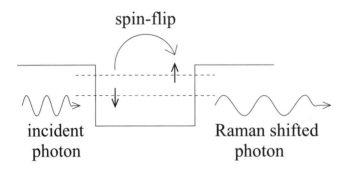

Figure 5.21   Schematic illustration of the spin–flip process

The application of a magnetic field to a heterostructure leads to a Zeeman splitting of the confined energy states, where in the case of an n-type material (i.e. material with donors present), two electron states are formed with the spin-up states having a higher energy than the spin-down states. At low temperatures, the carriers populate the lowest energy state, i.e. the spin-down, but can be excited by photons into the higher energy state. This excitation can be detected as a Raman shift [119], as illustrated schematically in Fig. 5.21.

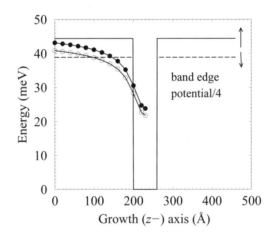

Figure 5.22   Energy of the spin-up (solid circles) and spin-down (open circles) states at a magnetic field of 8 T, for a range of donor positions across a 60 Å CdTe well surrounded by $Cd_{0.85}Mn_{0.15}Te$ barriers

This is a very powerful technique because the spectroscopy depends only upon one carrier type and theoretical modelling of the experimental data requires the parameter set of only one band. In this case, the parameters consist of the relative permittivity of the material, and the electron effective mass from bulk, together with the conduction

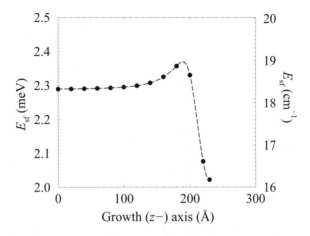

Figure 5.23   Spin-flip energy as a function of donor position

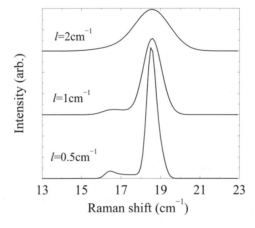

Figure 5.24   Intensity of Raman signal versus energy shift

band offset between the non-magnetic CdTe well and the magnetic $Cd_{1-x}Mn_xTe$ barriers [120].

The Raman shift is simply the energy difference between the two electron-donor spin states, $E_{sf} = E^\uparrow - E^\downarrow$. Fig. 5.22 displays the energy of these two spin states for a lightly n-doped $CdTe/Cd_{1-x}Mn_xTe$ system, calculated using the 3D trial wave function. The magnetic ion concentration in the barriers has been chosen as 15% in order to maximize the energy difference between the states. Fig. 5.23 displays the spin–flip energy for this system, again as a function of donor position.

If the donors are assumed to be uniformly distributed across the entire heterostructure, it is possible to represent the data of Fig. 5.23 in the form of an intensity I versus spin–flip energy E Raman spectrum, by assigning a Gaussian distribution to each point in energy space with a certain linewidth, i.e.

$$I = \sum_{r_d} \frac{1}{\sigma\sqrt{2\pi}} \exp\left[-\frac{(E - E_{sf}(r_d))^2}{2\sigma^2} \right]$$ (5.135)

where $E_{sf}(r_d)$ is the spin–flip energy of the donor at position r_d. The finite linewidth of any spin–flip signal will arise from microscopic fluctuations of the material parameters. For example, the well width, alloy fluctuations in the barrier, and the random nature of the donor distribution itself, will all give rise to a broadening of the signal from each of the donor positions r_d along the axis of the quantum well structure. The standard deviation σ is related to the linewidth l (full width at half-maximum) by:

$$\sigma = \frac{l}{2\sqrt{2\ln 2}}$$ (5.136)

In order to save considerable computational effort, the donor calculations were performed only at the points marked by circles on Fig. 5.23. A spline of this data was produced, given by the dashed line, and then this more detailed curve was used to produce the intensity versus energy data of Fig. 5.24, which also shows the effect of different linewidths on the predicted Raman spectrum of this single quantum well.

In the lowest curve ($l = 0.5$ cm^{-1}) it is possible, in principle, to resolve the donors in the well from those in the barrier, as observed experimentally [120]. Comparing Figs 5.23 and 5.24, it can be seen that the central peak at around 18.5 cm^{-1} corresponds to spin–flips of electrons bound to donors *in the barrier*. This peak in intensity is due entirely to the proportionately larger fraction of donors in the barrier compared with those in the well. In addition, spin–flips from electrons bound to donors in the well are clearly resolved as a small peak at about 16 cm^{-1}.

As the linewidth l of the signals is allowed to increase, the resolution and information in the simulated spectrum decreases. At a linewidth of 1 cm^{-1}, spin–flips due to donors in the well appear as just a shoulder on the larger 'barrier' peak, while by $l = 2$ cm^{-2} this information is lost altogether, and a broad central peak ensues.

In conclusion, the theoretical donor binding-energy calculations outlined above have been shown to be of direct relevance to simulating spin–flip Raman spectroscopy in diluted magnetic semiconductors. This allows the spatial distribution of donors to be investigated, although the importance of high quality samples has been shown to be significant.

It has been proposed that the magnetic behaviour of the first few monolayers of a dilute magnetic semiconductor adjacent to an interface with a non-magnetic semiconductor, could be significantly different from that of the bulk [121–123]. One contribution to this effect arises from a reduction in the number of antiferromagnetically coupled pairs due to a decrease in the number of nearest neighbour magnetic ions [121]. A single layer of donors, δ-doped into a quantum well structure in the region of the well–barrier interface could provide a useful probe of the magnetism—via the observations made from spin–flip Raman spectroscopy.

5.12 ALTERNATIVE APPROACH TO EXCITED IMPURITY STATES

The earlier method of calculating the energy levels of impurities within semiconductor heterostructures requires major mathematical analysis in order to extend to calculate excited states. In this section, a simpler, more general approach is developed that can handle the ground and excited states as well as the spatially degenerate states (2s, 2p$_x$, 2p$_z$) within a general heterostructure potential, that might also include an electric field (bias).

As in equation (5.4) the Hamiltonian of an impurity atom within a semiconductor heterostructure under the single-band effective mass and envelope function approximations is:

$$\mathcal{H} = -\frac{\hbar^2}{2}\frac{\partial}{\partial z}\frac{1}{m^*}\frac{\partial}{\partial z} + V(z) - \frac{e^2}{4\pi\epsilon r} \tag{5.137}$$

where, again, m^* is the effective mass of the charge carrier, $V(z)$ is the electrostatic potential which defines the heterostructure (and may include a bias $-eFz$) and r is the distance between the impurity and the charge carrier. Placing the x and y origins on the impurity atom, which is at a position r_i, then:

$$r^2 = x^2 + y^2 + (z - r_i)^2 \tag{5.138}$$

Taking a trial wave function of a charge carrier (electron or hole):

$$\Psi = \psi(z)\xi(r) \tag{5.139}$$

where $\psi(z)$ is the wave function of the electron (or hole) in the same heterostructure but *without* the impurity present and $\xi(r)$ is a hydrogenic like term describing the interaction between the electron (or hole) and the donor (or acceptor) ion.

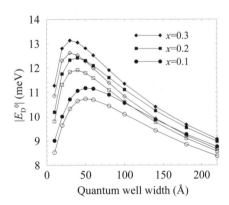

Figure 5.25 Comparison of the results of calculations using the approach developed in this section (open symbols) with the earlier method (closed symbols), presented in Fig. 5.13. The system is a donor at the centre of a GaAs quantum well surrounded by Ga$_{1-x}$Al$_x$As barriers.

For the 1s (ground) state of the impurity then the interaction term would be given by:

$$\xi(r) = \exp\left(-\frac{r}{\lambda}\right) \tag{5.140}$$

where λ is a variational parameter.

The variational calculation is implemented by adjusting λ in order to minimize the expectation value of the Hamiltonian operator (the total energy):

$$E = \frac{\langle \Psi | \mathcal{H} | \Psi \rangle}{\langle \Psi | \Psi \rangle} \qquad (5.141)$$

The energy E is evaluated for different values of λ by direct numerical integration of the numerator and denominator in the above equation. For example,

$$\langle \Psi | \mathcal{H} | \Psi \rangle = \int_0^\infty \int_0^\infty \int_0^\infty \Psi \left(-\frac{\hbar^2}{2} \frac{\partial}{\partial z} \frac{1}{m^*} \frac{\partial}{\partial z} + V(z) - \frac{e^2}{4\pi\epsilon r} \right) \Psi \; dx \, dy \, dz \qquad (5.142)$$

These integrals are calculated using a simple strip summation over a three-dimensional uniform mesh, with the differentials in the kinetic energy component evaluated using finite difference expansions.

5.13 DIRECT EVALUATION OF THE EXPECTATION VALUE OF THE HAMILTONIAN FOR THE GROUND STATE

Fig. 5.25 shows a comparison between the two numerical approaches of the ground state binding energy of a donor at the centre of a GaAs quantum well surrounded by $Ga_{1-x}Al_x As$ barriers. The closed symbols represent the data calculated using the three-dimensional trial wave function in Section 5.4, and presented in Fig. 5.13 while the open symbols show the results of calculations with the more direct numerical integration approach just developed.

It can be seen that for all three barrier heights (defined by three different Al concentrations, $x = 0.1$, 0.2 and 0.3), the function form of the binding energies with the quantum well width, is very similar. However, in all cases, the new data are around 0.5 meV below the earlier calculations.

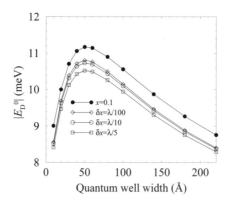

Figure 5.26   The effect of the numerical integration step length δx on the binding energy of a donor at the centre of a GaAs quantum well surrounded by $Ga_{0.9}Al_{0.1}As$ barriers.

Figs 5.26 and 5.28 explore the effect of two numerical parameters which could affect the accuracy of the calculations in this section.

In Fig. 5.26, the in-plane integration step length δx (equal to δy) was taken as 1/5th, 1/10th and 1/100th of the Bohr radius λ. The results of the calculations are compared to the $x = 0.1$ data from the original method. It can be seen that decreasing the size of the integration step length δx (which will increase the accuracy of any numerical integrations) does indeed move the new data closer to that of the earlier method. However, it also increases computational time, and given that this represents integration over a plane, increasing the number of steps by a factor of 10 (from $\lambda/10$ to $\lambda/100$, say), increases the computational time by a factor of 100. The step length $\delta x = \lambda/10$ is a good compromise and is adopted in all subsequent calculations.

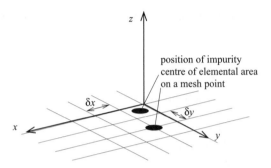

Figure 5.27   Illustration of the possible choices of the position of the impurity within the integration mesh.

The second computational parameter is the position of the impurity itself within the integration mesh. This would not normally be an issue, but in this case with impurities, the Coulomb potential $-e^2/(4\pi\epsilon r)$ has a singularity at $r = 0$. This can be avoided if the impurity is placed anywhere *within* the elemental area $\delta x \delta y$, and not on a mesh point itself, see Fig. 5.27.

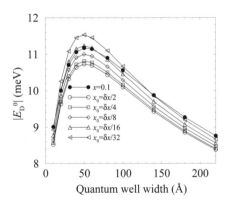

Figure 5.28   The effect of the position of the origin within the elemental volume on the binding energy of a donor at the centre of a GaAs quantum well surrounded by $Ga_{0.9}Al_{0.1}As$ barriers.

Fig. 5.28 shows the results of calculations of the effect of the origin (x_0,y_0) of the in-plane integration mesh with respect to the position of the impurity. It can be seen that the position of the origin can make up to nearly 1 meV difference in the binding energy, and as the impurity is brought closer and closer to a mesh point ($\delta x \to 0$) the binding energy increases. The latter is not surprising because the mesh points represent the points at which the deep Coulomb potential is sampled. The closer it is sampled to the impurity, then the deeper the potential and this value makes a greater contribution to the integral. In all subsequent calculations the mesh origin was chosen so that the impurity lay in the centre of an elemental area, this also has the additional justification of symmetry.

5.14   VALIDATION OF THE MODEL FOR THE POSITION DEPENDENCE OF THE IMPURITY

It has thus been established that this approach to the calculation of the ground state energy level of an impurity at the centre of a quantum well is in good agreement with the earlier technique. However, it is now important to validate the new approach for various impurity positions—not restricting the impurity to the highly symmetric position at the centre of a quantum well. This is particularly important because of the choice of the impurity wave function as the product of the envelope function without the impurity present with the hydrogenic term i.e. $\Psi = \psi(z)\xi(x, y, z, r_i)$, rather than the original method which recalculated the envelope function.

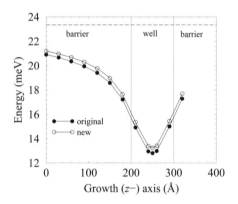

Figure 5.29   Comparison of the results of calculations using the approach developed in this section (open symbols) with the earlier method (closed symbols). The figure shows the total energy of an electron in a 100 Å GaAs quantum well surrounded by 200 Å $Ga_{0.9}Al_{0.1}As$ barriers, as a function of the position of the donor across the barrier–well–barrier system.

Fig. 5.29 shows the results of a series of calculations of the ground state donor energy for a range of donor positions across the barrier, well and barrier regions of a single 100 Å GaAs quantum well surrounded by 200 Å barriers of $Ga_{0.9}Al_{0.1}As$. The figure also shows the results from the earlier method and it can be seen that the difference in the energy of the donor is around 0.5 meV, which is quite acceptable in device design.

The calculations span the quantum well to show that the expected symmetry of the energy with position is reproduced. The horizontal dashed line near the top of the figure shows the energy of an electron in the same heterostructure but without a donor present.

5.15   EXCITED STATES

Thus, it now remains to move onto the initial purpose of this work, which is to develop
a model of the energy levels of *excited* impurity states. Taking the hydrogenic factor ξ
of the total impurity wave function as:

$$\xi = \left(1 - \frac{r}{\lambda}\right) \exp\left(-\frac{r}{\lambda}\right) \tag{5.143}$$

where $r = \sqrt{x^2 + y^2 + (z - r_i)^2}$, the energy level of the excited 2s state of a donor in the
centre of a GaAs quantum well can be calculated. Strictly speaking the total impurity
wave function $\Psi = \psi(z)\xi(x, y, z, r_i)$ for this excited 2s state is not orthogonal to the
corresponding 1s wave function and a more complicated approach incorporating a factor
'α' in ξ should be used. However, the results that follow indicate that the simple form
taken in equation (5.143) is a good approximation for obtaining the *energy* of excited
impurity states in heterostructures.

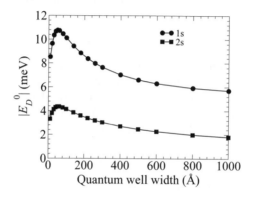

Figure 5.30   The well width dependence of the binding energies of the 1s and 2s states of
donors at the centre of GaAs quantum wells surrounded by $Ga_{1-x}Al_xAs$ barriers.

Fig. 5.30 shows the binding energy of the 1s and the 2s states as a function of the
width of the GaAs quantum well. It can be seen that the binding energy of the 1s state
is much larger than that of the 2s (as expected) and that the functional form of both
curves is the same. Fig. 5.31 shows the ratio of these two binding energies. Bohr theory
[4] gives the (binding) energy of the hydrogen atom as $E_n = E_1/n^2$, where E_1 is the
(binding) energy of the lowest ($n = 1$) state. Thus for a hydrogen atom the ratio of the
energy of the 1s state to the 2s state is exactly 4. In the figure, this number is much
smaller; however, as the width of the well is increased, and the system tends towards
bulk, this ratio does begin to move towards 4. This is supporting evidence in favour
of the application of this method to the calculation of the binding energy of excited
impurity states in heterostructures.

Fig. 5.32 shows the binding energies of the 2s and $2p_x$ impurity levels for a donor at
the centre of a GaAs quantum well, as a function of the quantum well width, again the

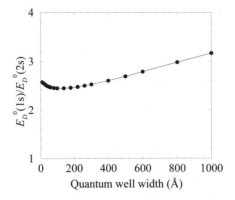

Figure 5.31   The well width dependence of the ratio of the binding energies of the 1s and 2s states of donors at the centre of GaAs quantum wells surrounded by $Ga_{1-x}Al_xAs$ barriers.

barrier material is $Ga_{0.9}Al_{0.1}As$. The hydrogenic factor for the $2p_x$ state was taken as:

$$\xi = x \exp\left(-\frac{r}{\lambda}\right) \qquad (5.144)$$

It can be seen that the binding energy of the $2p_x$ state is considerably less than that of the 2s state—only around half around the peak at 100 Å. However, as the width of the quantum well is increased, the binding energy of the $2p_x$ level converges towards that of the 2s level, as would be expected, because in the limit of an infinitely wide well (i.e. bulk material), these states should be degenerate.

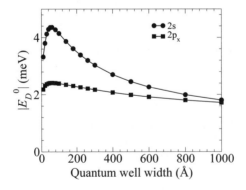

Figure 5.32   The well width dependence of the binding energies of the 2s and $2p_x$ states of donors at the centre of GaAs quantum wells surrounded by $Ga_{1-x}Al_xAs$ barriers.

Figs 5.33 shows the Bohr radii (the value of the variational parameter) λ for the 1s and $2p_x$ states. Fig. 5.34 shows the ratio of these two radii. It can be seen that it is very close to 2—as expected from Bohr theory. In fact, the $2p_x$ state is a very good one to calculate because near the impurity, where the singularity exists, the functional form of the hydrogenic term, i.e. $x \exp(-r/\lambda)$, naturally tends to zero, hence reducing any inaccuracies that might arise because of integrating over the singularity.

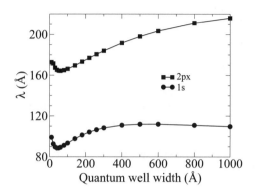

Figure 5.33   The well width dependence of the Bohr radius λ of the 1s and $2p_x$ states of donors at the centre of GaAs quantum wells surrounded by $Ga_{1-x}Al_xAs$ barriers.

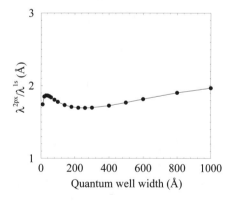

Figure 5.34   The well width dependence of the *ratio* of the Bohr radii of the 1s and $2p_x$ states of donors at the centre of GaAs quantum wells surrounded by $Ga_{1-x}Al_xAs$ barriers.

Fig. 5.35 shows the binding energy of the $2p_z$ impurity state, calculated using a hydrogenic term analagous as that for the $2p_x$ state, as in equation (5.144), but with the 'lobes' aligned along the growth axis. The lobes now overlap with the repulsive potentials of the heterostructure barriers and hence the energy of the impurity state is increased. This *decreases* the binding energy, and in fact, as the figure shows, the binding energy becomes negative, i.e. the state is unbound. This effect is reduced as the well becomes wider and the z-lobes can 'fit' inside the well. The binding energy of the $2\mathrm{p}_z$ state tends to that of the 2s and $2\mathrm{p}_x$ states as the well approaches 1000 Å.

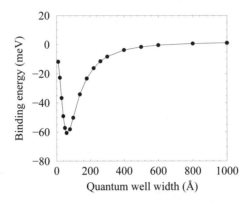

Figure 5.35   The well width dependence of the binding energy of the $2\mathrm{p}_z$ state of donors at the centre of GaAs quantum wells surrounded by $\mathrm{Ga}_{0.9}\mathrm{Al}_{0.1}\mathrm{As}$ barriers.

5.16 IMPURITY OCCUPANCY STATISTICS

Any book on solid-state physics, for example Kasap [124] (page 314) or Sze [44] (page 16), gives the number of free electrons introduced into a crystal due to doping as the integral over the entire conduction band of the density of states multiplied by the probability of occupation:

$$n = \int_{E_c}^{\infty} \rho(E) f(E) \ \mathrm{d}E \tag{5.145}$$

where the density of states $\rho(E)$ (derived earlier in Section 2.3) is written in relation to the conduction band edge E_c as:

$$\rho(E) = \frac{1}{2\pi^2} \left(\frac{2m^*}{\hbar^2} \right)^{\frac{3}{2}} (E - E_c)^{\frac{1}{2}} \tag{5.146}$$

and the Fermi–Dirac[†] distribution function $f(E)$ is approximated by a Boltzmann distribution function:

$$f(E) = \exp\left(-\frac{(E - E_F)}{kT} \right) \tag{5.147}$$

The number of free electrons (which is the same as the number of ionized donors) then follows as:

$$n = \frac{1}{2\pi^2} \left(\frac{2m^*}{\hbar^2} \right)^{\frac{3}{2}} \int_{E_c}^{\infty} (E - E_c)^{\frac{1}{2}} \exp\left(-\frac{(E - E_F)}{kT} \right) \ \mathrm{d}E \tag{5.148}$$

[†]This is the true Fermi–Dirac distribution function for a doped semiconductor and not the quasi-function introduced earlier to describe the distribution of electrons within a subband.

which leads to (see Kasap [124] (page 314)):

$$n = N_c \exp\left(-\frac{(E_c - E_F)}{kT}\right) \tag{5.149}$$

where N_c is known as the 'effective density of states' and is given by:

$$N_c = 2\left(\frac{2\pi m^* kT}{h^2}\right)^{\frac{3}{2}} \tag{5.150}$$

and the temperature-dependent Fermi energy for a donor volume density N_d and energy E_d is given by:

$$E_F = \frac{E_c + E_d}{2} + \frac{1}{2}kT \ln\left(\frac{N_d}{2N_c}\right) \tag{5.151}$$

see, for example, Kasap [124] (page 335). It can be seen from equation (5.151) that, at low temperatures, E_F would be halfway between the donor energy levels E_d and the conduction band edge E_c as expected.

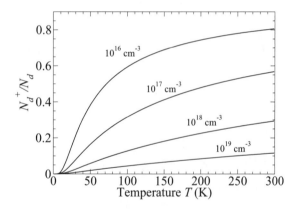

Figure 5.36   The proportion of ionized donors N_d^+ from a doping concentration N_d as a function of temperature for a bulk semiconductor, for several different dopant densities. The material system is GaAs with a donor energy 5.8 meV below the conduction band edge and an effective mass of 0.067 m_0.

Fig. 5.36 shows the results of calculations of the number of ionized donors N_d^+ ($=n$) as a function of the temperature T for several different doping densities N_d. It can be seen from the figure that as the doping density increases the proportion of donors ionized decreases at any given temperature. This is just a result of statistical mechanics—the effective density of states N_c remains constant, so as more donors are added to the system a smaller *proportion* of them can find empty states for their electrons to move to. It is interesting to note that at the typical doping density of 10^{18} cm^{-3} only around one-third of donors are ionized at room temperature.

The focus of this book is on low-dimensional systems so generalising to the two dimensions of a quantum well heterostructure, then the density of states would become the two-dimensional density of states again derived in Section 2.3:

$$\rho^{2D}(E) = \frac{m^*}{\pi \hbar^2} \tag{5.152}$$

hence:

$$n = \frac{m^*}{\pi \hbar^2} \int_{E_c}^{\infty} \exp\left(-\frac{(E - E_F)}{kT}\right) dE = \frac{m^*}{\pi \hbar^2} \left[-kT \exp\left(-\frac{(E - E_F)}{kT}\right)\right]_{E_c}^{\infty} \tag{5.153}$$

$$\therefore n = \frac{m^* kT}{\pi \hbar^2} \exp\left(-\frac{(E_c - E_F)}{kT}\right) \tag{5.154}$$

which in analogy to three dimensions, could also be written:

$$n = N_c^{2D} \exp\left(-\frac{(E_c - E_F)}{kT}\right) \tag{5.155}$$

where the two-dimensional effective density of states:

$$N_c^{2D} = \frac{m^* kT}{\pi \hbar^2} \tag{5.156}$$

and the Fermi energy would be equivalent to that in equation (5.151) but with N_c^{2D} instead of N_c and the donor density N_d would be taken as a *sheet density*.

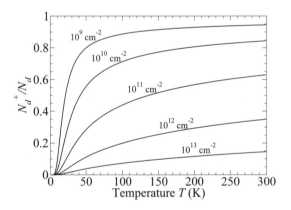

Figure 5.37   The proportion of ionized donors N_d^+ from a doping concentration N_d as a function of temperature for a two-dimensional (quantum well) system, for several different dopant densities. The material system is GaAs with a donor energy 5.8 meV below the conduction band edge and an effective mass of 0.067 m_0.

Fig. 5.37 shows the corresponding curve to Fig. 5.36 for a two-dimensional system. The generic behaviour is the same as in bulk, i.e. that the higher the temperature the greater proportion of donors are ionized and the greater the doping density the lower the proportion of ionized donors. Again, the number of free carriers generated by doping might seem surprisingly small.

Although this view has been recently reinforced [125], it is actually too simplistic and the proportion of carriers ionized is too small. The reason is that the model assumes that the impurity ionization energy (which in this section would be $E_c - E_d$) is left as a constant, whereas it was found that screening should see it reduce as the impurity concentration increases [126]. Pearson and Bardeen [126] argued that the ionization energy of an impurity would be reduced by an energy inversely proportional to the distance between the impurities, hence the ionization energy should be equal to:

$$|E_D^0| - aN_d^{\frac{1}{3}}$$

where E_D^0 is the ionization (binding) energy of the isolated impurity and N_d is the impurity (donor) concentration. The constant a depends on the impurity species and the host and was deduced experimentally by Pearson and Bardeen for boron and phosphorus in silicon.

CHAPTER 6

EXCITONS

6.1 EXCITONS IN BULK

If photons of energy comparable to the band gap are incident on a semiconductor, then they can be absorbed by the electrons forming atomic bonds between neighbouring atoms, and so provide them with enough energy to break free and move around in the body of the crystal. Within the band theory of solids, this would be described as 'exciting an electron from the valence band across the band gap into the conduction band'. If the energy of the photon is larger than the band gap, then a free electron is created *and* an empty state is left within the valence band (see 'high energy excitation' in Fig. 6.1).

The empty state within the valence band behaves very much like an air bubble in a liquid and rises to the top—the lowest energy state. This 'hole' behaves as though it were positively charged and hence often forms a bond with a conduction-band electron (see 'exciton formation' in Fig. 6.1). The attractive potential leads to a reduction (by an amount E_{X^0}) in the total energy of the electron and hole. This bound electron–hole pair is known as an 'exciton'. Photons of energy just below the band gap can be absorbed, thus creating excitons directly (see 'resonant excitation').

As the hole mass is generally much greater than the electron mass, then the two-body system resembles a hydrogen atom, with the negatively charged electron orbiting the positive hole. The exciton is quite stable and can have a relatively long lifetime, of the order of hundreds of ps to ns. Exciton recombination is an important feature of low

Quantum Wells, Wires and Dots, Third Edition. P. Harrison
©2009 John Wiley & Sons, Ltd.

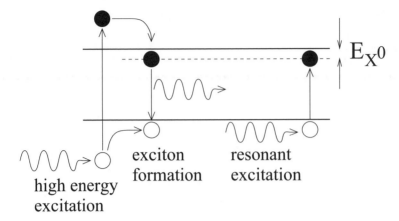

Figure 6.1   Schematic representation of the non-resonant and resonant generation of excitons

temperature photoluminescence, although as the binding energies are relatively low, i.e. a few meV to a few tens of meV, they tend to dissociate at higher temperatures.

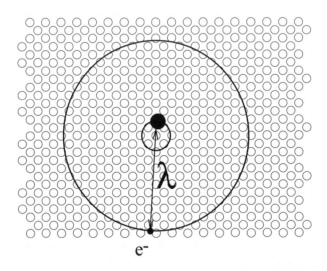

Figure 6.2   Schematic diagram of an exciton in bulk material, with the hole (filled circle near centre) and electron 'separated' by a Bohr radius λ, orbiting around the centre of mass

Therefore, in a similar manner to the hydrogenic impurities discussed in Chapter 5, the binding energy and orbital radius can be represented well by Bohr theory, *with* the correction for the finite mass of the central charge [3, 4]. This is implemented by exchanging the (in the case of a hydrogen atom, or donor) orbiting electron mass, with the reduced mass of the two-body system, in this case, the electron–hole pair. The reduced mass is given by:

$$\frac{1}{\mu} = \frac{1}{m_e^*} + \frac{1}{m_h^*}$$

(6.1)

Hence, the binding energy becomes:

$$E_{X^0} = -\frac{\mu e^4}{32\pi^2 \hbar^2 \epsilon_r^2 \epsilon_0^2} \tag{6.2}$$

and the Bohr radius follows as:

$$\lambda = \frac{4\pi \epsilon_r \epsilon_0 \hbar^2}{\mu e^2} \tag{6.3}$$

Taking typical values for bulk GaAs, i.e. the Γ valley electron and heavy-hole effective masses, $m_e^* = 0.067m_0$ and $m_{hh}^* = 0.62m_0$, respectively, then $\mu = 0.060m_0$. Using the static dielectric constant $\epsilon_r = 13.18$ [14], then the exciton binding energy and Bohr radius follow, respectively, as:

$$E_{X^0} = -4.7 \text{ meV} \quad \text{and} \quad \lambda = 115 \text{ Å} \tag{6.4}$$

which are exactly the same as the low-temperature measured values (see [14], p.420). Correspondingly for CdTe, $m_e^* = 0.096m_0$ and $m_{hh} = 0.6m_0$, therefore $\mu = 0.083m_0$, with $\epsilon_r = 10.6$, then:

$$E_{X^0} = -10.1 \text{ meV} \quad \text{and} \quad \lambda = 67 \text{ Å} \tag{6.5}$$

which again agree with experiment [127].

6.2 EXCITONS IN HETEROSTRUCTURES

In the same way as in bulk, excitons can be formed by the bonding of free electron–free hole pairs or through resonant excitation. Whereas in bulk, the total energy of the exciton is simply the energy of the free electron–free hole pair (i.e. the band gap) plus the exciton binding energy E_{X^0} , in a heterostructure there are additional components due to the electron and hole confinement energies, i.e.

$$E = E_g + E_{X^0} \text{ (bulk)} \quad E = E_g + E_e + E_h + E_{X^0} \text{ (heterostructure)} \tag{6.6}$$

The total exciton energy is clearly a function of structure because of the structural dependency of the confinement energies. In addition, it must be expected that the Coulombic potential energy, i.e. E_{X^0}, also depends upon the structure. This latter effect arises because the electron–hole separation can vary quite considerably between heterostructures. Fig. 6.3 illustrates this schematically; clearly the electron and hole separation is much smaller in the Type-I quantum well, where both particles are localized in the same layer of semiconductor, than in the Type-II system, where they are localized in different layers. Hence, the exciton binding energy will be larger in the former.

6.3 EXCITON BINDING ENERGIES

The Hamiltonian representing the interacting two-body electron–hole complex can be considered to be the sum of three terms

$$\mathcal{H} = \mathcal{H}_e + \mathcal{H}_h + \mathcal{H}_{e-h} \tag{6.7}$$

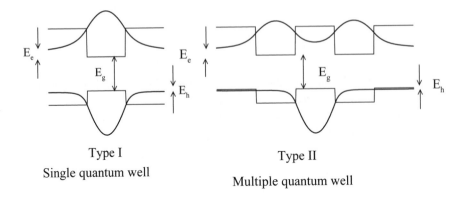

Type I

Single quantum well

Type II

Multiple quantum well

Figure 6.3   Electrons and holes in Type-I and Type-II systems

where $\mathcal{H}_e$ and $\mathcal{H}_h$ are the one-particle Hamiltonians appropriate to the conduction and valence bands, respectively, of the particular microstructure of interest (as in Chapter 2). However, in the context of the exciton a constant effective mass along the growth (z-) axis must be assumed, i.e. the one-particle Hamiltonians are written:

$$\mathcal{H}_e = -\frac{\hbar^2}{2m_e^*}\frac{\partial^2}{\partial z^2} + V_{\mathrm{CB}}(z) \quad ; \quad \mathcal{H}_h = -\frac{\hbar^2}{2m_h^*}\frac{\partial^2}{\partial z^2} + V_{\mathrm{VB}}(z) \tag{6.8}$$

The third term, $\mathcal{H}_{e-h}$ represents the electron–hole interaction, and is itself composed of two terms. One of these corresponds to the kinetic energy of the relative motion of the electron and hole in the x–y plane (perpendicular to the growth axis), while the other represents the Coulombic potential energy, i.e.

$$\mathcal{H}_{e-h} = \frac{\mathcal{P}_\perp^2}{2\mu_\perp} - \frac{e^2}{4\pi\epsilon r} \tag{6.9}$$

where $\mathcal{P}_\perp$ is the quantum mechanical momentum operator for the in-plane component of the relative motion. Now:

$$\mathcal{P} = -i\hbar\nabla = -i\hbar\left(\frac{\partial}{\partial x}\mathbf{i} + \frac{\partial}{\partial y}\mathbf{j} + \frac{\partial}{\partial z}\mathbf{k}\right) \tag{6.10}$$

$$\therefore \mathcal{P}_\perp^2 = -\hbar^2\left(\frac{\partial^2}{\partial x^2} + \frac{\partial^2}{\partial y^2}\right) \tag{6.11}$$

and r is simply the electron–hole separation, given by:

$$r^2 = (x_e - x_h)^2 + (y_e - y_h)^2 + (z_e - z_h)^2 \tag{6.12}$$

Note, that as the main motivation is concerned with the optical properties of excitons in quantum well systems, the Hamiltonian then contains no term for the motion of the centre of mass of the exciton in the x–y plane, i.e. the exciton is assumed to be at rest within the plane of the well.

The problem, as always, is to find the eigenfunctions Ψ and eigenvalues E of the Schrödinger equation:

$$\mathcal{H}\Psi = E\Psi \tag{6.13}$$

which describes the system. Following standard procedures, the two-body exciton wave function Ψ is chosen to be a product of three factors, as follows:

$$\Psi = \psi_e(z_e)\psi_h(z_h)\psi_r \tag{6.14}$$

where ψ_r represents the electron and hole relative motion. Following Hilton *et al.* [128, 129] and Harrison *et al.* [130, 131] ψ_r will be a variational wave function employed to minimize the total energy E of the system. The specific form of ψ_r will be discussed later. The other two factors, $\psi_e(z_e)$ and $\psi_h(z_h)$ are simply the eigenfunctions of the one-particle Hamiltonians of the heterostructure:

$$\mathcal{H}_e\psi_e(z_e) = E_e\psi_e(z_e) \quad ; \quad \mathcal{H}_h\psi_h(z_h) = E_h\psi_h(z_h) \tag{6.15}$$

One of the main advantages of this formalism is that it is independent of the form of the one-particle Hamiltonians $\mathcal{H}_e$ and $\mathcal{H}_h$, and indeed calculations can be performed on any system in which the standard electron and hole wave functions can be calculated [77, 132, 133].

Multiplying equation (6.13) on the left by Ψ and integrating over all space, then the total exciton energy follows simply as the expectation value:

$$E = \frac{\langle\Psi|\mathcal{H}|\Psi\rangle}{\langle\Psi|\Psi\rangle} \tag{6.16}$$

Now, whatever the form of the relative motion term ψ_r of equation (6.14), it will be a function of the electron–hole separation, which is quantified in terms of the three variables, $x = |x_e - x_h|$, $y = |y_e - y_h|$ and $a = |z_e - z_h|$. The denominator $\mathcal{D}$ in equation (6.16) can therefore be written as:

$$\mathcal{D} = \langle\Psi|\Psi\rangle = \int_{\text{all space}} |\psi_e(z_e)|^2|\psi_h(z_h)|^2|\psi_r|^2 \ \mathrm{d}x \ \mathrm{d}y \ \mathrm{d}z_e \ \mathrm{d}z_h \tag{6.17}$$

Since ψ_e and ψ_h are functions of z only, then the integrations over the x–y plane will involve only the relative motion term ψ_r. Correspondingly, the integrations over the electron and hole coordinates, z_e and z_h will yield a result which is a function of the electron and hole separation (a) only, i.e. it is possible to write:

$$\int |\psi_r|^2 \ \mathrm{d}x \ \mathrm{d}y = F(a) \tag{6.18}$$

Although the one-particle wave functions, ψ_e and ψ_h, in principle, extend to $\pm\infty$, in practice, from a computational viewpoint, it is necessary to define them between 'effective infinities'. As discussed in Chapter 3, these are defined so that:

$$\int_\mu^\nu \psi_e(z_e) \ \mathrm{d}z_e \approx \int_{-\infty}^{+\infty} \psi_e(z_e) \ \mathrm{d}z_e \tag{6.19}$$

and similarly for ψ_h. Using these results, the denominator $\mathcal{D}$ in equation (6.17) can be re-written as:

$$\mathcal{D} = \int_0^{\nu-\mu} \int_\mu^\nu \int_\mu^\nu |\psi_e(z_e)|^2 |\psi_h(z_h)|^2 F(a) \; dz_e \; dz_h$$

$$\times \left[\delta \left(z_e - z_h - a \right) + \delta \left(z_h - z_e - a \right) \right] \; da \tag{6.20}$$

where the Dirac δ-functions have been introduced in order to ensure that contributions are included from both $z_e - z_h = a$ and $z_h - z_e = a$. If the integration over z_h is performed first, then the first term only has a finite value when $z_h = z_e - a$ and the second term when $z_h = z_e + a$; hence:

$$\mathcal{D} = \int_0^{\nu-\mu} \int_{\mu+a}^\nu |\psi_e(z_e)|^2 |\psi_h(z_e - a)|^2 F(a) \; dz_e \; da$$

$$+ \int_0^{\nu-\mu} \int_\mu^{\nu-a} |\psi_e(z_e)|^2 |\psi_h(z_e + a)|^2 F(a) \; dz_e \; da \tag{6.21}$$

Making the substitution $z = z_e - a$ in the first term, and relabelling z as z_e in the second, finally gives the denominator in equation (6.16) as:

$$\mathcal{D} = \int_0^{\nu-\mu} p(a) F(a) \; da \tag{6.22}$$

where $p(a)$ represents the uncorrelated probability of finding the electron and hole separated by a distance a, i.e.

$$p(a) = \int_\mu^{\nu-a} |\psi_e(z+a)|^2 |\psi_h(z)|^2 + |\psi_e(z)|^2 |\psi_h(z+a)|^2 \; dz \tag{6.23}$$

Returning to the expression for the expectation value of the total exciton energy in equation (6.16) and using the three-term Hamiltonian of equation (6.7), the numerator $\mathcal{N}$ can then be written as:

$$\mathcal{N} = \langle \Psi | \mathcal{H} | \Psi \rangle = \langle \Psi | \mathcal{H}_e | \Psi \rangle + \langle \Psi | \mathcal{H}_h | \Psi \rangle + \langle \Psi | \mathcal{H}_{e-h} | \Psi \rangle \tag{6.24}$$

Labelling these three terms as $\mathcal{A}$, $\mathcal{B}$ and $\mathcal{C}$, respectively, i.e. $\mathcal{N} = \mathcal{A} + \mathcal{B} + \mathcal{C}$, then consider:

$$\mathcal{A} = \langle \Psi | T | \Psi \rangle + \langle \Psi | V | \Psi \rangle \tag{6.25}$$

where T and V are the kinetic and potential energy operators, respectively, of the electron one-particle Hamiltonian $\mathcal{H}_e$. Employing the standard constant-mass kinetic operator then (in an obvious notation) obtain:

$$\mathcal{A}_T = \int \psi_e^*(z_e) \psi_h^*(z_h) \psi_r^* \frac{-\hbar^2}{2m_e^*} \frac{\partial^2}{\partial z_e^2} \psi_e(z_e) \psi_h(z_h) \psi_r \; dz_e \; dz_h \; dx \; dy \tag{6.26}$$

On performing the differentiations:

$$\mathcal{A}_T = \int \psi_e^*(z_e) \psi_h^*(z_h) \psi_r^* \frac{-\hbar^2}{2m_e^*} [\psi_e''(z_e) \psi_h(z_h) \psi_r + 2\psi_e'(z_e) \psi_h(z_h) \psi_r' +$$

$$\psi_e(z_e)\psi_h(z_h)\psi_r''] \ \mathrm{d}z_e \ \mathrm{d}z_h \ \mathrm{d}x \ \mathrm{d}y \tag{6.27}$$

which can be written as follows:

$$\mathcal{A}_T = \int \frac{-\hbar^2}{2m_e^*} \left[\psi_e^*(z_e)\psi_e''(z_e)|\psi_h(z_h)|^2|\psi_r|^2 + \right.$$

$$\left. 2\psi_e^*(z_e)\psi_e'(z_e)|\psi_h(z_h)|^2\psi_r^*\psi_r' + |\psi_e(z_e)|^2|\psi_h(z_h)|^2\psi_r^*\psi_r''\right] \ \mathrm{d}z_e \ \mathrm{d}z_h \ \mathrm{d}x \ \mathrm{d}y \tag{6.28}$$

Note that the first term, when normalized by the denominator $\mathcal{D}$, merely represents the one-particle kinetic energy of the electron, which coupled together with $\langle \Psi|\mathcal{V}|\Psi \rangle$ in equation (6.25) gives the one-particle electron energy E_e as defined in equation (6.15). Furthermore, noting that, for the problems of interest, i.e. stationary states within semiconductor microstructures, the wave functions are real and that the chain rule gives:

$$\psi_e(z_e)\psi_e'(z_e) = \frac{1}{2}\frac{\partial \psi_e^2(z_e)}{\partial z_e} \tag{6.29}$$

Equation (6.25) therefore becomes:

$$\mathcal{A} = E_e\mathcal{D} + \int \frac{-\hbar^2}{2m_e^*} \left\{ |\psi_h(z_h)|^2 \left[\frac{\partial \psi_e^2(z_e)}{\partial z_e} \right] \psi_r\psi_r' + \right.$$

$$\left. |\psi_e(z_e)|^2|\psi_h(z_h)|^2\psi_r\psi_r'' \right\} \ \mathrm{d}z_e \ \mathrm{d}z_h \ \mathrm{d}x \ \mathrm{d}y \tag{6.30}$$

Integrate, by parts, over $\mathrm{d}z_e$ the first term in the integrand, i.e.:

$$\int |\psi_h(z_h)|^2 \left[\frac{\partial |\psi_e(z_e)|^2}{\partial z_e} \right] \psi_r\psi_r' \ \mathrm{d}z_e =$$

$$|\psi_h(z_h)|^2 \left[|\psi_e(z_e)|^2\psi_r\psi_r' \right]_\mu^\nu - \int \left(\psi_r'^2 + \psi_r\psi_r'' \right) |\psi_e(z_e)|^2|\psi_h(z_h)|^2 \ \mathrm{d}z_e \tag{6.31}$$

Hence, substituting back into equation (6.30) gives:

$$\mathcal{A} = E_e\mathcal{D} + \int \frac{-\hbar^2}{2m_e^*} |\psi_h(z_h)|^2 \left\{ \left[|\psi_e(z_e)|^2\psi_r\psi_r' \right]_\mu^\nu \right.$$

$$\left. - \int \left[|\psi_e(z_e)|^2\psi_r'^2 + |\psi_e(z_e)|^2\psi_r\psi_r'' - |\psi_e(z_e)|^2\psi_r\psi_r'' \right] \ \mathrm{d}z_e \right\} \mathrm{d}z_h \ \mathrm{d}x \ \mathrm{d}y \tag{6.32}$$

The last two terms in the integration over z_e obviously cancel out, and in addition, examination of the first term on the right-hand side, shows that this term will not contribute, since whatever the form of the eigenfunction ψ_e, it will undoubtedly vanish at the effective infinities μ and ν. Therefore:

$$\mathcal{A} = E_e\mathcal{D} + \int_{\text{all space}} \frac{\hbar^2}{2m_e^*} |\psi_h(z_h)|^2|\psi_e(z_e)|^2\psi_r'^2 \ \mathrm{d}z_e \ \mathrm{d}z_h \ \mathrm{d}x \ \mathrm{d}y \tag{6.33}$$

Adopting a similar argument as above, the integral over the plane can then be written as:

$$G(a) = \int \left| \frac{\partial \psi_r}{\partial z_e} \right|^2 \ \mathrm{d}x \ \mathrm{d}y \tag{6.34}$$

and hence equation (6.33) gives the final expression for $\mathcal{A}$ as follows:

$$\mathcal{A} = E_e \mathcal{D} + \frac{\hbar^2}{2m_e^*} \int_0^{\nu-\mu} p(a)G(a) \ da \tag{6.35}$$

An analogous expression exists for $\mathcal{B}$ with the subscripts e and h, labelling electron and hole, respectively, being interchanged.

Now consider $\mathcal{C} = \langle \Psi | \mathcal{H}_{e-h} | \Psi \rangle$ as defined in equation (6.24). As mentioned earlier, $\mathcal{H}_{e-h}$ is composed of two terms which represent the in-plane kinetic energy of the relative motion and the Coulombic potential energy between the electron and hole, i.e.

$$\mathcal{C} = \mathcal{C}_T + \mathcal{C}_V \tag{6.36}$$

By using the kinetic energy operator defined above in equation (6.11) then the first of these terms can then be written as:

$$\mathcal{C}_T = \int \psi_e^*(z_e)\psi_h^*(z_h)\psi_r^* \frac{-\hbar^2}{2\mu_\perp} \left(\frac{\partial^2}{\partial x^2} + \frac{\partial^2}{\partial y^2} \right) \psi_e(z_e)\psi_h(z_h)\psi_r \ dz_e \ dz_h \ dx \ dy \tag{6.37}$$

where the in-plane reduced mass $\mu_\perp$ is given by:

$$\frac{1}{\mu_\perp} = \frac{1}{m_e^\perp} + \frac{1}{m_h^\perp} \tag{6.38}$$

Again, whatever the functional form of the relative motion term ψ_r, it is apparent that the kinetic energy operator in $\mathcal{C}_T$ will act only upon ψ_r, as the one-particle electron and hole wave functions, ψ_e and ψ_h, respectively, are not functions of x or y. Furthermore, the integration over the plane will result in an entity that is a function of a only, i.e. in the spirit of equation (6.35):

$$\mathcal{C}_T = -\frac{\hbar^2}{2\mu_\perp} \int_0^{\nu-\mu} p(a)J(a) \ da \tag{6.39}$$

where

$$J(a) = \int \psi_r \left(\frac{\partial^2}{\partial x^2} + \frac{\partial^2}{\partial y^2} \right) \psi_r \ dx \ dy \tag{6.40}$$

since ψ_r is real. In a similar manner, the potential energy term $\mathcal{C}_V$ can be written as:

$$\mathcal{C}_V = -\frac{e^2}{4\pi\epsilon} \int_0^{\nu-\mu} p(a)K(a) \ da \tag{6.41}$$

where

$$K(a) = \int \frac{1}{r}\psi_r^2 \ dx \ dy \tag{6.42}$$

with $r^2 = x^2 + y^2 + a^2$.

To summarize then, the total exciton energy can be written as:

$$E = \frac{\mathcal{A} + \mathcal{B} + \mathcal{C}}{\mathcal{D}} \tag{6.43}$$

with $\mathcal{A}$ given in equation (6.35), $\mathcal{B}$ following analogously, $\mathcal{C}$ given by equation (6.36) and $\mathcal{D}$ by equation (6.22). Evaluation of the entities, $F(a)$, $G(a)$, $J(a)$ and $K(a)$, necessitates the introduction of a specific form for the relative motion term ψ_r. The exciton binding energy E_{X^0} follows simply from:

$$E = E_e + E_h + E_{X^0} \tag{6.44}$$

6.4   1S EXCITON

In line with other work in the literature, the choice of wave function representing the electron–hole interaction that is to be employed is a hydrogenic type, given by:

$$\psi_r = \exp\left(-\frac{r'}{\lambda}\right) \tag{6.45}$$

where the Bohr radius λ will be used as a parameter and systematically varied in order to minimize the total energy E of the system (which is equivalent to maximizing the exciton binding energy E_{X^0}). However, where this work differs is in the precise form of the choice of the relative coordinate r'. Using the knowledge and experience derived from the considerations of impurities in Chapter 5, a variable symmetry-type relative motion term is chosen:

$$r'^2 = (x_e - x_h)^2 + (y_e - y_h)^2 + \zeta^2 (z_e - z_h)^2 \tag{6.46}$$

or by using $a = |z_e - z_h|$ as above, then:

$$r'^2 = r_\perp^2 + \zeta^2 a^2 \tag{6.47}$$

The second variational parameter, ζ, allows the exciton to assume any shape of wave function that is energetically favourable. Traditionally, the case with $\zeta = 0$ has become known as the two-dimensional exciton [134–138], and $\zeta = 1$, as the three-dimensional exciton [134, 139]. Cases where ζ is allowed to take values other than 0 and 1 are rarely found in the literature because of the increased complexity in the resulting mathematics of the problem [140–143].

In the present approach, ζ will be allowed to take all values from 0 upwards; however, due to complications in the mathematics, in particular deriving the limits of integration for the evaluation of $K(a)$ (see later), it is necessary to introduce the following transformations:

$$\text{for} \ \ 0 \leq \zeta \leq 1, \ \ \zeta^2 \ = \ 1 - \beta^2, \ \ \text{where} \ \ 0 \leq \beta \leq 1 \tag{6.48}$$
$$\text{for} \ \ 1 \leq \zeta \leq \infty, \ \ \zeta^2 \ = \ 1 + \eta^2, \ \ \text{where} \ \ 0 \leq \eta \leq \infty \tag{6.49}$$

In order to illustrate the technique, attention will be focused initially on the former of the two cases. First, consider evaluation of the entity $F(a)$ as defined in equation (6.18); using this particular choice of ψ_r and moving into plane polar coordinates gives:

$$F(a) = 2\pi \int_0^\infty \exp\left(-\frac{2r'}{\lambda}\right) r_\perp \ dr_\perp \tag{6.50}$$

Using equation (6.47) to substitute $r_\perp$ with r', noting that $r'dr' = r_\perp dr_\perp$, and accounting for the change in limits, i.e. when:

$$r_\perp = 0, \ \ r' = \ \sqrt{1 - \beta^2}a \tag{6.51}$$
$$r_\perp = \infty, \ \ r' = \ \infty \tag{6.52}$$

then obtain:

$$F(a) = 2\pi \int_{\sqrt{1-\beta^2}a}^\infty \exp\left(-\frac{2r'}{\lambda}\right) r' \ dr' \tag{6.53}$$

and by integrating by parts:

$$F(a) = 2\pi \left(\frac{\lambda\sqrt{1-\beta^2}a}{2} + \frac{\lambda^2}{4} \right) \exp\left(-\frac{2\sqrt{1-\beta^2}a}{\lambda} \right) \tag{6.54}$$

Next consider $G(a)$ as defined in equation (6.34), then:

$$G(a) = 2\pi \int_0^\infty \left| \frac{\partial}{\partial z_e} \exp\left(-\frac{r'}{\lambda} \right) \right|^2 r_\perp \, dr_\perp \tag{6.55}$$

and so performing the differentiation and substituting r' for $r_\perp$ as above, then:

$$G(a) = 2\pi \int_{\sqrt{1-\beta^2}a}^\infty \left| -\frac{1}{\lambda} \exp\left(-\frac{r'}{\lambda} \right) \frac{(1-\beta^2)a}{r'} \right|^2 r' \, dr' \tag{6.56}$$

Now let $r' = \sqrt{1-\beta^2}a\cosh\theta$, then:

$$G(a) = 2\pi \int_0^\infty \frac{1}{\lambda^2}(1-\beta^2)^2 a^2 \exp\left(-\frac{2\sqrt{1-\beta^2}a\cosh\theta}{\lambda} \right) \frac{\sqrt{1-\beta^2}a\sinh\theta}{\sqrt{1-\beta^2}a\cosh\theta} \, d\theta \tag{6.57}$$

Making the further substitution, $w = \exp(-\theta)$, then $d\theta = -dw/w$ and noting that:

$$\cosh\theta = \frac{1}{2}\left(e^\theta + e^{-\theta} \right) = \frac{1}{2}\left(\frac{1}{w} + w \right), \quad \text{and} \quad \sinh\theta = \frac{1}{2}\left(\frac{1}{w} - w \right) \tag{6.58}$$

then finally:

$$G(a) = 2\pi \int_0^1 \frac{(1-\beta^2)^2 a^2}{\lambda^2} \exp\left[-\frac{\sqrt{1-\beta^2}a}{\lambda}\left(\frac{1}{w} + w \right) \right] \left(\frac{1-w^2}{w(1+w^2)} \right) \, dw \tag{6.59}$$

An analogous argument shows that $G(a)$ appearing in $\mathcal{B}$ is exactly equal to this form.

Next consider evaluation of the integral $J(a)$ as defined in equation (6.40). With this aim, note that:

$$\frac{\partial r'}{\partial x} = \frac{x}{r'}, \quad \text{and} \quad \frac{\partial^2 r'}{\partial x^2} = \frac{1}{r'} - \frac{x^2}{r'^3} \tag{6.60}$$

This then gives the following:

$$\frac{\partial^2 \psi_r}{\partial x^2} = \left(\frac{-1}{r'\lambda} + \frac{x^2}{r'^3\lambda} + \frac{x^2}{r'^2\lambda^2} \right) \psi_r \tag{6.61}$$

and hence:

$$\left(\frac{\partial^2}{\partial x^2} + \frac{\partial^2}{\partial y^2} \right) \psi_r = \left[\frac{1}{\lambda^2} - \frac{(1-\beta^2)a^2}{\lambda r'^3} - \frac{1}{r'\lambda} - \frac{(1-\beta^2)a^2}{r'^2\lambda^2} \right] \psi_r \tag{6.62}$$

Therefore, moving from Cartesian into plane polar coordinates:

$$J(a) = 2\pi \int_0^\infty \left[\frac{1}{\lambda^2} - \frac{(1-\beta^2)a^2}{\lambda r'^3} - \frac{1}{r'\lambda} - \frac{(1-\beta^2)a^2}{r'^2\lambda^2} \right] \exp\left(-\frac{2r'}{\lambda} \right) r_\perp \, dr_\perp \tag{6.63}$$

It is standard practice in the literature [144] to expand the expressions involving $(r')^{-n}$ as a power series in $r_\perp$ and a, and then to perform the integration numerically. This involves summing over a series of terms, each of which must be integrated over a range from 0 to ∞; this procedure, however, can be avoided.

Writing equation (6.63) as $J(a) = J_1 + J_2 + J_3 + J_4$, where J_i represents the first, second, etc, terms, respectively, then:

$$J_1 + J_3 = 2\pi \int_0^\infty \left[\frac{1}{\lambda^2} - \frac{1}{\lambda r'} \right] \exp\left(-\frac{2r'}{\lambda} \right) r_\perp \, dr_\perp \tag{6.64}$$

Again, substituting r' for $r_\perp$:

$$J_1 + J_3 = 2\pi \int_{\sqrt{1-\beta^2}a}^\infty \left[\frac{1}{\lambda^2} - \frac{1}{\lambda r'} \right] \exp\left(-\frac{2r'}{\lambda} \right) r' \, dr' \tag{6.65}$$

and therefore:

$$J_1 + J_3 = 2\pi \left\{ \left[\frac{1}{2} \exp\left(-\frac{2r'}{\lambda} \right) - \frac{r'}{2\lambda} \exp\left(-\frac{2r'}{\lambda} \right) \right]_{\sqrt{1-\beta^2}a}^\infty \right.$$

$$\left. + \frac{1}{2\lambda} \int_{\sqrt{1-\beta^2}a}^\infty \exp\left(-\frac{2r'}{\lambda} \right) dr' \right\} \tag{6.66}$$

which gives:

$$J_1 + J_3 = 2\pi \left(\frac{\sqrt{1-\beta^2}a}{2\lambda} - \frac{1}{4} \right) \exp\left(-\frac{2\sqrt{1-\beta^2}a}{\lambda} \right) \tag{6.67}$$

The two remaining terms of equation (6.63) give:

$$J_2 + J_4 = 2\pi \int_0^\infty \left[-\frac{(1-\beta^2)a^2}{\lambda r'^3} - \frac{(1-\beta^2)a^2}{\lambda^2 r'^2} \right] \exp\left(-\frac{2r'}{\lambda} \right) r_\perp \, dr_\perp \tag{6.68}$$

Again substituting r' for $r_\perp$, then:

$$J_2 + J_4 = 2\pi \int_{\sqrt{1-\beta^2}a}^\infty (1-\beta^2)a^2 \left(-\frac{1}{\lambda r'^2} - \frac{1}{\lambda^2 r'} \right) \exp\left(-\frac{2r'}{\lambda} \right) dr' \tag{6.69}$$

Making a further scale change of $r' = \sqrt{1-\beta^2}a\cosh\theta$, then:

$$J_2 + J_4 = 2\pi \int_0^\infty \left(-\frac{1}{\lambda\cosh^2\theta} - \frac{\sqrt{1-\beta^2}a}{\lambda^2\cosh\theta} \right) \exp\left(-\frac{2\sqrt{1-\beta^2}a\cosh\theta}{\lambda} \right)$$

$$\times \sqrt{1-\beta^2}a\sinh\theta \, d\theta \tag{6.70}$$

With the final substitution of $w = \exp -\theta$, then $d\theta = -dw/w$, therefore:

$$J_2 + J_4 = 2\pi \int_0^1 \left[-\frac{1}{\frac{\lambda}{4}\left(\frac{1}{w}+w\right)^2} - \frac{\sqrt{1-\beta^2}a}{\frac{\lambda^2}{2}\left(\frac{1}{w}+w\right)} \right]$$

$$\times \exp\left[-\frac{\sqrt{1-\beta^2}a}{\lambda}\left(\frac{1}{w}+w\right)\right]\sqrt{1-\beta^2}\frac{a}{2}\left(\frac{1}{w^2}-1\right)\,dw \tag{6.71}$$

This last equation illustrates the advantage of the present formalism, namely that the computationally difficult integral of equation (6.63), which has hitherto been expanded into a infinite series and integrated to infinity, has been replaced with a simple integral over the range from 0 to 1. Even if the integrand had a finite number of singularities, this would still pose no problem in its evaluation.

Finally consider $K(a)$, as defined in equation (6.42), i.e.

$$K(a) = 2\pi \int_0^\infty \frac{1}{r}\exp\left(-\frac{2r'}{\lambda}\right)\,r_\perp\,dr_\perp \tag{6.72}$$

Recalling that $r^2 = r_\perp^2 + a^2$, then:

$$K(a) = 2\pi \int_a^\infty \exp\left(-\frac{2r'}{\lambda}\right)\,dr \tag{6.73}$$

The form of r', i.e. $r'^2 = r^2 - \beta^2 a^2$, suggests the substitution $r = \beta a\cosh\theta$, which gives:

$$K(a) = 2\pi \int_{\cosh^{-1}\frac{1}{\beta}}^\infty \beta a\sinh\theta\exp\left(-\frac{2\beta a\sinh\theta}{\lambda}\right)\,d\theta \tag{6.74}$$

Again making use of the substitution $w = \exp-\theta$, this then necessitates evaluating w corresponding to $\theta = \cosh^{-1}\frac{1}{\beta}$, i.e. $\cosh\theta = \frac{1}{\beta} = \frac{1}{2}(w+\frac{1}{w})$; this then yields the quadratic equation:

$$w^2 - \frac{2w}{\beta} + 1 = 0 \tag{6.75}$$

Since the product of the two roots of this equation is unity, one root must correspond to $\exp(-\theta)$ and the other to $\exp\theta$. It is readily ascertained that the $\exp(-\theta)$ root is as follows:

$$\exp(-\theta) = \frac{1}{\beta} - \frac{1}{\beta}\sqrt{1-\beta^2} \tag{6.76}$$

This follows since the limits on β are 0 and 1. Hence:

$$K(a) = 2\pi\frac{\beta a}{2}\int_0^{\frac{1}{\beta}-\frac{1}{\beta}\sqrt{1-\beta^2}} \exp\left[-\frac{\beta a}{\lambda}\left(\frac{1}{w}-w\right)\right]\left(\frac{1}{w^2}-1\right)\,dw \tag{6.77}$$

In a similar manner, the second form of ψ_r, with $r' = \sqrt{r_\perp^2 + (1+\eta^2)a^2}$, gives the same expressions for $F(a)$, $G(a)$ and $J(a)$ as above, but with the simple substitution, $1+\eta^2$, in place of $1-\beta^2$. Only the expression for the last of the 'a' functions differs, in particular:

$$K(a) = 2\pi\frac{\eta a}{2}\int_0^{\frac{1}{\eta}\sqrt{1+\eta^2}-\frac{1}{\eta}} \exp\left[-\frac{\eta a}{\lambda}\left(\frac{1}{w}+w\right)\right]\left(\frac{1}{w^2}+1\right)\,dw \tag{6.78}$$

6.5 THE TWO-DIMENSIONAL AND THREE-DIMENSIONAL LIMITS

It is *always* worthwhile performing convergence tests, i.e. taking the theoretical-computational model to established, often analytical, limits. The idea is to increase confidence in the theory and, as ever, to demonstrate that the previous theories are limits of the new. For example, classical mechanics is recovered from relativistic mechanics, in the low-velocity limit.

Although not as grand an example, there exist two limits which the analysis above can be compared with. In the limit of very wide quantum wells, the exciton should look like a bulk exciton, in both its binding energy and Bohr radius. In addition, in the limit of very narrow wells, the exciton should become two-dimensional in nature.

The bulk, or three-dimensional limiting case of hydrogenic two-body systems, such as impurities and excitons, has been discussed and used already. A transparent treatise of this Bohr model of the hydrogen atom is given by Weidner and Sells [4]. This approach can easily be adapted to the case of an electron orbiting a positively charged central infinite mass, with the orbit restricted to a single plane, i.e. it is two-dimensional (2D). The Schrödinger equation is then written as:

$$
-\frac{\hbar^2}{2m}\left(\frac{\partial^2 \psi}{\partial x^2} + \frac{\partial^2 \psi}{\partial y^2}\right) - \frac{e^2}{4\pi\epsilon_r\epsilon_0 r_\perp}\psi = E\psi \tag{6.79}
$$

where the electron–proton separation

$$
r_\perp = \sqrt{x^2 + y^2} \tag{6.80}
$$

For solution, the Cartesian coordinates need to be converted into the plane polar coordinate $r_\perp$, with this aim, note:

$$
\frac{\partial \psi}{\partial x} = \frac{\partial \psi}{\partial r_\perp}\frac{\partial r_\perp}{\partial x} = \frac{x}{r_\perp}\frac{\partial \psi}{\partial r_\perp} \tag{6.81}
$$

and so:

$$
\frac{\partial^2 \psi}{\partial x^2} = \frac{\partial}{\partial x}\left(\frac{x}{r_\perp}\frac{\partial \psi}{\partial r_\perp}\right) = \frac{1}{r_\perp}\frac{\partial \psi}{\partial r_\perp} - \frac{x}{r_\perp^2}\frac{\partial r_\perp}{\partial x}\frac{\partial \psi}{\partial r_\perp} + \frac{x}{r_\perp}\frac{\partial^2 \psi}{\partial r_\perp^2}\frac{\partial r_\perp}{\partial x} \tag{6.82}
$$

$$
\therefore \frac{\partial^2 \psi}{\partial x^2} = \frac{1}{r_\perp}\frac{\partial \psi}{\partial r_\perp} - \frac{x^2}{r_\perp^3}\frac{\partial \psi}{\partial r_\perp} + \frac{x^2}{r_\perp^2}\frac{\partial^2 \psi}{\partial r_\perp^2} \tag{6.83}
$$

Thus:

$$
\frac{\partial^2 \psi}{\partial x^2} + \frac{\partial^2 \psi}{\partial y^2} = \frac{2}{r_\perp}\frac{\partial \psi}{\partial r_\perp} - \frac{x^2 + y^2}{r_\perp^3}\frac{\partial \psi}{\partial r_\perp} + \frac{x^2 + y^2}{r_\perp^2}\frac{\partial^2 \psi}{\partial r_\perp^2} \tag{6.84}
$$

$$
\therefore \frac{\partial^2 \psi}{\partial x^2} + \frac{\partial^2 \psi}{\partial y^2} = \frac{1}{r_\perp}\frac{\partial \psi}{\partial r_\perp} + \frac{\partial^2 \psi}{\partial r_\perp^2} \tag{6.85}
$$

Therefore the Schrödinger equation, (equation 6.79), in plane polar coordinates becomes:

$$
\frac{1}{r_\perp}\frac{\partial \psi}{\partial r_\perp} + \frac{\partial^2 \psi}{\partial r_\perp^2} + \frac{2m}{\hbar^2}\left(E + \frac{e^2}{4\pi\epsilon_r\epsilon_0 r_\perp}\right)\psi = 0 \tag{6.86}
$$

The standard technique for solving the hydrogen atom is to choose a spherically symmetric wave function of the form:

$$\psi = \exp\left(-\frac{r}{\lambda_{3D}}\right) \tag{6.87}$$

In analogy to this, the 2D equivalent is:

$$\psi = \exp\left(-\frac{r_\perp}{\lambda_{2D}}\right) \tag{6.88}$$

By calculating the derivatives:

$$\frac{\partial\psi}{\partial r_\perp} = -\frac{1}{\lambda_{2D}}\psi \quad \text{and} \quad \frac{\partial^2\psi}{\partial r_\perp^2} = \frac{1}{\lambda_{2D}^2}\psi \tag{6.89}$$

and substituting into equation (6.86), then obtain:

$$\left(\frac{1}{\lambda_{2D}^2} + \frac{2mE}{\hbar^2}\right)\psi + \frac{1}{r_\perp}\left(\frac{2me^2}{\hbar^2 4\pi\epsilon_r\epsilon_0} - \frac{1}{\lambda}\right)\psi = 0 \tag{6.90}$$

The standard argument is that as this equation must be valid for all values of $r_\perp$, then

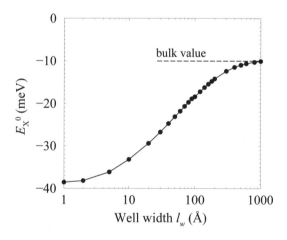

Figure 6.4   Exciton binding energy in an infinitely deep quantum well

divergence as $r_\perp \to 0$ can only be avoided if the second term is zero; this implies:

$$\lambda_{2D} = \frac{4\pi\epsilon_r\epsilon_0\hbar^2}{2me^2} = \frac{\lambda_{3D}}{2} \tag{6.91}$$

i.e. the Bohr radius of the 2D (planar) two-body system, is half that of the 3D (spherical) system. Using this form for the Bohr radius, the first term then yields the energy of the 2D system as follows:

$$E^{2D} = -4\frac{me^4}{32\pi^2\hbar^2\epsilon_r^2\epsilon_0^2} = 4E^{3D} \tag{6.92}$$

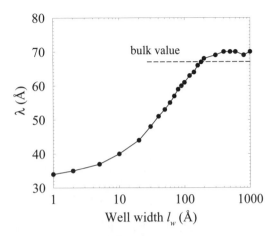

Figure 6.5 Bohr radius in an infinitely deep quantum well

As mentioned above, and as already utilized for neutral impurities in Chapter 5, the 3D (or bulk) limit can be approached by increasing the width of a *finite* quantum well. However, the finite depth of such a quantum well is not conducive to the 2D limit, for as the well width decreases, the one-particle electron and hole wave functions begin to 'spill out' over the top of the well. The 2D limit can only be approached hypothetically in an infinitely deep quantum well. Fig. 6.4 illustrates the results of calculations of the exciton binding energy as a function of the width of an infinitely deep CdTe quantum well. The magnitude of the bulk exciton binding energy was calculated above in Section 6.1 as 10.1 meV. The negative values on the graph illustrate that it is indeed a bound state.

Clearly, both limits are obeyed, i.e.

$$\lim_{l_w \to \infty} E_{X^0} = E_{X^0}^{3D} \quad \text{and} \quad \lim_{l_w \to 0} E_{X^0} = 4E_{X^0}^{3D} \tag{6.93}$$

Fig. 6.5 displays the corresponding Bohr radii λ for the energies of Fig. 6.4. Remembering that the Bohr radius in bulk, $\lambda_{3D} = 67$ Å, then the 2D limit, i.e.

$$\lim_{l_w \to 0} \lambda = \frac{\lambda_{3D}}{2} \tag{6.94}$$

is satisfied. The 3D limit *is* obeyed, although the data on the graph show a slight scatter around the bulk radii of 67 Å. The source of this discrepancy is numerical accuracy. At the larger well widths, the wave function needs to be known at many points in order to calculate the binding energy to very high tolerances (thus leading to long computational times), as required here in these convergence tests. Despite this high accuracy, the Bohr radius still shows some deviation from the bulk, although greater numerical accuracy has reduced this towards the value of 67 Å.

The calculations also showed that at the very narrow (5–40 Å) well widths, the second parameter ζ, representing the symmetry of the relative motion term was close to zero, i.e. a 2D wave function. As the width increased, ζ increased steadily towards unity,

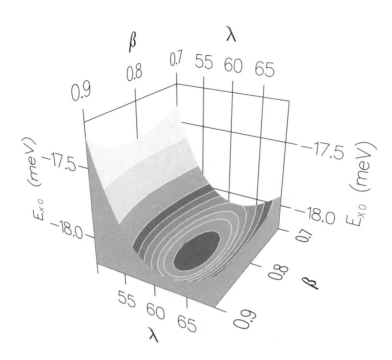

Figure 6.6   The exciton binding energy E_{X^0} in the two-dimensional λ–β parameter space

i.e. the 3D (spherical or bulk-like) exciton. This point that ζ generally lies in the limits $0 < \zeta < 1$ will be reinforced further by the use of several different example systems in the following discussions. For this reason, discussion of the symmetry of the wave function will use ζ and $\beta = \sqrt{1 - \zeta^2}$ interchangeably.

Fig. 6.6 illustrates the variational technique in the case of the infinite well of width l_w=100 Å. It can be seen that there is just one minimum in the exciton energy in the two-dimensional λ–β space. This is almost always the case. However, there is an exception to this rule [131]. In double quantum well systems where the one-particle electron wave function has a significant component in each of the wells, there exist two local energy minima within λ–β space. This implies that there are two possible exciton states which can be formed from the same one-particle wave functions. The two states can be thought of as originating from the localized hole binding separately with both of the 'lobes' of the electron wave function.

It is worthwhile just looking at the form of the function $p(a)$, which plays a significant rôle in the theoretical analysis. As mentioned at the point of its introduction, $p(a)$ represents the *uncorrelated* (i.e. calculated without taking account of the electron–hole Coulombic interaction) probability of finding the electron and hole separated by a distance a along the growth (z-) axis. While $p(a)$ is not a physical observable, often the form of it can help in understanding and provide an insight into the nature of the exciton. One particular example of that would be in the 'twin' exciton states of the double quantum well discussed above. Fig. 6.7 shows $p(a)$ for this simple case of

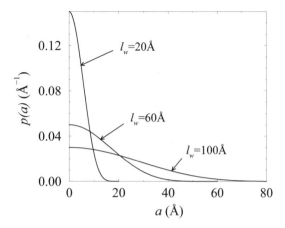

Figure 6.7   The uncorrelated probability $p(a)$ of finding the electron and hole separated by a distance a in the infinite quantum well, shown for different well widths

the infinite quantum well, while later figures will compare it with other more complex heterostructures.

In conclusion, the validity of the technique presented above has been substantiated by the calculation of exciton binding energies and wave functions for the full width range of infinite quantum wells. In the sections that follow, examples of calculations will be presented that demonstrate the versatility of the method. In addition, some of the results are expected to be of interest in their own right.

6.6   EXCITONS IN SINGLE QUANTUM WELLS

Only in a very few circumstances are the band offsets in both the conduction and valence bands large enough that the system can be approximated with an infinite quantum well model. The vast majority of cases demand finite potentials. Reiterating, the approach developed above depends only upon the one-particle electron and hole wave functions, ψ_e and ψ_h, respectively, and no further knowledge of the system is required (except for the basic material parameters of electron and hole effective masses and dielectric constants).

To illustrate this, Figs. 6.8–6.10 display the results of calculations of exciton binding energies in finite GaAs single quantum wells surrounded by $Ga_{1-x}Al_xAs$ barriers. Unlike the infinite well case, the exciton binding energy E_{X^0} is a non-monotonic function of well width. As in the impurity binding energy case of Chapter 5, this is due to the effect of the well width on the electron and hole confinement. At very narrow well widths, the one-particle states are 'squeezed' up the well to reside at energies just below the top of the barrier. The wave function tends to 'spill' over the top, thus leading to a reduced probability of the particle being within the region of the quantum well. This non-monotonic behaviour has been observed in experiment [145]. In addition, increasing Al concentration in the barrier (both the conduction and valence band offsets

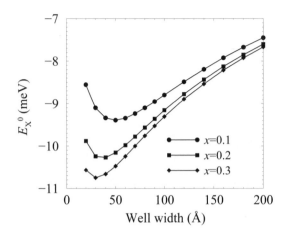

Figure 6.8   The exciton binding energy E_{X^0} in a GaAs quantum well surrounded by $Ga_{1-x}Al_xAs$ barriers

are proportional to x) leads to an increase in the magnitude of the exciton binding energy for *all* well widths. However, there are two points to note regarding this. First, the effect of the barrier height is reduced at larger well widths. This substantiates the fact that a particular choice of barrier height (infinity) was chosen to illustrate the 3D limit in the previous section. It is clear from Fig. 6.8 that the exciton binding energy is tending towards its bulk value of 4.7 meV for *all* barrier heights. Secondly, the effect of an increasing barrier height is largest at smaller well widths, and for any given well width, further increases in the barrier height lead to smaller increases in the binding energy.

Figs 6.9 and 6.10 summarize the excitonic wave function corresponding to the minimized variational energies of Fig. 6.8. The binding energy is essentially calculated over a range of discrete values of both λ and (in this case) β, hence the accuracy of these parameters is known only to the resolution of the mesh. The most important result of the variational calculation is the energy, which around the minima is a weak function of the parameters, and thus even a coarse mesh would have little consequences for the binding energy; Fig. 6.6 illustrates this nicely. The minima can be determined to 0.001 meV with a λ resolution of 1 Å and a β resolution of 0.01. The symbols in Fig. 6.9 are drawn open and of a size in order to illustrate the accuracy in the determination of the Bohr radius λ. In many ways the functional dependencies of λ on well width and barrier height mirror that of the binding energy E_{X^0} —it is a non-monotonic function of well width, decreases with increasing barrier height and tends towards its bulk value of 115 Å at large well widths.

The symmetry parameter is displayed in Fig. 6.10 as both β (as calculated) and as its transformation into ζ ($= \sqrt{1 - \beta^2}$). Recalling that the trial wave function of the

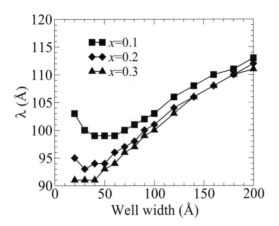

Figure 6.9   The exciton Bohr radius λ in a GaAs quantum well surrounded by $Ga_{1-x}Al_xAs$ barriers

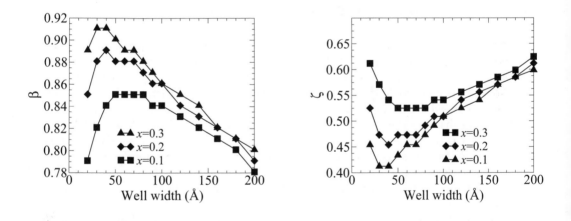

Figure 6.10   The symmetry parameters β and ζ ($= \sqrt{1 - \beta^2}$) in a GaAs quantum well surrounded by $Ga_{1-x}Al_xAs$ barriers

relative motion is given by:

$$\psi_r = \exp\left[-\frac{\sqrt{r_\perp^2 + \zeta^2(z_e - z_h)^2}}{\lambda}\right]$$

(6.95)

then $\zeta \sim 0.5$ implies that:

$$\psi_r \sim \exp\left[-\frac{\sqrt{r_\perp^2 + \frac{(z_e - z_h)^2}{4}}}{\lambda}\right] \tag{6.96}$$

which is roughly midway between the two-dimensional case, i.e.

$$\psi_r \sim \exp\left(-\frac{r_\perp}{\lambda}\right) \tag{6.97}$$

and the three-dimensional case:

$$\psi_r \sim \exp\left[-\frac{\sqrt{r_\perp^2 + (z_e - z_h)^2}}{\lambda}\right] \tag{6.98}$$

While the implications for the binding energy, by using either one of the simpler wave functions above, might be relatively small, i.e. probably less than 1 meV, the fact remains that excitons in quantum wells do have mixed symmetry. In this example the ζ value implies that 'equi-surfaces' of ψ_r are prolate spheroids, i.e. they look like rugby footballs (American footballs), with the major axis along the growth (z-) direction of the heterostructure.

6.7   EXCITONS IN MULTIPLE QUANTUM WELLS

Semiconductor heterostructures are often more complex than single quantum wells. Perhaps the next stage in complexity of design is to incorporate many identical wells within the same grown layer. As discussed previously, if the barriers separating the wells are such that the one-particle wave functions in each well overlap, then the system is called a superlattice, while if they do not overlap, it is called a multiple quantum well system.

Fig. 6.11 displays the electron and heavy-hole one-particle wave functions for a 5-period 50 Å GaAs/50 Å $Ga_{0.9}Al_{0.1}As$ multiple quantum well. Clearly, this system is a finite superlattice. The form of the wave functions in these finite superlattices has been discussed earlier in Chapter 3; however, a simple comparison of the two, indicates that the heavy-hole wave function is more localized than that of the electron on account of its larger mass.

Although it serves no *real* purpose, it is interesting to compare the functional form of $p(a)$ with that of the infinite well of earlier. The troughs in Fig. 6.12 indicate that there are electron–hole separations, which are less favoured than others. These troughs at 50, 150 and 250 Å correspond to separations where one particle would be located in a well and the other in a barrier. As in the infinite well case, the electron–hole separation (a) with the highest probability is zero.

The exciton binding energy is displayed in Fig. 6.13 for a series of 5-period multiple quantum wells as a function of the equal well (l_w) and barrier (l_b) widths. In contrast to the finite well, the magnitude of E_{X^0} passes through a *minimum*, which in this material system is at around 50–60 Å, before increasing again. In the GaAs-$Ga_{1-x}Al_xAs$ material system here, larger well and barrier widths reduced the interaction between the states in adjacent wells to such an extent that the separation between the symmetric ground

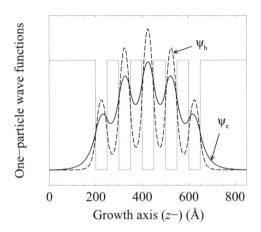

Figure 6.11   The electron and heavy-hole one-particle wave functions in a 5-period 50 Å GaAs/50 Å $Ga_{0.9}Al_{0.1}As$ multiple quantum well

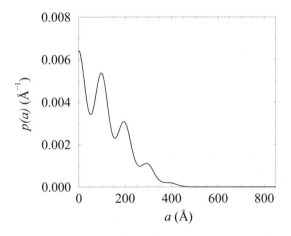

Figure 6.12 $p(a)$ for the 5 period GaAs-$Ga_{0.9}Al_{0.1}As$ multiple quantum well

state (as displayed in Fig. 6.11) and the first excited state was less than 10^{-6} meV, and could not be resolved; hence the truncation of the data at 60 and 80 Å. Such calculations have been performed at larger well and barrier widths in the CdTe-$Cd_{1-x}Mn_xTe$ system [131], where it was shown that beyond the minimum in the magnitude of E_{X^0} there is a maximum and then the binding energy of the multiple quantum well tends towards that of the finite well, as would be expected.

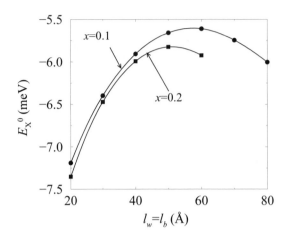

Figure 6.13   The exciton binding energy E_{X^0} as a function of well width in a 5-period GaAs-Ga$_{1-x}$Al$_x$As multiple quantum well

6.8   STARK LADDERS

As discussed previously (Chapter 3, in the context of unipolar (single-charge-carrier)), systems, when an electric field is applied along the growth (z-) axis of a (finite or infinite) superlattice, the eigenstates which previously extended over the whole system, begin to localize. As the field increases, the miniband breaks up and a localized state forms in each quantum well. At this point, the energy separation between the states in adjacent wells is proportional to the electric field and is given simply by the difference in potential energy of the well centres, i.e.

$$E_{n+1} - E_n = eFL \tag{6.99}$$

where F is the electric field strength and L is the superlattice period. The conduction and valence band states thus form a series of steps similar to a 'ladder'.

In a bipolar system the hole wave functions tend to localize more rapidly than those of the electron's, hence a regime exists where a hole wave function centred in one particular well has a significant overlap with the electron wave functions from adjacent wells, in addition to its own. Therefore, the photoluminescence emission and absorption spectra show a series of lines, which represent exciton transitions from the hole in well n with electrons in well n, and in wells $n-1$ and $n+1$, as displayed in Fig. 6.14.

Given that the electron and hole energy separations between adjacent wells are proportional to the field, then the electron–hole separation (ignoring the exciton binding energy for now) will also be proportional to the field, as displayed in Fig. 6.15 (left). The addition of the exciton binding energy in this simplistic explanation complicates the situation slightly (see Fig. 6.15 (right)), but by the time that moderate electric fields are present, the spectral lines are equally spaced and radiating from the zero-field point. Mendez *et al.* [146] have observed very similar behaviour in a related system, namely a 30 Å GaAs/35 Å Ga$_{0.65}$Al$_{0.35}$As superlattice.

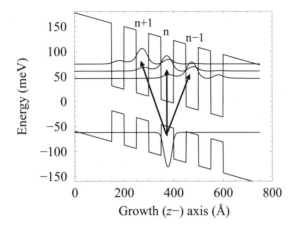

Figure 6.14 Illustration of the field (14 kVcm^{-1})-localized heavy-hole state recombining with the three closest electron states, which characterize the Stark ladder

Fig. 6.15 (right), however, does not tell the whole story. While it does indicate the position of the spectral lines, it does not give any information about their intensities. The oscillator strength (defined as follows) gives a measure of the expected intensities of the photoluminescence emission and photoluminescence excitation absorption lines:

$$O_s \propto \frac{\langle \psi_e | \psi_h \rangle^2}{\langle \Psi | \Psi \rangle} = \frac{O^2}{\mathcal{D}} \tag{6.100}$$

Fig. 6.16 displays the oscillator strength as a function of the applied electric field for the three excitonic transitions of interest. At zero field, all eigenstates have definite parity and hence the $|e_{n+1}\rangle \rightarrow |h_n\rangle$ and $|e_{n-1}\rangle \rightarrow |h_n\rangle$ transitions are forbidden; this result

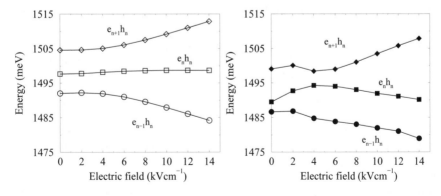

Figure 6.15 The excitonic transition energies characteristic of a Stark Ladder; (left) without exciton binding energies; (right) with exciton binding energies

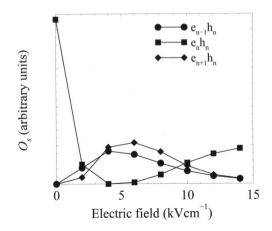

Figure 6.16   The oscillator strength O_s as function of electric field, for three of the Stark ladder transitions

arises naturally from the analysis and is demonstrated numerically in the figure by a zero oscillator strength. As the field is increased, the system passes through the region of 'miniband breakup', with the effect of the symmetry breaking induced by the field being to allow the previously forbidden transitions. At larger fields still, the one-particle wave functions will tend towards complete localization in a single well, and hence the oscillator strength of the inter-well transitions is expected (and confirmed by the calculations) to peak and then reduce. Meanwhile, the intra-well exciton transition $|e_n\rangle \rightarrow |h_n\rangle$ initially has the largest oscillator strength (obviously), which then temporarily decreases at low fields, before increasing steadily to dominate in the Stark ladder regime.

To date, no excitonic states have been calculated with the variable symmetry parameter $\zeta > 1$. However, for the work discussed in this present section, some of the Stark ladder excitons have β values tending to zero, i.e. $\zeta = 1$.

6.9   SELF-CONSISTENT EFFECTS

In the analysis put forward in this present chapter and for the majority of the relevant literature, the exciton binding energy has been calculated by using a variational approach, based on the choice of trial wave function of the following form:

$$\Psi = \psi_e(z_e)\psi_h(z_h)\psi_r(x, y, z_e - z_h) \tag{6.101}$$

where most importantly, ψ_e and ψ_h were simply the one-particle eigenstates of the one-dimensional Schrödinger equation of the heterostructure; this means that ψ_e was calculated without the presence of the hole and ψ_h was calculated without the presence of the electron.

Attempts to improve upon this began with the simple idea of the Coulombic-hole model. This has been implemented [147] in the dilute magnetic semiconductor system CdTe-Cd$_{1-x}$Mn$_x$Te (see Section 5.11.1), which is thought to have a magnetic-field-induced Type-I to Type-II transition. It is generally accepted that such corrections to the one-particle wave functions are more important in Type-II systems, as a well-confined electron may be able to 'suck in' the hole from the adjacent layer, thus having a large effect on the exciton binding energy.

A fully self-consistent exciton model has been proposed by Warnock *et al.* [148] and later extended to include the variable symmetry relative motion term of this analysis [149]. The self-consistent correction to the one-particle eigenfunctions, ψ_e and ψ_h, is incorporated by solving a new Schrödinger equation:

$$\left(\int_{\text{all space}} \psi_h^* \psi_r^* \mathcal{H} \psi_h \psi_r \ dz_h \ dx \ dy \right) \psi_e$$

$$= E \left(\int_{\text{all space}} \psi_h^* \psi_r^* \psi_h \psi_r \ dz_h \ dx \ dy \right) \psi_e \qquad (6.102)$$

where $\mathcal{H}$ is the total Hamiltonian describing the system, as in equation (6.7). This yields the first improved iteration to the electron wave function, $\psi_e^{(1)}$, say. The latter is then employed to improve the hole wave function, as follows:

$$\left(\int_{\text{all space}} \psi_e^{(1)*} \psi_r^* \mathcal{H} \psi_h \psi_r \ dz_h \ dx \ dy \right) \psi_h$$

$$= E \left(\int_{\text{all space}} \psi_e^{(1)*} \psi_r^* \psi_h \psi_r \ dz_h \ dx \ dy \right) \psi_h \qquad (6.103)$$

Following this, both $\psi_e^{(1)}$ and $\psi_h^{(1)}$ are utilized in the standard exciton-binding-energy calculation to give the new relative motion term, characterized entirely in terms of the parameters λ and ζ. The procedure is repeated until the total energy of the exciton is minimized and λ and ζ have converged.

Piorek *et al.* [149] demonstrated that for Type-I systems, a variable-symmetry relative motion term is *more important* than self-consistency with simple 2D or 3D wave functions.

The self-consistent correction proved itself to be important in flat-band (i.e. a zero offset in the valence band) and Type-II systems. The simple one-particle solutions are usually localized in different semiconductor layers, thus leading to a quite small exciton binding energy. However, the addition of self-consistency allowed the electron and hole eigenfunctions to move closer together, thus increasing the binding energy and, most importantly, reducing the total energy of the exciton.

6.10 SPONTANEOUS SYMMETRY BREAKING

The aim of this book is to discuss the theoretical methods necessary to analyse semiconductor heterostructures. In addition, the computational implementation of these methods is paramount. This present section gives an example of purely computational work giving rise to new science—something that could not have been predicted from theory.

In calculating the self-consistent exciton energy in series of multiple quantum wells, Piorek *et al.* [150] found that a 4-period multiple quantum well could have a larger exciton energy than the equivalent 3-period system. This seemed a contradiction as a 4-well system can be thought of as a 3-well system but with some of the potential barrier *removed*. Therefore it should have a lower exciton energy. This seemed to be a general result, i.e. that $2n$-period multiple quantum wells had a larger energy than $2n - 1$ systems. Inspection of the self-consistently iterated electron and hole wave functions, $\psi_e^{(m)*}$ and $\psi_h^{(m)*}$, respectively, showed that in systems with odd numbers of wells, the electron and hole localized in the central well leading to a large binding energy. However, in systems with an even number of wells, the electron and hole localized in the *two* central wells, thus giving a smaller binding energy, and so leading to a higher total energy.

However, for certain calculations, exceptions to the above rule were identified. It was found that if there was the slightest asymmetry in the original electron and hole wave functions, ψ_e and ψ_h, then the self-consistent iteration repeatedly increased it. Such asymmetries in ψ_e and ψ_h needed only to be minute in the first instance, and could arise merely from a coarse finite difference mesh; in fact, a difference in the wave function maxima of 1 part in 10^{10} was enough to have an effect. Given these initial conditions, the iteration forcibly broke the symmetry of the electron and hole, thus forcing them to localize in the second well of a four-well system, for example.

A systematic study of the original multiple quantum wells found that, if a very small asymmetry was introduced into the potential profile of the heterostructure (perhaps just one monolayer within the well, having an alloy concentration of 10^{-5} rather than zero), then it was always enough to spontaneously break the symmetry of the exciton [150], with the result being that the exciton energy *always* decreased as the number of wells in the system increased. A point to note here is the size of the asymmetry necessary in the potential to bring about localization. This is very small and will always be present in even the highest quality semiconductor layers. However, in addition, there are many other mechanisms which could bring about the small asymmetry, e.g. the presence of a single impurity atom, a phonon, or another charge carrier, will all be sufficient to induce a tiny perturbation in either the electron or the hole wave function, which will then influence the final exciton state. Once the exciton begins to form, it 'pulls itself up by its own bootstraps' until it is fully localized within one well. Therefore, in conclusion, excitons within multiple quantum well systems will (in general) spontaneous localize into just one of the wells.

It must be noted for the record, however, that the spontaneous symmetry-breaking of excitons in multiple quantum wells has recently been disputed in the literature [151].

6.11   2S EXCITON

In a similar manner to both the cases of the hydrogen atom and impurities, excitons can also have excited states. While the electron and hole one-particle states remain unchanged, the corresponding choice for the relative motion factor of the 2s excited state is given by:

$$\psi_r^{2s} = \left(1 - \frac{\alpha r'}{\lambda_{2s}}\right) \exp\left(-\frac{r'}{\lambda_{2s}}\right) \tag{6.104}$$

where λ_{2s} is stated specifically to imply the Bohr radius of the 2s state and α is a (as yet undefined) parameter, chosen to ensure orthogonality of the 1s and 2s eigenstates, say, for example, as follows:

$$\mathcal{E} = \left\langle \Psi^{1s} | \Psi^{2s} \right\rangle = 0 \tag{6.105}$$

In the spirit of the above, this can be written as:

$$\mathcal{E} = \int_0^{\nu - \mu} p(a)L(a) \; da \qquad (6.106)$$

where, in a similar manner to equation (6.18):

$$L(a) = \int \psi_r^{1s} \psi_r^{2s} \; dx \; dy \qquad (6.107)$$

Using the specific forms for ψ_r^{1s} and ψ_r^{2s} as above, then:

$$L(a) = 2\pi \left\{ \sqrt{1 - \beta^2} a\lambda' + \lambda'^2 - \frac{\alpha}{\lambda_{2s}} \left[(1 - \beta^2)a^2\lambda' + 2\sqrt{1 - \beta^2}a\lambda'^2 + 2\lambda'^3 \right] \right\}$$

$$\times \exp\left(-\frac{\sqrt{1 - \beta^2}a}{\lambda'} \right) \qquad (6.108)$$

where $1/\lambda' = 1/\lambda_{1s} + 1/\lambda_{2s}$. Writing this as $L(a) = L_1 - \alpha L_2$, then obtain:

$$\alpha = \frac{\int_0^{\nu - \mu} p(a)L_1(a) \; da}{\int_0^{\nu - \mu} p(a)L_2(a) \; da} \qquad (6.109)$$

The remaining functions, $F(a)$, $G(a)$, $J(a)$ and $K(a)$, can all be derived by using similar procedures to the 1s case, and after lengthy manipulation are given by:

$$F(a) = 2\pi \left\{ \frac{\sqrt{1 - \beta^2}a\lambda_{2s}}{2} + \frac{\lambda_{2s}^2}{4} \right.$$

$$- \frac{2\alpha}{\lambda_{2s}} \left[\frac{(1 - \beta^2)a^2\lambda_{2s}}{2} + \frac{\sqrt{1 - \beta^2}a\lambda_{2s}^2}{2} + \frac{\lambda_{2s}^3}{4} \right]$$

$$\left. + \frac{\alpha^2}{\lambda_{2s}^2} \left(\frac{(1 - \beta^2)^{\frac{3}{2}}a^3\lambda_{2s}}{2} + \frac{3(1 - \beta^2)a^2\lambda_{2s}^2}{4} + \frac{6\sqrt{1 - \beta^2}a\lambda_{2s}^3}{8} + \frac{6\lambda_{2s}^4}{16} \right) \right\}$$

$$\times \exp\left(-\frac{2\sqrt{1 - \beta^2}a}{\lambda_{2s}} \right) \qquad (6.110)$$

In addition, $G(a) = G_1 + G_2$ where:

$$G_1 = 2\pi \int_0^1 (\alpha + 1)^2 \frac{(1 - \beta^2)^2 a^2}{\lambda_{2s}^2} \exp\left[-\frac{\sqrt{1 - \beta^2}a}{\lambda_{2s}} \left(\frac{1}{w} + w \right) \right]$$

$$\times \left[\frac{1 - w^2}{w(1 + w^2)} \right] \; dw \qquad (6.111)$$

and:

$$G_2 = 2\pi \left[-\frac{\alpha(\alpha + 1)}{\lambda_{2s}^2} + \frac{\alpha^2}{\lambda_{2s}^3} \left(\frac{\sqrt{1 - \beta^2}a}{2} + \frac{\lambda_{2s}}{4} \right) \right]$$

$$\times \exp\left(-\frac{2\sqrt{1-\beta^2}a}{\lambda_{2s}}\right)(1-\beta^2)^2 a^2 \qquad (6.112)$$

Continuing for the remaining functions:

$$J(a) = 2\pi \int_0^1 \left[1 - \frac{\alpha\sqrt{1-\beta^2}a\left(\frac{1}{w}+w\right)}{2\lambda_{2s}}\right]$$

$$\times \left\{\left[\frac{2\alpha+1}{\lambda_{2s}^2} - \frac{\alpha\sqrt{1-\beta^2}a\left(\frac{1}{w}+w\right)}{2\lambda_{2s}^3}\right]\left(\frac{1-w^2}{1+w^2}\right)^4\right.$$

$$\left. - \left[\frac{2(\alpha+1)w}{\lambda_{2s}\sqrt{1-\beta^2}a(1+w^2)} - \frac{\alpha}{\lambda_{2s}^2}\right]\left[1 + \frac{4w^2}{(1+w^2)^2}\right]\right\}$$

$$\times \exp\left[-\frac{\sqrt{1-\beta^2}a\left(\frac{1}{w}+w\right)}{\lambda_{2s}}\right]\left[\frac{(1-\beta^2)a^2}{4}\right]\left(\frac{1}{w}-w\right)\left(\frac{1}{w^2}+1\right)\,dw \qquad (6.113)$$

and finally:

$$K(a) = 2\pi\frac{\beta a}{2}\int_0^{\left(\frac{1}{\beta}-\frac{1}{\beta}\sqrt{1-\beta^2}\right)}\left[1 - \frac{\alpha\beta a}{2\lambda_{2s}}\left(\frac{1}{w}-w\right)\right]^2$$

$$\times \exp\left[-\frac{\beta a}{\lambda_{2s}}\left(\frac{1}{w}-w\right)\right]\left(\frac{1}{w^2}-1\right)\,dw \qquad (6.114)$$

CHAPTER 7

STRAINED QUANTUM WELLS

contributed by V. D. Jovanović

7.1 STRESS AND STRAIN IN BULK CRYSTALS

A mechanical force acting on a crystal lattice changes the relative positions of the lattice points (sites) i.e. the positions of the atoms forming the crystal structure. This can be characterized by a vector $\vec{u}$, which defines the relative displacement of an atom into a new position $\vec{r}' = \vec{r} + \vec{u}$ in some arbitrary Cartesian coordinate system. Different crystal lattice points can have different relative displacements making the vector $\vec{u}$ coordinate dependent i.e. $\vec{u} = \vec{u}(\vec{r})$. If the lattice points return to their original positions after the force is removed, then the deformation is described as elastic.

Knowing the relative displacements of each lattice site, the state of the crystal deformation can be described by strain components defining a second-rank tensor (for more about application of tensors in crystals, see Nye [152]) as:

$$\epsilon_{ij} = \frac{1}{2}\left(\frac{\partial u_i}{\partial x_j} + \frac{\partial u_j}{\partial x_i}\right), \qquad i,j = 1,2,3 \tag{7.1}$$

where u_1, u_2, and u_3 are the relative displacements of the crystal lattice points along the x_1, x_2 and x_3 axes, respectively. The diagonal components represent extensions per unit length along the x_1, x_2 and x_3 directions and are usually referred as 'stretches' (see

Quantum Wells, Wires and Dots, Third Edition. P. Harrison
©2009 John Wiley & Sons, Ltd.

Fig. 7.1), while the off-diagonal components ϵ_{ij} are related to 'rotations' e.g. the term $\partial u_i / \partial x_j$ represents a rotation about the x_3 axis toward x_1 of a line element parallel to x_2. As such, the angle between the two line elements parallel to x_1 and x_2 changes from $\pi/2$ before deformation to $\pi/2 - \epsilon_{ij}$ after *. The strain tensor is symmetrical ($\epsilon_{ij} = \epsilon_{ji}$) and can be written as:

$$\epsilon = \begin{pmatrix} \epsilon_{11} & \epsilon_{12} & \epsilon_{31} \\ \epsilon_{12} & \epsilon_{22} & \epsilon_{23} \\ \epsilon_{31} & \epsilon_{23} & \epsilon_{33} \end{pmatrix} \tag{7.2}$$

If the shear strain components are zero, the diagonal elements also determine the change in the crystal volume (also known as 'dilation') as:

$$\frac{\Delta V}{V} = \text{Tr}(\epsilon) = \epsilon_{11} + \epsilon_{22} + \epsilon_{33} \tag{7.3}$$

This can be explained in terms of the unity cube strained by the diagonal components. Assuming the strains are small, which is true in the limits of linear strain theory as well as in most applications, the distorted volume will then be $(1 + \epsilon_{11})(1 + \epsilon_{22})(1 + \epsilon_{33})$ giving a total change in the cube volume (to first-order) of $\epsilon_{11} + \epsilon_{22} + \epsilon_{33}$ (see Fig. 7.1).

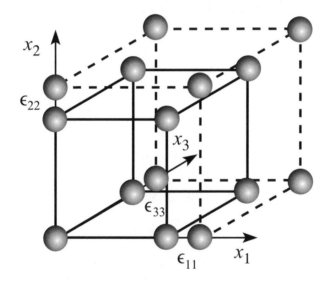

Figure 7.1   Schematic illustration of the influence of the diagonal strain components or stretches (ϵ_{11}, ϵ_{22}, ϵ_{33}) on the crystal lattice.

It is also important to define the properties of the force causing the strain in the crystal lattice. If a crystal is acted on by an external force or if part of a crystal is applying a force on a neighbouring part, then the crystal is said to be in the state of 'stress', where the stress is usually defined as the force per unit area of the crystal. If a unit cube is considered, then normal stress components can be defined as σ_{ii} e.g. σ_{11}, σ_{22}, σ_{33}, etc. and the shear stress components as σ_{ij} e.g. σ_{12}, σ_{21}, σ_{13} etc. as shown in

---

*Note that in the literature an intuitive notation is occasionally seen replacing the indices 1, 2 and 3 with x, y and z, e.g. $\epsilon_{11} \rightarrow \epsilon_{xx}$.

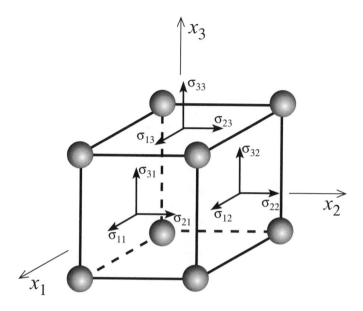

Figure 7.2   Directions of the stress components acting on a crystal lattice that are usually assumed.

Fig. 7.2. In the homogeneous case when the force is independent of the position on the crystal surface, the stress components form a symmetric second-rank tensor ($\sigma_{ij} = \sigma_{ji}$) as:

$$\sigma = \begin{pmatrix} \sigma_{11} & \sigma_{12} & \sigma_{31} \\ \sigma_{12} & \sigma_{22} & \sigma_{23} \\ \sigma_{31} & \sigma_{23} & \sigma_{33} \end{pmatrix} \qquad (7.4)$$

The diagonal elements of the stress tensor are of special importance in applications. If only σ_{11} is a non-zero stress component, then the stress is said to be uniaxial, while if both σ_{11} and σ_{22} are not equal to zero the stress is defined as 'biaxial'. The latter is the case common in quantum well structures as elaborated on in the following section.

In the limit of linear elastic theory the stress and the strain are connected by Hooke's law which reads:

$$\epsilon_{ij} = \sum_{k=1}^{3} \sum_{l=1}^{3} S_{ijkl} \sigma_{kl} \qquad (7.5)$$

where S_{ijkl} are the 'elastic compliance constants' which form a fourth-rank tensor. In the literature, stress is commonly presented as a function of strain and Hooke's law takes the similar form:

$$\sigma_{ij} = \sum_{k=1}^{3} \sum_{l=1}^{3} C_{ijkl} \epsilon_{kl} \qquad (7.6)$$

where C_{ijkl} are the 'elastic stiffness constants' and also form a fourth-rank tensor.

A more concise, matrix (or Voigt's) notation is commonly used in the literature which reduces the number of tensor indices and transforms the strain, stress and elastic stiffness components as follows:

$$
\begin{pmatrix}
\epsilon_{11} & \epsilon_{12} & \epsilon_{31} \\
\epsilon_{12} & \epsilon_{22} & \epsilon_{23} \\
\epsilon_{31} & \epsilon_{23} & \epsilon_{33}
\end{pmatrix}
\rightarrow
\begin{pmatrix}
\epsilon_1 & \frac{1}{2}\epsilon_6 & \frac{1}{2}\epsilon_5 \\
\frac{1}{2}\epsilon_6 & \epsilon_2 & \frac{1}{2}\epsilon_4 \\
\frac{1}{2}\epsilon_5 & \frac{1}{2}\epsilon_4 & \epsilon_3
\end{pmatrix}
\tag{7.7}
$$

$$
\begin{pmatrix}
\sigma_{11} & \sigma_{12} & \sigma_{31} \\
\sigma_{12} & \sigma_{22} & \sigma_{23} \\
\sigma_{31} & \sigma_{23} & \sigma_{33}
\end{pmatrix}
\rightarrow
\begin{pmatrix}
\sigma_1 & \sigma_6 & \sigma_5 \\
\sigma_6 & \sigma_1 & \sigma_4 \\
\sigma_5 & \sigma_4 & \sigma_3
\end{pmatrix}
\tag{7.8}
$$

$$
C_{ijkl} \rightarrow C_{mn} \qquad i,j,k,l = 1,2,3; \ m,n = 1,\ldots 6
\tag{7.9}
$$

The $\frac{1}{2}$ terms have been introduced in order to give a clear relationship between the stress, strain and other parameters used. Hooke's law given by equation (7.6) in the matrix notation takes a simplified form and reads:

$$
\sigma_i = \sum_{k=1}^{6} C_{ik}\epsilon_k
\tag{7.10}
$$

or equivalently:

$$
\begin{bmatrix}
\sigma_1 \\ \sigma_2 \\ \sigma_3 \\ \sigma_4 \\ \sigma_5 \\ \sigma_6
\end{bmatrix}
=
\begin{bmatrix}
C_{11} & C_{12} & C_{13} & C_{14} & C_{15} & C_{16} \\
C_{21} & C_{22} & C_{23} & C_{24} & C_{25} & C_{26} \\
C_{31} & C_{32} & C_{33} & C_{34} & C_{35} & C_{36} \\
C_{41} & C_{42} & C_{43} & C_{44} & C_{45} & C_{46} \\
C_{51} & C_{52} & C_{53} & C_{54} & C_{55} & C_{56} \\
C_{61} & C_{62} & C_{63} & C_{64} & C_{65} & C_{66}
\end{bmatrix}
\begin{bmatrix}
\epsilon_1 \\ \epsilon_2 \\ \epsilon_3 \\ \epsilon_4 \\ \epsilon_5 \\ \epsilon_6
\end{bmatrix}
\tag{7.11}
$$

where C_{ik} is now the elastic stiffness matrix, which is related to the elastic compliance matrix as $[C] = [S]^{-1}$ and defines the unique correlation between the elastic constants.

In order to define the elastic properties of the crystal, 36 independent elastic stiffness constants are required. However, crystal lattices commonly exhibit certain symmetries, which can be employed to reduce the number of constants necessary to describe their elastic behaviour. The number of independent matrix components for the most used semiconductors are: 3 for the cubic (e.g. GaAs) and 5 for the hexagonal (e.g. wurtzite GaN) crystal geometries. Hence, the C matrices for these semiconductors are given as:

$$
\begin{bmatrix}
C_{11} & C_{12} & C_{12} & 0 & 0 & 0 \\
C_{12} & C_{11} & C_{12} & 0 & 0 & 0 \\
C_{12} & C_{12} & C_{11} & 0 & 0 & 0 \\
0 & 0 & 0 & C_{44} & 0 & 0 \\
0 & 0 & 0 & 0 & C_{44} & 0 \\
0 & 0 & 0 & 0 & 0 & C_{44}
\end{bmatrix}
\tag{7.12}
$$

and

$$
\begin{bmatrix}
C_{11} & C_{12} & C_{13} & 0 & 0 & 0 \\
C_{12} & C_{11} & C_{13} & 0 & 0 & 0 \\
C_{13} & C_{13} & C_{33} & 0 & 0 & 0 \\
0 & 0 & 0 & C_{44} & 0 & 0 \\
0 & 0 & 0 & 0 & C_{44} & 0 \\
0 & 0 & 0 & 0 & 0 & \frac{1}{2}(C_{11} - C_{12})
\end{bmatrix}
\tag{7.13}
$$

for the cubic and the hexagonal crystals, respectively.

Finally, to deform the crystal lattice, a certain amount of energy is needed. The sum of the work done by the stress components acting on the crystal defines the strain energy as another important parameter influencing the crystal strain state. In terms of the strain components with respect to the validity of Hooke's law, the strain energy per volume unit can be expressed in the matrix notation as:

$$W = \frac{1}{2} \sum_{i=1}^{6} \sum_{j=1}^{6} C_{ij} \epsilon_i \epsilon_j \tag{7.14}$$

The strain energy density always has to be greater than zero. Furthermore, the crystal system reaches the most stable strain state for the minimal value of the strain energy—a condition which is used in the derivation of strain balancing in Section 7.3.

For cubic semiconductors, equations (7.12) and (7.14) give the elastic energy as:

$$W_{\text{cubic}} = \frac{1}{2} C_{11} \left(\epsilon_1^2 + \epsilon_2^2 + \epsilon_3^2 \right) + \frac{1}{2} C_{44} \left(\epsilon_4^2 + \epsilon_5^2 + \epsilon_6^2 \right) + C_{12} \left(\epsilon_1 \epsilon_2 + \epsilon_2 \epsilon_3 + \epsilon_3 \epsilon_1 \right) \tag{7.15}$$

whilst for hexagonal crystals it takes the form:

$$W_{\text{hex}} = \frac{1}{2} C_{11} \left(\epsilon_1^2 + \epsilon_2^2 \right) + \frac{1}{2} C_{33} \epsilon_3^2 + C_{12} \epsilon_1 \epsilon_2 + C_{13} \epsilon_3 \left(\epsilon_1 + \epsilon_2 \right)$$

$$+ \frac{1}{2} C_{44} \left(\epsilon_4^2 + \epsilon_5^2 \right) + \frac{1}{4} (C_{11} - C_{12}) \epsilon_6^2 \tag{7.16}$$

7.2 STRAIN IN QUANTUM WELLS

The effects of strain are of particular interest in quantum well structures. If a thin epitaxial layer is deposited on a much thicker substrate (usually assumed to be infinitely thick in comparison to the epitaxial layer), then the lattice constant (a_l) in the growth plane (perpendicular to the growth direction) of the layer will be forced to change to try and equal the lattice constant of the substrate (a_0). As a consequence, the crystal lattice is under biaxial stress along the growth interface and, while no force is applied along the growth direction, the crystal is able to relax freely along that direction. Therefore, the stress has only two diagonal components σ_1 and σ_2 while σ_3 and the shear components are zero. Hence, in the matrix notation the stress in the epitaxial (quantum well) layer can be written as:

$$\sigma = \begin{pmatrix} \sigma_1 & 0 & 0 \\ 0 & \sigma_2 & 0 \\ 0 & 0 & 0 \end{pmatrix} \tag{7.17}$$

The existence of biaxial stress results in the appearance of an in-plane strain. Growth which allows the lattice constant of the epitaxial layer to fully equal (match) the substrate is usually referred as pseudomorphic growth. In such cases, the in-plane strain can be easily calculated as:

$$\epsilon_{\parallel} = \frac{a_0 - a_l}{a_l} \tag{7.18}$$

In quantum well systems the in-plane strain is usually of the order of 1%, e.g. for an AlN layer grown on a GaN substrate $\epsilon_{\parallel} \approx 2\%$. However, some materials and their

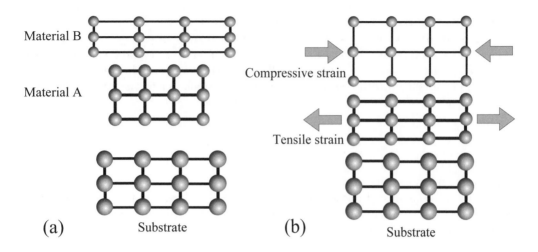

Figure 7.3 Schematic illustration of a substrate with two mismatched layers, (a) shows the free-standing unstrained layers with lattice constants smaller $a_l < a_0$ (material A) and larger $a_l > a_0$ (material B) than the lattice constant of the substrate and (b) shows the situation when either of the materials is grown on the substrate, i.e. material A is under tensile strain and material B is under compressive strain as their lattice constants are forced to be equal to the lattice constant of the substrate.

alloys have very similar lattice constants such as the GaAs/AlGaAs system for which $\epsilon_\parallel < 0.1\%$ allowing them to be considered as unstrained.

The lattice constant of the epitaxial layer can be either larger or smaller than the lattice constant of the substrate (see Fig. 7.3) defining a compressive ($\epsilon_\parallel < 0$) or a tensile ($\epsilon_\parallel > 0$) strain, respectively. A compressive strain will therefore force the lattice constant in the plane to shrink (see material B in Fig. 7.3) conversely tensile strain will force it to expand (see material A in Fig. 7.3). It is also reasonable to assume that the strain in the thin semiconductor layers is homogeneous and therefore constant throughout the layer.

If the epitaxial layer width is larger than some critical thickness, the layer relaxes in the plane (perpendicular to the growth direction) restoring its original lattice constant. This is a violent process producing a large number of defects and imperfections (cracks) in the growth surface. The formation of defects and the existence of a critical thickness can be understood in term of the elastic energy. The strained layer system possesses a certain additional elastic energy which is a function of the layer thickness (approximately a product of the elastic energy density, the area and the layer width). If, for a specific layer width, the strain energy exceeds the energy required for the generation of defects, then the system will tend to relax to a new state with lower strain energy forming imperfections in the growth plane. The width of the layer for which this relaxation process occurs is commonly referred to as the 'critical thickness' (see [153]). Therefore, pseudomorphic growth is a necessary condition for the fabrication of good quality layers with a small number of intrinsic defects.

Though no stress exists ($\sigma_3 = 0$) in the growth direction, the lattice constant is still forced to change due to the 'Poisson effect' (see Fig. 7.3). If the compressive strain forces the in-plane lattice constant to reduce, then the lattice constant in the growth direction will increase and vice versa for tensile strain. Hence, strain exists in the growth direction as well. The ratio that determines the increase or decrease of the lattice constant due to the in-plane stress is called Poisson's ratio (ν) and it connects the in-plane and the perpendicular strains as:

$$\epsilon_3 = -\nu\epsilon_1 \tag{7.19}$$

For the commonly used cubic semiconductor materials grown along the [001] direction, the relationship between the stress and the strain in the epitaxial (quantum well) layers is given by Hooke's law (equations (7.6) and (7.12)) and for the biaxial stress assumed reads:

$$
\begin{bmatrix} \sigma_1 \\ \sigma_2 \\ \sigma_3 \\ 0 \\ 0 \\ 0 \end{bmatrix} =
\begin{bmatrix}
C_{11} & C_{12} & C_{12} & 0 & 0 & 0 \\
C_{12} & C_{11} & C_{12} & 0 & 0 & 0 \\
C_{12} & C_{12} & C_{11} & 0 & 0 & 0 \\
0 & 0 & 0 & C_{44} & 0 & 0 \\
0 & 0 & 0 & 0 & C_{44} & 0 \\
0 & 0 & 0 & 0 & 0 & C_{44}
\end{bmatrix}
\begin{bmatrix} \epsilon_1 \\ \epsilon_2 \\ \epsilon_3 \\ \epsilon_4 \\ \epsilon_5 \\ \epsilon_6 \end{bmatrix}
\tag{7.20}
$$

The matrix equation defines a system of two linear independent equations. Knowing that the in-plane strain is actually equal to ϵ_1, the strains in the epitaxial layers are given as:

$$\epsilon_1 = \epsilon_2 = \epsilon_{\parallel}$$

$$\epsilon_3 = -2\frac{C_{12}}{C_{11}}\epsilon_1 \tag{7.21}$$

or in matrix form:

$$
\epsilon = \begin{pmatrix}
\epsilon_1 & 0 & 0 \\
0 & \epsilon_1 & 0 \\
0 & 0 & -2\frac{C_{12}}{C_{11}}\epsilon_1
\end{pmatrix}
\tag{7.22}
$$

where the factor $\nu = -2\frac{C_{12}}{C_{11}}$ represents Poisson's ratio for cubic semiconductors in the [001] direction. Typical values of the strain components and the strain distribution in cubic InGaAs/AlGaAs quantum wells are presented in Fig. 7.4.

Knowing the strain components makes it possible to determine the relative change in the lattice volume (or dilation) using equation (7.3) which now reads:

$$\Theta = \frac{\Delta V}{V} = 2\epsilon_1\left(1 - \frac{C_{12}}{C_{11}}\right) \tag{7.23}$$

For quantum wells with the layers based on the hexagonal crystal geometry grown in the [0001] direction a similar approach can be applied, modifying Hooke's law as:

$$
\begin{bmatrix} \sigma_1 \\ \sigma_2 \\ \sigma_3 \\ 0 \\ 0 \\ 0 \end{bmatrix} =
\begin{bmatrix}
C_{11} & C_{12} & C_{13} & 0 & 0 & 0 \\
C_{12} & C_{11} & C_{13} & 0 & 0 & 0 \\
C_{13} & C_{13} & C_{33} & 0 & 0 & 0 \\
0 & 0 & 0 & C_{44} & 0 & 0 \\
0 & 0 & 0 & 0 & C_{44} & 0 \\
0 & 0 & 0 & 0 & 0 & \frac{1}{2}(C_{11} - C_{12})
\end{bmatrix}
\begin{bmatrix} \epsilon_1 \\ \epsilon_2 \\ \epsilon_3 \\ \epsilon_4 \\ \epsilon_5 \\ \epsilon_6 \end{bmatrix}
\tag{7.24}
$$

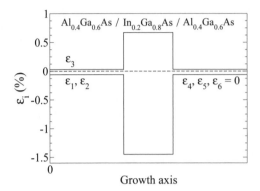

Figure 7.4 Values of the strain components in an $In_{0.2}Ga_{0.8}As$ quantum well surrounded by $Al_{0.4}Ga_{0.6}As$ barriers.

which implies the solutions for the strain components are:

$$\epsilon_1 = \epsilon_2 = \epsilon_\parallel \quad \text{and} \quad \epsilon_3 = -2\frac{C_{12}}{C_{11}}\epsilon_1 \qquad (7.25)$$

or in matrix form:

$$\epsilon = \begin{pmatrix} \epsilon_1 & 0 & 0 \\ 0 & \epsilon_1 & 0 \\ 0 & 0 & -2\frac{C_{13}}{C_{33}}\epsilon_1 \end{pmatrix} \qquad (7.26)$$

Similarly as for the cubic crystal geometry, for epitaxial layers based on hexagonal materials the relative change of the volume can be expressed as:

$$\Theta = \frac{\Delta V}{V} = 2\epsilon_1\left(1 - \frac{C_{13}}{C_{33}}\right) \qquad (7.27)$$

7.3   STRAIN BALANCING

In order to ensure pseudomorphic growth of multiple layers of quantum wells, a technique called 'strain-balancing' is commonly employed. This implies a careful choice of the epitaxial layer compositions and widths as well as in the choice of the substrate (or buffer layer) parameters if possible, with the aim of engineering the strain state of the entire structure, and therefore preventing the appearance of misfit dislocations and other defects. However, the available margins for such parameter manipulation are quite small as the requirements of the quantum well structure enforce tight constraints.

The general method for evaluating a 'strain-balancing condition' is based on the fact that the stable strain state corresponds to the minimum of the elastic energy in each layer of the quantum well structure. However, when the multilayer system is considered, such a state needs to be reached across the entire structure and not only in a single layer so that a favourable strain distribution is reached overall and prevents the relaxation process. Such a condition can be met by minimizing the average elastic energy with respect to the in-plane strain (known as the 'zero-stress' condition) [154] given as:

$$\bar{W} = \frac{\sum\limits_{k=1}^{n} W_k l_k}{\sum\limits_{k=1}^{n} l_k} \tag{7.28}$$

where l_k is the width and W_k the strain energy density of the kth layer.

Recalling the elastic energy density definition stated in equation (7.14), the energy density of the kth layer can be written as

$$W_k = \frac{1}{2} \sum_{i=1}^{6} \sum_{j=1}^{6} C_{ij}^{(k)} \epsilon_i^{(k)} \epsilon_j^{(k)} \tag{7.29}$$

where $C_{ij}^{(k)}$ is the elastic stiffness constant and $\epsilon_i^{(k)}$ is the strain component (in matrix notation) of the kth layer.

For the commonly used cubic and hexagonal semiconductors, the elastic energy densities are given by equations (7.15) and (7.16). If the nature of the strain in the epitaxial layers is assumed to be due to a biaxial stress, then the equations can be written in a simpler form by substituting equation (7.12) or (7.13) and equation (7.21) into equation (7.14) for the cubic and equivalently for the hexagonal crystals. The elastic energy density then reads:

$$W_k = A_k \cdot [\epsilon_1^{(k)}]^2 \tag{7.30}$$

where:

$$A_k = C_{11} + C_{12} - 2\frac{C_{13}^2}{C_{33}} \tag{7.31}$$

for the cubic and:

$$A_k = C_{11} + C_{12} - 2\frac{C_{13}^2}{C_{33}} \tag{7.32}$$

for the hexagonal crystal geometries.

With respect to equation (7.28), equation (7.29) can then written as:

$$\bar{W} = \frac{\sum\limits_{k=1}^{n} A_k [\epsilon_1^{(k)}]^2 l_k}{\sum\limits_{k=1}^{n} l_k} \tag{7.33}$$

The pseudomorphic condition implies that the modified lattice constants (a_k and a_{k-1}) of adjacent layers must be equal to that of the substrate. The in-plane strains in each of the layers can then be written as:

$$\epsilon_1^{(k)} = \frac{a_0 - a_k}{a_k} \tag{7.34}$$

and:

$$\epsilon_1^{(k-1)} = \frac{a_0 - a_{k-1}}{a_{k-1}} \tag{7.35}$$

which means that the in-plane strains in adjacent layers ($\epsilon_1^{(k)}$ and $\epsilon_1^{(k-1)}$) can be related as:

$$\epsilon_1^{(k)} = \frac{a_{k-1}}{a_k}\epsilon_1^{(k-1)} + \frac{a_{k-1} - a_k}{a_k} \tag{7.36}$$

and the average in-plane stress can follow as:

$$\bar{\sigma} = \frac{\partial \bar{W}}{\partial \epsilon_1^{(1)}} = \frac{2}{l_1 + l_2 + l_3 + \dots} \times$$
$$\left\{ A_1\epsilon_1^{(1)}l_1 + A_2\epsilon_1^{(2)}l_2\frac{\partial\epsilon_1^{(2)}}{\partial\epsilon_1^{(1)}} + A_3\epsilon_1^{(3)}l_3\frac{\partial\epsilon_1^{(3)}}{\partial\epsilon_1^{(1)}} + \dots \right\} \tag{7.37}$$

The zero-stress (or strain-balance) condition implies that the *in-plane* stress is zero ($\bar{\sigma} = 0$), and using $\partial\epsilon_1^{(k)}/\partial\epsilon_1^{(1)} = a_1/a_k$ ($k = 2, 3, \dots$) then equation (7.37) gives:

$$A_1\epsilon_1^{(1)}l_1 + A_2\epsilon_1^{(2)}l_2\frac{a_1}{a_2} + A_3\epsilon_1^{(3)}l_3\frac{a_1}{a_3} + \dots = 0 \tag{7.38}$$

which delivers the lattice constant of the substrate (or suitably grown buffer layer) necessary for strain-balancing the quantum well stack as:

$$a_0 = \frac{\sum\limits_{k=1}^{n} A_kl_k/a_k}{\sum\limits_{k=1}^{n} A_kl_k/a_k^2} \tag{7.39}$$

If this lattice constant is not equal to that of any readily available substrate, then it can be achieved by growth of a suitable buffer layer. For example, in the $Si_{1-x}Ge_x/Si$ material system, strain-balancing is achieved by growing a buffer layer with the appropriate Ge composition, i.e. $Si_{1-y}Ge_y$. This is usually linearly graded from the pure silicon composition of the substrate to the required composition y.

If a multilayer structure has periodicity, the above expression should be understood in terms of the single period, i.e. it should be applied to the n layers constituting a single period, and the whole structure will then clearly be strain balanced.

The previous derivation takes into account the difference in elastic properties of the layers (i.e. elastic stiffness constants). If the elastic constants are similar for all layers of the structure, then the stain-balancing condition can be simplified by taking into account only the difference of the lattice constants (also known as the 'average lattice method') as:

$$a_0 = \frac{\sum\limits_{k=1}^{n} l_ka_k}{\sum\limits_{k=1}^{n} l_k} \tag{7.40}$$

If the substrate is fixed and cannot be engineered, the quantum well structure itself has to be 'tailored' to match the substrate, the complexity of which depends on the number of different layers per period and the boundaries imposed by the desired application. In something like a quantum well infrared photodetector (QWIP) this could be achieved by altering the width of the thick barrier layer separating the quantum well absorbing regions, as provided the thickness of this layer is above some minimum (to limit the dark current) it's precise value is not too important. In a quantum cascade laser the injector region offers some flexibility to allow a design to be created which also satisfies the strain-balancing condition.

7.4   EFFECT ON THE BAND PROFILE OF QUANTUM WELLS

The strain in epitaxial layers acts to change the crystal lattice geometry, i.e. to perturb its size and symmetry. This results in a change of the electronic structure in particular a modification of the conduction and valence band edges and therefore a corresponding shift in the energy levels. In order to give a quantitative measure of this further effect of strain, a general case of the strained crystal is considered under the framework of deformation potential theory. This section will consider the simple case of the conduction band Γ minimum, while more detailed analysis exploring the valence band will be presented in Chapter 14.

If the strain is assumed to be small, which is the case in semiconductor quantum wells ($\sim 1\%$), first-order perturbation theory can be used to calculate the band shift. The Hamiltonian under strain can be expressed as the sum of the unperturbed Hamiltonian ($\mathcal{H}_0$) and the strain induced contribution ($\mathcal{H}_\epsilon$), i.e.

$$\mathcal{H} = \mathcal{H}_0 + \mathcal{H}_\epsilon \tag{7.41}$$

where the dependence of $\mathcal{H}_\epsilon$ on the strain is:

$$\mathcal{H}_\epsilon = \sum_{i,j} \frac{\partial V}{\partial \epsilon_{ij}} \epsilon_{ij} \tag{7.42}$$

and V is the original (unperturbed) crystal potential. From perturbation theory it follows that the energy shift due to strain can be expressed in terms of the deformation potential components representing the matrix elements of the relative band shift due to strain (the $\frac{\partial V}{\partial \epsilon_{ij}}$ term in equation (7.42)) as:

$$\delta E_c = \sum_{i,j} D_{ij} \epsilon_{ij} \tag{7.43}$$

Note that the D_{ij} forms a second-rank tensor though the matrix representation is commonly used. The number of non-zero D_{ij} components is dependent on the crystal symmetry (similar to strain theory presented earlier) and the type of band minimum.

For the conduction band Γ point in cubic semiconductors all the off-diagonal deformation potential components vanish ($D_{ij} = 0$) leaving only three equal diagonal constants $D_{11} = D_{22} = D_{33} = a_c$ (the isotropic case) which are usually determined by fitting to experimental data. Equation (7.43), giving the shift of the conduction band energy due to strain, now reads:

$$\delta E_c^{\text{cubic}} = a_c(\epsilon_1 + \epsilon_2 + \epsilon_3) \tag{7.44}$$

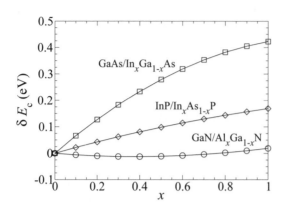

Figure 7.5   Conduction band energy shifts for strained layers based on different materials: AlGaN grown on GaN substrate (circles), InAsP grown on InP (diamonds) and InGaAs grown on GaAs (squares), as a function on the layer composition.

Its clear from the equation that the energy band shift depends only on the strain state of the crystal given via the diagonal strain components.

Hexagonal crystals exhibit much less symmetry than cubic ones. Consequently, a certain anisotropic behaviour can be expected and this is reflected in the different deformation potential constants. For example, GaN in the growth plane has identical deformation potential constants $D_{11} = D_{22} = a_{c\perp}$ whilst in the growth direction the constant $D_{33} = a_{c\parallel}$ is almost three times larger. The total conduction band shift in GaN then follows as:

$$\delta E_c^{\text{hex}} = a_{c\perp}(\epsilon_1 + \epsilon_2) + a_{c\parallel}\epsilon_3 \tag{7.45}$$

The previous analysis can be applied to determine the energy shift of the conduction band edge in the vicinity of the Γ minimum in the epitaxial layer of the quantum well structure. For the layers based on cubic semiconductor material grown in the [001] direction, substituting the strain components given by equation (7.21) the band edge energy shift becomes:

$$\delta E_c^{\text{cubic}} = 2a_c \left(1 - \frac{C_{12}}{C_{11}}\right)\epsilon_1 \tag{7.46}$$

where ϵ_1 is the in-plane strain due to lattice mismatch with the substrate and C_{ij} are the elastic stiffness constants as previously defined. Similarly, for the hexagonal crystal geometry in the [0001] direction the energy shift is given as:

$$\delta E_c^{\text{hex}} = 2 \left(a_{c\perp} - a_{c\parallel}\frac{C_{13}}{C_{33}}\right)\epsilon_1 \tag{7.47}$$

The conduction band shift depends significantly on the material deformation potential constants and the lattice mismatch, and can be up to a few hundreds of meV as shown in Fig. 7.5. It is interesting to see that, although the wurtzite AlGaN layer grown on

GaN is highly strained in the growth plane, because of the anisotropy of the hexagonal crystals the overall shift is smaller.

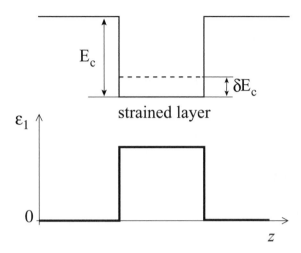

Figure 7.6   Schematic diagram of the conduction band edge of a quantum well with a strain induced energy band shift (δE_c). The original conduction band discontinuity is E_c. The lower diagram shows the corresponding in-plane strain distribution across the quantum well.

Consider now the influence of the conduction band edge shift on the electronic levels in a single quantum well. For simplicity it is assumed that the well material is strained while the barriers are matched to the substrate (Fig. 7.6). The conduction band edge shift implies the equivalent change in the barrier height, defining the new quantum well potential as:

$$U(z)_{str} = U(z) + \delta E_c(z) \tag{7.48}$$

which results in a shift of the discrete energy levels localized within the quantum well.

In order to illustrate the importance of strain in quantum wells, Fig. 7.7 gives the energy of the ground state (with respect to the quantum well band edge) with and without the strain perturbation, as a function of the alloy composition (x) in the $InP/In_xAs_{1-x}P$ quantum well strained on an InP substrate. The calculation shows an average energy shift of around 10%, which can significantly influence the designs and applications. In addition, strain will influence the energy difference between subbands so introducing a shift in intersubband transition energies. This implies that strain 'engineering' could be used to reach desired emission and detection wavelengths otherwise unattainable (see, for example, Faist *et al.* [155]).

7.5 THE PIEZOELECTRIC EFFECT

Certain types of crystal materials exhibit a behaviour such that, under stress, an extra electric charge gathers on their surfaces. The effect is called 'piezoelectricity' and is a consequence of a non-compensated electric polarization generated in the volume of the crystal. Piezoelectric behaviour is exhibited in quartz, Rochelle Salt and Tourmaline, as well as in cubic and hexagonal semiconductors such as InGaAs and GaN. The changes

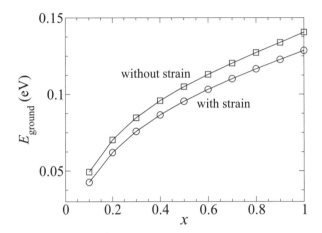

Figure 7.7   Energy of the ground state (with respect to the conduction band edge in the well) in the InP/In$_x$As$_{1-x}$P system with and without strain, as a function of the alloy composition (x). The well width is set to 50 Å and the barrier width to 60 Å.

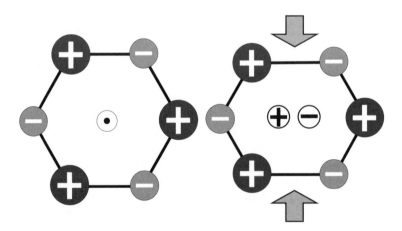

Figure 7.8   Schematic illustration of the charge separation in GaN grown along the [0001] direction which causes the appearance of an uncompensated polarization. The left hand diagram shows the unstrained crystal lattice, and on the right the atomic positions when under strain.

in the crystal lattice geometry due to stress act to separate the charges and break the local electrostatic neutrality in the crystal, a schematic diagram illustrating this effect for wurtzite GaN is given in Fig. 7.8. The uncompensated positive and negative charges induce a global polarization vector, the properties of which are defined by the material and the nature of the applied stress.

If an arbitrary stress is acting on the crystal, then the total induced polarization (in matrix notation similar to that in Sec. 7.1) can be written as:

$$P_i = \sum_{k=1}^{6} d_{ik}\sigma_k \tag{7.49}$$

where d_{ik} is the piezoelectric modulus. The piezoelectric polarization is a vector which is described by three components, i.e. P_1 and P_2 in the growth plane and P_3 along the growth direction, which in turn define a 3×6 matrix of the piezoelectric moduli (d). As for the case of the elastic constants, the symmetry in the crystals reduces the number of independent elements of the d matrix. For cubic and hexagonal crystal geometries, the number of independent moduli relies on the crystal class, e.g. one modulus for InGaAs and three moduli for wurtzite GaN.

So it is stress induced symmetry breaking in a crystal that leads to a piezoelectric polarization, hence the underlying symmetry of the crystal is important. A good example of this is InGaAs. When grown on a GaAs substrate in the [111] direction it exhibits a piezoelectric behaviour. However, layers grown along the [001] direction are polarization free, while those grown along the [110] direction exhibit a piezoelectric polarization in the plane of the layers.

In the following, the piezoelectric properties of epitaxial layers will be considered in more detail for the example of the increasingly more widespread wurtzite GaN. In order to determine the piezoelectric polarization for GaN grown in the [0001] direction consider equation (7.49) which assuming a biaxial stress can be written as:

$$\begin{bmatrix} P_1 \\ P_2 \\ P_3 \end{bmatrix} = \begin{bmatrix} 0 & 0 & 0 & 0 & \frac{1}{2}d_{15} & 0 \\ 0 & 0 & 0 & \frac{1}{2}d_{15} & 0 & 0 \\ d_{31} & d_{31} & d_{33} & 0 & 0 & 0 \end{bmatrix} \begin{bmatrix} \sigma_1 \\ \sigma_2 \\ 0 \\ 0 \\ 0 \\ 0 \end{bmatrix} \tag{7.50}$$

Note that the factor $\frac{1}{2}$ is a consequence of adopting the matrix notation as discussed in Section 7.1. The matrix equation can be decoupled into three linear equations defining the polarization vector components, the solution of which gives a single non-zero component of the piezoelectric polarization in the growth direction as:

$$P_3 = d_{31}(\sigma_1 + \sigma_2) = 2d_{31}\sigma_1 \tag{7.51}$$

Using equations (7.20) and (7.21), the stress component σ_1 can be expressed in terms of the strain as $\sigma_1 = \epsilon_1 \left(C_{11} + C_{12} - 2\frac{C_{13}^2}{C_{33}} \right)$. Hence, the final form of the in-plane piezoelectric polarization of GaN reads:

$$P_3 = 2d_{31}\epsilon_1 \left(C_{11} + C_{12} - 2\frac{C_{13}^2}{C_{33}} \right) \tag{7.52}$$

where C_{ij} are the elastic stiffness constants. As the term $C_{11} + C_{12} - 2C_{13}^2/C_{33} > 0$ is always greater than zero, the sign of the polarization is determined by the sign of the strain. Hence, P_3 is always positive for layers under biaxial compressive strain and negative for layers under tensile strain.

A more often used approach for describing piezoelectric properties of crystals than the moduli is that of the piezoelectric constants defined as:

$$e_{kl} = \sum_{j=1}^{6} d_{kj} C_{jl} \tag{7.53}$$

where C_{jl} are the elastic stiffness constants. Then equation (7.49) for the piezoelectric polarization can be rewritten as:

$$P_i = \sum_{k=1}^{6} e_{ik} \epsilon_k \tag{7.54}$$

In terms of these piezoelectric constants, the polarization in the wurtzite GaN layer under biaxial strain now reads:

$$P_3 = \epsilon_1 e_{31} + \epsilon_2 e_{32} + \epsilon_3 e_{33} = 2\epsilon_1 e_{31} + \epsilon_3 e_{33} = 2\epsilon_1 \left(e_{31} - e_{33} \frac{C_{13}}{C_{33}} \right) \tag{7.55}$$

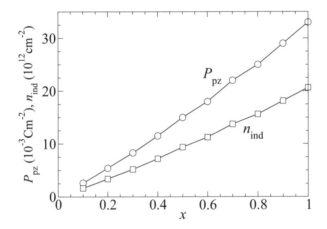

Figure 7.9   The piezoelectric polarization and induced charge at the interface in a single $Al_x Ga_{1-x}N$ layer grown on a GaN substrate as a function of the layer composition x.

Fig. 7.9 shows the calculated piezoelectric polarization and the induced charge ($n_{ind} = P_{pz}/q$) at the growth interface for a single AlGaN layer grown strained on a GaN substrate. For example, for 50% of Al in the epitaxial layer the sheet electron density is around $1 \times 10^{13} cm^{-2}$. This is a common case in GaN-based HEMTs where the induced charge can significantly influence the electronic properties of devices. Furthermore, in quantum well structures the piezoelectric polarization can have a significant effect on the band profile through the electric fields induced.

7.6   INDUCED PIEZOELECTRIC FIELDS IN QUANTUM WELLS

Consider a single quantum well structure based on piezoelectric active material. The different piezoelectric and strain properties of the well and the barrier materials will result in different polarizations. Hence, at the interfaces a gradient of the piezoelectric polarization appears, which induces a fixed charge which gathers in the vicinity of the interface. The charge density is given by the equation:

$$\rho_p = -\nabla \vec{P} \tag{7.56}$$

Constant strain across the epitaxial layer implies a constant polarization as well (see equation (7.55)). Hence, the change of the piezoelectric polarization is abrupt and the induced charge can be found as the difference between the polarizations in the adjacent layers. Furthermore, this accumulated charge at the interfaces induces an electric field in order to satisfy Gauss's law.

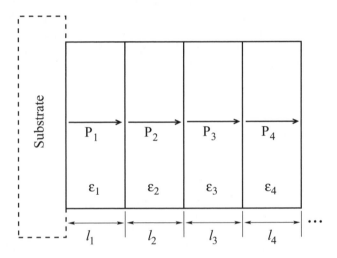

Figure 7.10   Schematic diagram of a multilayer structure based on piezoelectric material giving the notation used in text.

Consider now the general case of the multilayer structure in Fig. 7.10 in order to estimate the intrinsic electric fields in the layers. The electrostatic boundary conditions of the displacement vector $(\vec{D})$ at the adjacent interfaces are applied as:

$$\vec{D}_i = \vec{D}_{i+1} \tag{7.57}$$

where linear material properties[†] are assumed, i.e. $\vec{D}_i = \varepsilon_i \vec{E}_i + \vec{P}_i$. Though the piezoelectric polarization vector can have an arbitrary direction only the component along the growth direction will influence the electronic properties of a quantum well structure and therefore only this component will be considered here. For an arbitrary n-layer

[†] ε is used to represent the permittivity in this chapter to distinguish it from ϵ, which is the commonly accepted symbol for the components of strain.

structure, equation (7.57) gives a system of $n - 1$ linear equations:

$$\varepsilon_1 F_1 + P_1 = \varepsilon_2 F_2 + P_2$$
$$\varepsilon_2 F_2 + P_2 = \varepsilon_3 F_2 + P_3$$
$$\vdots$$
$$\varepsilon_{i-1} F_{i-1} + P_{i-1} = \varepsilon_i F_i + P_i$$
$$\vdots$$
$$\varepsilon_{n-1} F_{n-1} + P_{n-1} = \varepsilon_n F_n + P_n \quad (7.58)$$

which can be solved provided appropriate boundary conditions across the structure are defined. Though there are no constraints on the choice of the boundary condition, the natural choice are the 'hard wall' boundary conditions, which enforce the total potential drop across the structure to be zero. All other possibilities would imply that the quantum well structure acts as a voltage source in the closed circuit. For a periodic structure (e.g. a superlattice) the hard wall and the periodic boundary conditions are equivalent.

Knowing the boundary condition, the necessary nth equation of the system is defined and reads:

$$\sum_{i=1}^{n} F_i l_i = 0 \quad (7.59)$$

where L_i is the width of the ith layer.

The solution of the system gives a rather simple equation:

$$F_j = \frac{\sum\limits_k (P_k - P_j) \frac{l_k}{\varepsilon_k}}{\varepsilon_j \sum\limits_k \frac{l_k}{\varepsilon_k}} \quad (7.60)$$

from which the intrinsic electric field can be easily calculated. It is interesting to see that in the quantum wells the electric field exists even in the layers with zero piezoelectric polarization, this is a consequence of the charge induced at the interfaces with the adjacent piezoelectric layers.

In the case of wurtzite GaN-based quantum wells, the piezoelectric field due to the large strain and piezoelectric constants, can be stronger than 1 MV/cm [156]. Fig. 7.11 shows such an example, where the piezoelectric field in the well and barrier layers of a single GaN/AlGaN quantum well structure grown on a GaN substrate is calculated as a function of the barrier layer composition. The electric field in an unstrained 20 Å wide layer is as large as 5 MV/cm, while in a thicker 60 Å strained AlGaN layer it is proportionally smaller but still around 1 MV/cm. The piezoelectric field in GaN alloys is an order of magnitude larger than in another widely used material—cubic InGaAs which exhibits piezoelectric fields of around ~ 100 kV/cm [157].

7.7   EFFECT OF PIEZOELECTRIC FIELDS ON QUANTUM WELLS

As shown in the previous section, the piezoelectric effect manifests itself as an electric field in the layers of the quantum well structure. To explore the influence of such fields on the band profile and the electronic structure, the simplest case of a single quantum

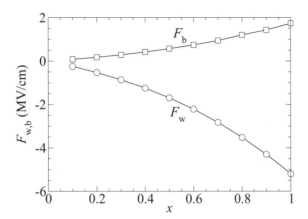

Figure 7.11   The induced piezoelectric field in a GaN well and $Al_xGa_{1-x}N$ barrier layers grown on a GaN substrate as a function of the barrier layer composition x.

well will be considered. Also, without loss of generality, a uniform dielectric constant can be assumed across the structure. Hence, the magnitude of the piezoelectric fields in the quantum well, assuming identical barriers, follows from equation (7.60) as:

$$F_{w,b} = \frac{(P_{b,w} - P_{w,b})\, l_{b,w}}{\varepsilon(l_w + l_b)} \tag{7.61}$$

where $P_{w,b}$ are the polarization magnitudes in the well and the barrier respectively and ε is the constant permittivity. The electric fields in the well and the barrier layer must be of opposite sign, as the overall potential drop across the quantum well has to be zero. This can be expressed analytically using equation (7.59) as:

$$F_w l_w + F_b l_b = 0 \tag{7.62}$$

This implies that the distribution of the electric field across the structure is proportional to the ratio of the layer widths. Consequently, the thinner epitaxial layer will have a higher electric field than that of the thicker.

The piezoelectric field alters the quantum well band profile, which can be represented as a step-linear potential (with respect to the coordinate system as in Fig. 7.12):

$$V(z) = \begin{cases} eF_b z + V_b, & \text{for } z < 0 \\ eF_w z, & \text{for } 0 < z < l_w \\ eF_b z + F_w l_w + V_b & \text{for } z > l_w \end{cases} \tag{7.63}$$

where l_w is the well width and V_b is the barrier height. The previous potential definition is equivalent to a biased quantum well structure and the solutions can be found as a linear combination of Airy's functions as discussed in Chapter 2.

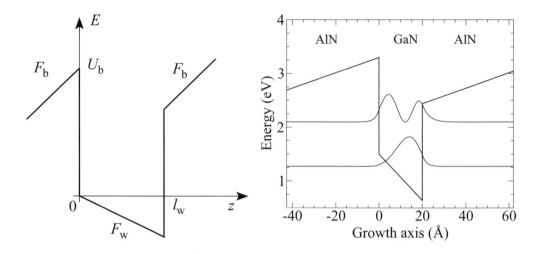

Figure 7.12   On the left-hand side is a schematic diagram of typical induced electric fields in a quantum well based on piezoelectric materials and on the right-hand side is the band profile of a GaN/AlN quantum well structure with wave function moduli of the first two bound states.

Another important effect is that the piezoelectric field can be screened by free electrons. The linear potential causes electrons to gather close to the interfaces, therefore inducing space-charges. If the carrier density in the quantum well is comparable to the density of the piezoelectric induced charges the electrostatic field due to the free electrons can suppress the piezoelectric field. However, as the induced charge at the interfaces is usually much larger than the quantum well doping limit this screening effect is generally not pronounced in applications.

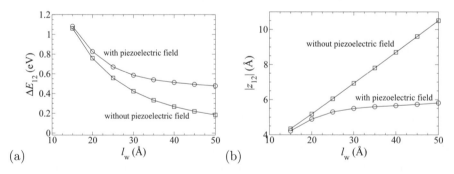

Figure 7.13   Illustration of the influence of the piezoelectric field on the electronic and optical properties of GaN/AlN quantum wells, (a) the energy difference and (b) the dipole matrix element, between the first two bound states with and without the piezoelectric field.

The change of the band profile due to a piezoelectric field can have a strong influence on the energies of intersubband transitions. This is illustrated in Fig. 7.13(a), which shows a comparison of the transition energies between the lowest two subbands in GaN/AlGaN quantum wells with and without the internal electric field included. A substantial increase (blue shift) of the intersubband transition energies is clearly observed and has been observed experimentally [158].

The piezoelectric field can also considerably modify the localization properties of the wave functions, increasing or decreasing the overlap between the ground and other excited states. This can reflect on the dipole matrix element (defined as $z_{ij} = \langle \Psi_i | z | \Psi_j \rangle$) and consequently on the intersubband optical properties of quantum well structures such as the intersubband absorption ($A_{ij} \sim |z_{ij}|^2$). For GaN/AlGaN quantum wells, the relative change in the dipole matrix element z_{12} is shown in Fig. 7.13(b). Clearly, the estimated decrease of up to 50% in z_{ij} is enough to significantly deteriorate the performance of a device in a potential application. A similar problem is present in quantum well lasers based on interband transitions between the conduction and the valence bands (see [159]).

The previously mentioned effects on optical transition energies and absorption strengths highlight the importance of a thorough understanding of the piezoelectric field properties for the modelling and design of optoelectronic devices.

CHAPTER 8

SIMPLE MODELS OF QUANTUM WIRES AND DOTS

8.1 FURTHER CONFINEMENT

It has already been shown that the reduction in dimensionality produced by confining electrons (or holes) to a thin semiconductor layer leads to a dramatic change in their behaviour. This principle can be developed by further reducing the dimensionality of the electron's environment from a two-dimensional quantum well to a one-dimensional quantum wire and eventually to a zero-dimensional quantum dot. In this context, of course, the dimensionality refers to the number of degrees of freedom in the electron momentum; in fact, within a quantum wire, the electron is confined across two directions, rather than just the one in a quantum well, and, so, therefore, reducing the degrees of freedom to one. In a quantum dot, the electron is confined in all three dimensions, thus reducing the degrees of freedom to zero. If the number of degrees of freedom are labelled as $\mathcal{D}_f$ and the number of directions of confinement are labelled as $\mathcal{D}_c$, then clearly:

$$\mathcal{D}_f + \mathcal{D}_c = 3 \tag{8.1}$$

for all solid state systems. These values are highlighted for the four possibilities shown in Table 8.1. Tradition has determined that the reduced-dimensionality systems are labelled by the remaining degrees of freedom in the electron motion, i.e. $\mathcal{D}_f$, rather than the number of directions with confinement $\mathcal{D}_c$.

Fig. 8.1 gives a simple outline of how quantum wires *might* be fabricated, although note that there is more than one method and the interested reader should refer to a

Quantum Wells, Wires and Dots, Third Edition. P. Harrison
©2009 John Wiley & Sons, Ltd.

Table 8.1   The number of degrees of freedom $\mathcal{D}_f$ in the electron motion, together with the extent of the confinement $\mathcal{D}_c$, for the four basic dimensionality systems

System	$\mathcal{D}_c$	$\mathcal{D}_f$
Bulk	0	3
Quantum well	1	2
Quantum wire	2	1
Quantum dot	3	0

specialist growth treatise for further details. A standard quantum well layer can be patterned with photolithography or perhaps electron-beam lithography, and etched to leave a free standing strip of quantum well material; the latter may or may not be filled in with an overgrowth of the barrier material (in this case, $Ga_{1-x}Al_xAs$). Any charge carriers are still confined along the heterostructure growth (z-) axis, as they were in the quantum well, but in addition (provided the strip is narrow enough) they are now confined along an additional direction, either the x- or the y-axis, depending on the lithography.

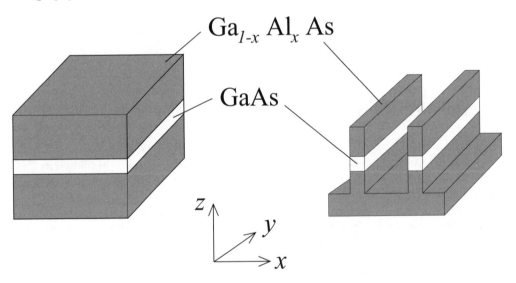

Figure 8.1   Fabrication of quantum wires

Fig. 8.2 shows an expanded view of a single quantum wire, where clearly the electron (or hole) is free to move in only one direction, in this case along the y-axis. Within the effective-mass approximation the motion along the axis of the wire can still be described by a parabolic dispersion, i.e.

$$E = \frac{\hbar^2 k^2}{2m^*} \tag{8.2}$$

just as in bulk and for the in-plane motion within a quantum well.

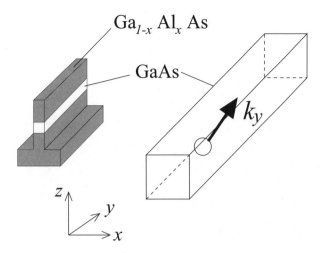

Figure 8.2   A single wire and an expanded view showing schematically the single degree of freedom in the electron momentum

Another class of quantum wire can be formed by patterning the substrate *before* growth. This leads to the formation of so-called *V-grooved quantum wires*, see e.g. [7], p. 35; the solution of these has been dealt with by Gangopadhyay and Nag [160], and will also be touched upon later in Section 17.4.

Quantum dots (see [8] for an introduction to their applications) *can* again be formed by further lithography and etching, e.g. if a quantum well sample is etched to leave pillars rather than wires, then a charge carrier can become confined in all three dimensions, as illustrated in Fig. 8.3.

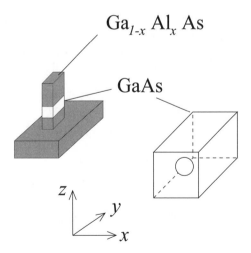

Figure 8.3   A single free standing pillar containing a quantum dot and an expanded view which shows schematically the removal of all degrees of freedom for the electron momentum

Under certain growth conditions, when a thin layer of a semiconductor is grown on top of a substrate which has a quite different lattice constant, then in an attempt to minimize the total strain energy between the bonds, the thin layer spontaneously orders, or *self-assembles* into quantum dots. Microscopy has shown the dots to take the shape of pyramids, or square based 'tetrahedron' [161, 162].

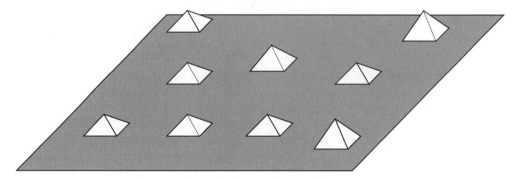

Figure 8.4   Schematic representation of the pyramidal shape of self-assembled quantum dots in highly lattice mismatched systems

In the following sections example solutions will be derived for some of the simpler geometries; some of the more complex ones will have to wait until Chapter 17.

8.2 SCHRÖDINGER'S EQUATION IN QUANTUM WIRES

The general three-dimensional Schrödinger equation for constant effective mass is:

$$-\frac{\hbar^2}{2m^*}\nabla^2\psi(x,y,z) + V(x,y,z)\psi(x,y,z) = E\psi(x,y,z) \tag{8.3}$$

In analogy to the in-plane dispersion discussed in Section 2.2, in a quantum wire it is possible to decouple the motion along the length of the wire. Taking the axis of the wire along x, then the total potential $V(x,y,z)$ can always be written as the sum of a two-dimensional confinement potential plus the potential along the wire (which happens to be zero in this case), i.e.

$$V(x,y,z) = V(x) + V(y,z) \tag{8.4}$$

The eigenfunction can then be written as a product of two components:

$$\psi(x,y,z) = \psi(x)\psi(y,z) \tag{8.5}$$

Substituting both equation (8.4) and equation (8.5) into equation (8.3), then:

$$-\frac{\hbar^2}{2m^*}\left(\frac{\partial^2}{\partial x^2} + \frac{\partial^2}{\partial y^2} + \frac{\partial^2}{\partial z^2}\right)\psi(x)\psi(y,z) + (V(x) + V(y,z))\,\psi(x)\psi(y,z)$$

$$= E\psi(x)\psi(y,z) \tag{8.6}$$

Writing the energy as a sum of terms associated with the two components of the motion, then:

$$-\frac{\hbar^2}{2m^*}\left[\psi(y,z)\frac{\partial^2\psi(x)}{\partial x^2} + \psi(x)\frac{\partial^2\psi(y,z)}{\partial y^2} + \psi(x)\frac{\partial^2\psi(y,z)}{\partial z^2}\right]$$

$$+\psi(y,z)V(x)\psi(x) + \psi(x)V(y,z)\psi(y,z) = (E_x + E_{y,z})\psi(x)\psi(y,z) \qquad (8.7)$$

It is now possible to associate distinct kinetic and potential energies on the left-hand side of equation (8.7), with the components E_x and $E_{y,z}$ on the right-hand side, thus giving two *decoupled* equations, as follows:

$$-\frac{\hbar^2}{2m^*}\psi(y,z)\frac{\partial^2\psi(x)}{\partial x^2} + \psi(y,z)V(x)\psi(x) = \psi(y,z)E_x\psi(x) \qquad (8.8)$$

$$-\frac{\hbar^2}{2m^*}\left[\psi(x)\frac{\partial^2\psi(y,z)}{\partial y^2} + \psi(x)\frac{\partial^2\psi(y,z)}{\partial z^2}\right]$$

$$+\psi(x)V(y,z)\psi(y,z) = \psi(x)E_{y,z}\psi(y,z) \qquad (8.9)$$

In the above $\psi(y,z)$ is not acted upon by any operator in the first equation, and similarly for $\psi(x)$ in the second equation, and thus they can be divided out. In addition, as mentioned above, the potential component along the axis of the wire $V(x) = 0$, thus giving the final decoupled equations of motion as follows:

$$-\frac{\hbar^2}{2m^*}\frac{\partial^2\psi(x)}{\partial x^2} = E_x\psi(x) \qquad (8.10)$$

$$-\frac{\hbar^2}{2m^*}\left[\frac{\partial^2\psi(y,z)}{\partial y^2} + \frac{\partial^2\psi(y,z)}{\partial z^2}\right] + V(y,z)\psi(y,z) = E_{y,z}\psi(y,z) \qquad (8.11)$$

Clearly, the first of these equations is satisfied by a plane wave of the form $\exp(ik_x x)$, thus giving the standard dispersion relationship:

$$E_x = \frac{\hbar^2 k_x^2}{2m^*} \qquad (8.12)$$

The second of these equations of motion, equation (8.11), is merely the Schrödinger equation for the two-dimensional confinement potential characterizing a quantum wire. For a general cross-sectional wire, equation (8.11) should really be solved by using a full two-dimensional solution, which lies beyond the scope of this present work. For the purpose here, special cases of the solution of equation (8.11), for the relevant commonly found geometries, will be illustrated. Such particular solutions rely upon the ability to further decouple the motion into independent components.

8.3   INFINITELY DEEP RECTANGULAR WIRES

Perhaps the simplest quantum wire geometry would be a rectangular cross-section surrounded by infinite barriers. This is illustrated schematically in Fig. 8.5 and can be considered to be the two-dimensional analogy to the one-dimensional confinement potential of the standard infinitely deep quantum well.

Within the quantum wire, the potential is zero, while outside the wire it is infinite; thus in the latter case the wave function is zero. Hence, the Schrödinger equation is only defined within the wire for the motion in the two confined y- and z-directions, i.e. equation (8.11) becomes:

$$-\frac{\hbar^2}{2m^*}\left[\frac{\partial^2\psi(y,z)}{\partial y^2} + \frac{\partial^2\psi(y,z)}{\partial z^2}\right] = E_{y,z}\psi(y,z) \qquad (8.13)$$

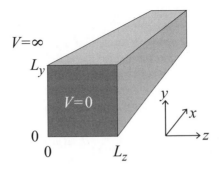

Figure 8.5   The infinitely deep rectangular cross-section quantum wire

The form of the potential in this Schrödinger equation (i.e. zero!) allows the motions to be decoupled further, writing:

$$\psi(y, z) = \psi(y)\psi(z) \tag{8.14}$$

and then:

$$-\frac{\hbar^2}{2m^*}\psi(z)\frac{\partial^2\psi(y)}{\partial y^2} - \frac{\hbar^2}{2m^*}\psi(y)\frac{\partial^2\psi(z)}{\partial z^2} = E_{y,z}\psi(y)\psi(z) \tag{8.15}$$

Again, it is possible to associate the individual kinetic energy terms on the left-hand side of equation (8.15) with separate energy components, i.e. by writing $E_{y,z} = E_y + E_z$, then:

$$-\frac{\hbar^2}{2m^*}\psi(z)\frac{\partial^2\psi(y)}{\partial y^2} - \frac{\hbar^2}{2m^*}\psi(y)\frac{\partial^2\psi(z)}{\partial z^2} = \psi(z)E_y\psi(y) + \psi(y)E_z\psi(z) \tag{8.16}$$

The decoupling is completed with the following:

$$-\frac{\hbar^2}{2m^*}\psi(z)\frac{\partial^2\psi(y)}{\partial y^2} = \psi(z)E_y\psi(y) \tag{8.17}$$

$$-\frac{\hbar^2}{2m^*}\psi(y)\frac{\partial^2\psi(z)}{\partial z^2} = \psi(y)E_z\psi(z) \tag{8.18}$$

Dividing the first of this pair of equations by $\psi(z)$ and the second by $\psi(y)$ gives:

$$-\frac{\hbar^2}{2m^*}\frac{\partial^2\psi(y)}{\partial y^2} = E_y\psi(y) \tag{8.19}$$

$$-\frac{\hbar^2}{2m^*}\frac{\partial^2\psi(z)}{\partial z^2} = E_z\psi(z) \tag{8.20}$$

Given that the potential outside the wire is infinite, then the standard boundary condition of continuity in the wave function implies that *both* $\psi(y)$ and $\psi(z)$ are zero at the edges of the wire. Thus, equations (8.19) and (8.20) are identical to those of the one-dimensional infinitely deep quantum well—the two-dimensional Schrödinger equation has been decoupled into two one-dimensional equations. Given the origin in a 'corner', and the wire dimensions L_y and L_z, as in Fig. 8.5, then the solutions follow as:

$$\psi(y) = \sqrt{\frac{2}{L_y}}\sin\left(\frac{\pi n_y y}{L_y}\right) \tag{8.21}$$

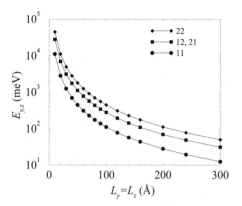

Figure 8.6   The confinement energy in an infinitely deep quantum wire with a square cross-section

and

$$\psi(z) = \sqrt{\frac{2}{L_z}} \sin\left(\frac{\pi n_z z}{L_z}\right) \tag{8.22}$$

which give the components of energy as:

$$E_y = \frac{\hbar^2 \pi^2 n_y^2}{2m^* L_y^2} \tag{8.23}$$

and

$$E_z = \frac{\hbar^2 \pi^2 n_z^2}{2m^* L_z^2} \tag{8.24}$$

Thus, the total energy due to confinement, $E_{y,z} = E_y + E_z$, is:

$$E_{y,z} = \frac{\hbar^2 \pi^2}{2m^*} \left(\frac{n_y^2}{L_y^2} + \frac{n_z^2}{L_z^2} \right) \tag{8.25}$$

The confined states of a quantum wire are therefore described by the two principle quantum numbers n_y and n_z, in contrast to the sole number required for the one-dimensional confinement potential in quantum wells.

Fig. 8.6 displays the confinement energies $E_{y,z}$ for (n_y,n_z) equal to (1,1), (1,2), (2,1), and (2,2) as a function of the side length $L_y = L_z$ for a *square* cross-section infinitely deep quantum wire. In this case of a square cross-sectional wire, the confinement energies of the (1,2) and (2,1) states are equal; however, clearly this will not be the case for a rectangular cross-section wire, which has $L_y \neq L_z$. Just as in the quantum wells met previously, the confinement energy decreases as the size of the system increases.

The wave function for a stationary (wave function along the length of the wire, independent of position) confined particle within the wire is real, and hence the charge density is simply $[\psi(y)\psi(z)]^2$. This is plotted in Fig. 8.7 for the four lowest confined states over a cross-sectional plane. The spatial distribution of the charge density is dependent upon the principle quantum numbers n_y and n_z (as would be expected), and the number of anti-nodes (local maxima) is equal to $n_y n_z$.

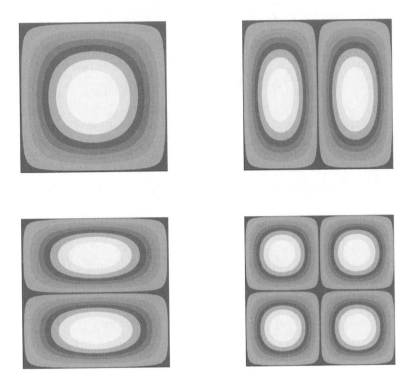

Figure 8.7   The charge densities of the four lowest energy confined states within a infinitely deep quantum wire; (top left) $n_y=1$, $n_z=1$; (top right) $n_y=1$, $n_z=2$; (bottom left) $n_y=2$, $n_z=1$; (bottom right) $n_y=2$, $n_z=2$

8.4 SIMPLE APPROXIMATION TO A FINITE RECTANGULAR WIRE

More relevant to real devices would be a rectangular cross-sectional quantum wire, but with finite height barriers, as would be fabricated with a post-etch overgrowth. Fig. 8.8 (left) illustrates the two-dimensional confinement potential $V(y, z)$ for this system. With this configuration it is not possible to write the potential $V(y, z)$ as a sum of two independent potentials $V(y)$ and $V(z)$, and thus it is not possible to separate the y- and z-motions.

However, a very loose approximation may be to write the potential as in Fig. 8.8 (right). With this form, $V(y, z)$ does equal $V(y) + V(z)$, where $V(y)$ and $V(z)$ are independent finite well potentials, as in Sections 2.5 and 2.6. The approximation occurs in the 'corner regions' outside of the wire where the two quantum well potential barrier heights V sum to give $2V$. This is in areas which are not expected to be sampled too much by the eigenfunctions, particularly those in wide wires and the lower energy states.

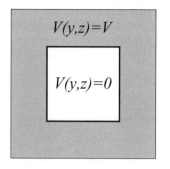
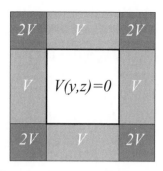

Figure 8.8   The rectangular cross-section quantum wire with finite barriers (left) and an approximate form for the potential (right), suitable for decoupling the motion

So proceeding, the Schrödinger equation for the y–z cross-sectional confined motion in a general quantum wire is given by equation (8.11), i.e.

$$-\frac{\hbar^2}{2m^*}\left[\frac{\partial^2\psi(y,z)}{\partial y^2} + \frac{\partial^2\psi(y,z)}{\partial z^2}\right] + V(y,z)\psi(y,z) = E_{y,z}\psi(y,z) \tag{8.26}$$

Therefore, writing the potential $V(y,z) = V(y) + V(z)$ and the wave function $\psi(y,z)$ as $\psi(y)\psi(z)$, gives:

$$-\frac{\hbar^2}{2m^*}\psi(z)\frac{\partial^2\psi(y)}{\partial y^2} - \frac{\hbar^2}{2m^*}\psi(y)\frac{\partial^2\psi(z)}{\partial z^2}$$

$$+ [V(y) + V(z)]\,\psi(y)\psi(z) = E_{y,z}\psi(y)\psi(z) \tag{8.27}$$

Again, by writing the energy $E_{y,z}$ as a sum of two components associated with the y- and z-motions, equation (8.27) can then be split into two, giving:

$$-\frac{\hbar^2}{2m^*}\psi(z)\frac{\partial^2\psi(y)}{\partial y^2} + \psi(z)V(y)\psi(y) = \psi(z)E_y\psi(y) \tag{8.28}$$

and

$$-\frac{\hbar^2}{2m^*}\psi(y)\frac{\partial^2\psi(z)}{\partial z^2} + \psi(y)V(z)\psi(z) = \psi(y)E_z\psi(z) \tag{8.29}$$

Dividing the first of these equations by $\psi(z)$, and the second by $\psi(y)$, gives familiar equations for straightforward one-dimensional potentials, i.e.

$$-\frac{\hbar^2}{2m^*}\frac{\partial^2\psi(y)}{\partial y^2} + V(y)\psi(y) = E_y\psi(y) \tag{8.30}$$

and

$$-\frac{\hbar^2}{2m^*}\frac{\partial^2\psi(z)}{\partial z^2} + V(z)\psi(z) = E_z\psi(z) \tag{8.31}$$

In this derivation, the kinetic energy operator appropriate for a constant effective mass has been employed, but equally well as there is nothing which depends upon this form, the variable effective mass kinetic energy operator, i.e. $(\partial/\partial z)(1/m^*(z))(\partial/\partial z)$, could

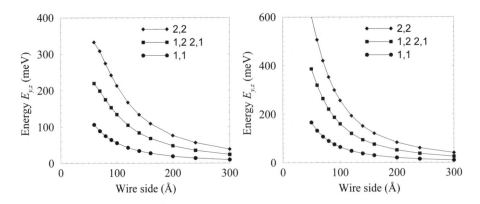

Figure 8.9   The confinement energy, $E_{y,z} = E_y + E_z$, in a square cross-section GaAs quantum wire surrounded by finite height barriers of $Ga_{0.8}Al_{0.2}As$ (left) and $Ga_{0.6}Al_{0.4}As$ (right)

have been used. The final outcome is the same, namely that as the independent potentials $V(y)$ and $V(z)$ are simply those of a finite quantum well, then the solutions for both the wave functions and the confinement energies follow as before (Sections 2.5 and 2.6).

Fig. 8.9 shows the equivalent of Fig. 8.6 but with finite barriers, corresponding to Al concentrations of 20 and 40% in $Ga_{1-x}Al_xAs$. The solutions were derived by combining the eigenvalues from two independent one-dimensional quantum well calculations, including the effective mass mismatch at the well (in this case wire) interface. The eigenvalues labelled '1,1' correspond to the ground state of the y-motion combined with the ground state of the z-motion. The others represent the combinations formed between the lowest two energy states in each direction. The behaviour of the energy with the length of the wire side is qualitatively similar to the infinite barrier case, i.e. there is degeneracy between the 1,2 and the 2,1 solutions, the ordering of the energy levels is the same and the eigenvalues decrease with increasing wire side.

The corresponding charge densities $[\psi(y,z)\psi(y,z)]$ are plotted in Fig. 8.10 for a cross-section of the wire for these same four eigenstates. Again, the behaviour is similar to the infinitely deep wire case, with the same distribution of maxima and minima. However, the main difference is, of course, that the finite barrier height allows for a significant 'leakage' of the wave function into the surrounding material. It is this interaction with the 'potential pillars' located outside of the wire at each corner, which originate from the approximated potential in Fig. 8.8, which limits the applicability or accuracy of this simple model.

For this relatively narrow (100 Å×100 Å) wire, the scale next to the ground state charge density in Fig. 8.10 (top left), implies that, roughly speaking, probably around 80–90% of the charge is confined within the wire. Thus the effect of the approximate potential outside of the wire will not be too substantial. However, for the higher-energy states more of the 'lighter shade' is outside the wire, and thus it would be expected that this approximation would be worse. The same effect would be true when reducing the length of the wire side.

This simple approach to the finite quantum wire is just a way of understanding the basic physics and being able to predict *qualitatively* how the electronic properties change

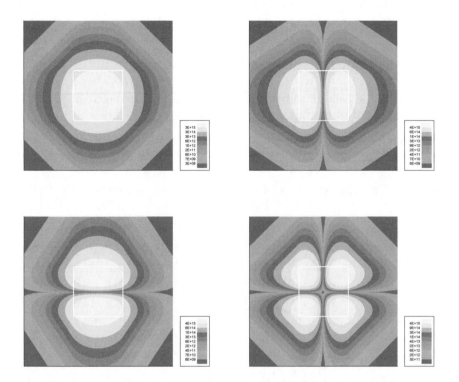

Figure 8.10   The charge densities of the four lowest energy confined states of a GaAs quantum wire of side 100 Å with finite $Ga_{0.8}Al_{0.2}As$ barriers, (top left) $n_y = 1$, $n_z = 1$; (top right) $n_y = 1$, $n_z = 2$; (bottom left) $n_y = 2$, $n_z = 1$; (bottom right) $n_y = 2$, $n_z = 2$; the edges of the wire are indicated by the boxes

with the system parameters. For a more complete description, a full two-dimensional Schrödinger solution may be required (see [160, 163]), or alternatively the empirical pseudopotential approach (see Chapter 17) may be employed.

The energy eigenvalues $E_{y,z}$ *could* be improved by considering a perturbation on the two-dimensional system which removed the '$2V$' potential pillars. Using first-order perturbation theory, the change in energy of a level would be given by:

$$\Delta E = \langle \psi(y,z)|V'(y,z)|\psi(y,z)\rangle \tag{8.32}$$

i.e.

$$\Delta E = \int_{-\infty}^{+\infty} \int_{-\infty}^{+\infty} \psi^*(y)\psi^*(z)V'(y,z)\psi(y)\psi(z) \ dy\,dz \tag{8.33}$$

The perturbation to the potential $V'(y,z)$ would be negative and of magnitude V, and employing the fourfold symmetry of this *square* cross-sectional wire, would give:

$$\Delta E = -4V \int_{L_y}^{+\infty} \psi^*(y)\psi(y) \ dy \int_{L_z}^{+\infty} \psi^*(z)\psi(z) \ dz \tag{8.34}$$

which is relatively straightforward to evaluate. Califano and Harrison [164] have demonstrated that this can be a quite useful approach to the solution of finite barrier quantum wires and dots.

8.5 CIRCULAR CROSS-SECTION WIRE

Consider again the Schrödinger equation for the motion in the confined cross-sectional plane of a quantum wire, as given earlier in equation (8.11), i.e.

$$-\frac{\hbar^2}{2m^*}\left[\frac{\partial^2\psi(y,z)}{\partial y^2} + \frac{\partial^2\psi(y,z)}{\partial z^2}\right] + V(y,z)\psi(y,z) = E_{y,z}\psi(y,z) \qquad (8.35)$$

Given the cylindrical symmetry of the quantum wire, as shown in Fig. 8.11, it would

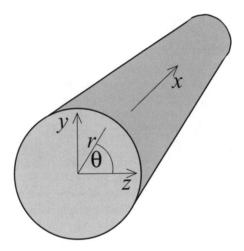

Figure 8.11   Schematic illustration of a circular cross-section quantum wire

seem advantageous to move into polar coordinates for the cross-sectional motion. With the definitions of the modulus r and angle θ as in the figure, the Cartesian coordinates then follow as:

$$y = r\sin\theta \quad \text{and} \quad z = r\cos\theta \qquad (8.36)$$

$$\therefore r = \sqrt{y^2 + z^2} \qquad (8.37)$$

The wave function $\psi(y,z)$ can clearly be written in terms of the new variables r and θ; however, given the circular symmetry the wave functions should not have a dependence on the angle θ. Thus, the wave function can actually be written as $\psi(r)$, and the Schrödinger equation therefore becomes:

$$-\frac{\hbar^2}{2m^*}\left(\frac{\partial^2}{\partial y^2} + \frac{\partial^2}{\partial z^2}\right)\psi(r) + V(r)\psi(r) = E_r\psi(r) \qquad (8.38)$$

where the index on E_r just indicates that this eigenvalue is associated with the confined cross-sectional motion, as opposed to the unconfined motion along the axis of the wire.

In addition, the circular symmetry of the potential that defines the wire can be written explicitly as $V(r)$. Now:

$$\frac{\partial}{\partial y}\psi(r) = \frac{\partial}{\partial r}\psi(r) \times \frac{\partial r}{\partial y} \tag{8.39}$$

Differentiating both sides of equation (8.37) with respect to y, gives:

$$\frac{\partial r}{\partial y} = \frac{1}{2}\left(y^2 + z^2\right)^{-\frac{1}{2}} \times 2y = \frac{y}{r} \tag{8.40}$$

Hence:

$$\frac{\partial}{\partial y}\psi(r) = \frac{\partial}{\partial r}\psi(r) \times \frac{y}{r} \tag{8.41}$$

The second derivative is then:

$$\frac{\partial}{\partial y}\frac{\partial}{\partial y}\psi(r) = \frac{\partial}{\partial y}\left[\frac{\partial}{\partial r}\psi(r) \times \frac{y}{r}\right] \tag{8.42}$$

$$\therefore \frac{\partial^2}{\partial y^2}\psi(r) = \frac{\partial^2}{\partial r^2}\psi(r) \times \frac{\partial r}{\partial y}\frac{y}{r} + \frac{\partial}{\partial r}\psi(r)\frac{\partial}{\partial y}\left(\frac{y}{r}\right) \tag{8.43}$$

and thus:

$$\frac{\partial^2}{\partial y^2}\psi(r) = \frac{y^2}{r^2}\frac{\partial^2}{\partial r^2}\psi(r) + \frac{\partial}{\partial r}\psi(r)\left(\frac{1}{r} - \frac{y}{r^2}\frac{\partial r}{\partial y}\right) \tag{8.44}$$

Finally:

$$\frac{\partial^2}{\partial y^2}\psi(r) = \frac{1}{r}\frac{\partial}{\partial r}\psi(r) - \frac{y^2}{r^3}\frac{\partial}{\partial r}\psi(r) + \frac{y^2}{r^2}\frac{\partial^2}{\partial r^2}\psi(r) \tag{8.45}$$

and similarly for z, hence:

$$\left(\frac{\partial^2}{\partial y^2} + \frac{\partial^2}{\partial z^2}\right)\psi(r) = \frac{2}{r}\frac{\partial}{\partial r}\psi(r) - \frac{(y^2 + z^2)}{r^3}\frac{\partial}{\partial r}\psi(r) + \frac{(y^2 + z^2)}{r^2}\frac{\partial^2}{\partial r^2}\psi(r) \tag{8.46}$$

Recalling that $y^2 + z^2 = r^2$, then:

$$\left(\frac{\partial^2}{\partial y^2} + \frac{\partial^2}{\partial z^2}\right)\psi(r) = \frac{1}{r}\frac{\partial}{\partial r}\psi(r) + \frac{\partial^2}{\partial r^2}\psi(r) \tag{8.47}$$

Substituting into equation (8.38) gives the final form for the Schrödinger equation as follows:

$$-\frac{\hbar^2}{2m^*}\left(\frac{1}{r}\frac{\partial}{\partial r} + \frac{\partial^2}{\partial r^2}\right)\psi(r) + V(r)\psi(r) = E_r\psi(r) \tag{8.48}$$

In this case, reliance has been made on the specific form of the kinetic energy operator, unlike the earlier example of the rectangular cross-section quantum wire, and hence this Schrödinger equation *is* only valid for a constant effective mass.

One numerical approach for solving equation (8.48) would be to follow a similar procedure to that employed in Section 3.1, i.e. expand the derivatives in terms of the standard finite differences equivalents:

$$\frac{\partial}{\partial r}\psi(r) = \frac{\psi(r + \delta r) - \psi(r - \delta r)}{2\delta r} \tag{8.49}$$

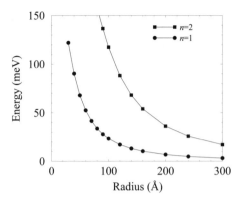

Figure 8.12   The confinement energy in a finite barrier circular cross-section quantum wire

$$\frac{\partial^2}{\partial r^2}\psi(r) = \frac{\psi(r+\delta r) - 2\psi(r) + \psi(r-\delta r)}{(\delta r)^2} \tag{8.50}$$

and then rearrange to obtain a shooting iterative equation. This is achieved by substituting the above into equation (8.48) to give:

$$\frac{1}{r}\left[\frac{\psi(r+\delta r) - \psi(r-\delta r)}{2\delta r}\right] + \frac{\psi(r+\delta r) - 2\psi(r) + \psi(r-\delta r)}{(\delta r)^2}$$

$$= \frac{2m^*}{\hbar^2}\left[V(r) - E_r\right]\psi(r) \tag{8.51}$$

Multiplying both sides by $2r(\delta r)^2$ gives:

$$\left[\psi(r+\delta r) - \psi(r-\delta r)\right]\delta r + 2r\left[\psi(r+\delta r) - 2\psi(r) + \psi(r-\delta r)\right]$$

$$= 2r(\delta r)^2\frac{2m^*}{\hbar^2}\left[V(r) - E_r\right]\psi(r) \tag{8.52}$$

Gathering terms in $\psi(r+\delta r)$, $\psi(r)$, and $\psi(r-\delta r)$, then:

$$(2r + \delta r)\,\psi(r+\delta r)$$

$$= 2r\left\{(\delta r)^2\frac{2m^*}{\hbar^2}\left[V(r) - E_r\right] + 2\right\}\psi(r) + (-2r + \delta r)\,\psi(r-\delta r) \tag{8.53}$$

Then finally, obtain:

$$\psi(r+\delta r) = \frac{2r\left\{2m^*(\delta r/\hbar)^2\left[V(r) - E_r\right] + 2\right\}\psi(r) + (-2r + \delta r)\,\psi(r-\delta r)}{2r + \delta r} \tag{8.54}$$

which is an iterative shooting equation, similar to those met earlier in Chapter 3. It can be solved according to the standard boundary condition, $\psi(r) \to 0$ as $r \to \infty$, as before. Now in regions of constant potential, the wave functions are, in general, continuous, and therefore for the particular case of a straight line perpendicular to and through the wire (x-) axis, the wave function must also be continuous. Thus, when crossing the wire axis, the radial component of the wave function $\psi(r)$ must have a derivative of zero, i.e. a local maxima or minima. This allows the iterative starting conditions to be chosen as:

$$\psi(0) = 1 \quad \text{and} \quad \psi(\delta r) = 1 \tag{8.55}$$

These can look a little simplistic, but in the limit of decreasing step length δr, the maximum or minimum is clearly flat.

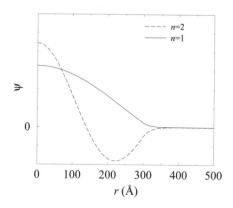

Figure 8.13   The radial component of the wave function $\psi(r)$ for the lowest two eigenstates in a finite-barrier quantum wire with radius 300 Å of circular cross-section

Fig. 8.12 displays the results of calculations of the electron confinement energy versus the wire radius, for a GaAs wire surrounded by $Ga_{0.8}Al_{0.2}As$, *for constant effective mass*. As expected, the confinement energy decreases with increasing radius and the odd-parity eigenstate is of higher energy than the even. This latter point is highlighted in Fig. 8.13, which plots the radial motion $\psi(r)$ for the 300 Å radius wire. The even- $(n = 1)$ and odd- $(n = 2)$ parity nature of the eigenstates can clearly be seen.

8.6 QUANTUM BOXES

Cuboid quantum dots, perhaps more specifically designated as *quantum boxes* can be thought of as simply a generalisation of the rectangular cross-section quantum wires, in which there is now additional confinement along the remaining x-axis. This additional confinement removes the remaining degree of freedom in the particle's momentum and localises it in all directions. Thus the energy levels can no longer be referred to as *subbands* and are now known as *sublevels*.

Considering the case of an infinite potential separating the inside of the box from the outside, then the three-dimensional Schrödinger equation within the box is simply:

$$-\frac{\hbar^2}{2m^*}\left(\frac{\partial^2}{\partial x^2} + \frac{\partial^2}{\partial y^2} + \frac{\partial^2}{\partial z^2}\right)\psi(x, y, z) = E_{x,y,z}\psi(x, y, z) \tag{8.56}$$

Again, writing the total energy $E_{x,y,z}$ as a sum of the three terms E_x, E_y, and E_z, then this single three-dimensional equation can be decoupled into three one-dimensional equations:

$$-\frac{\hbar^2}{2m^*}\frac{\partial^2}{\partial x^2}\psi(x) = E_x\psi(x) \tag{8.57}$$

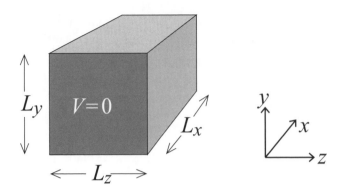

Figure 8.14   Schematic illustration of a quantum box with side L_x, L_y, and L_z

$$-\frac{\hbar^2}{2m^*}\frac{\partial^2}{\partial y^2}\psi(y) = E_y\psi(y) \tag{8.58}$$

$$-\frac{\hbar^2}{2m^*}\frac{\partial^2}{\partial z^2}\psi(z) = E_z\psi(z) \tag{8.59}$$

which for the infinitely deep barrier case, are just the solutions to infinitely deep quantum wells, of widths L_x, L_y, and L_z respectively, i.e. in analogy to the infinitely deep quantum wire confinement energy of equation (8.25), the confinement energy within this quantum box follows as:

$$E_{x,y,z} = \frac{\hbar^2\pi^2}{2m^*}\left(\frac{n_x^2}{L_x^2} + \frac{n_y^2}{L_y^2} + \frac{n_z^2}{L_z^2}\right) \tag{8.60}$$

The three-dimensional nature of the confinement thus requires three *quantum numbers*, i.e. n_x, n_y, and n_z to label each state.

Finite-barrier quantum boxes could be gleaned from three decoupled one-dimensional quantum well calculations in a similar manner to the finite barrier quantum wire discussed in Section 8.4. However, in this case any perturbative correction would have to account for eight corner-cubes of additional potential '$2V$' and twelve edge-cuboids of additional potential 'V'. Alternatively, a full three-dimensional solution can be constructed by expanding the wave function as a linear combination of infinite well solutions (see Gangopadhyay and Nag [165]).

8.7   SPHERICAL QUANTUM DOTS

It is perhaps easier to deal with a finite barrier quantum dot with spherical rather than cuboid symmetry. The approach is rather similar to that derived earlier for the circular cross-section quantum wire.

Given the spherical symmetry of the potential, then the wave function would also be expected to have spherical symmetry, hence the Schrödinger equation for a *constant* effective mass could be written:

$$-\frac{\hbar^2}{2m^*}\left(\frac{\partial^2}{\partial x^2} + \frac{\partial^2}{\partial y^2} + \frac{\partial^2}{\partial z^2}\right)\psi(r) + V(r)\psi(r) = E_r\psi(r) \tag{8.61}$$

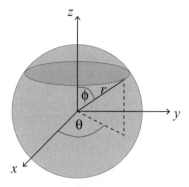

Figure 8.15   Schematic illustration of a spherical quantum dot

where the index on E_r has been added just to indicate that this energy is associated with the confinement along the radius. In this case:

$$r = \sqrt{x^2 + y^2 + z^2} \tag{8.62}$$

The transition can be made from Cartesian (x, y, z) to spherical polar coordinates, in effect just r, in the same way as detailed in Section 8.5. Using equation (8.45), each of the three Cartesian axes gives an equation of the following form:

$$\frac{\partial^2}{\partial x^2}\psi(r) = \frac{1}{r}\frac{\partial}{\partial r}\psi(r) - \frac{x^2}{r^3}\frac{\partial}{\partial r}\psi(r) + \frac{x^2}{r^2}\frac{\partial^2}{\partial r^2}\psi(r) \tag{8.63}$$

Therefore, the complete $\nabla^2\psi(r)$ is given by:

$$\left(\frac{\partial^2}{\partial x^2} + \frac{\partial^2}{\partial y^2} + \frac{\partial^2}{\partial z^2}\right)\psi(r) =$$

$$\frac{3}{r}\frac{\partial}{\partial r}\psi(r) - \frac{(x^2 + y^2 + z^2)}{r^3}\frac{\partial}{\partial r}\psi(r) + \frac{(x^2 + y^2 + z^2)}{r^2}\frac{\partial^2}{\partial r^2}\psi(r) \tag{8.64}$$

$$\therefore \left(\frac{\partial^2}{\partial x^2} + \frac{\partial^2}{\partial y^2} + \frac{\partial^2}{\partial z^2}\right)\psi(r) = \frac{2}{r}\frac{\partial}{\partial r}\psi(r) + \frac{\partial^2}{\partial r^2}\psi(r) \tag{8.65}$$

(as, for example, in [4], p. 188).

Substituting into the Schrödinger equation then:

$$-\frac{\hbar^2}{2m^*}\left(\frac{2}{r}\frac{\partial}{\partial r} + \frac{\partial^2}{\partial r^2}\right)\psi(r) + V(r)\psi(r) = E_r\psi(r) \tag{8.66}$$

Such spherically symmetric Schrödinger equations have been investigated before (see, for example [29], p. 76). As an alternative to such a well-established approach, and with the impetus in this work on simple numerical schemes, a shooting technique similar to that described for the circular cross-section quantum wire is sought. With this aim, expanding the first and second derivatives in terms of finite differences gives:

$$\frac{2}{r}\left[\frac{\psi(r + \delta r) - \psi(r - \delta r)}{2\delta r}\right] + \frac{\psi(r + \delta r) - 2\psi(r) + \psi(r - \delta r)}{(\delta r)^2}$$

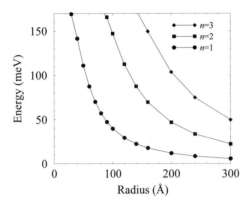

Figure 8.16   The confinement energy in a spherical GaAs quantum dot surrounded by a $Ga_{0.8}Al_{0.2}As$ barrier

$$= \frac{2m^*}{\hbar^2}\left[V(r) - E_r\right]\psi(r) \qquad (8.67)$$

Multiplying both sides by $r(\delta r)^2$ gives:

$$\left[\psi(r + \delta r) - \psi(r - \delta r)\right]\delta r + r\left[\psi(r + \delta r) - 2\psi(r) + \psi(r - \delta r)\right]$$

$$= r(\delta r)^2 \frac{2m^*}{\hbar^2}\left[V(r) - E_r\right]\psi(r) \qquad (8.68)$$

Gathering terms in $\psi(r + \delta r)$, $\psi(r)$, and $\psi(r - \delta r)$ then:

$$(r + \delta r)\,\psi(r + \delta r)$$

$$= r\left\{(\delta r)^2 \frac{2m^*}{\hbar^2}\left[V(r) - E_r\right] + 2\right\}\psi(r) + (-r + \delta r)\,\psi(r - \delta r) \qquad (8.69)$$

and finally:

$$\psi(r + \delta r) = \frac{r\left\{2m^*(\delta r/\hbar)^2\left[V(r) - E_r\right] + 2\right\}\psi(r) + (-r + \delta r)\,\psi(r - \delta r)}{r + \delta r} \qquad (8.70)$$

which is again an iterative equation which can be solved with a numerical shooting technique according to the same boundary conditions as discussed in Section 8.5.

Fig. 8.16 shows the results of calculations of the three lowest energy levels of a spherical GaAs quantum dot surrounded by a finite barrier composed of $Ga_{0.8}Al_{0.2}As$, with a *sharp* boundary. In fact, the formalism above, as that of the circular cross-section quantum wire, is applicable for any radial potential profile $V(r)$, e.g. it is also valid for diffused interfaces. Again, the behaviour of the energies as a function of the spatial dimension, as shown in Fig. 8.16, is as expected in confined systems, namely the confinement energy decreases as the size of the system increases.

Fig. 8.17 displays the corresponding radial components of the wave functions. It can be seen that they all have a maximum at the centre of the potential and that as the principle quantum number n increases, then the number of nodes increases. The nature of the states is perhaps better illustrated by considering the charge density. Given

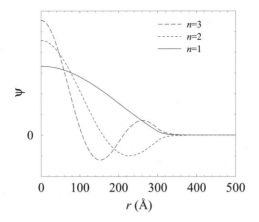

Figure 8.17   The wave functions of the three lowest energy states in the 300 Å spherical quantum dot

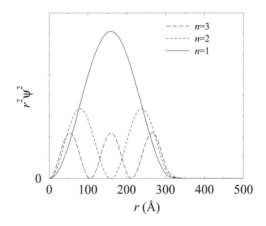

Figure 8.18   The probability density of the three lowest energy states in the 300 Å spherical quantum dot

that successive increases in the radial coordinate r lead to spherical shells of volume proportional to r^2, then the charge density between the radii r and $r + \delta r$ is proportional to $r^2 \psi^*(r) \psi(r)$; this is plotted in Fig. 8.18. The lowest energy state, i.e. $n = 1$, has a single anti-node which is close to half the radius of the potential, while the second has two maxima in the charge density, etc.

8.8   NON-ZERO ANGULAR MOMENTUM STATES

The states deduced in the previous section were of the form $\psi(r)$, and therefore were analogous to the 's' orbitals akin to hydrogen atoms. However, in atoms there exist other types of state, with the next simplest being that of the 'p' orbitals. These have wave functions of the following form:

$$\psi_{p_x} = x\psi(r) \tag{8.71}$$

$$\psi_{p_y} = y\psi(r) \tag{8.72}$$

$$\psi_{p_z} = z\psi(r) \tag{8.73}$$

(see any quantum theory book, e.g. [4], p. 190). These states are clearly not spherically symmetric, and have non-zero angular momenta.

Consider the first of these states in the (spherical polar coordinates) Schrödinger equation (equation (8.66)). As the orientation of the axes is purely arbitrary, then the radius r in any shooting equation *could* be aligned along the x–axis. With this simplification, the Schrödinger equation would become:

$$-\frac{\hbar^2}{2m^*}\left(\frac{2}{r}\frac{\partial}{\partial r} + \frac{\partial^2}{\partial r^2}\right)r\psi(r) + V(r)r\psi(r) = E_r r\psi(r) \tag{8.74}$$

Dropping the indices on ψ and V, then:

$$-\frac{\hbar^2}{2m^*}\left[\frac{2}{r}\left(\psi + r\frac{\partial\psi}{\partial r}\right) + \frac{\partial}{\partial r}\left(\psi + r\frac{\partial\psi}{\partial r}\right)\right] + Vr\psi = E_r r\psi \tag{8.75}$$

$$\therefore -\frac{\hbar^2}{2m^*}\left(\frac{2}{r}\psi + 2\frac{\partial\psi}{\partial r} + \frac{\partial\psi}{\partial r} + \frac{\partial\psi}{\partial r} + r\frac{\partial^2\psi}{\partial r^2}\right) + Vr\psi = E_r r\psi \tag{8.76}$$

and

$$-\frac{\hbar^2}{2m^*}\left(4\frac{\partial\psi}{\partial r} + r\frac{\partial^2\psi}{\partial r^2}\right) + \left(Vr - \frac{\hbar^2}{2m^*}\frac{2}{r}\right)\psi = E_r r\psi \tag{8.77}$$

Dividing through by r gives:

$$\frac{\partial^2\psi}{\partial r^2} + \frac{4}{r}\frac{\partial\psi}{\partial r} + \left[\frac{2m^*}{\hbar^2}(E_r - V) + \frac{2}{r^2}\right]\psi = 0 \tag{8.78}$$

(which resembles equation (15.4) of Schiff [29]). The extra potential term $(1/r^2)$ represents the contribution of the angular momentum to the energy, see [29], p. 81.

This current section has shown a possible way forward for deducing non-spherical eigenstates from a *general* spherically symmetric potential, as may be encountered in a quantum dot. Clearly, this work is incomplete and remains left open for the interested reader.

8.9   APPROACHES TO PYRAMIDAL DOTS

There has been a drive to simplify the fabrication of quantum dots, and in a particular kind of material system this has been achieved with the aid of a process known as *self-assembly*. This occurs when a thin (perhaps even sub-mono) layer of one material is deposited on top of a substrate which has a quite different lattice constant. In such a system, the strain energy in the bonds forms a substantial fraction of the total energy, and in an effort to minimize this the deposited atoms rearrange themselves, moving from a thin layer into 'clumps'. This method has become known as the 'Stranski-Krastanov' growth mode [166]. Microscopy has shown that these small groups of atoms form pyramidal dots, as illustrated earlier in Fig. 8.4.

It appears that the pyramidal dot requires a fully three-dimensional solution of Schrödinger's equation, as given by Cusack *et al.* [167]. However, a little thought allows the problem to be simplified into a combination of a series of decoupled Schrödinger equations and some perturbation theory.

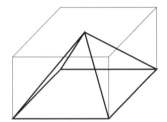 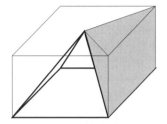

Figure 8.19   A possible perturbative approach for small pyramidal dots

For example, a pyramidal dot could be thought of as a finite-barrier cuboid dot, but with four additional perturbing potentials, as shown in Fig. 8.19 (right) (shaded area). These also take the shape of square based pyramids, with the base, in this example, vertical. However, these pyramids are not symmetric, having the apex centred on the upper edge rather than over the centre of the base.

If the finite dot was solved with decoupled one-dimensional wave functions, then for relatively small systems (< 100 Å) where the unperturbed wave function penetrates considerably into the barrier of the cuboid quantum dot, there is the possibility of a partial cancellation of the '$-2V$' and '$-V$' potentials at the corners and along the edges (see Section 8.6), with these '$+V$' non-symmetric pyramidal potentials. Such a partial cancellation may be vital to ensure convergence of the perturbed energy levels. Recent work employing this approach has derived connection rules between the energy of a pyramidal dot with a cubic quantum dot of a different volume [164].

As mentioned above, lattice mismatch is what drives the self-assembly process, hence the atoms constituting self-assembled quantum dots are under the influence of significant amounts of strain. The effect of this strain is to alter the conduction and valence band profiles so they are no longer constant either inside or outside the quantum dot. The method for determining the strain field and the effect on the conduction and valence band edges has been summarised by Califano and Harrison [168] for both $In_{1-x}Ga_xAs$ on GaAs and $Si_{1-x}Ge_x$ on silicon quantum dots.

8.10   MATRIX APPROACHES

Alternative solutions are being sought to this problem by using an extension of the method of Gangopadhyay and Nag cited earlier [165]. This involves expanding the wave function as a Fourier series of infinite quantum well solutions and has proved itself to be a workable and reliable technique [169, 170].

The basic approach is to expand the two- or three-dimensional wave function of the quantum wire or quantum dot as a linear combination of some basis functions. The basis functions are usually sine or plane waves of a larger box that encompasses the quantum wire or dot and has an infinite exterior potential. This leads to a formulation where the energy solutions are the eigenvales of a matrix equation. There is a thorough introduction to this method and a description on how to apply it to the solution of Schrödinger's equation in two and three dimensions in the book by Harrison [171] (Chapter 3).

8.11   FINITE DIFFERENCE EXPANSIONS

Although less important with the advent of desktop computers with more and more memory, the main problem with matrix methods for the solution of Schrödinger's equation in multiple dimensions is the sheer size of the expansion set. If in one-dimension it was found that ten basis states were needed to reproduce the ground state energy and wave function of some perturbed potential, then in two dimensions for similar accuracy the basis set would have 10×10=100 components. In three dimensions this would become 10×10×10=1000 states. Thus the matrix would be of order 100 or 1000, respectively and this is for the relatively small number of basis functions, in reality 20 or more in each direction may be necessary.

An alternative approach [172, 173] which is less demanding on memory is to return to the idea of a finite difference expansion of the derivatives in Schrödinger's equation, similar to that developed for the shooting method for one dimension in Chapter 3. Recalling the Schrödinger equation for the motion in the confined cross-sectional plane of a quantum wire, as given earlier in equation (8.11) and used again in equation (8.35), i.e.

$$-\frac{\hbar^2}{2m^*}\left[\frac{\partial^2\psi(y,z)}{\partial y^2} + \frac{\partial^2\psi(y,z)}{\partial z^2}\right] + V(y,z)\psi(y,z) = E_{y,z}\psi(y,z) \tag{8.79}$$

Expanding the two derivatives in terms of finite differences, as in equation (3.8), gives:

$$-\frac{\hbar^2}{2m^*}\left[\frac{\psi(y+\delta y,z) - 2\psi(y,z) + \psi(y-\delta y,z)}{(\delta y)^2}\right.$$

$$\left. + \frac{\psi(y,z+\delta z) - 2\psi(y,z) + \psi(y,z-\delta z)}{(\delta z)^2}\right] + V(y,z)\psi(y,z) = E_{y,z}\psi(y,z) \tag{8.80}$$

Multiplying through by $(\delta y)^2(\delta z)^2$:

$$(\delta z)^2\left[\psi(y+\delta y,z) - 2\psi(y,z) + \psi(y-\delta y,z)\right]$$

$$+(\delta y)^2\left[\psi(y,z+\delta z) - 2\psi(y,z) + \psi(y,z-\delta z)\right]$$

$$= -\frac{2m^*}{\hbar^2}\left[E_{y,z} - V(y,z)\right](\delta y\delta z)^2\,\psi(y,z) \tag{8.81}$$

Gathering terms together:

$$(\delta z)^2\left[\psi(y+\delta y,z) + \psi(y-\delta y,z)\right] + (\delta y)^2\left[\psi(y,z+\delta z) + \psi(y,z-\delta z)\right]$$

$$+\left\{\frac{2m^*}{\hbar^2}\left[E_{y,z} - V(y,z)\right](\delta y\delta z)^2 - 2\left[(\delta y)^2 + (\delta z)^2\right]\right\}\psi(y,z) = 0 \tag{8.82}$$

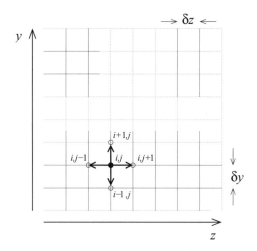

Figure 8.20   The two-dimensional mesh for the finite difference approach to the quantum wire

Thus the wave function at the general point $\psi(y, z)$ depends upon the values at 4 neighbouring points, see Fig. 8.20. Equation (8.82) can be written more succinctly in terms of the indices given in Fig. 8.20:

$$(\delta y)^2(\psi_{i,j+1} + \psi_{i,j-1}) + (\delta z)^2(\psi_{i+1,j} + \psi_{i-1,j}) + k\psi_{i,j} = 0 \tag{8.83}$$

where:

$$k = \frac{2m^*}{\hbar^2}[E_{y,z} - V(y, z)](\delta y \delta z)^2 - 2[(\delta y)^2 + (\delta z)^2] \tag{8.84}$$

Such an equation exists for each grid point on the mesh and to solve simultaneously they have to be written in the form of a matrix equation and solved according to the standard boundary conditions of the wave function and its first derivative tending to zero as the spatial coordinates tend to infinity. The memory saving occurs because the matrix is sparse. El-Moghraby *et al.* [172] map out this method in detail and apply it to rectangular and triangular cross-section quantum wires and pyramidal self-assembled quantum dots. In a later work El-Moghraby *et al.* [173] apply the method to vertically aligned coupled quantum dots.

8.12   DENSITY OF STATES

Just as there is a change in the density of states moving from the bulk (3D) crystal to a quantum well (2D), there is a further change in the density of states on moving to quantum wires (1D) and quantum dots (0D). Recall from Section 2.3 that the density of states is defined as the number of states per unit energy per unit volume of real space, which was expressed mathematically in equation (2.33) as:

$$\rho(E) = \frac{dN}{dE} \tag{8.85}$$

In the bulk crystal, the three degrees of freedom for the electron momentum mapped out a sphere in **k**-space, while in a quantum well the electron momenta fill successively larger circles. Continuing this argument for a quantum wire with just one degree of freedom, the electron momenta then fill states along a line.

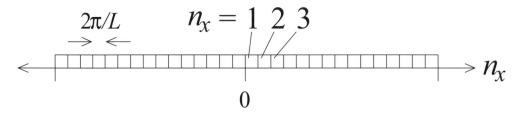

Figure 8.21   The occupation of states in **k**-space along a one-dimensional quantum wire

Therefore, proceeding with the same argument as before, the total number of states N is then equal to the length of the line in **k**-space $(2k)$, divided by the length occupied by one state (again $2\pi/L$), and divided by the *length* in real space (see Fig. 8.21), i.e.

$$N^{1D} = 2 \times 2k \frac{1}{2\pi/L} \frac{1}{L} \tag{8.86}$$

where again the '2' prefactor accounts for spin degeneracy. Therefore:

$$N^{1D} = \frac{4k}{2\pi} \tag{8.87}$$

and for later:

$$\frac{dN^{1D}}{dk} = \frac{2}{\pi} \tag{8.88}$$

In analogy to both the bulk and quantum-well cases, define the density of states for a one-dimensional wire as:

$$\rho^{1D}(E) = \frac{dN^{1D}}{dE} = \frac{dN^{1D}}{dk} \frac{dk}{dE} \tag{8.89}$$

As the along-axis dispersion curve can still be described as a parabola (as given by equation (8.12)), then reuse can be made of equation (2.39), i.e.

$$\frac{dk}{dE} = \left(\frac{2m^*}{\hbar^2}\right)^{\frac{1}{2}} \frac{E^{-\frac{1}{2}}}{2} \tag{8.90}$$

This, taken together with the result in equation (8.88), then gives:

$$\rho^{1D}(E) = \frac{2}{\pi} \left(\frac{2m^*}{\hbar^2}\right)^{\frac{1}{2}} \frac{E^{-\frac{1}{2}}}{2} \tag{8.91}$$

$$\therefore \rho^{1D}(E) = \left(\frac{2m^*}{\hbar^2}\right)^{\frac{1}{2}} \frac{1}{\pi E^{\frac{1}{2}}} \tag{8.92}$$

where the energy E is measured upwards from a subband minimum. Therefore, comparing the density of states for bulk (3D), quantum wells (2D) and quantum wires (1D), as summarized in Table 8.2, it can be seen that successive reductions in degrees of freedom

Table 8.2   The density of states for reduced dimensionality systems, rewritten in a standard form

Dimensionality	$\rho(E)$
3D	$\frac{1}{2\pi^2}\left(\frac{2m^*}{\hbar^2}\right)^{\frac{3}{2}} E^{\frac{1}{2}}$
2D	$\frac{1}{2\pi}\left(\frac{2m^*}{\hbar^2}\right)^{1} E^{0}$
1D	$\frac{1}{\pi}\left(\frac{2m^*}{\hbar^2}\right)^{\frac{1}{2}} E^{-\frac{1}{2}}$

for the electron motion, lead to reductions in the functional form of $\rho(E)$ by factors of $E^{\frac{1}{2}}$.

If there are many (n) confined states within the quantum wire with subband minima E_i, then the density of states at any particular energy is the sum over all the subbands below that point, which can be written as:

$$\rho^{1D}(E) = \sum_{i=1}^{n} \left(\frac{2m^*}{\hbar^2}\right)^{\frac{1}{2}} \frac{1}{\pi(E - E_i)^{\frac{1}{2}}} \Theta(E - E_i) \tag{8.93}$$

Fig. 8.22 gives an example of the 1D density of states, for a 60×70 Å rectangular cross-section GaAs quantum wire surrounded by infinite barriers. In contrast to the bulk and 2D cases displayed earlier in Fig. 2.7, quantum wires show maxima in the density of states at around the subband minima, i.e. at around the point at which charge would be expected to accumulate. Therefore, interband (electron–hole) recombination will have a narrower linewidth than that of the 2D or 3D cases.

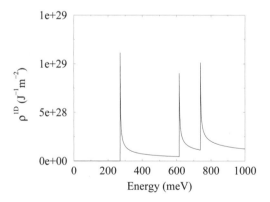

Figure 8.22   The density of states in a one-dimensional quantum wire

The situation for quantum dots is quite different. As the particles are confined in all directions, then there are no dispersion curves, and thus the density of states is just dependent upon the number of confined levels. One single isolated dot would therefore offer just two (spin-degenerate) states at the energy of each confined level, and a plot of the density of states versus energy would be a series of δ-functions.

CHAPTER 9

QUANTUM DOTS

contributed by M. Califano

9.1 0-DIMENSIONAL SYSTEMS AND THEIR EXPERIMENTAL REALIZATION

As discussed in the previous chapter, structures in which the the motion of the carriers is confined further by increasing the value of $\mathcal{D}_c$ to 3 (i.e. confinement in all three dimensions) are called *quantum dots* (QDs). Because the number of degrees of freedom $\mathcal{D}_f$ in these structures is 0 (see equation (8.1) and Table 8.1) they are also known as *zero-dimensional* systems.

A quantum dot can be thought of as an artificial atom, as its density of states consists of a series of very sharp peaks, and its physical properties resemble, in many respects, those of an atom in a cage. However these artificial atoms are also expected to have some significantly different properties compared with real atoms, as they can be filled with both electrons and holes.

Historically, the first techniques employed to realize semiconductor heterostructures that provided 3D confinement for the carriers were lithographic patterning and etching of quantum well structures. There are, in practice, different lithographic techniques (optical lithography and holography, X-ray lithography, electron and focused ion beam lithography, scanning tunnelling lithography, etc.); the fundamental steps involved are, however, essentially the same: (i) the growth of a layered structure followed by the

251

Quantum Wells, Wires and Dots, Third Edition. P. Harrison
©2009 John Wiley & Sons, Ltd.

imposition of a further structure on that, or (ii) the formation of a three-dimensional pattern during growth.

These techniques involve serial processes and are therefore very slow. Nevertheless, they also have several advantages that still make them appealing, such as the possibility of realizing dots of almost arbitrary lateral shape, size and arrangement, depending on the resolution of the particular lithographic technique used, and their general compatibility with modern very large-scale integrated semiconductor technology.

Another related method worth mentioning consists of the creation (with lithographic techniques) of miniature electrodes on the surface of a quantum well. A spatially modulated electric field, that localizes the electrons within a small region, is then produced by applying a suitable voltage to the electrodes. Such electrostatically confined dots are also called *parabolic dots*, due to the fact that the shape of the confining potential can be approximated reasonably well by a parabola [174].

A different approach to the creation of QD structures is self-assembly. This can be achieved in two ways: by chemical or by epitaxial synthesis.

Chemical synthetic processes provide a cheap and fast technique by which almost perfectly crystalline clusters ranging from few hundreds to tens of thousands of atoms can be obtained in a variety of shapes (spheres, rods, arrows, tetrapods, trees and so on) and materials. These so-called *nanocrystals* or *colloidal dots* are synthesized in solution and are covered by organic molecules (ligands) that allow them to be soluble and prevent aggregation. They can be dried and deposited onto substrates made of other materials, incorporated into devices such as photovoltaic cells as sensitisers or used as markers in biological applications.

In the epitaxial deposition on each other of semiconductor materials that differ slightly in lattice constant, the resulting lattice strain is exploited to obtain arrays of three-dimensional islands or quantum dots. In fact, when a material is grown on top of another material with a different lattice constant, as is the case for InGaAs on GaAs, it can strain to conform to the substrate in the plane of the junction (it is assumed that the substrate is so thick that it cannot be distorted significantly). Thus its lattice constant in that plane is reduced (for InGaAs) while it is extended along the direction of growth (the opposite occurs for Si on SiGe) due to the elastic response of the material (as illustrated in Fig. 7.3). This distortion of the active layer causes the build-up of elastic energy, which is relaxed when a critical thickness τ, which depends on the particular heterostructure, is reached, and the two-dimensional growth changes into a three-dimensional one, as shown schematically in Fig. 9.1 for the case of SiGe: coherent [175] (i.e. defect free) islands of the deposited material form spontaneously, with a thin wetting layer left under the islands. The quantum dot islands are then covered (capped, in technical terms) with a layer of the substrate material.

Islands of various sizes and shapes have been reported, depending on the growth conditions: square-based pyramids with typical base length and height of 120 and 60 Å [177], or 240 and 30 Å [178] respectively, lens shaped (hemispherical) islands with base diameters and heights of 200 and 50 Å [179], or 200 and 7 Å [180], respectively. Other islands have been observed [181] to have a square base pyramidal shape before GaAs regrowth, and a lens shape after that, or to become truncated pyramids with an In-rich core having an inverted-triangle [182] or an inverted-cone shape after capping.

It is worth stressing that both shape and size are important parameters variations in which can dramatically affect the electronic structure of the island in both conduction

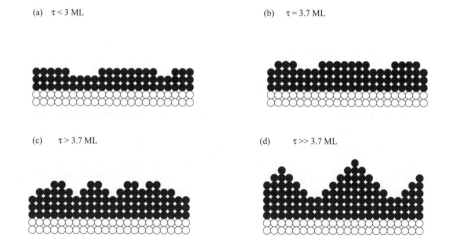

Figure 9.1   Schematic illustration of the growth of Ge (solid circles) on a Si substrate (open circles) [176]. The initial (a) 2D layer-by-layer growth continues up to the critical thickness for the island formation (τ=3.7 monolayers (ML)) (b), after which small islands nucleate on the flat 3.7 ML Ge (c), which turn into 3D islands (d) leaving 3.0 ML of 2D Ge.

and valence bands. Therefore, their knowledge is essential in order for a theoretical model to give an accurate description of the electronic and optical properties of a single QD.

The complexity of the problem increases further, as it is now possible, by using suitable growth conditions, to fabricate arrays and stacks of small self-assembled quantum dots ($\sim$ 10 nm), ordered in size and shape, with high area density ($> 10^{11}$ cm^{-2}) and high optical quality, suitable for creating devices such as low threshold lasers [183]. Furthermore, the presence of external electric and/or magnetic fields is often required to control the number of charge carriers inside these structures.

Calculating the energy levels of a *single* quantum dot alone is already a formidable computational task and the Schrödinger equation must often be solved in three dimensions by means of a numerical method, where in most of the cases, recourse is made to approximations and simplifying hypotheses such as the assumption of equal and isotropic elastic properties of both dot and matrix, and of an effective mass variable only with composition but not with position [184] (the position dependence of the strain components leads, in fact, to a position dependent effective mass [169]), invariance under strain of the matrix elements in the calculation of the electron effective masses [185], decoupling of valence and conduction bands in the calculation of the energies and wave functions [186], neglect of spin-orbit split-off band coupling effects [185, 186], etc. Considered under this light, the problem of solving the Schrödinger equation for an ensemble of dots in a complex structure becomes almost hopeless if all sorts of contributions have to be considered properly. The need therefore arises for as simple a model as possible for describing the electronic structure of a single dot.

9.2   CUBOIDAL DOTS

One of the easiest dot shapes to consider in Cartesian coordinates is the cube. A cuboidal dot can, in fact, be thought of as a generalization of the rectangular cross-section quantum wire treated in Chapter 8, where the motion of the carriers is further confined along the x-axis. Quantum boxes were treated in Section 8.6; however, due to their importance for what follows, their solutions will be presented here as well.

In the simplest case of an infinite confining potential, within the dot the three-dimensional Schrödinger equation (8.6) assumes the form of equation (8.56) The eigenstates of this system will be exploited later on in this chapter as a basis set for the expansion of the solution of the more general problem of a dot with arbitrary shape in a finite potential. By writing the total wave function as a product, and the energy as a sum of three terms:

$$\psi(x, y, z) = \psi(x)\psi(y)\psi(z) \tag{9.1}$$

$$E_{x,y,z} = E_x + E_y + E_z \tag{9.2}$$

equation (8.56) can be decoupled into a set of three independent one-dimensional equations, following the same procedure used in Section 8.3, yielding equations (8.57) (8.58) and (8.59), each of which is identical to equation (2.6), relative to an infinitely deep quantum well (here the three wells are orientated along the three perpendicular directions x, y and z). The total solutions of equation (8.3) will therefore be products of three terms of the form of equation (2.15) in all directions:

$$\psi_{lmn} = \sqrt{\frac{2}{L_x}} \sin\left[\frac{l\pi x}{L_x}\right] \sqrt{\frac{2}{L_y}} \sin\left[\frac{m\pi y}{L_y}\right] \sqrt{\frac{2}{L_z}} \sin\left[\frac{n\pi z}{L_z}\right], \tag{9.3}$$

and the energies of the confined states will be as in equation (8.60), i.e. sums of three terms like equation (2.13)

$$E_{x,y,z} = \frac{\hbar^2 \pi^2}{2m^*} \left(\frac{n_x^2}{L_x^2} + \frac{n_y^2}{L_y^2} + \frac{n_z^2}{L_z^2}\right). \tag{9.4}$$

9.3   DOTS OF ARBITRARY SHAPE

In the more interesting case of a realistic (i.e. finite) potential, $V(x, y, z)$ cannot be expressed as a sum of three independent potentials $V(x)$, $V(y)$ and $V(z)$, therefore the three-dimensional equation of motion cannot be decoupled into three independent one-dimensional equations. Also, as was mentioned at the beginning of this chapter, the shape of the dots synthesised experimentally may vary greatly from the idealized cube treated in the previous section. Still, it is possible to develop a simple model to describe their electronic and optical properties, within the framework of the effective mass approximation using a single-band envelope function technique.

Gershoni et al. [187] first developed a similar approach in 2D, in which they expressed the envelope function Ψ of a rectangular quantum wire with finite confining potential in terms of a linear combination of periodic functions ψ (called basis functions), solutions of a rectangular wire with an infinite barrier height and suitably chosen dimensions:

$$\Psi(x, y, z) = \sum_{lmn} a_{lmn} \psi_{lmn}(x, y, z). \tag{9.5}$$

This expansion is mathematically equivalent to the function Ψ, as the basis functions $\{\psi\}$ form a complete and orthonormal set (COS). However, the choice of the basis set is not unique, and many different COSs can be used. The selection of a specific basis is usually based on symmetry considerations, aiming to minimize the number of functions required to describe a specific system (the larger this number, the longer the calculation will take to run and the larger the memory requirement will be!). In fact, as any COS $\{\psi\}$ is infinite, for practical purposes (i.e. to be able to perform any calculation), only a limited number of basis functions is used and the set is truncated at some value of l, m, n corresponding to a cutoff wave vector $\mathbf{k}_{cut} = (k_x, k_y, k_z)$ in reciprocal space, see equation (2.12). The error made in this approximation, needs, however to be assessed by running convergence tests in which energy eigenvalues and other relevant quantities (e.g. dipole matrix elements) are calculated (see below). The accuracy can be increased by increasing the number of states included in the expansion, equation (9.5).

The advantage of this technique is that, in contrast to the case of the finite quantum well presented in Section 2.5, here there is no need to explicitly match the wave functions across the boundary between barrier and dot materials, as the basis functions extend over the whole system. This method can therefore be applied to structures of arbitrary shape. Moreover, all the matrix elements can be calculated analytically, allowing considerable savings in computational resources.

Gangopadhyay and Nag [188] extended this method to study 3D confined structures such as parallelepipeds and cylinders, whereas Califano and Harrison [189] applied it to pyramidal structures.

In the effective mass approximation the Schrödinger equation for the envelope function of an arbitrarily shaped dot can be written as:

$$-\frac{\hbar^2}{2} \left(\nabla \frac{1}{m^*(x, y, z)} \nabla \right) \Psi(x, y, z) + V(x, y, z) \Psi(x, y, z) = E\Psi(x, y, z) \tag{9.6}$$

This form ensures, amongst other things, that (i) the Hamiltonian is Hermitian, (ii) the wave functions are orthogonal and (iii) the probability current is conserved at the interface of the heterojunction. The envelope function of the dot, $\Psi(x, y, z)$, can then be expanded in terms of a COS of solutions of a structure with infinite confining potential and suitable symmetry (sphere, cylinder, cube, etc.) to match that of the dot. The following example will use as a basis set the solutions of the *cuboidal* problem with infinite barrier height deduced above, i.e.

$$\psi_{lmn} = \sqrt{\frac{2}{L_x}} \sin\left[l\pi \left(\frac{1}{2} - \frac{x}{L_x} \right) \right] \sqrt{\frac{2}{L_y}} \sin\left[m\pi \left(\frac{1}{2} - \frac{y}{L_y} \right) \right] \sqrt{\frac{2}{L_z}} \sin\left[n\pi \frac{z}{L_z} \right] \tag{9.7}$$

where the domains $[-L_x/2, L_x/2]$, $[-L_y/2, L_y/2]$ were chosen for the variation of x and y, and $[0, L_z]$ for that of z (hence the slight difference in the argument of the first two sine functions from those in equation (9.3) above).

Care has to be taken to place the boundaries L_x, L_y, L_z away from the dot surface, so that the energy eigenvalues are independent of their choice. This can be achieved by solving equation (9.6) for different (increasing) sizes of the outer box until the energies obtained converge (see Section 9.3.1 below). As mentioned before, one of the advantages of this approach is that, because there is no need to explicitly match wave functions across the boundary between barrier and dot, it can easily be applied to an arbitrary confining potential. Substituting expression (9.5) into equation (9.6) one obtains:

$$-\frac{\hbar^2}{2}\nabla\left(\frac{1}{m^*(x,y,z)}\nabla\sum_{lmn}a_{lmn}\psi_{lmn}(x,y,z)\right)$$

$$+V(x,y,z)\sum_{lmn}a_{lmn}\psi_{lmn}(x,y,z) = E\sum_{lmn}a_{lmn}\psi_{lmn}(x,y,z) \qquad (9.8)$$

Multiplying on the left by $\psi^*_{l'm'n'}$, and integrating over the cuboid $L_xL_yL_z$ yields:

$$\sum_{lmn}a_{lmn}\left[-\frac{\hbar^2}{2}\int\psi^*_{l'm'n'}\nabla\left(\frac{1}{m^*(x,y,z)}\nabla\psi_{lmn}\right)dx\,dy\,dz\right.$$

$$\left.+\int\psi^*_{l'm'n'}V\psi_{lmn}\,dx\,dy\,dz - E\int\psi^*_{l'm'n'}\psi_{lmn}\,dx\,dy\,dz\right] = 0 \qquad (9.9)$$

which can be expressed in the form of a matrix equation as:

$$(M_{lmnl'm'n'} - E\delta_{mm'}\delta_{ll'}\delta_{nn'})a_{lmn} = 0 \qquad (9.10)$$

where use has been made of the orthonormality of the wave functions:

$$\int\psi^*_{l'm'n'}\psi_{lmn}\,dx\,dy\,dz = \delta_{mm'}\delta_{ll'}\delta_{nn'} \qquad (9.11)$$

The matrix elements $M_{lmnl'm'n'}$ are given by:

$$M_{lmnl'm'n'} = -\frac{\hbar^2}{2}\int\psi^*_{l'm'n'}\nabla\left(\frac{1}{m^*(x,y,z)}\nabla\psi_{lmn}\right)dx\,dy\,dz$$

$$+\int\psi^*_{l'm'n'}V\psi_{lmn}\,dx\,dy\,dz \qquad (9.12)$$

Carrying out the differentiation, the first integral of equation (9.12) becomes:

$$-\frac{\hbar^2}{2}\left[\int\psi^*_{l'm'n'}\left(\nabla\frac{1}{m^*(x,y,z)}\right)(\nabla\psi_{lmn})\,dx\,dy\,dz\right.$$

$$\left.+\int\psi^*_{l'm'n'}\frac{1}{m^*(x,y,z)}\nabla(\nabla\psi_{lmn})\,dx\,dy\,dz\right] \qquad (9.13)$$

The second integral of equation (9.13) is then integrated by parts:

$$\int\psi^*_{l'm'n'}\frac{1}{m^*(x,y,z)}\nabla(\nabla\psi_{lmn})\,dx\,dy\,dz =$$

$$\psi^*_{l'm'n'}\frac{1}{m^*(x,y,z)}\nabla\psi_{lmn}\bigg|_{cuboid} - \int\nabla\left(\psi^*_{l'm'n'}\frac{1}{m^*(x,y,z)}\right)\nabla\psi_{lmn}\,dx\,dy\,dz$$

$$(9.14)$$

The first term (the non-integral one) of equation (9.14) vanishes (remember that the wave functions vanish at the boundaries of the cuboid), and one is left with:

$$-\int \frac{1}{m^*(x,y,z)} \left(\nabla\psi^*_{l'm'n'}\right)\left(\nabla\psi_{lmn}\right) dx\,dy\,dz$$

$$-\int \psi^*_{l'm'n'} \left(\nabla\frac{1}{m^*(x,y,z)}\right)\left(\nabla\psi_{lmn}\right) dx\,dy\,dz$$

the second term of which cancels the first integral of equation (9.13), so that only one term is left (besides the one containing the potential in equation (9.12)). One finally gets:

$$M_{lmnl'm'n'} = \frac{\hbar^2}{2}\int \frac{1}{m^*(x,y,z)}\nabla\psi^*_{l'm'n'}\nabla\psi_{lmn}\,dx\,dy\,dz$$

$$+\int \psi^*_{l'm'n'}V\psi_{lmn}\,dx\,dy\,dz \tag{9.15}$$

The problem here is still the spatial dependence of both the potential and the effective mass in the integrals. Whilst in principle even in the framework of the effective mass approximation such dependence has a complex form (mainly due to the inhomogeneous strain profile), in order to simplify the calculation, V and m^* can be considered constant within each material (a more rigorous justification for that will be given in Section 9.4), i.e. their value exhibits a step-like discontinuity in passing from the barrier region into the dot and another, opposite to the previous, going from the dot into the barrier region. To handle this discontinuity computationally, each integral is split into three parts, within each of which the potential and the effective mass are constant: first one takes an integral with $m^* = m_B$ ($V = V_B$) over the whole cuboid (i.e. barrier plus dot regions), second the integral with $m^* = m_B$ ($V = V_B$) over the dot region is subtracted and third the integral with $m^* = m_D$ ($V = V_D$) over the dot region is added:

$$M_{lmnl'm'n'} = \frac{\hbar^2}{2}\left[\frac{1}{m_B}\int_B \nabla\psi^*_{l'm'n'}\nabla\psi_{lmn}\,dx\,dy\,dz\right.$$

$$-\frac{1}{m_B}\int_D \nabla\psi^*_{l'm'n'}\nabla\psi_{lmn}\,dx\,dy\,dz$$

$$\left.+\frac{1}{m_D}\int_D \nabla\psi^*_{l'm'n'}\nabla\psi_{lmn}\,dx\,dy\,dz\right] + V_B\int_B \psi^*_{l'm'n'}\psi_{lmn}\,dx\,dy\,dz$$

$$-V_B\int_D \psi^*_{l'm'n'}\psi_{lmn}\,dx\,dy\,dz + V_D\int_D \psi^*_{l'm'n'}\psi_{lmn}\,dx\,dy\,dz \tag{9.16}$$

where the subscripts B and D in the integrals mean that the integration is over the barrier and dot region, respectively. The product in the first integral (I_1) of equation (9.16) gives three similar terms. Consider first the term with the derivation in x:

$$I_1^x = \int_B \frac{\partial\psi_{l'm'n'}}{\partial x}\cdot\frac{\partial\psi_{lmn}}{\partial x}\,dx\,dy\,dz$$

$$= \frac{2}{L_x}\pi^2\frac{ll'}{L_x^2}\int_{-L_x/2}^{L_x/2}\cos\left[l'\pi\left(\frac{1}{2}-\frac{x}{L_x}\right)\right]\cos\left[l\pi\left(\frac{1}{2}-\frac{x}{L_x}\right)\right]dx\,\delta_{mm'}\delta_{nn'}$$

Using the relation $2\cos z_1 \cos z_2 = \cos(z_1 - z_2) + \cos(z_1 + z_2)$, the product of the cosines can be expressed as a sum:

$$
I_1^x = \frac{2}{L_x} \frac{ll'}{L_x^2} \pi^2 \delta_{mm'} \delta_{nn'} \frac{1}{2} \int_{-L_x/2}^{L_x/2} \left\{ \cos\left[\pi(l'-l) \left(\frac{1}{2} - \frac{x}{L_x} \right) \right] \right.
$$
$$
\left. + \cos\left[\pi(l'+l) \left(\frac{1}{2} - \frac{x}{L_x} \right) \right] \right\} \, dx \tag{9.17}
$$

which is easily integrated, giving:

$$
I_1^x = \frac{2}{L_x} \frac{ll'}{L_x^2} \pi^2 \delta_{mm'} \delta_{nn'} \frac{1}{2} \left\{ -\frac{\sin\left[\pi(l'-l) \left(\frac{1}{2} - \frac{x}{L_x} \right) \right]}{\frac{\pi(l'-l)}{L_x}} \right|_{-L_x/2}^{L_x/2}
$$
$$
\left. - \frac{\sin\left[\pi(l'+l) \left(\frac{1}{2} - \frac{x}{L_x} \right) \right]}{\frac{\pi(l'+l)}{L_x}} \right|_{-L_x/2}^{L_x/2} \right\}. \tag{9.18}
$$

The second term in the braces vanishes for all l and l'. The first term, instead, is zero for all $l \neq l'$, i.e. is equal to $\delta_{ll'}$. In fact, it vanishes for all l and l' when evaluated at $L_x/2$ (it behaves like $(\sin x^2)/x$ for $x \to 0$), but it's equal to 1 for $l = l'$, when evaluated at $-L_x/2$ (as the usual limit $(\sin x)/x$ for $x \to 0$). It follows that:

$$
I_1^x = \pi^2 \frac{ll'}{L_x^2} \delta_{mm'} \delta_{nn'} \delta_{ll'}. \tag{9.19}
$$

The same holds for the terms containing the derivatives along y and z. The first integral over the barrier region gives therefore:

$$
\int_B \nabla \psi_{l'm'n'}^* \nabla \psi_{lmn} \, dx \, dy \, dz = \pi^2 \left(\frac{ll'}{L_x^2} + \frac{mm'}{L_y^2} + \frac{nn'}{L_z^2} \right) \delta_{ll'} \delta_{mm'} \delta_{nn'} \tag{9.20}
$$

The other integral over the barrier region is simply the normalization of the wave functions (equation (9.11)), i.e. $\delta_{ll'} \delta_{mm'} \delta_{nn'}$, and the only integrals left in the final expression are those over the dot:

$$
M_{lmnl'm'n'} = \left[\frac{\hbar^2 \pi^2}{2} \frac{1}{m_B} \left(\frac{ll'}{L_x^2} + \frac{mm'}{L_y^2} + \frac{nn'}{L_z^2} \right) + V \right] \delta_{ll'} \delta_{mm'} \delta_{nn'}
$$
$$
+ \frac{\hbar^2}{2} \left(\frac{1}{m_D} - \frac{1}{m_B} \right) \int_D \nabla \psi_{l'm'n'}^* \nabla \psi_{lmn} \, dx \, dy \, dz
$$
$$
- V \int_D \psi_{l'm'n'}^* \psi_{lmn} \, dx \, dy \, dz \tag{9.21}
$$

where the confining potential in the dot region is zero ($V_D = 0$) and $V_B = V$. Whilst the calculation described so far applies to dots with any shape, the evaluation of these two remaining integrals requires a mathematical expression for the dot boundaries. In other words, the shape of the dot needs to be specified at this point.

A relevant feature of this method is that all the integrals in equation (9.21) can be performed analytically. Since the integration limits for two of the three variables (x and

y are chosen here) depend on the third one (z, along the height), those two integrations can be performed independently.

$$\int_D \psi^*_{l'm'n'} \psi_{lmn} \, dx \, dy \, dz =$$

$$\frac{8}{L_x L_y L_z} \int_z \int_{x(z)} \sin\left[l'\pi\left(\frac{1}{2} - \frac{x}{L_x}\right)\right] \sin\left[l\pi\left(\frac{1}{2} - \frac{x}{L_x}\right)\right] dx$$

$$\times \int_{y(z)} \sin\left[m'\pi\left(\frac{1}{2} - \frac{y}{L_y}\right)\right] \sin\left[m\pi\left(\frac{1}{2} - \frac{y}{L_y}\right)\right] dy$$

$$\times \sin\left[n'\pi\frac{z}{L_z}\right] \sin\left[n\pi\frac{z}{L_z}\right] dz \qquad (9.22)$$

The general expressions for the integral of a product of sine functions are (see [190]):

$$\int \sin(ax + b)\sin(cx + d) \, dx = \frac{\sin\left[(a - c)x + b - d\right]}{2(a - c)} - \frac{\sin\left[(a + c)x + b + d\right]}{2(a + c)}$$

$$\int \sin(ax + b)\sin(ax + d) \, dx = \frac{x}{2}\cos(b - d) - \frac{\sin(2ax + b + d)}{4a}$$

where the first expression holds for $[a^2 \neq c^2]$.

In the present case $a = -h\pi/L_i$, $b = h\pi/2$, $c = -h'\pi/L_i$ and $d = h'\pi/2$ (where $h = l, m$ and $i = x, y$). Four different situations therefore arise, depending on whether $h = h'$ or not:

(i) $l = l'$ and $m = m'$;

(ii) $l = l'$ and $m \neq m'$;

(iii) $l \neq l'$ and $m = m'$;

(iv) $l \neq l'$ and $m \neq m'$.

Within each of these, there are two more distinct cases, when the integration over z is finally considered:

(a) $n = n'$;

(b) $n \neq n'$.

The situation is similar for the integral containing the product of the gradients of the wave functions.

Once calculated, the 16 different analytical expressions for the matrix elements, equation (9.10) can be solved by direct diagonalization using standard mathematical software such as LAPACK [191].

9.3.1 Convergence tests

Once a specific structure has been identified and the geometry of the problem has been determined (i.e. the shape and composition of both dot and matrix material have been selected), two important parameters must be chosen: the size of the barrier region (which is also known as the supercell) and the number of basis states that will be used in the calculation (which corresponds to the value of the cutoff wave vector k_{cut} in reciprocal space). These quantities, which are mutually interdependent (i.e. the choice of one of them affects that of the other, as, for example, a larger supercell requires a larger number of basis states to achieve a given degree of accuracy in the calculation of the eigenvalues), have the largest influence on the calculation of the electronic structure of any system.

It has been previously mentioned that the boundaries L_x, L_y, L_z must be chosen so that the energy eigenvalues are essentially independent on their choice. The L_i (where $i = x, y, z$) can be defined in terms of the barrier dimensions, i.e. the smallest distance from the dot to the outer box along the three axes, b_x, b_y and b_z, and the dot dimensions l_x, l_y and l_z (where l_x and l_y are the base diameters, and $l_z = h$ is the height), as $L_i = 2b_i + l_i$.

If the convergence tests are performed setting naively $b_x = b_y = b_z$ (a reasonable choice aimed at minimizing the number of independent parameters), problems arise for some matrix elements (i.e. their denominator vanishes) whenever one of the conditions $il_h/L_h = jl_k/L_k$ occurs (where i and j are integers and $h, k = x, y, z$). These problems can be overcome by multiplying the barrier dimensions along x, y and z by different constants, c_x, c_y and c_z (all ≈ 1, so that, from the point of view of the energy calculations, the b_i can be considered equal), which avoids the occurrence of the divergence conditions. A set of calculations has then to be performed for each value of the parameters L_i, using different numbers of basis states.

An example of such a calculation for the ground state energy eigenvalue in the conduction band of an InAs square-based pyramid with $b = 120$ Å and $h = 60$ Å embedded in a GaAs matrix is shown in Fig. 9.2, where b_i values range from 25 Å to 300 Å, and the total number of basis states used is $n_{tot} = n_{wf}^3$, as n_{wf} (varying from 7 to 23) has been chosen to be the same along all three directions x, y and z for simplicity (the material parameters used are listed in Table 9.1 below).

As expected, the number of wave functions required to achieve convergence (defined as the achievement of a minimum for the $y = f(x)$ curve, whose value does not change by increasing the value of x. In the present case $E = f(n_{wf})$), increases with increasing b_i (n_{wf} =15, 19, 19 and 21 for $b_i = 25, 50, 100$ and 200 Å, respectively). However, from Fig. 9.2 another important feature also emerges: if b_i is too small compared with the dot size (e.g. $b_i = 25$ Å), the calculated energy eigenvalues will be too high, no matter how many basis states are used in the expansion of the dot envelope function. In this case the energies are said to be unconverged with respect to the supercell size ($E = f(b_i)$). An indication that convergence in this respect is being achieved is provided by the fact that the difference in the energies calculated with $50 \leq b_i \leq 300$ is only 2 meV, for $n_{wf} = 23$. (A further effect, not shown in Fig. 9.2, of increasing b_i is the decrease in the spacing between excited states energies.)

The choice $b_i = 200$ Å is found to ensure convergence within 1 meV for a wide range of pyramid sizes when a basis set of 6859 (i.e. 19^3) wave functions is used (i.e. using $b_i = 200$ Å, the difference from the converged energy eigenvalue is only 1 meV, as in the example in Fig. 9.2), while giving the lowest excited states energies, compatible with that number of wave functions and the condition of convergence of the ground state energy. Unlike in Fig. 9.2, where $E(b_i = 100) = E(b_i = 200)$ for $n_{wf} = 19$, a barrier dimension $b_i = 100$ Å proves instead too small to yield convergence in large pyramids.

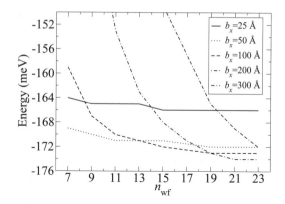

Figure 9.2   Electronic ground state energy levels (with respect to the GaAs conduction band), as a function of the number of wave functions along each direction (i.e. $l, m, n = 1, \ldots, 23$ in equation (9.7): the total number of wave function used is n_{wf}^3, and n_{wf}^3 is the order of the matrix), for an InAs square-based pyramidal QD with $b = 120$ Å and $h = 60$ Å, and different barrier dimensions $b_x (= b_y = b_z)$: $b_x = 25$ Å, dashed line; $b_x = 50$ Å, dot–dashed line; $b_x = 100$ Å, dotted line; $b_x = 200$ Å, solid line; $b_x = 300$ Å, long dashed line.

In order to improve accuracy further, one would need to increase the value of b_i; however this comes at the cost of increasing n_{wf}: more than 12,167 (i.e. 23^3) wave functions are needed for $b_i = 300$ Å in order to obtain a similar accuracy (1 meV), which translates into a memory requirement of ~ 1 Gb to store the matrix. The final choice of the parameters is therefore determined as a trade-off between accuracy and computational demands. For many practical purposes, an accuracy of 1 meV is more than adequate, whereas computers with memories in excess of 1 Gb can be quite expensive.

9.3.2   Efficiency

Given the memory constraints just discussed, a legitimate question to ask is whether the choice of basis states, equation (9.7), was the most efficient. In other words, is it possible by using a different basis set (still solutions of the infinite cube) to decrease the number of states used in the expansion of the envelope function, retaining at the same time the same level of accuracy? A popular choice in the literature is the use of plane waves $\exp{(i k_j x_j)}$ (where $j = 1, 2, 3$, x_j are the spatial coordinates x, y and z, $k_j = 2\pi h / L_j$ and $h = l, m, n$). In this basis, the envelope function could be expressed in a compact form as:

$$\Psi(x, y, z) = \sum_{lmn} \frac{a_{lmn}}{\sqrt{L_x L_y L_z}} \exp\left(2\pi i \left(\frac{l}{L_x} + \frac{m}{L_y} + \frac{n}{L_z} \right) \right)$$

$$= \sum_{\mathbf{k}} \frac{a_{\mathbf{k}}}{\sqrt{\Omega}} \exp(i \mathbf{k} \cdot \mathbf{r}) \tag{9.23}$$

The two basis sets are compared here in order to determine their efficiency in terms of:

- converged energy values, i.e. do both methods give the same set of eigenvalues?

- number of wave functions required to achieve convergence, i.e. how many wave functions are needed in each method to obtain convergence?

- memory requirements, i.e. how much memory is needed to perform the calculations?

- run time, i.e. how long do the calculations take?

The mathematical expressions of the matrix elements for the plane wave calculation are much easier to calculate and more compact to write (they take about one-quarter of the number of lines of code, compared with the sine wave formulation). Nevertheless, when the results of equivalent calculations (i.e. calculations where the same sets of parameters and the same number of wave functions n_{wf} were used), are compared, it is found that the sine waves method gives quicker convergence, in terms of both the number of wave functions employed and the computational time. The latter depends on the order of the matrix and on the size of each element, therefore the more matrix elements to be computed (i.e. the larger n_{wf}), the longer the time required to diagonalize the matrix (see Fig. 9.3).

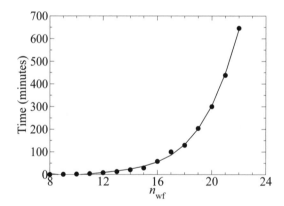

Figure 9.3   Run time (in minutes) as a function of the cube root of the order of the matrix to be diagonalized (i.e. the number of wave functions n_{wf} along each direction). Note the line is a quartic polynomial fit, but is merely there to guide the eye.

And also, considering matrices of the same order, the larger the size of the single matrix element (in terms of memory), the longer the diagonalization time. The matrix elements in the plane wave calculation are *complex* quantities, and therefore occupy twice as much memory as the (*real*) elements of the sine wave method.

Fig. 9.4 compares the ground state energies calculated with the two methods, for a pyramid with $b = 120$ Å and $h = 60$ Å (the barrier dimension is $b_i = 200$ Å): in this case, the plane wave calculation gives an energy at least 7 meV larger than that obtained with the sine wave calculation, for all the n_{wf} considered. This means that, in order to obtain the same accuracy, the plane wave calculation requires more wave functions (i.e.

bigger matrices) and takes much longer to run than the sine wave method. This proves that sine waves are, from all points of view, a more efficient basis set than plane waves for the expansion of the envelope function, when performing such calculations.

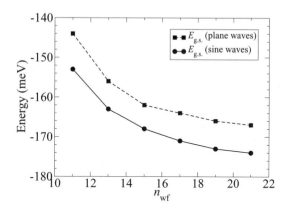

Figure 9.4 Electronic ground state energy levels as a function of the equivalent number of wave functions along each direction (the total number of wave function used is n^3), for an InAs square-based pyramidal QD with $b = 120$ Å and $h = 60$ Å, with respect to the GaAs conduction band. Solid line: sine wave calculations; dashed line: plane wave calculations

It is, perhaps, worth mentioning that, although Barker *et al.* [184] claim that only 11 plane waves along the in-plane directions and 17 along the growth direction are enough to ensure convergence, for a pyramid of the same dimensions considered here, they chose a separation between neighbouring dots equal to the dot dimension, which means a barrier dimension of half this quantity (i.e. $b = 60$ Å). As already discussed above for the sine wave calculation, this choice ensures a quicker convergence (i.e. less wave functions are required to achieve it), but the converged eigenvalue is higher than that obtained when larger barriers are used (as seen in Fig. 9.2). In order to confirm this, another set of plane wave calculations is performed, using a barrier dimension of 60 Å: the converged ground state energy (obtained with $n_{wf} = 19$ along all directions) is found to be $E_{g.s.} = -170$ meV, which is only 1 meV lower than the value obtained with 11 plane waves along the in-plane directions and 17 along the growth direction, but it is 5 meV higher than the value found with the sine wave calculation and a barrier of 200 Å, and 2 meV higher than the energy obtained with the sine wave method and the same barrier of 60 Å.

9.3.3 Optimization

Having proved the superior efficiency of the sine wave formulation compared with the full plane wave approach, it would now be worthwhile exploring possible simple strategies to optimise the code performance, exploiting the distinctive characteristics of the specific system treated. Fig. 9.3 shows the calculation run time as a function of the cube root of the order of the matrix ($n_{wf}^3 \times n_{wf}^3$), or, in other words, the number of wave functions n_{wf} along each direction (where $l, m, n = 1, \ldots, n_{wf}$ in equation (9.7)). So far, the same

n_{wf} has been used along the three directions x, y, z, in the expansion of the envelope function).

A simple general strategy that can be adopted in order to reduce the calculation time is the reduction of the *accuracy* of the quantities used in the calculations from *double* precision to *single* precision (i.e. double→float in C, real*8 → real*4 in Fortran), reducing their size from 8 to 4 bytes. This means that the matrix requires less memory to be stored (it occupies a quarter of the memory needed by a matrix made of *doubles*), and less time (about 25% less) to be diagonalized. The *accuracy* of the eigenvalues, however, is not affected in this case by this change of *computational accuracy*.

Another improvement in terms of run time can be obtained by exploiting the symmetries of the system: in a pyramid, for example, the height is the symmetry axis, therefore the potential is symmetric along x and y, and the in-plane (xy) wave functions can be of either even or odd parity with respect to that axis. If, in the calculations only even (odd) parity functions are considered along x and y, and all, even and odd parity functions, along z (to take into account that the potential is not symmetric along that direction), one might expect to obtain eigenvalues and eigenvectors of only even (odd) parity states. Unfortunately, this conclusion proves not to be completely correct: the calculations with even parity wave functions yield the energy of the ground state (as expected) and of some excited states (fourth, fifth, sixth, etc.), whereas the use of odd parity functions produces the third excited state energy and some higher ones. No calculation is, however, able to reproduce the energies of the first two excited states, which implies that they must contain a superposition of both even- and odd-parity wave functions. Nevertheless, if only the ground state energy is needed, this method proves to be very efficient and reliable, with very low computational demands: it reproduces the results obtained with $n_{wf} = 19$ in the full calculation, using only 10 wave functions along x and y and 19 along z (i.e. a matrix 13 times smaller), in less than 4% of the run time of the unoptimized code.

9.4 APPLICATION TO REAL PROBLEMS

It is now instructive, from the point of view of the aims of this book, to give an example of how such a very simple computational method can be developed into a simple physical model with a surprising predicting power in real-world applications. To show its potential, it will be applied to the study of the electronic and optical properties of InAs self-assembled quantum dots in a GaAs matrix and its results compared both with the predictions of much more complex theoretical approaches and with experimental data relative to a few different samples. The outcome will prove truly remarkable.

9.4.1 InAs/GaAs self-assembled quantum dots

As discussed in Chapter 7 for the more general case of a quantum well, the difference in lattice constant between InAs and GaAs (where $a_{InAs} > a_{GaAs}$), which provides the key ingredient for the formation of the dots, leads to the build-up of a strain field in the system, which affects both the confining potential and the effective mass of the carriers. The confining potential becomes a piecewise continuous function of position and differs from the square well formed by the difference in the absolute energy of the conduction or valence band edges in the bulk dot and barrier material. In the dot, the compressive stress alters the curvature of the bulk bands, causing also the effective masses to differ from the unstrained ones. The presence of shear strain, moreover, leads to the appearance of polarization charges with an associated piezoelectric potential (see Sections 7.5–7.7). Finally, the charge carriers interact via Coulomb giving rise to additional terms in the Hamiltonian (see equation (6.7)).

9.4.2 Working assumptions

This complex picture seems to undermine the simplicity of the computational method illustrated in Section 9.3; however, the complexity of the problem can be considerably reduced, without compromising the integrity of the physical description of the system, by introducing the following approximations:

(i) the motion of electrons and holes is decoupled;

(ii) the effects of strain on both the confining potentials and effective masses are accounted for *on average*;

(iii) piezoelectric effects and Coulomb interactions are neglected.

In this framework (the physical model), therefore, the method can be easily applied, as: (i) the energy levels of electrons and holes are calculated with separate one-band Hamiltonians within the envelope function effective-mass approximation; (ii) the confining potentials and effective masses are assumed to be constant (and equal to their average *strained* value) throughout the dot and the matrix for both carriers and for all dot sizes (see Table 9.1); (iii)(a) piezoelectric effects are disregarded, as in single dots of realistic sizes they generally affect the energies of the levels involved in optical transitions by less than 1 meV [192] (although the effects may be considerably larger in systems of closely spaced dots, since their amplitude is different in each dot: theoretical calculations show [193] that piezoelectricity is one of the most important factors contributing to the level splitting in stacked QDs); (iii)(b) the Coulomb interaction is also neglected, since the QDs considered are in the *strong confinement regime* [194], where the size quantization represents the main part of the carrier energy (their effective radius $r = b/(8\pi)^{1/3}$, where b is the pyramid base, is small compared with the bulk exciton Bohr radius $a_{0,\mathrm{InAs}} = 340$ Å).

Table 9.1 Calculation parameters: m_B barrier region effective mass; m_D dot region effective mass (all in units of the bare electron mass m_0); V_0 carrier confining potential (in meV).

	Electron			Heavy hole	
m_B	m_D	V_0	m_B	m_D	V_0
0.0665^a	0.040^b	450^a	0.3774^a	0.341^a	316^c

[a] [192], [b] [195], [c] [189]

9.4.3 Results

The ground state energy eigenvalues for both electrons and heavy holes, calculated using this simple model, are presented in Fig. 9.5 and compared with the results of more sophisticated theoretical approaches [186, 192] which take into account band mixing, piezoelectric effects and the spatial variation of the confining potentials due to the strain distribution. The agreement is very good for both the conduction and the valence band, especially considering the complex nature of the latter.

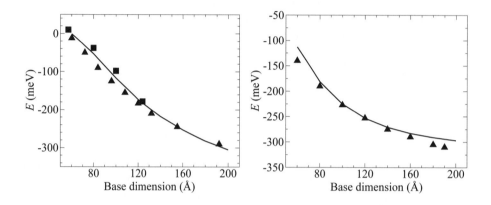

Figure 9.5   Electron (left-hand side) and heavy-hole (right-hand side) ground state energy levels as a function of the base dimension, for InAs square based pyramidal QDs with aspect ratio 1, with respect to the unstrained GaAs conduction and valence bands, respectively. Comparison of the results of the present model (solid lines) with those of Ref. [186] (squares) and of [192] (triangles).

The model predicts no bound electron states for base lengths smaller than about 60 Å, one bound state for 60 Å $< b <$ 100 Å and up to ten for $b =$ 200 Å. Interestingly, the supposedly more accurate (and certainly more elaborate) approach of Ref. [192] predicts only one electron state for this size. The importance of this feature will appear more clearly later on when the results of the simple model will be compared with experimental data. The prediction of three bound states in QDs with $b =$ 120 Å, five for $b =$ 160 Å and seven for $b =$ 180 Å is in excellent agreement with calculations reported by Stier *et al.* [196] who found 4 and 6 bound electron states in structures with $b =$ 136 Å and 170 Å, respectively, by modelling strained QDs with the much more sophisticated 8-band $\mathbf{k} \cdot \mathbf{p}$ theory. As in [186], the first and second excited electron states are found to be degenerate, as expected, due to the C_{4v} symmetry of the pyramidal dot. Fig. 9.6 shows the electronic wave functions $\Psi(x, y, z)$ (calculated at the base of the dot), for the ground and the first excited state (higher excited states are not bound), in an InAs pyramid with $b =$ 120 Å and $h =$ 60 Å.

The predictions of the model are next compared with several photoluminescence (PL) spectra, relative to samples with different sizes [197–200], in Tables 9.2, 9.3 and 9.4, (and in the text), where the identification of the peaks, in terms of transitions between specific

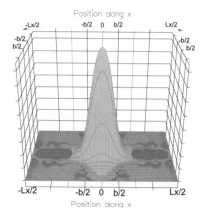 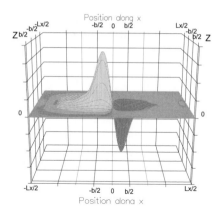

Figure 9.6 Electronic wave function Ψ (calculated at the base of the dot), for the two lowermost conduction states in an InAs pyramid with $b = 120$ Å and $h = 60$ Å.

electron and hole single-particle states, follows the attribution proposed in the original papers.

Table 9.2 Transition energies: comparison between experimental data (deduced from PL spectra reported in Ref. [198]), and theoretical results for a square based pyramid with $b = 110$ Å and $h = 60$ Å. Since the model predicts the first and second excited hole states to be degenerate, the third peak has been attributed to a transition to the third excited hole level.

Transition	$E_{\text{exp.}}$ (eV)	$E_{\text{th.}}$ (eV)	$|\delta E|$ (eV)
$e_0 \rightarrow h_0$	1.10	1.112	0.012
$e_0 \rightarrow h_{1(2)}$	1.17	1.176	0.006
$e_0 \rightarrow h_3$	1.24	1.224	0.016

The emission energy for the transition between states e_i in the conduction band and h_j in the valence band are computed as:

$$E(e_i \rightarrow h_j) = E_g(\text{GaAs}) + E(e_i) + E(h_j)$$

where $E_g(\text{GaAs})$ is the energy gap of bulk GaAs (1.519 eV) and $E(e_i), E(h_j)$ are, respectively, the electron and hole energies relative to the unstrained GaAs band edges (which, as in the example in Fig. 9.5 assume negative values if relative to bound states).

The agreement with both the number and the position of the observed spectral peaks over a wide range of dot sizes and aspect ratios is astonishingly good, especially considering the simplicity of the model: the error δE between calculated and measured energies is always much smaller than the broadening of the PL peaks whose full width at half maximum typical values are of the order of 60–70 meV.

Table 9.3 Transition energies: comparison between experimental data (deduced from PL spectra of Ref. [197]), and theoretical results for an InAs pyramidal dot with $b = 200$ Å and $h = 70$ Å.

Transition	$E_{\text{exp.}}$ (eV)	$E_{\text{th.}}$ (eV)	$|\delta E|$ (eV)
$e_0 \rightarrow h_0$	1.01	0.979	0.022
$e_{1,2} \rightarrow h_{1,2}$	1.10	1.112	0.012
$e_3 \rightarrow h_3$	1.17	1.214	0.044
$e_4 \rightarrow h_4$	1.23	1.237	0.007
$e_5 \rightarrow h_5$	1.29	1.264	0.026

It is interesting to note that, for an InAs pyramidal dot with $b = 200$ Å and $h = 70$ Å, the simple model predicts five different electron states (actually six bound states are predicted, the first and second excited states being degenerate so that there are only five distinct energy levels), whereas the more complex approaches of [186,192,201] are unable to reproduce this fundamental feature. The importance of this result becomes clear when considering that the PL spectrum of such a sample grown by Schmidt *et al.* [197] shows precisely five peaks (see Table 9.3), which were attributed to transitions between electron and hole states with the same quantum numbers and are therefore consistent with the existence of five different electronic energy levels in the QD. Furthermore, the double degeneracy of the first electron and hole excited states predicted by the simple model is also mentioned in Ref. [197]. The calculated energy splitting $\Delta E^{\text{th.}}_{h0,h1} = 29$ meV between the ground and first excited heavy-hole states in this structure is in excellent agreement with the experimentally estimated value of $\Delta E^{\text{exp.}}_{h0,h1} = 27$ meV, obtained by combining capacitance and PL measurements [197]. The predicted electron energy splitting $\Delta E^{\text{th.}}_{e0,e1}$ however, is about 50% larger (104 meV) than the experimental value deduced from the separation between the first two PL peaks (i.e. $\Delta E_{e0,e1} \approx E(e_1 \rightarrow h_1) - E(e_0 \rightarrow h_0) - \Delta E_{h0,h1} = 63$ meV).

These results are surprising, considering that the constant confining potential approximation used here is expected to give a better description of the conduction than the valence band [192]. In fact, whereas the conduction band offset is almost constant throughout the dot, the use of such a simplistic assumption in the valence band is expected to affect the hole levels alignment because of the more complex shape of the confinement's real profile in structures with an aspect ratio of 1 (see [169, 202]). This approximation should be more suitable for flatter structures with high aspect ratios, where both the electrons and holes confining potentials have almost a square well shape [203]. As discussed in Section 9.4.4 below, the suggestion by Nishiguchi and Yoh [204], of including an energy dependence for the effective mass to reduce the overestimate of the energy level separation in the conduction band would not improve the agreement with the experimental data.

For a rectangular based pyramidal QD of height 30 Å and base dimensions 250 and 300 Å along the [110] and [1$\bar{1}$0] directions, respectively, the model predicts six bound

Table 9.4   Transition energies: comparison between experimental data (deduced from PL and PLE spectra of Ref. [199]), and theoretical results for a rectangular-based pyramidal QD of height 30 Å and base dimensions 250 and 300 Å along the [110] and [1$\bar{1}$0] directions, respectively. In this case the model predicts the first and second excited levels not to be degenerate, because of the different symmetry of the system (i.e. the pyramid is not square based).

Transition	$E_{\text{exp.}}$ (eV)	$E_{\text{th.}}$ (eV)	$|\delta E|$ (eV)
$e_0 \rightarrow h_0$	1.220	1.168	0.052
$e_1 \rightarrow h_1$	1.270	1.261	0.009
$e_2 \rightarrow h_2$	1.284	1.280	0.004
$e_3 \rightarrow h_3$	1.332	1.338	0.006
$e_4 \rightarrow h_4$	1.340	1.341	0.001
$e_5 \rightarrow h_5$	1.380	1.378	0.002
$e_6 \rightarrow h_6$	1.412	/	/

electron states (i.e. the ground state and five excited states), thus the absence of any value for the transition between the sixth excited electron and heavy-hole energy levels in Table 9.4. However, Noda *et al.* [199] mention that this last peak may be due to the wetting layer signal as well.

Another spectral position well reproduced by the model is that obtained by Toda *et al.* [200] in near-field magneto-optical spectroscopy measurements of single self-assembled QDs. The structures they investigate have lateral size of ~200 Å and height of ~ 20 Å as indicated by atomic force microscope (AFM) studies of uncapped layers. The typical magnetic field dependence of the peak energies from a single QD they show has a value of about 1347 meV for zero magnetic field. The value calculated with the simple model is 1321 meV. Moreover, only one bound electron state is predicted for such structures.

9.4.4   Concluding remarks

It is, perhaps, worthwhile stressing once more that, although the Schrödinger equation used in the present approach does not explicitly include the strain, its effects are nevertheless accounted for by the choice of average strained values for effective masses and confining potentials (for both electrons and heavy-holes). The choice of these parameters is based on a set of calculations of the strain distribution in pyramidal structures with aspect ratio Q ranging from 1 to 4.5 performed by Califano and Harrison [169], using a method based on the Green's function technique [205], where the anisotropy of the elastic properties was taken into account as well. The carriers' strained confining potentials were calculated as a function of position along the growth direction, in the framework of the 8-band $\mathbf{k} \cdot \mathbf{p}$ theory. The results showed that the confining potentials of both electrons and holes are almost constant along the dot axis for $Q \geq 2$ (in agreement with previously published data [203]), with values in the centre of the island ranging from about 414 to about 463 meV (for the electrons), and from 253 to 332 meV (for the heavy holes), for the experimental structures considered above, proving the constant-confining-potential approximation as a reasonable choice.

Nevertheless, the high degree of agreement obtained by this simple model with the transition energies of so many different experimental samples might appear hard to explain. It seems to strongly indicate that the *details* of the strain distribution are not as important for the transition energy in InAs QDs as they are commonly believed to be.

Califano and Harrison [169] found that the strain distribution for a pyramid depends very weakly on the dot volume (for a given aspect ratio Q), but is more sensitive to a change in Q. This variation, however, principally affects the biaxial component (and is more accentuated near the tip of the dot), whereas the hydrostatic strain is only slightly modified. The strain-induced shift of both conduction and *average* valence band energies, however, is due only to the hydrostatic component, i.e.:

$$\Delta E_c = a_c \epsilon_h \tag{9.24}$$

$$\Delta E_{hh} = a_v \epsilon_h - \frac{b}{2} \epsilon_b \tag{9.25}$$

where ΔE_c, ΔE_{hh}, ϵ_h and ϵ_b are the energy shifts of conduction and heavy-hole bands, and the hydrostatic and biaxial components of the strain, respectively (a_c, a_v and b are the deformation potentials). Therefore, only the biaxial-component-induced heavy-hole energy shift is substantially affected by the variation of the dot aspect ratio. However, whilst this effect is most pronounced towards the pyramid tip, the main contribution to the integral in the matrix element of the heavy-hole confining potential that appears in the Hamiltonian comes from the bottom of the pyramid, where the heavy-hole wave function is localized (see for example [184, 192, 196]), and depends on the pyramid size. Therefore, it is the size quantization together with the (less size-sensitive) hydrostatic strain distribution (which determines the shift of the *average* band position) that represents the dominating part of the heavy-hole energy. This could explain why the transition energies calculated using the same values for the *average* strained hole confining potential agree so well with experiment for such a wide range of dot sizes, and the excellent agreement obtained for the heavy-holes energy splitting $\Delta E_{h0,h1}$.

The situation is a bit more complex for the electron effective mass tensor, the expression of which depends on the particular approximation adopted. Generally, one can say that, because it is a function of the carriers' band edge energies, a similar argument applies as for the heavy-hole band edge, except that in this case the energy shifts of light hole and spin-orbit split-off bands (if their important contribution is included into the effective mass expression) also depend on the biaxial strain. Due to the more complex dependence on the biaxial component of strain however, the resultant effect of a variation in the aspect ratio is less predictable. Furthermore, in the case of the electron the wave function is localized above that of the heavy-holes, yielding a higher value for the matrix element in the region where the difference (of the strain distribution between dots with different values of the aspect ratio) is bigger.

In this case, a compensation mechanism in which a positive difference in the confining potential (i.e. between the assumed average strained value of 450 meV and the average strained value calculated for the specific sample) is compensated by a negative difference in the value of the effective mass (so that when the confining potential is < 450 meV, the effective mass is > 0.04), and vice versa, could be responsible for the agreement in the transition energies.

Finally, a few words on excitonic effects are perhaps appropriate here as well. The simple model discussed so far calculates the total energy of the electron-hole system (the exciton) as the sum of its *single-particle* energies, i.e. the energy levels of electrons and holes are considered to be independent from the presence of the other particle as if, instead of 'living together' under the same 'pyramidal roof' they were infinitely far apart. However, as in reality the electron and the hole are confined within a space that is much smaller than the exciton Bohr radius in the bulk (the radius of an exciton in bulk InAs is 34 nm), there is considerable Coulomb (i.e. electrostatic) interaction between them due to their opposite charge. This effect can be taken into account perturbatively, and the exciton energy can be written within first order as:

$$E(h_i, e_j) = (\varepsilon_{e_j} - \varepsilon_{h_i}) - (J_{e_j, h_i} - K_{e_j, h_i}) \tag{9.26}$$

where the first term contains the energies of the (non-interacting) electron and hole, respectively, in levels j and i, and the last term represents direct and exchange Coulomb energies given by:

$$J_{i,j} = \iint \frac{|\psi_i(\boldsymbol{r}_1)|^2 |\psi_j(\boldsymbol{r}_2)|^2}{\epsilon(\boldsymbol{r}_1, \boldsymbol{r}_2)|\boldsymbol{r}_1 - \boldsymbol{r}_2|} \mathrm{d}\boldsymbol{r}_1 \mathrm{d}\boldsymbol{r}_2 \tag{9.27}$$

and

$$K_{i,j} = \iint \frac{\psi_i^*(\boldsymbol{r}_1)\psi_j^*(\boldsymbol{r}_1)\psi_i(\boldsymbol{r}_2)\psi_j(\boldsymbol{r}_2)}{\epsilon(\boldsymbol{r}_1, \boldsymbol{r}_2)|\boldsymbol{r}_1 - \boldsymbol{r}_2|} \mathrm{d}\boldsymbol{r}_1 \mathrm{d}\boldsymbol{r}_2 \tag{9.28}$$

where ϵ is the dielectric constant of the dot. The largest contribution in the second term comes from the direct Coulomb interaction and was calculated to be of the order of a few tens of meV [206] for a wide range of dot shapes and has therefore been neglected for simplicity in the present approach (Incidentally this is also the order of magnitude of the discrepancies between theory and experiment found above, see Tables 9.2–9.4.) The exchange energy, was found [206] to be an order of magnitude smaller than the direct Coulomb term and was similarly neglected. Other many-body effects such as correlations, even though responsible for the excitonic 'fine structure' [207], are not relevant at the level of resolution discussed above, as they are expected to have even smaller contributions to the total exciton energy.

9.5   A MORE COMPLEX MODEL IS NOT ALWAYS A BETTER MODEL

This simple model can be modified to include a more detailed account of the strain-induced spatial variation of the crucial parameters V and $m_{\mathrm{eff.}}$. As mentioned in the previous section, the strain distribution in self-assembled pyramidal $In_{1-x}Ga_xAs/GaAs$ quantum dots can be calculated using a method based on the Green's function technique [205], accounting for the dependence of the biaxial and hydrostatic components on the quantum dot volume, aspect ratio and composition. The position dependence of the strained carriers' confining potentials and electron effective mass (in-plane and perpendicular components) can then be calculated in the framework of eight-band $\mathbf{k} \cdot \mathbf{p}$ theory [169]. In this respect, the inclusion of the coupling with the spin-orbit split-off band in the calculation of the effective mass tensor has been shown [169] to be crucial for a correct description of the effects of the strain.

The spatial dependence of V and $m_{eff.}$, however, makes it impossible to keep an analytical form for the matrix elements in the Hamiltonian, and resort to a time consuming numerical integration is needed in the calculation of the electronic energy levels. Unfortunately, this considerable increase in the complexity of the model does not lead to a comparable increase in the accuracy of its results. Calculations performed within the framework of the effective mass approximation showed [202] that the inclusion of such an accurate account of the effect of strain, resulting in position-dependent parameters, does not induce any appreciable variation, at least in the ground state electronic energies, compared with the constant potential and effective mass choice detailed in Table 9.1.

This result seems to confirm the suggestion made above that *details* of the strain distribution are apparently not as important for the transition energies in InAs QDs as they are commonly believed to be, and that simple models that correctly capture the physics of the system can successfully predict its energies too.

CHAPTER 10

CARRIER SCATTERING

10.1 FERMI'S GOLDEN RULE

If a charge carrier, i.e. an electron or a hole, is moving within the body of a perfect crystal lattice which is free from all defects and with all atoms stationary, then it will continue in that state *ad infinitum*. Of course, such a situation is never reached, which implies that the charge carrier will change its state—a process which is known as *scattering*.

Quantum mechanical scattering is usually summarized in terms of *Fermi's Golden Rule* [208] which states the following: if an electron (or hole) in a state $|i\rangle$ of energy E_i experiences a time-dependent perturbation $\tilde{\mathcal{H}}$ which could scatter (transfer) it into any one of the final states $|f\rangle$ of energy E_f, then the lifetime of the carrier in state $|i\rangle$ is given by:

$$\frac{1}{\tau_i} = \frac{2\pi}{\hbar} \sum_f \left| \langle f | \tilde{\mathcal{H}} | i \rangle \right|^2 \delta(E_f - E_i) \tag{10.1}$$

10.2 PHONONS

The massive atoms that constitute semiconductor crystals are all connected together by chemical bonds which are nominally covalent, although in compounds can have a degree of ionicity. These atoms are always in a state of continual motion, which because of the definite crystal lattice structure, is vibrational around an equilibrium position. The

Quantum Wells, Wires and Dots, Third Edition. P. Harrison
©2009 John Wiley & Sons, Ltd.

atoms vibrate even at the hypothetical zero of absolute temperature—the so-called *zero point energy* (see Section 3.5). In some ways, the vibrations of these interconnected quantum particles (atoms) resembles a classical (macroscopic) system of a series of masses connected by springs. There are basically four different modes of vibration, as illustrated in Figs. 10.1 and 10.2, each one of which is referred to as a *phonon*.

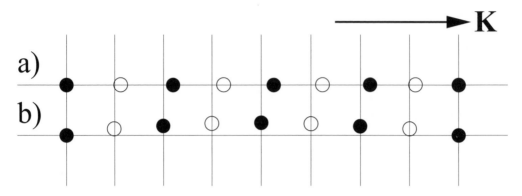

Figure 10.1   Schematic illustrations of the atomic displacements in (a) longitudinal acoustic (LA) and (b) transverse acoustic (TA) phonon modes

The acoustic modes shown in Fig. 10.1 are characterized by the neighbouring atoms being in phase. In the longitudinal mode the atomic displacements are in the same direction as the direction of energy transfer, while in the transverse mode the atomic displacements are perpendicular to this direction.

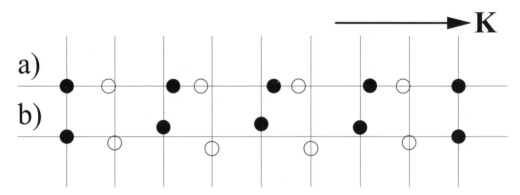

Figure 10.2   Schematic illustrations of the atomic displacements in (a) longitudinal optic (LO) and (b) transverse optic (TO) phonon modes

The longitudinal and transverse definitions also apply to the two types of optic phonon modes as illustrated in Fig. 10.2. However, in this type of lattice vibration the displacements of neighbouring atoms are in opposite phase.

The wave-like nature of the lattice vibrations allows them to be described, say, by an angular frequency ω and a wave vector $\mathbf{K}$. Thus the energy of a phonon is $\hbar\omega$—the same as a photon of light. In addition, and in analogy to propagating electrons, the momentum of a phonon is said to be quantized and of value $\hbar\mathbf{K}$. Furthermore, phonons are diffracted by the crystal lattice just like electrons and holes, and thus a Brillouin zone type summary of the energy–momentum curves can be employed.

Fig. 10.3 shows schematically just such a set of phonon dispersion curves for a typical semiconductor. The form of the curves are reasonably similar, although with differing energy scales for the common semiconductors that are of interest in this work (see, for example [7], p. 14); note, however, that the coupling (interaction) between the different phonon modes with charge carriers does differ between the various materials.

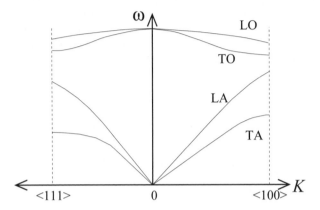

Figure 10.3   Phonon energy ($\hbar\omega$) versus momentum ($\hbar K$) curves for a typical semiconductor

Phonons are bosons, and hence their number per unit volume is given by the Bose–Einstein factor (see [209], p. 391 or [1], p. 454):

$$N_0 = \frac{1}{\exp\left(\hbar\omega/kT\right) - 1} \tag{10.2}$$

As the phonons themselves represent the motion of atoms that are centres of electric charge, then they also represent time-dependent perturbations $\tilde{\mathcal{H}}$ of the crystal potential and can therefore scatter charge carriers. Many authors have previously considered scattering via phonons (see for example [210–212]); however, the spirit of this present work is to provide a *fully* documented derivation.

With this aim, consider a simple wave function of a phonon in bulk material*:

$$\phi = C(\mathbf{K})\mathrm{e}^{-i\mathbf{K}\cdot\mathbf{r}} \tag{10.3}$$

Therefore, the electric field, which is the derivative of this wave function, can be described by the following relationship:

$$\mathbf{E} = \nabla\phi = -i\mathbf{K}\phi \tag{10.4}$$

and hence:

$$\mathbf{E}^*\mathbf{E} = |C(\mathbf{K})|^2\,\mathbf{K}\cdot\mathbf{K} \tag{10.5}$$

The normalization condition is therefore:

$$\frac{1}{2}\hbar\omega(\mathbf{K}) = \frac{1}{2}\omega(\mathbf{K})\left.\frac{\partial\epsilon}{\partial\omega}\right|_{\mathbf{K}}\int\mathbf{E}^*\mathbf{E}\;\mathrm{d}\tau \tag{10.6}$$

*The author would like to thank Paul Kinsler for this contribution.

If V is the volume of the crystal, then as the integrand is independent of position (see equation (10.5)):

$$\frac{1}{2}\hbar\omega(\mathbf{K}) = \frac{1}{2}\omega(\mathbf{K}) \left.\frac{\partial\epsilon}{\partial\omega}\right|_{\mathbf{K}} V \left|C(\mathbf{K})\right|^2 \mathbf{K}_\bullet\mathbf{K} \tag{10.7}$$

The majority of the interest here lies with heterostructures made from compound semiconductors. These materials are polar as the different electronegativities of the constituent atoms lead to a degree of ionicity in the chemical bonds (see Chapter 15). In such materials the dominant electron–phonon interaction (scattering) is with the *longitudinal optic phonon*, often referred to as the LO phonon. The LO phonon dispersion curve (see Fig. 10.3), is relatively flat, and hence it is possible to approximate it as being *dispersionless*, which gives $\omega\partial\epsilon/\partial\omega = 2/P$ [213], where:

$$P = \frac{1}{\epsilon_\infty} - \frac{1}{\epsilon_s} \tag{10.8}$$

with ϵ_∞ and ϵ_s being, respectively, the high- and low-frequency permittivities of the material. Therefore:

$$\frac{1}{2}\hbar\omega(\mathbf{K}) = \frac{1}{2}\frac{2}{P}V \left|C(\mathbf{K})\right|^2 \mathbf{K}_\bullet\mathbf{K} \tag{10.9}$$

The normalization coefficients $C(\mathbf{K})$ of the phonon wave functions are therefore given by:

$$\left|C(\mathbf{K})\right|^2 = \frac{\hbar\omega P}{2V\left|\mathbf{K}\right|^2} \tag{10.10}$$

i.e. the normalized wave function of a single dispersionless phonon is:

$$\phi = \left(\frac{\hbar\omega P}{2\left|\mathbf{K}\right|^2}\right)^{\frac{1}{2}} \frac{e^{-i\mathbf{K}_\bullet\mathbf{r}}}{V^{\frac{1}{2}}} \tag{10.11}$$

The total *phonon interaction term* is thus obtained by summing over all phonon wave vectors, i.e.

$$\tilde{\mathcal{H}} = e\sum_{\mathbf{K}}\phi \tag{10.12}$$

and therefore, by using equation (10.12), obtain:

$$\tilde{\mathcal{H}} = e\sum_{\mathbf{K}}\left(\frac{\hbar\omega P}{2\left|\mathbf{K}\right|^2}\right)^{\frac{1}{2}} \frac{e^{-i\mathbf{K}_\bullet\mathbf{r}}}{V^{\frac{1}{2}}} \tag{10.13}$$

10.3   LONGITUDINAL OPTIC PHONON SCATTERING OF BULK CARRIERS

Although not the main emphasis of this present work, it is worthwhile deducing here the *scattering rates* $(1/\tau_i)$ of electrons in bulk bands with these longitudinal optic (LO) phonons, as many of the mathematical techniques used will be required for the quantum well system that follows.

In a bulk crystal, the electron wave functions are simply given by:

$$|i\rangle = \psi_i = \frac{e^{-i\mathbf{k}_i_\bullet\mathbf{r}}}{V^{\frac{1}{2}}} \tag{10.14}$$

Therefore, substituting for the electron wave function and the phonon interaction of equation (10.13) into Fermi's Golden Rule (equation (10.1)) gives:

$$\frac{1}{\tau_i} = \frac{2\pi}{\hbar} \sum_{\mathbf{k}_f} \left| \int \frac{e^{i\mathbf{k}_f \cdot \mathbf{r}}}{V^{\frac{1}{2}}} e \sum_{\mathbf{K}} \left(\frac{\hbar \omega P}{2|\mathbf{K}|^2} \right)^{\frac{1}{2}} \frac{e^{-i\mathbf{K} \cdot \mathbf{r}}}{V^{\frac{1}{2}}} \frac{e^{-i\mathbf{k}_i \cdot \mathbf{r}}}{V^{\frac{1}{2}}} \, d\mathbf{r} \right|^2 \delta(E_f^t - E_i^t) \qquad (10.15)$$

The additional index on the initial and final state energies has been introduced in order to specify total energy, which could be the sum of a band minimum and the electron kinetic energy.

The above basically represents the lifetime for scattering for a total phonon population of 1, and although the sum over all of the phonon wave vectors should introduce a population term, it is simpler just to add it here manually, i.e.

$$\frac{1}{\tau_i} = \frac{2\pi e^2 \hbar \omega P}{2\hbar V} \left(N_0 + \frac{1}{2} \mp \frac{1}{2} \right) \sum_{\mathbf{k}_f} \left| \int \sum_{\mathbf{K}} \frac{e^{-i(\mathbf{k}_i - \mathbf{k}_f + \mathbf{K}) \cdot \mathbf{r}}}{V|\mathbf{K}|} \, d\mathbf{r} \right|^2 \delta(E_f^t - E_i^t) \qquad (10.16)$$

The factor $(N_0 + \frac{1}{2} \mp \frac{1}{2})$ represents the phonon density within the crystal. The upper sign of the $\mp$ represents absorption, which reduces the phonon population from $(N_0 + 1)$ to N_0, while the lower sign represents emission of a phonon which increases the number of phonons from N_0 to $(N_0 + 1)$. Just for convenience, absorb this factor into P and rewrite as P', i.e.

$$P' = \left(\frac{1}{\epsilon_\infty} - \frac{1}{\epsilon_s} \right) \left(N_0 + \frac{1}{2} \mp \frac{1}{2} \right) \qquad (10.17)$$

Then the integral over all space, specified by the $d\mathbf{r}$, of the exponential function $\exp\left[-i\left(\mathbf{k}_i - \mathbf{k}_f + \mathbf{K}\right) \cdot \mathbf{r}\right]$, can be converted into a δ-function by taking the surfaces of the volume V to be effectively at infinity, thus giving a factor of 2π per dimension, i.e. the lifetime of the carrier in state i is now given by:

$$\frac{1}{\tau_i} = \frac{\pi e^2 \omega P'}{V} \sum_{\mathbf{k}_f} \left| \sum_{\mathbf{K}} \frac{(2\pi)^3}{V} \delta \left(\mathbf{k}_i - \mathbf{k}_f + \mathbf{K} \right) \frac{1}{|\mathbf{K}|} \right|^2 \delta(E_f^t - E_i^t) \qquad (10.18)$$

The sum over $\mathbf{K}$ can be converted into an integral, which introduces a factor of $L/2\pi$ per dimension (see Section 2.3); over all three dimensions, this introduces a factor of $V/(2\pi)^3$, which cancels with the existing $(2\pi)^3/V$, and hence:

$$\frac{1}{\tau_i} = \frac{\pi e^2 \omega P'}{V} \sum_{\mathbf{k}_f} \left| \int \frac{\delta \left(\mathbf{k}_i - \mathbf{k}_f + \mathbf{K} \right)}{|\mathbf{K}|} \, d\mathbf{K} \right|^2 \delta(E_f^t - E_i^t) \qquad (10.19)$$

In a similar manner, the sum over $\mathbf{k}_f$ can also be changed into an integral, again introducing a factor of $V/(2\pi)^3$:

$$\therefore \frac{1}{\tau_i} = \frac{\pi e^2 \omega P'}{V} \int \frac{V}{(2\pi)^3} \left| \int \frac{\delta \left(\mathbf{k}_i - \mathbf{k}_f + \mathbf{K} \right)}{|\mathbf{K}|} \, d\mathbf{K} \right|^2 \delta(E_f^t - E_i^t) \, d\mathbf{k}_f \qquad (10.20)$$

Expanding the modulus squared gives:

$$\frac{1}{\tau_i} = \frac{\pi e^2 \omega P'}{(2\pi)^3} \int \left[\int \frac{\delta \left(\mathbf{k}_i - \mathbf{k}_f + \mathbf{K} \right)}{|\mathbf{K}|} \, d\mathbf{K} \right]$$

$$\times \left[\int \frac{\delta \left(\mathbf{k}_i - \mathbf{k}_f + \mathbf{K}' \right)}{|\mathbf{K}'|} \ d\mathbf{K}' \right] \delta(E_f^t - E_i^t) \ d\mathbf{k}_f \tag{10.21}$$

Changing the order of integration, then:

$$\therefore \frac{1}{\tau_i} = \frac{\pi e^2 \omega P'}{(2\pi)^3} \int \int \int \frac{1}{|\mathbf{K}||\mathbf{K}'|} \delta \left(\mathbf{k}_i - \mathbf{k}_f + \mathbf{K} \right) \delta \left(\mathbf{k}_i - \mathbf{k}_f + \mathbf{K}' \right)$$

$$\times \delta(E_f^t - E_i^t) \ d\mathbf{K}' \ d\mathbf{k}_f \ d\mathbf{K} \tag{10.22}$$

Now the integral over $\mathbf{K}'$ only gives anything when $\mathbf{k}_i - \mathbf{k}_f + \mathbf{K}' = 0$, i.e. when $\mathbf{K}' = -\mathbf{k}_i + \mathbf{k}_f$, and therefore:

$$\frac{1}{\tau_i} = \frac{\pi e^2 \omega P'}{(2\pi)^3} \int \int \frac{1}{|\mathbf{K}|| - \mathbf{k}_i + \mathbf{k}_f|} \delta \left(\mathbf{k}_i - \mathbf{k}_f + \mathbf{K} \right) \delta(E_f^t - E_i^t) \ d\mathbf{k}_f \ d\mathbf{K} \tag{10.23}$$

Similarly, the integral over $\mathbf{k}_f$ only makes a contribution when $\mathbf{k}_i - \mathbf{k}_f + \mathbf{K} = 0$, i.e. when $\mathbf{k}_f = \mathbf{k}_i + \mathbf{K}$, and therefore:

$$\frac{1}{\tau_i} = \frac{\pi e^2 \omega P'}{(2\pi)^3} \int \frac{1}{|\mathbf{K}|| - \mathbf{k}_i + \mathbf{k}_i + \mathbf{K}|} \delta(E_f^t - E_i^t) \ d\mathbf{K} \tag{10.24}$$

which gives:

$$\frac{1}{\tau_i} = \frac{\pi e^2 \omega P'}{(2\pi)^3} \int \frac{1}{|\mathbf{K}|^2} \delta(E_f^t - E_i^t) \ d\mathbf{K} \tag{10.25}$$

Now the total energy E_i^t of the system before the scattering event is equal to the energy of the charge carrier plus (or minus) the energy ($\hbar \omega$) of the phonon for absorption (or emission). In addition, the carrier energy can be split into its potential and kinetic components, i.e. for parabolic bands, the total energy of a carrier is equal to the energy of the band minimum plus a component proportional to the momentum squared, i.e.

$$E_i^t = E_i + \frac{\hbar^2 \mathbf{k}_i^2}{2m^*} \pm \hbar \omega \tag{10.26}$$

$$E_f^t = E_f + \frac{\hbar^2 \mathbf{k}_f^2}{2m^*} \tag{10.27}$$

where E_i and E_f represent the energy band minima of the initial (i) and final (f) states, respectively. A note of caution must be added here; the assumption of parabolic bands is fundamental, the derivation beyond this point is dependent upon it. While this is generally true for typical situations met in n-type material, care must be taken for the valence band, where non-parabolicity is paramount.

The optical branches of the phonon dispersion curves in typical III–V materials are quite flat, which means that the energy $\hbar \omega$ of the LO phonon is only a weak function of the phonon wave vector $\mathbf{K}$ (see for example [1], p. 435 and [14] p. 70). Hence, as mentioned before, this derivation will follow the standard assumption that the phonon energy $\hbar \omega$ can be approximated well with a constant value (taken as 36 meV in GaAs ([14], p. 92)).

Substituting into equation (10.25) gives:

$$\frac{1}{\tau_i} = \frac{\pi e^2 \omega P'}{(2\pi)^3} \int \frac{1}{|\mathbf{K}|^2} \delta \left(E_f + \frac{\hbar^2 \mathbf{k}_f^2}{2m^*} - \left(E_i + \frac{\hbar^2 \mathbf{k}_i^2}{2m^*} \pm \hbar \omega \right) \right) \ d\mathbf{K} \tag{10.28}$$

$$\therefore \frac{1}{\tau_i} = \frac{\pi e^2 \omega P'}{(2\pi)^3} \int \frac{1}{|\mathbf{K}|^2} \delta \left(E_f - E_i + \frac{\hbar^2 \mathbf{k}_f^2}{2m^*} - \frac{\hbar^2 \mathbf{k}_i^2}{2m^*} \mp \hbar\omega \right) \, d\mathbf{K} \qquad (10.29)$$

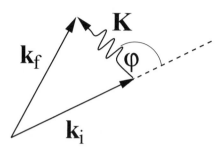

Figure 10.4   Momentum conservation in phonon scattering

As this is scattering of electrons (carriers) in bulk material, then consider the simplest case where the carrier remains within the same *band*, i.e. $E_i = E_f$, then:

$$\frac{1}{\tau_i} = \frac{\pi e^2 \omega P'}{(2\pi)^3} \int \frac{1}{|\mathbf{K}|^2} \delta \left(\frac{\hbar^2 \mathbf{k}_f^2}{2m^*} - \frac{\hbar^2 \mathbf{k}_i^2}{2m^*} \mp \hbar\omega \right) \, d\mathbf{K} \qquad (10.30)$$

where the argument of the δ-function can be factorized, thus giving:

$$\frac{1}{\tau_i} = \frac{\pi e^2 \omega P'}{(2\pi)^3} \int \frac{1}{|\mathbf{K}|^2} \delta \left(\frac{\hbar^2}{2m^*} \left(\mathbf{k}_f^2 - \mathbf{k}_i^2 \mp \frac{2m^* \omega}{\hbar} \right) \right) \, d\mathbf{K} \qquad (10.31)$$

Then, employing the rule $\delta(ax) = \delta(x)/a$, with the constant $a = \hbar^2/2m^*$, obtain:

$$\frac{1}{\tau_i} = \frac{2\pi m^* e^2 \omega P'}{\hbar^2 (2\pi)^3} \int \frac{1}{|\mathbf{K}|^2} \delta \left(\mathbf{k}_f^2 - \mathbf{k}_i^2 \mp \frac{2m^* \omega}{\hbar} \right) \, d\mathbf{K} \qquad (10.32)$$

Consider now the conservation of momentum between the two states of the charge carrier, labelled with their initial $\mathbf{k}_i$ and final $\mathbf{k}_f$ momenta, respectively, together with the momentum of the phonon $\mathbf{K}$, as illustrated in Fig. 10.4. The aim is to remove the dependency of the integrand on the final momentum state $\mathbf{k}_f$. This can be achieved by defining the angle between the initial momentum $\mathbf{k}_i$ and the phonon momentum $\mathbf{K}$ as ϕ, and then by using the cosine rule and the shorthand notation $|\mathbf{K}| = K$:

$$\mathbf{k}_f^2 = \mathbf{k}_i^2 + \mathbf{K}^2 - 2k_i K \cos(\pi - \phi) \qquad (10.33)$$

which gives:

$$\mathbf{k}_f^2 = \mathbf{k}_i^2 + \mathbf{K}^2 + 2k_i K \cos\phi \qquad (10.34)$$

Substituting for $\mathbf{k}_f^2$ in equation (10.32) then gives:

$$\frac{1}{\tau_i} = \frac{m^* e^2 \omega P'}{\hbar^2 (2\pi)^2} \int \frac{1}{|\mathbf{K}|^2} \delta \left(\mathbf{K}^2 + 2k_i K \cos\phi \mp \frac{2m^* \omega}{\hbar} \right) \, d\mathbf{K} \qquad (10.35)$$

Converting the integral over all of the phonon momentum states $\mathbf{K}$ into spherical polar coordinates, with the Cartesian z-axis along the initial momentum $\mathbf{k}_i$ in order to conserve the angle ϕ, then the elemental volume becomes $K^2 \, dK \, \sin\phi \, d\phi \, d\theta$, therefore:

$$\frac{1}{\tau_i} = \frac{m^* e^2 \omega P'}{\hbar^2 (2\pi)^2} \int_0^{2\pi} \int_0^{\pi} \int_0^{\infty} \frac{1}{|\mathbf{K}|^2} \delta \left(\mathbf{K}^2 + 2k_i K \cos\phi \mp \frac{2m^* \omega}{\hbar} \right)$$

$$\times K^2 \; dK \; \sin\phi \; d\phi \; d\theta \tag{10.36}$$

Now clearly, $|\mathbf{K}|^2 = K^2$, as the latter is just a shorthand version, and in addition, the integral over the angle θ only introduces a factor of 2π as the integrand is independent of θ; therefore:

$$\frac{1}{\tau_i} = \frac{m^* e^2 \omega P'}{\hbar^2 2\pi} \int_0^\pi \int_0^\infty \delta\left(K^2 + 2k_i K \cos\phi \mp \frac{2m^*\omega}{\hbar}\right) \; dK \; \sin\phi \; d\phi \tag{10.37}$$

Write the constant prefactor as:

$$\Upsilon' = \frac{m^* e^2 \omega P'}{2\pi\hbar^2} = \frac{m^* e^2 \omega P}{2\pi\hbar^2}\left(N_0 + \frac{1}{2} \mp \frac{1}{2}\right) \tag{10.38}$$

and by considering the absorption process first (i.e. the upper sign in the $\mp$) then:

$$\frac{1}{\tau_i} = \Upsilon' \int_0^\pi \int_0^\infty \delta\left(K^2 + 2k_i K \cos\phi - \frac{2m^*\omega}{\hbar}\right) \; dK \; \sin\phi \; d\phi \tag{10.39}$$

Following the approach of Hagston, Piorek, and Harrison (for more details of this and related work see the PhD thesis of Piorek [214]), the argument of the δ-function can be factorized uniquely into:

$$(K - \alpha_1)(K + \alpha_2) = K^2 + (\alpha_2 - \alpha_1)K - \alpha_1\alpha_2 \tag{10.40}$$

where the constants α_1 and α_2 are real and positive—this follows since the product $\alpha_1\alpha_2$ must be negative. Hence, since K, which is the modulus of the phonon wave vector, must be greater than zero, then there is only one contribution to the δ-function, i.e.

$$\frac{1}{\tau_i} = \Upsilon' \int_0^\pi \int_0^\infty \delta\left((K - \alpha_1)(K + \alpha_2)\right) \; dK \; \sin\phi \; d\phi \tag{10.41}$$

Around the solution $K = \alpha_1$ the other factor $(K + \alpha_2)$ is clearly finite (and nearly constant!) and hence can be brought outside the δ-function, again by using the relation $\delta(ax) = \delta(x)/a$, thus giving:

$$\frac{1}{\tau_i} = \Upsilon' \int_0^\pi \int_0^\infty \frac{1}{K + \alpha_2}\delta(K - \alpha_1) \; dK \; \sin\phi \; d\phi \tag{10.42}$$

On performing the integration over K, the only contribution occurs when the argument of the δ-function is equal to zero, i.e. when $K = \alpha_1$, which results in:

$$\frac{1}{\tau_i} = \Upsilon' \int_0^\pi \frac{1}{\alpha_1 + \alpha_2} \sin\phi \; d\phi \tag{10.43}$$

Recalling the definition in equation (10.40) then it is worth noting for later that $\alpha_2 > \alpha_1$ for $0 < \phi < \pi/2$ and $\alpha_1 > \alpha_2$ for $\pi/2 < \phi < \pi$. The quadratic equation for K is then:

$$K^2 + 2k_i K \cos\phi - \frac{2m\omega}{\hbar} = 0 \tag{10.44}$$

Hence, application of the standard formula for the solution of quadratics gives:

$$K = -k_i \cos\phi \pm \sqrt{k_i^2 \cos^2\phi + \frac{2m\omega}{\hbar}} \tag{10.45}$$

so therefore:

$$K = -k_i \cos\phi + \sqrt{k_i^2 \cos^2\phi + \frac{2m\omega}{\hbar}} \tag{10.46}$$

and

$$K = -k_i \cos\phi - \sqrt{k_i^2 \cos^2\phi + \frac{2m\omega}{\hbar}} \tag{10.47}$$

Now the constants α_1 and α_2 are equal to the root (K) and the negative root $(-K)$ respectively, and therefore their sum is:

$$\alpha_1 + \alpha_2 = 2\sqrt{k_i^2 \cos^2\phi + \frac{2m\omega}{\hbar}} \tag{10.48}$$

Substituting this result into equation (10.43) gives:

$$\frac{1}{\tau(p)} = \Upsilon' \int_0^\pi \frac{1}{2\sqrt{k_i \cos^2\phi + \frac{2m\omega}{\hbar}}} \sin\phi \ d\phi \tag{10.49}$$

Thus, the problem has been reduced to a single integral over the variable ϕ. One way to proceed from here is to introduce a new variable φ, defined as:

$$\varphi = -k_i \cos\phi + \sqrt{k_i^2 \cos^2\phi + \frac{2m\omega}{\hbar}} \tag{10.50}$$

where the limits of φ follow as:

$$\varphi_{\min} = \varphi(\phi = 0) = -k_i + \sqrt{k_i^2 + \frac{2m\omega}{\hbar}} \tag{10.51}$$

and

$$\varphi_{\max} = \varphi(\phi = \pi) = +k_i + \sqrt{k_i^2 + \frac{2m\omega}{\hbar}} \tag{10.52}$$

Using equation (10.50) then:

$$\varphi + k_i \cos\phi = \sqrt{k_i^2 \cos^2\phi + \frac{2m\omega}{\hbar}} \tag{10.53}$$

and squaring both sides:

$$\varphi^2 + 2\varphi k_i \cos\phi + k_i^2 \cos^2\phi = k_i^2 \cos^2\phi + \frac{2m\omega}{\hbar} \tag{10.54}$$

$$\therefore k_i \cos\phi = \frac{m\omega}{\hbar\varphi} - \frac{\varphi}{2} \tag{10.55}$$

Substituting $k_i \cos\phi$ back into equation (10.53) gives:

$$\varphi + \frac{m\omega}{\hbar\varphi} - \frac{\varphi}{2} = \sqrt{k_i^2 \cos^2\phi + \frac{2m\omega}{\hbar}} \tag{10.56}$$

$$\therefore \frac{\varphi}{2} + \frac{m\omega}{\hbar\varphi} = \sqrt{k_i^2 \cos^2\phi + \frac{2m\omega}{\hbar}} \tag{10.57}$$

Using equation (10.55) then:

$$\cos\phi = \frac{m\omega}{\hbar k_i \varphi} - \frac{\varphi}{2k_i} \tag{10.58}$$

and by differentiating both sides with respect to φ:

$$\sin\phi \; d\phi = -\left(\frac{m\omega}{\hbar k_i \varphi^2} + \frac{1}{2k_i}\right) d\varphi \tag{10.59}$$

Using equations (10.57) and (10.59) in equation (10.49) gives:

$$\frac{1}{\tau_i} = \frac{\Upsilon'}{2}\int_{\varphi_{\min}}^{\varphi_{\max}} \frac{1}{\frac{\varphi}{2} + \frac{m\omega}{\hbar\varphi}}\left(\frac{m\omega}{\hbar k_i \varphi^2} + \frac{1}{2k_i}\right) d\varphi \tag{10.60}$$

$$\therefore \frac{1}{\tau_i} = \frac{\Upsilon'}{2}\int_{\varphi_{\min}}^{\varphi_{\max}} \frac{1}{\frac{\varphi}{2} + \frac{m\omega}{\hbar\varphi}} \times \frac{1}{k_i \varphi}\left(\frac{m\omega}{\hbar\varphi} + \frac{\varphi}{2}\right) d\varphi \tag{10.61}$$

and therefore:

$$\frac{1}{\tau_i} = \frac{\Upsilon'}{2}\int_{\varphi_{\min}}^{\varphi_{\max}} \frac{1}{k_i \varphi}\, d\varphi \tag{10.62}$$

which finally gives:

$$\frac{1}{\tau_i} = \frac{\Upsilon'}{2}\frac{1}{k_i}\,[\ln\varphi]_{\varphi_{\min}}^{\varphi_{\max}} \tag{10.63}$$

Now for emission, equation (10.37) becomes:

$$\frac{1}{\tau_i} = \Upsilon'\int_0^\pi \int_0^\infty \delta\left(K^2 + 2Kk_i\cos\phi + \frac{2m^*\omega}{\hbar}\right)\, dK \; \sin\phi \; d\phi \tag{10.64}$$

In analogy with absorption, the quadratic argument of the δ-function can be factorized. Considering the integration over the phonon momentum K first, then for $0 < \phi < \pi/2$, where $\cos\phi > 0$, the only possible factors are $(K + \alpha_1)(K + \alpha_2)$. These both imply roots for K which are less than zero and hence unphysical, and thus the argument of the δ-function in this ϕ domain is never zero and there are therefore no contributions to the integral.

However, when $\pi/2 < \phi < \pi$, $\cos\phi$ is negative and the quadratic in K can be factorized with two possible roots: i.e.

$$K = -k_i\cos\phi \pm \sqrt{k_i^2\cos^2\phi - \frac{2m\omega}{\hbar}} \tag{10.65}$$

For the phonon momentum K to remain real, the argument of the square root function must be greater than or equal to zero, i.e.

$$k_i^2\cos^2\phi - \frac{2m\omega}{\hbar} > 0 \tag{10.66}$$

$$\therefore k_i^2\cos^2\phi > \frac{2m\omega}{\hbar} \tag{10.67}$$

Thus, there is a minimum value for ϕ, given by:

$$\cos\phi_{\min} = -\sqrt{\frac{2m\omega}{\hbar k_i^2}} \tag{10.68}$$

where the negative square root has been specified as $\phi_{min} > \pi/2$. Given that the angle ϕ belongs to the domain $\pi/2 < \phi < \pi$, the coefficient of K within the δ-function is negative, and hence the argument can be factorized into the form $(K - \alpha_1)(K - \alpha_2)$, where as before α_1 and α_2 are assumed to be real and positive. Therefore, equation (10.64) becomes:

$$\frac{1}{\tau_i} = \Upsilon' \int_{\phi_{min}}^{\pi} \int_0^{\infty} \delta\left((K - \alpha_1)(K - \alpha_2)\right) \, dK \, \sin\phi \, d\phi \tag{10.69}$$

Equation (10.65) implies that the constants α_1 and α_2 are distinct, and thus there are two contributions to the integral. Choosing $\alpha_1 > \alpha_2$, then $\alpha_2 - \alpha_1$ will be negative, and hence around α_2:

$$\delta((K - \alpha_1)(K - \alpha_2)) = \frac{\delta(K - \alpha_2)}{|K - \alpha_1|} \tag{10.70}$$

i.e.

$$\frac{1}{\tau_i} = \Upsilon' \int_{\phi_{min}}^{\pi} \int_0^{\infty} \left[\frac{\delta(K - \alpha_1)}{K - \alpha_2} + \frac{\delta(K - \alpha_2)}{|K - \alpha_1|}\right] \, dK \, \sin\phi \, d\phi \tag{10.71}$$

Completing the integration over the phonon momentum K then gives:

$$\frac{1}{\tau_i} = \Upsilon' \int_{\phi_{min}}^{\pi} \left(\frac{1}{\alpha_1 - \alpha_2} + \frac{1}{|\alpha_2 - \alpha_1|}\right) \sin\phi \, d\phi \tag{10.72}$$

Again in analogy to absorption, introduce a variable φ, which differs only in the sign of the $2m\omega/\hbar$ term, i.e.

$$\varphi = -k_i \cos\phi + \sqrt{k_i^2 \cos^2\phi - \frac{2m\omega}{\hbar}} \tag{10.73}$$

As the integration over the angle ϕ is to be replaced with an integral over the new variable φ, then the new limits of integration have to be deduced. Now the minimum value of φ occurs when ϕ is at its minimum, as deduced above in equation (10.66); this occurs when the argument of the square root is zero, i.e.

$$\varphi_{min} = \varphi(\phi_{min}) = -k_i \cos\phi_{min} \tag{10.74}$$

Using $\cos\phi_{min}$ from equation (10.68), then:

$$\varphi_{min} = -k_i \times -\sqrt{\frac{2m\omega}{\hbar k_i^2}} = \sqrt{\frac{2m\omega}{\hbar}} \tag{10.75}$$

Correspondingly, the maximum value of φ occurs when $\cos\phi_{min}$ is at its maximum, i.e.

$$\varphi_{max} = \varphi(\phi = \pi) = k_i + \sqrt{k_i^2 - \frac{2m\omega}{\hbar}} \tag{10.76}$$

Following a similar argument to that used for absorption eventually gives the scattering rate for emission as:

$$\frac{1}{\tau_i} = \Upsilon' \frac{1}{k_i} [\ln\varphi]_{\varphi_{min}}^{\varphi_{max}} \tag{10.77}$$

which is equivalent to that given by Lundstrom ([211], equation (2.76)) and can be manipulated further to give the more familiar form (for example [211], equation (2.79)).

10.4   LO PHONON SCATTERING OF TWO-DIMENSIONAL CARRIERS

Again, following the method of Kinsler [213], the phonon interaction term is as for the case of bulk phonon modes, as defined in equation (10.13). However, in order to make use of the symmetry of a general heterostructure, it can be split into components along the growth (z-) axis and in the (x–y) plane of the layers, i.e.

$$\tilde{\mathcal{H}} = e \sum_{\mathbf{K}} \left(\frac{\hbar \omega P}{2|\mathbf{K}|^2} \right)^{\frac{1}{2}} \frac{e^{-i\mathbf{K} \cdot \mathbf{r}}}{V^{\frac{1}{2}}} \tag{10.78}$$

which becomes:

$$\tilde{\mathcal{H}} = e \sum_{\mathbf{K}_{xy}} \sum_{K_z} \left(\frac{\hbar \omega P}{2 \left(|\mathbf{K}_{xy}|^2 + |K_z|^2 \right)} \right)^{\frac{1}{2}} \frac{e^{-i\mathbf{K}_{xy} \cdot \mathbf{r}_{xy}}}{A^{\frac{1}{2}}} \frac{e^{-iK_z z}}{L^{\frac{1}{2}}} \tag{10.79}$$

Now the electron (or hole) wave functions in a heterostructure are a product of an envelope along the growth axis and an in-plane travelling wave (see Section 2.2), i.e.

$$\psi = \psi(z) \frac{e^{-i\mathbf{k} \cdot \mathbf{r}_{xy}}}{A^{\frac{1}{2}}} \tag{10.80}$$

Therefore, substituting this new form for the phonon interaction and the electron wave function into Fermi's Golden Rule (equation 10.1), gives:

$$\frac{1}{\tau_i} = \frac{2\pi}{\hbar} \sum_{\mathbf{k}_f} \left| \int \int \psi_f^*(z) \frac{e^{i\mathbf{k}_f \cdot \mathbf{r}_{xy}}}{A^{\frac{1}{2}}} e \sum_{\mathbf{K}_{xy}} \sum_{K_z} \left(\frac{\hbar \omega P}{2 \left(|\mathbf{K}_{xy}|^2 + |K_z|^2 \right)} \right)^{\frac{1}{2}} \right.$$

$$\left. \times \frac{e^{-i\mathbf{K}_{xy} \cdot \mathbf{r}_{xy}}}{A^{\frac{1}{2}}} \frac{e^{-iK_z z}}{L^{\frac{1}{2}}} \psi_i(z) \frac{e^{-i\mathbf{k}_i \cdot \mathbf{r}_{xy}}}{A^{\frac{1}{2}}} \, dz \, d\mathbf{r}_{xy} \right|^2 \delta(E_f^t - E_i^t) \tag{10.81}$$

where the electron wave vectors, $\mathbf{k}_i$ and $\mathbf{k}_f$ of the initial and final states, respectively, are taken explicitly to lie in the plane of the quantum wells only, i.e. $\mathbf{k} = \mathbf{k}(x, y)$ only.

  This is a good place to introduce the phonon population factor ($N_0 + \frac{1}{2} \mp \frac{1}{2}$); again incorporating it within the factor P to give P' as in equation (10.17), and rearranging, then gives:

$$\frac{1}{\tau_i} = \frac{2\pi e^2 \hbar \omega P'}{2\hbar A} \sum_{\mathbf{k}_f} \left| \sum_{K_z} \frac{1}{L^{\frac{1}{2}}} \int \psi_f^*(z) e^{-iK_z z} \psi_i(z) \, dz \right.$$

$$\left. \times \sum_{\mathbf{K}_{xy}} \frac{1}{\left(|\mathbf{K}_{xy}|^2 + |K_z|^2 \right)^{\frac{1}{2}}} \frac{1}{A} \int e^{-i(\mathbf{k}_i - \mathbf{k}_f + \mathbf{K}_{xy}) \cdot \mathbf{r}_{xy}} \, d\mathbf{r}_{xy} \right|^2 \delta(E_f^t - E_i^t) \tag{10.82}$$

  As in the previous section, converting the integral over the x–y plane (denoted by $d\mathbf{r}_{xy}$) into a δ-function, then gives a factor of 2π per dimension, provided that the limits of integration are effectively at infinity. This results in the following:

$$\frac{1}{\tau_i} = \frac{\pi e^2 \omega P'}{AL} \sum_{\mathbf{k}_f} \left| \sum_{K_z} G_{if}(K_z) \right.$$

$$\times \sum_{\mathbf{K}_{xy}} \frac{1}{(|\mathbf{K}_{xy}|^2 + |K_z|^2)^{\frac{1}{2}}} \frac{(2\pi)^2}{A} \delta(\mathbf{k}_i - \mathbf{k}_f + \mathbf{K}_{xy}) \Bigg|^2 \delta(E_f^t - E_i^t) \tag{10.83}$$

where $G_{if}(K_z)$ is known as the *form factor* and is given by:

$$G_{if}(K_z) = \int \psi_f^*(z) e^{-iK_z z} \psi_i(z) \, dz \tag{10.84}$$

Often this form factor is normalized by dividing by some length, or square root of a length (see, for example, Lundstrom [211], equation (2.102)). However, it suits the purpose better here to leave it unnormalized as above.

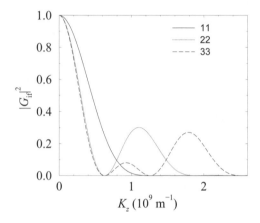

Figure 10.5   Intrasubband form factors

Fig. 10.5 shows this form factor for the case of *intrasubband* scattering events, i.e. transitions where the carrier remains within the same quantum well subband. For this particular case, the example system is that of an electron in a 100 Å GaAs infinitely deep quantum well. As the form factor $G_{if}(K_z)$ always appears as a modulus squared, i.e. $|G_{if}(K_z)|^2$ (see later), and as the initial- and final-state wave functions, ψ_i and ψ_f, in these confined systems are real, then $|G_{if}(K_z)|^2$ is symmetric about $K_z = 0$ and hence it is plotted for positive K_z only.

The figure shows that $|G_{if}(K_z)|^2$ can have more than one maxima, with the overall maximum always for the zone-centre ($K_z = 0$) phonon, which implies that this scattering event is most likely to occur. The physical interpretation is that as the initial and final states of the carrier are the same, then no momentum change along the growth (z-) axis is required, and hence the dominance of the zero-momentum transition. There are additional local maxima, the number of which depend on the number of antinodes in the wave functions. The form factor squared for the 2→2 event, for example, has a total of two maxima, i.e. one at the zone centre and another local maximum. This second local maximum *may* be thought of as corresponding to a scattering event from one of the wave function nodes to the other.

Fig. 10.6 shows the corresponding case for *intersubband* transitions, i.e. scattering events where the carrier changes from one confined level to another. These form factors are quite different in their behaviour to those of the intrasubband case in that there is no maximum at the zone centre—rather it is at a finite wave vector. The particular value corresponds to the difference in momenta along the z-axis of the carrier states. Thus, as 'i' increases, the wave vector of the antinode of the i→1 transition also increases.

Converting the summations over the phonon wave vectors $\mathbf{K}_{xy}$ and K_z into integrals introduces factors of $A/(2\pi)^2$ and $L/(2\pi)$, respectively, from the density of states, and therefore:

$$\frac{1}{\tau_i} = \frac{\pi e^2 \omega P'}{AL} \sum_{\mathbf{k}_f} \left| \frac{L}{2\pi} \int \int G_{if}(K_z) \frac{\delta(\mathbf{k}_i - \mathbf{k}_f + \mathbf{K}_{xy})}{(|\mathbf{K}_{xy}|^2 + |K_z|^2)^{\frac{1}{2}}} \, dK_z \, d\mathbf{K}_{xy} \right|^2 \delta(E_f^t - E_i^t) \quad (10.85)$$

Finally, by changing the summation over the final in-plane electron wave vector $\mathbf{k}_f$ into an integral, and also introducing a factor of $A/(2\pi)^2$ obtain:

$$\frac{1}{\tau_i} = \frac{\pi e^2 \omega P' L}{A(2\pi)^2} \frac{A}{(2\pi)^2} \int \left| \int \int G_{if}(K_z) \frac{\delta(\mathbf{k}_i - \mathbf{k}_f + \mathbf{K}_{xy})}{(|\mathbf{K}_{xy}|^2 + |K_z|^2)^{\frac{1}{2}}} \, dK_z \, d\mathbf{K}_{xy} \right|^2 \, d\mathbf{k}_f$$

$$\times \, \delta(E_f^t - E_i^t) \quad (10.86)$$

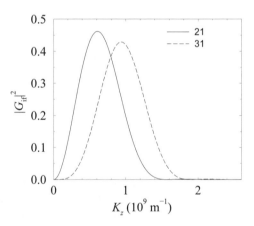

Figure 10.6   Intersubband form factors

The procedure followed is similar to the previous case for bulk electrons, expanding the modulus squared doubles the number of integrals over the phonon wave vectors, i.e.

$$\frac{1}{\tau_i} = \frac{\pi e^2 \omega P' L}{(2\pi)^4} \int \left(\int \int G_{if}^*(K_z) \frac{\delta(\mathbf{k}_i - \mathbf{k}_f + \mathbf{K}_{xy})}{(|\mathbf{K}_{xy}|^2 + |K_z|^2)^{\frac{1}{2}}} \, dK_z \, d\mathbf{K}_{xy} \right)$$

$$\times \left(\int \int G_{if}(K_z') \frac{\delta(\mathbf{k}_i - \mathbf{k}_f + \mathbf{K}_{xy}')}{(|\mathbf{K}_{xy}'|^2 + |K_z'|^2)^{\frac{1}{2}}} \, dK_z' \, d\mathbf{K}_{xy}' \right) \, d\mathbf{k}_f \, \delta(E_f^t - E_i^t) \quad (10.87)$$

Again, as before change the order of the integrations in such a way as to perform the integration over $\mathbf{K}_{xy}'$ first, i.e.

$$\frac{1}{\tau_i} = \frac{\pi e^2 \omega P' L}{(2\pi)^4} \int \int \int G_{if}^*(K_z) G_{if}(K_z') \int \int \frac{\delta(\mathbf{k}_i - \mathbf{k}_f + \mathbf{K}_{xy})}{(|\mathbf{K}_{xy}|^2 + |K_z|^2)^{\frac{1}{2}}}$$

$$\times \frac{\delta(\mathbf{k}_i - \mathbf{k}_f + \mathbf{K}'_{xy})}{\left(|\mathbf{K}'_{xy}|^2 + |K'_z|^2\right)^{\frac{1}{2}}} \; d\mathbf{K}'_{xy} \, d\mathbf{k}_f \, dK'_z \, dK_z \, d\mathbf{K}_{xy} \, \delta(E^t_f - E^t_i) \tag{10.88}$$

So, by performing this integration over $\mathbf{K}'_{xy}$, the second δ-function then limits the contribution to the point where $\mathbf{k}_i - \mathbf{k}_f + \mathbf{K}'_{xy} = 0$, i.e. when $\mathbf{K}'_{xy} = -\mathbf{k}_i + \mathbf{k}_f$, thus giving:

$$\frac{1}{\tau_i} = \frac{\pi e^2 \omega P' L}{(2\pi)^4} \int \int \int G^*_{if}(K_z) G_{if}(K'_z) \int \frac{\delta(\mathbf{k}_i - \mathbf{k}_f + \mathbf{K}_{xy})}{\left(|\mathbf{K}_{xy}|^2 + |K_z|^2\right)^{\frac{1}{2}}}$$

$$\times \frac{1}{\left(|-\mathbf{k}_i + \mathbf{k}_f|^2 + |K'_z|^2\right)^{\frac{1}{2}}} \; d\mathbf{k}_f \, dK'_z \, dK_z \, d\mathbf{K}_{xy} \, \delta(E^t_f - E^t_i) \tag{10.89}$$

Now when performing the integration over the final electron in-plane wave vector $\mathbf{k}_f$, as denoted by the $d\mathbf{k}_f$, then again its contribution is limited by the remaining δ-function to the point where $\mathbf{k}_i - \mathbf{k}_f + \mathbf{K}_{xy} = 0$, i.e. when $\mathbf{k}_f = \mathbf{k}_i + \mathbf{K}_{xy}$, thus giving:

$$\frac{1}{\tau_i} = \frac{\pi e^2 \omega P' L}{(2\pi)^4} \int \int \int G^*_{if}(K_z) G_{if}(K'_z) \frac{1}{\left(|\mathbf{K}_{xy}|^2 + |K_z|^2\right)^{\frac{1}{2}}}$$

$$\times \frac{1}{\left(|-\mathbf{k}_i + \mathbf{k}_i + \mathbf{K}_{xy}|^2 + |K'_z|^2\right)^{\frac{1}{2}}} \; dK'_z \, dK_z \, d\mathbf{K}_{xy} \, \delta(E^t_f - E^t_i) \tag{10.90}$$

and therefore:

$$\frac{1}{\tau_i} = \frac{\pi e^2 \omega P' L}{(2\pi)^4} \int \int \int \frac{G^*_{if}(K_z) \, G_{if}(K'_z)}{\left(|\mathbf{K}_{xy}|^2 + |K_z|^2\right)^{\frac{1}{2}} \left(|\mathbf{K}_{xy}|^2 + |K'_z|^2\right)^{\frac{1}{2}}} \; dK'_z \, dK_z \, d\mathbf{K}_{xy}$$

$$\times \; \delta(E^t_f - E^t_i) \tag{10.91}$$

It is now necessary to perform a little mathematical trickery in order to force the problem through and obtain the *accepted* result (see Lundstrom [211], for example). Such steps are questionable and the possibility remains that the scattering rate should depend on the 'square of an integral' rather than the 'integral of a square'. Proceeding, however, rearrange equation (10.91) to give:

$$\frac{1}{\tau_i} = \frac{\pi e^2 \omega P' L}{(2\pi)^4} \int \int \frac{G^*_{if}(K_z)}{\left(|\mathbf{K}_{xy}|^2 + |K_z|^2\right)^{\frac{1}{2}}} \; dK_z \int \frac{G_{if}(K'_z)}{\left(|\mathbf{K}_{xy}|^2 + |K'_z|^2\right)^{\frac{1}{2}}} \; dK'_z$$

$$\times \; d\mathbf{K}_{xy} \, \delta(E^t_f - E^t_i) \tag{10.92}$$

then:

$$\frac{1}{\tau_i} = \frac{\pi e^2 \omega P' L}{(2\pi)^4} \int \int \frac{G^*_{if}(K_z)}{\left(|\mathbf{K}_{xy}|^2 + |K_z|^2\right)^{\frac{1}{2}}}$$

$$\times \left[\int \frac{G_{if}(K_z)}{\left(|\mathbf{K}_{xy}|^2 + |K_z|^2\right)^{\frac{1}{2}}} \; \delta(K_z - K'_z)\right] \; dK_z \, dK'_z \, d\mathbf{K}_{xy} \, \delta(E^t_f - E^t_i) \tag{10.93}$$

Note that the δ-function ensures that the term in the square brackets ([]) only gives a contribution when $K_z = K'_z$, and when it does this term reverts to the integral over K'_z, as before. Changing the order of integration:

$$\frac{1}{\tau_i} = \frac{\pi e^2 \omega P' L}{(2\pi)^4} \int \int \frac{G^*_{if}(K_z) G_{if}(K_z)}{\left(|\mathbf{K}_{xy}|^2 + |K_z|^2\right)^{\frac{1}{2}} \left(|\mathbf{K}_{xy}|^2 + |K_z|^2\right)^{\frac{1}{2}}}$$

$$\times \int \delta(K_z - K'_z) \; \mathrm{d}K'_z \; \mathrm{d}K_z \; \mathrm{d}\mathbf{K}_{xy} \; \delta(E_\mathrm{f}^\mathrm{t} - E_\mathrm{i}^\mathrm{t}) \tag{10.94}$$

Performing the integration over K'_z first, then the δ-function has to yield a constant term with the units of a wave vector, e.g. $2\pi/L$, in order to satisfy dimensionality arguments; furthermore, introduction the notation $|\mathbf{K}| = K$ gives:

$$\frac{1}{\tau_\mathrm{i}} = \frac{\pi e^2 \omega P' L}{(2\pi)^4} \int \int \frac{2\pi}{L} \frac{|G_\mathrm{if}(K_z)|^2}{K_{xy}^2 + K_z^2} \; \mathrm{d}K_z \; \mathrm{d}\mathbf{K}_{xy} \delta(E_\mathrm{f}^\mathrm{t} - E_\mathrm{i}^\mathrm{t}) \tag{10.95}$$

$$\therefore \frac{1}{\tau_\mathrm{i}} = \frac{\pi e^2 \omega P'}{(2\pi)^3} \int \int \frac{|G_\mathrm{if}(K_z)|^2}{K_{xy}^2 + K_z^2} \; \mathrm{d}K_z \; \mathrm{d}\mathbf{K}_{xy} \delta(E_\mathrm{f}^\mathrm{t} - E_\mathrm{i}^\mathrm{t}) \tag{10.96}$$

As in the bulk case, the total electron (or hole) energies are a sum of a band minimum and the kinetic energy within the band, the only difference being here that, instead of the three-dimensional bulk bands, the carriers are in the two-dimensional subbands (with minima labelled E_f and E_i) of some general quantum well system. Thus:

$$E_\mathrm{i}^\mathrm{t} = E_\mathrm{i} + \frac{\hbar^2 \mathbf{k}_\mathrm{i}^2}{2m^*} \pm \hbar\omega \tag{10.97}$$

and

$$E_\mathrm{f}^\mathrm{t} = E_\mathrm{f} + \frac{\hbar^2 \mathbf{k}_\mathrm{f}^2}{2m^*} \tag{10.98}$$

where the upper sign in the $\pm\hbar\omega$ term accounts for scattering processes involving the absorption of a phonon and the lower sign represents emission. Again, it must be noted that this assumption of parabolic subbands could limit the range of applicability for the case of holes in the valence band. Exploiting the conservation of energy, as specified explicitly in the δ-function, by substituting E_f^t and E_i^t into equation (10.96) then gives:

$$\frac{1}{\tau_\mathrm{i}} = \frac{\pi e^2 \omega P'}{(2\pi)^3} \int \int \frac{|G_\mathrm{if}(K_z)|^2}{K_{xy}^2 + K_z^2}$$

$$\times \delta\left(E_\mathrm{f} + \frac{\hbar^2 \mathbf{k}_\mathrm{f}^2}{2m^*} - \left(E_\mathrm{i} + \frac{\hbar^2 \mathbf{k}_\mathrm{i}^2}{2m^*} \pm \hbar\omega \right) \right) \; \mathrm{d}K_z \; \mathrm{d}\mathbf{K}_{xy} \tag{10.99}$$

By labelling $E_\mathrm{f} - E_\mathrm{i} \mp \hbar\omega$ as Δ (say), then:

$$\frac{1}{\tau_\mathrm{i}} = \frac{\pi e^2 \omega P'}{(2\pi)^3} \int \int \frac{|G_\mathrm{if}(K_z)|^2}{K_{xy}^2 + K_z^2}$$

$$\times \delta\left(\frac{\hbar^2 \mathbf{k}_\mathrm{f}^2}{2m^*} - \frac{\hbar^2 \mathbf{k}_\mathrm{i}^2}{2m^*} + \Delta \right) \; \mathrm{d}K_z \; \mathrm{d}\mathbf{K}_{xy} \tag{10.100}$$

The physical requirement of conservation of energy has therefore advanced the mathematical derivation, and now it is the turn of the conservation of momentum. Consider again the application of the cosine rule to Fig. 10.4, with the additional feature that as the initial and final momenta of the carrier are in-plane, then the phonon momentum $\mathbf{K}$ must also be in the plane. This latter feature has been specified up until now with the additional index xy; thus, in analogy to equation (10.34):

$$\mathbf{k}_\mathrm{f}^2 = \mathbf{k}_\mathrm{i}^2 + \mathbf{K}_{xy}^2 + 2k_\mathrm{i} K_{xy} \cos\phi \tag{10.101}$$

Substituting for $\mathbf{k}_f^2$ in the δ-function therefore gives:

$$
\frac{1}{\tau_i} = \frac{\pi e^2 \omega P'}{(2\pi)^3} \int \int \frac{|G_{if}(K_z)|^2}{K_{xy}^2 + K_z^2}
$$

$$
\times \; \delta \left(\frac{\hbar^2}{2m^*} \left(\mathbf{k}_i^2 + \mathbf{K}_{xy}^2 + 2k_i K_{xy} \cos\phi \right) - \frac{\hbar^2 k_i^2}{2m^*} + \Delta \right) \; \mathrm{d}K_z \, \mathrm{d}\mathbf{K}_{xy} \tag{10.102}
$$

Making use of the fact that $\mathbf{k}^2 = k^2$, then:

$$
\frac{1}{\tau_i} = \frac{\pi e^2 \omega P'}{(2\pi)^3} \int \int \frac{|G_{if}(K_z)|^2}{K_{xy}^2 + K_z^2}
$$

$$
\times \; \delta \left(\frac{\hbar^2 K_{xy}^2}{2m^*} + \frac{\hbar^2 k_i K_{xy} \cos\phi}{m^*} + \Delta \right) \; \mathrm{d}K_z \, \mathrm{d}\mathbf{K}_{xy} \tag{10.103}
$$

Using $\delta(ax) = \delta(x)/a$, then:

$$
\frac{1}{\tau_i} = \frac{\pi e^2 \omega P'}{(2\pi)^3} \int \int \frac{|G_{if}(K_z)|^2}{K_{xy}^2 + K_z^2}
$$

$$
\times \frac{2m^*}{\hbar^2} \delta \left(K_{xy}^2 + 2k_i K_{xy} \cos\phi + \frac{2m^* \Delta}{\hbar^2} \right) \; \mathrm{d}K_z \, \mathrm{d}\mathbf{K}_{xy} \tag{10.104}
$$

Now the two-dimensional integral over the Cartesian in-plane phonon wave vector, as denoted by $\mathrm{d}\mathbf{K}_{xy}$, can be changed into polar coordinates, with the radius as the modulus K_{xy} and an angle ϕ between it and the initial carrier momentum state. The new elemental area is $K_{xy} \, \mathrm{d}K_{xy} \, \mathrm{d}\phi$, and therefore:

$$
\frac{1}{\tau_i} = \frac{m^* e^2 \omega P'}{(2\pi)^2 \hbar^2} \int_0^{2\pi} \int_0^{\infty} \int_{-\infty}^{+\infty} \frac{|G_{if}(K_z)|^2}{K_{xy}^2 + K_z^2}
$$

$$
\times \; \delta \left(K_{xy}^2 + 2k_i K_{xy} \cos\phi + \frac{2m^* \Delta}{\hbar^2} \right) K_{xy} \; \mathrm{d}K_z \, \mathrm{d}K_{xy} \, \mathrm{d}\phi \tag{10.105}
$$

Kinsler [213] has continued the analysis by introducing a new variable $y = \cos\phi$; however, in this work we will again follow the method of Hagston, Piorek, and Harrison as documented in [214]. This will involve factorising the argument of the δ-function; for now, however, consider the integral over ϕ first.

As $\cos\phi$ is an even function, then the integral from 0 to 2π is clearly twice the integral from 0 to π, and in addition, changing the order of integration gives:

$$
\frac{1}{\tau_i} = \frac{2m^* e^2 \omega P'}{(2\pi)^2 \hbar^2} \int_{-\infty}^{+\infty} \int_0^{\infty} \int_0^{\pi} \frac{|G_{if}(K_z)|^2}{K_{xy}^2 + K_z^2}
$$

$$
\times \; \delta \left(K_{xy}^2 + 2k_i K_{xy} \cos\phi + \frac{2m^* \Delta}{\hbar^2} \right) K_{xy} \; \mathrm{d}\phi \, \mathrm{d}K_{xy} \, \mathrm{d}K_z \tag{10.106}
$$

Write the constant prefactor as:

$$
\Upsilon'' = \frac{2m^* e^2 \omega P'}{(2\pi)^2 \hbar^2} \tag{10.107}
$$

and, in addition, consider the case of a positive Δ. When $0 < \phi < \pi/2$, then $\cos\phi > 0$, which implies that the argument of the δ-function can only be factorized into the form $(K_{xy} + \alpha_1)(K_{xy} + \alpha_2)$, where α_1 and α_2 are real and positive. As K_{xy} is the *magnitude* of the phonon wave vector in the plane of the quantum wells, then it is always positive; thus, the argument of the δ-function is never zero and no contributions to the integral are made.

However, when $\pi/2 < \phi < \pi$, the factor $\cos\phi$ is negative. Consider the substitution $\phi' = \pi - \phi$, and then $\cos\phi$ becomes $-\cos\phi'$ and $d\phi = -d\phi'$. When $\phi = \pi/2$, $\phi' = \pi/2$ and when $\phi = \pi$, $\phi' = 0$; therefore:

$$\frac{1}{\tau_i} = \Upsilon'' \int_{-\infty}^{+\infty} \int_0^\infty \int_0^{\pi/2} \frac{|G_{if}(K_z)|^2}{K_{xy}^2 + K_z^2}$$

$$\times \ \delta\left(K_{xy}^2 - 2k_i K_{xy} \cos\phi' + \frac{2m^*\Delta}{\hbar^2}\right) K_{xy} \ d\phi' \ dK_{xy} \ dK_z \tag{10.108}$$

Now the argument of the quadratic can be factorized into the form $(K_{xy}-\alpha_1)(K_{xy}-\alpha_2)$, where it is also specified that $\alpha_1 > \alpha_2$. Thus two solutions for the in-plane phonon wave vector exist; this is illustrated for a general case in Fig. 10.7. For each absorption or emission process, two possible scattering events can occur, each one of which conserves both energy and momentum, where the latter are satisfied by mediating via two different phonon momenta, K_{xy}.

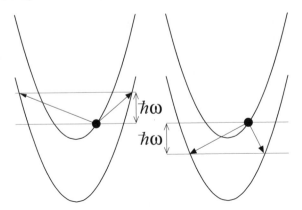

Figure 10.7   The two possible scattering events for (left) phonon absorption and (right) phonon emission

Using this factorized form in equation (10.108) gives:

$$\frac{1}{\tau_i} = \Upsilon'' \int_{-\infty}^{+\infty} \int_0^\infty \int_0^{\pi/2} \frac{|G_{if}(K_z)|^2}{K_{xy}^2 + K_z^2}$$

$$\times \ \delta\left((K_{xy} - \alpha_1)(K_{xy} - \alpha_2)\right) K_{xy} \ d\phi' \ dK_{xy} \ dK_z \tag{10.109}$$

Now around the solution $K_{xy} = \alpha_1$, the factor $(K_{xy} - \alpha_2)$ is a constant, and vice versa, and hence :

$$\frac{1}{\tau_i} = \Upsilon'' \int_{-\infty}^{+\infty} \int_0^\infty \int_0^{\pi/2} \frac{|G_{if}(K_z)|^2}{K_{xy}^2 + K_z^2}$$

$$\times \left[\frac{\delta(K_{xy} - \alpha_1)}{K_{xy} - \alpha_2} + \frac{\delta(K_{xy} - \alpha_2)}{|K_{xy} - \alpha_1|} \right] K_{xy} \; d\phi' \; dK_{xy} \; dK_z \tag{10.110}$$

Hence, completing the integration over K_{xy}:

$$\frac{1}{\tau_i} = \Upsilon'' \int_{-\infty}^{+\infty} \int_0^{\pi/2} |G_{if}(K_z)|^2$$

$$\times \left(\frac{1}{\alpha_1^2 + K_z^2} \frac{\alpha_1}{\alpha_1 - \alpha_2} + \frac{1}{\alpha_2^2 + K_z^2} \frac{\alpha_2}{\alpha_1 - \alpha_2} \right) \; d\phi' \; dK_z \tag{10.111}$$

Recalling the quadratic for K_{xy} in the δ-function argument of equation (10.108) then:

$$K_{xy} = k_i \cos \phi' \pm \sqrt{k_i^2 \cos^2 \phi' - \frac{2m^*\Delta}{\hbar^2}} \tag{10.112}$$

where it is known that as the roots for K_{xy} are real and distinct, then the argument of the square root function is greater than zero. Furthermore, as it has been specified that $\alpha_1 > \alpha_2$, then:

$$\alpha_1 = k_i \cos \phi' + \sqrt{k_i^2 \cos^2 \phi' - \frac{2m^*\Delta}{\hbar^2}} \tag{10.113}$$

and

$$\alpha_2 = k_i \cos \phi' - \sqrt{k_i^2 \cos^2 \phi' - \frac{2m^*\Delta}{\hbar^2}} \tag{10.114}$$

Therefore:

$$\alpha_1 + \alpha_2 = 2k_i \cos \phi' \tag{10.115}$$

and

$$\alpha_1 \alpha_2 = \frac{2m^*\Delta}{\hbar^2} \tag{10.116}$$

which also come from the 'sum of the roots$=-b/a$' and 'product of the roots$=c/a$'. In addition:

$$\alpha_1 - \alpha_2 = 2\sqrt{k_i^2 \cos^2 \phi' - \frac{2m^*\Delta}{\hbar^2}} \tag{10.117}$$

Hence, if equation (10.111) can be manipulated to only contain the roots in the form of these simple constructions, then a reasonably compact expression may be obtained. Consider the term in parentheses in equation (10.111), i.e.

$$\frac{1}{\alpha_1^2 + K_z^2} \frac{\alpha_1}{\alpha_1 - \alpha_2} + \frac{1}{\alpha_2^2 + K_z^2} \frac{\alpha_2}{\alpha_1 - \alpha_2} = \frac{(\alpha_2^2 + K_z^2)\alpha_1 + (\alpha_1^2 + K_z^2)\alpha_2}{(\alpha_1^2 + K_z^2)(\alpha_2^2 + K_z^2)(\alpha_1 - \alpha_2)}$$

$$= \frac{\alpha_1 \alpha_2^2 + \alpha_1 K_z^2 + \alpha_1^2 \alpha_2 + \alpha_2 K_z^2}{(\alpha_1 - \alpha_2)(\alpha_1^2 \alpha_2^2 + \alpha_1^2 K_z^2 + \alpha_2^2 K_z^2 + K_z^4)} \tag{10.118}$$

$$= \frac{(\alpha_1 + \alpha_2)(K_z^2 + \alpha_1 \alpha_2)}{(\alpha_1 - \alpha_2)\{(\alpha_1 \alpha_2)^2 + [(\alpha_1 + \alpha_2)^2 - 2\alpha_1 \alpha_2]K_z^2 + K_z^4\}} \tag{10.119}$$

Using the forms for $\alpha_1 + \alpha_2$, $\alpha_1 \alpha_2$, and $\alpha_1 - \alpha_2$ in equations (10.115), (10.116), and (10.117), then:

$$\frac{1}{\alpha_1^2 + K_z^2} \frac{\alpha_1}{\alpha_1 - \alpha_2} + \frac{1}{\alpha_2^2 + K_z^2} \frac{\alpha_2}{\alpha_1 - \alpha_2}$$

$$
= \frac{2k_i \cos\phi' \left[K_z^2 + \left(\frac{2m^*\Delta}{\hbar^2}\right)\right]}{2\sqrt{k_i^2 \cos^2\phi' - \frac{2m^*\Delta}{\hbar^2}}\left[\left(\frac{2m^*\Delta}{\hbar^2}\right)^2 + \left(4k_i^2 \cos^2\phi' - 2\frac{2m^*\Delta}{\hbar^2}\right)K_z^2 + K_z^4\right]}
$$

$$
= \frac{k_i \cos\phi' \left[K_z^2 + \left(\frac{2m^*\Delta}{\hbar^2}\right)\right]}{\sqrt{k_i^2 \cos^2\phi' - \frac{2m^*\Delta}{\hbar^2}}\left[\left(K_z^2 - \frac{2m^*\Delta}{\hbar^2}\right)^2 + 4k_i^2 K_z^2 \cos^2\phi'\right]} \tag{10.120}
$$

Substituting back into equation (10.111) gives:

$$
\frac{1}{\tau_i} = \Upsilon'' \int_{-\infty}^{+\infty} \int_0^{\pi/2} |G_{if}(K_z)|^2 k_i \left[K_z^2 + \left(\frac{2m^*\Delta}{\hbar^2}\right)\right]
$$

$$
\times \frac{\cos\phi'}{\sqrt{k_i^2 \cos^2\phi' - \frac{2m^*\Delta}{\hbar^2}}\left[\left(K_z^2 - \frac{2m^*\Delta}{\hbar^2}\right)^2 + 4k_i^2 K_z^2 \cos^2\phi'\right]} \, d\phi' \, dK_z \tag{10.121}
$$

Thus far, the derivation has followed fairly standard methods; however, it is possible to proceed further with the analytical work and evaluate the integral over the angle ϕ' by using the innovative approach of Hagston, Piorek and Harrison [214]. The remainder of this section is dedicated to this procedure, but if only the result is required, then skip to equation (10.151).

For now though, consider this integral over the angle ϕ', where there is clearly a maximum value for ϕ' which occurs when the argument of the square root function in equation (10.112) becomes zero, which is given by $k_i \cos\phi'_{\max} = \sqrt{2m^*\Delta/\hbar^2}$. Labelling as '$I$', this is then of the form:

$$
I = \int_0^{\phi'_{\max}} \frac{\cos\phi'}{\sqrt{a\cos^2\phi' - b(c + d\cos^2\phi')}} \, d\phi' \tag{10.122}
$$

where:

$$
a = k_i^2; \quad b = \frac{2m^*\Delta}{\hbar^2}; \quad c = \left(K_z^2 - \frac{2m^*\Delta}{\hbar^2}\right)^2; \quad d = 4k_i^2 K_z^2 \tag{10.123}
$$

Then:

$$
I = \int_0^{\phi'_{\max}} \frac{\sin\phi' \cos\phi'}{\sin\phi' \sqrt{a\cos^2\phi' - b(c + d\cos^2\phi')}} \, d\phi' \tag{10.124}
$$

Putting $x = \cos^2\phi'$, then $dx = -2\sin\phi'\cos\phi' \, d\phi'$, and therefore:

$$
I = \int_1^{\cos^2\phi'_{\max}} \frac{-\frac{1}{2}}{\sqrt{1-x}\sqrt{ax - b(c + dx)}} \, dx \tag{10.125}
$$

$$
\therefore I = \frac{1}{2} \int_{\cos^2\phi'_{\max}}^1 \frac{1}{\sqrt{-ax^2 + (a+b)x - b(c + dx)}} \, dx \tag{10.126}
$$

which is the same as:

$$
I = \frac{1}{2\sqrt{a}} \int_{\cos^2\phi'_{\max}}^1 \frac{1}{\sqrt{\frac{(a-b)^2}{4a^2} - \left(x - \frac{a+b}{2a}\right)^2 (c + dx)}} \, dx \tag{10.127}
$$

Consider the substitution:

$$x - \frac{a+b}{2a} = \frac{a-b}{2a} \cos\alpha \tag{10.128}$$

and so then:

$$dx = -\left(\frac{a-b}{2a}\right) \sin\alpha \ d\alpha \tag{10.129}$$

when $x = \cos^2\phi'_{max} = b/a$; therefore:

$$\frac{b}{a} - \frac{a+b}{2a} = \frac{a-b}{2a} \cos\alpha \tag{10.130}$$

$$\therefore 2b - a - b = (a-b)\cos\alpha \tag{10.131}$$

i.e. $\cos\alpha = -1$, which implies that the lower limit of the integral becomes π. In addition, when $x = 1$, $\cos\alpha = 1$ which implies that the upper limit is 0. Substituting into equation (10.127), then obtain:

$$I = \frac{1}{2\sqrt{a}} \int_\pi^0 \frac{-\left(\frac{a-b}{2a}\right)\sin\alpha}{\sqrt{\left(\frac{a-b}{2a}\right)^2 - \left(\frac{a-b}{2a}\cos\alpha\right)^2}\left(c + d\left(\frac{a+b}{2a} + \frac{a-b}{2a}\cos\alpha\right)\right)} \ d\alpha \tag{10.132}$$

Now, the factor $\left(\frac{a-b}{2a}\right)^2$ can be taken out of the square root, and cancelled with the factor in the numerator; the remainder of the argument then becomes $\sqrt{1 - \cos^2\alpha}$, which also cancels with the numerator, and hence:

$$I = \frac{1}{2\sqrt{a}} \int_0^\pi \frac{1}{c + d\left(\frac{a+b}{2a} + \frac{a-b}{2a}\cos\alpha\right)} \ d\alpha \tag{10.133}$$

Writing $e = c + d(a+b)/(2a)$ and $f = d(a-b)/(2a)$, then (for later):

$$e + f = c + d \quad \text{and} \quad e - f = c + d\frac{b}{a} \tag{10.134}$$

Substituting for e and f gives:

$$I = \frac{1}{2\sqrt{a}} \int_0^\pi \frac{1}{e + f\cos\alpha} \ d\alpha \tag{10.135}$$

Consider the substitution $t = \tan(\alpha/2)$, then:

$$dt = \frac{1}{2}\sec^2\left(\frac{\alpha}{2}\right) d\alpha = \frac{1}{2}\left[1 + \tan^2\left(\frac{\alpha}{2}\right)\right] d\alpha = \frac{1}{2}(1 + t^2) \ d\alpha \tag{10.136}$$

Making use of the tan half-angle formula (see [38], p. 72):

$$\cos\alpha = \frac{1 - t^2}{1 + t^2} \tag{10.137}$$

and changing the integral limits, equation (10.135) then becomes:

$$I = \frac{1}{2\sqrt{a}} \int_0^\infty \frac{1}{e + f\left(\frac{1-t^2}{1+t^2}\right)} \frac{2}{1 + t^2} \ dt \tag{10.138}$$

Multiplying the top and bottom of the above equation by $(1+t^2)$ then gives the following:

$$I = \frac{1}{2\sqrt{a}} \int_0^\infty \frac{2}{e(1+t^2)+f(1-t^2)}\, dt \tag{10.139}$$

$$I = \frac{1}{\sqrt{a}} \int_0^\infty \frac{1}{e+f+(e-f)t^2}\, dt \tag{10.140}$$

This is a standard form, and given that the coefficient of t^2 is greater than zero, the result is then given by the $(\Delta = 4(e+f)(e-f) > 0)$ component of equation (2.172) in [23], i.e.

$$I = \frac{1}{\sqrt{a}} \left[\frac{2}{\sqrt{4(e+f)(e-f)}} \arctan\left(\frac{2(e-f)t}{\sqrt{4(e+f)(e-f)}} \right) \right]_0^\infty \tag{10.141}$$

which when evaluated gives:

$$I = \frac{1}{\sqrt{a}} \frac{2}{\sqrt{4(e+f)(e-f)}} \left(\frac{\pi}{2} - 0 \right) \tag{10.142}$$

Recalling the forms for a, b, c, and d given in equation (10.123) and substituting these into equations (10.134) then gives the following:

$$e + f = \left(K_z^2 - \frac{2m^*\Delta}{\hbar^2} \right)^2 + 4k_i^2 K_z^2 \tag{10.143}$$

and

$$e - f = \left(K_z^2 - \frac{2m^*\Delta}{\hbar^2} \right)^2 + 4K_z^2 \frac{2m^*\Delta}{\hbar^2} = \left(K_z^2 + \frac{2m^*\Delta}{\hbar^2} \right)^2 \tag{10.144}$$

Substituting for both $(e+f)$ and $(e-f)$ in equation (10.142), and recalling that $a = k_i^2$, then:

$$I = \frac{1}{2k_i} \frac{\pi}{\sqrt{\left(K_z^2 - \frac{2m^*\Delta}{\hbar^2} \right)^2 + 4k_i^2 K_z^2 \left(K_z^2 + \frac{2m^*\Delta}{\hbar^2} \right)}} \tag{10.145}$$

$$\therefore I = \frac{1}{2k_i} \frac{\pi}{\sqrt{K_z^4 + 2K_z^2 \left(2k_i^2 - \frac{2m^*\Delta}{\hbar^2} \right) + \left(\frac{2m^*\Delta}{\hbar^2} \right)^2 \left(K_z^2 + \frac{2m^*\Delta}{\hbar^2} \right)}} \tag{10.146}$$

With this analytical form for the integral over the angle ϕ', the original equation, i.e. equation (10.121), becomes:

$$\frac{1}{\tau_i} = \frac{\Upsilon''}{2} \int_{-\infty}^{+\infty} \frac{\pi |G_{if}(K_z)|^2}{\sqrt{K_z^4 + 2K_z^2 \left(2k_i^2 - \frac{2m^*\Delta}{\hbar^2} \right) + \left(\frac{2m^*\Delta}{\hbar^2} \right)^2}}\, dK_z \tag{10.147}$$

This last equation represents the lifetime of a carrier in an initial subband 'i' with *any* in-plane wave vector k_i, strictly speaking only those k_i that satisfy energy conservation can have a lifetime. This information was really lost when the integration over the in-plane phonon wave vector K_{xy} was performed to remove the second (energy conservation) δ-function. It can be put back in with a Heaviside unit step function:

$$\Theta\left(k_i^2 - \frac{2m^*\Delta}{\hbar^2} \right)$$

Recalling that $\Delta = E_f - E_i \mp \hbar\omega$ then the Heaviside function ensures that there are only finite lifetimes τ_i when:

$$\frac{\hbar^2 k_i}{2m^*} > E_f - E_i \mp \hbar\omega \qquad (10.148)$$

Remembering that the upper sign describes phonon absorption then this would imply:

$$E_i + \frac{\hbar^2 k_i}{2m^*} + \hbar\omega > E_f \qquad (10.149)$$

and for emission:

$$E_i + \frac{\hbar^2 k_i}{2m^*} > E_f + \hbar\omega \qquad (10.150)$$

which are clearly the desired results, hence finally:

$$\frac{1}{\tau_i} = \frac{\Upsilon''}{2}\Theta\left(k_i^2 - \frac{2m^*\Delta}{\hbar^2}\right)\int_{-\infty}^{+\infty}\frac{\pi|G_{if}(K_z)|^2}{\sqrt{K_z^4 + 2K_z^2\left(2k_i^2 - \frac{2m^*\Delta}{\hbar^2}\right) + \left(\frac{2m^*\Delta}{\hbar^2}\right)^2}}\,dK_z \qquad (10.151)$$

The case for a negative Δ follows a similar route and leads to the same end result. Therefore, equation (10.151) represents the final form for the lifetime τ_i of a carrier in a subband 'i' with an in-plane wave vector k_i before scattering by an LO phonon. The information regarding whether such an event is with respect to absorption of a phonon or emission, and the final state of the carrier, is incorporated within the variables Δ and P' (which is within Υ'').

This result is a particularly powerful expression because it is applicable to all two-dimensional carrier distributions, regardless of the particular form for the wave functions. Such information is wrapped up in the form factor $G_{if}(K_z)$, and thus the carrier–LO phonon scattering rate can be calculated for any semiconductor heterostructure simply by evaluating a one-dimensional integral.

10.5   APPLICATION TO CONDUCTION SUBBANDS

In the previous section, the carrier–LO phonon scattering rate, which is the reciprocal of the lifetime, was derived for a two-dimensional distribution, as found in the subbands formed in quantum well systems. In this present section, this result, as summarized in equation (10.151), will be applied to a variety of examples in order to gain an intuitive understanding of this important phenomenon.

Fig. 10.8 displays the intersubband scattering rate as a function of the total initial energy E_i^t, as defined in equation (10.97), for an electron in the second subband of an infinitely deep quantum well with respect to LO phonon emission and scattering into the ground state. (Note here the initial carrier energy domain is from the subband minimum upwards.) The scattering rate increases as the carrier approaches the subband minimum, which in this case is around 220 meV. This is a general result and occurs for subband separations greater than the LO phonon energy.

If the quantum well width is allowed to increase, then the energy separation between the initial and final subbands decreases. Fig. 10.9 illustrates the effect that such a series of calculations has on the scattering rate. It can be seen that in all cases the scattering rate increases; however, for quantum well widths greater than 300 Å, there is a small

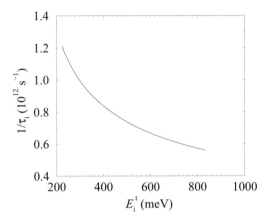

Figure 10.8 The scattering rate via LO phonon emission for an electron initially in the second subband and finally in the ground state, of a 100 Å GaAs infinitely deep quantum well at a lattice temperature of 77 K

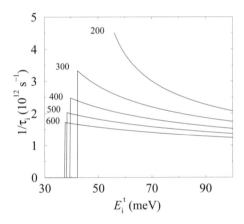

Figure 10.9 The scattering rate as given in Fig. 10.8, but for a variety of well widths: 200, 300, 400, 500 and 600 Å, as indicated on the figure

region where the scattering rate falls very rapidly to zero. This is indicative of the subband separation, $E_2 - E_1$, being less than the LO phonon energy.

The 'cut-off' in the scattering rate occurs because the electrons in the upper subband have not sufficient energy to emit an LO phonon and hence are unable to scatter. This feature is illustrated schematically in Fig. 10.10. Moving from the upper right down the curve, electrons have sufficient kinetic energy which, when combined with the potential energy from being in the upper subband, allows them to emit a phonon; however, the third electron represents the minimum kinetic energy for scattering. Below this point, the electrons are less than an LO phonon away from the energy minimum of the *complete* system and hence can not scatter.

Fig. 10.11 shows the corresponding intrasubband scattering rate for an electron in the second subband. It can be seen that the behaviour of the rate with the initial energy is qualitatively similar to the intersubband case shown in Fig. 10.8. However, in

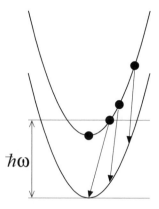

Figure 10.10   The effect of the LO phonon energy 'cut-off' on intersubband electron scattering

addition there is a cut-off energy at 260 meV, which is an LO phonon energy above the subband minima. This is always the case for intrasubband scattering—when a carrier is within a phonon energy of the subband minima it can not emit a phonon. Note that the scattering rate is almost an order of magnitude higher for the intrasubband case than for the intersubband case. This actually represents something of a minimum, as generally the intrasubband rate is between one and two orders of magnitude higher than the intersubband rate. This is because the overlap of the wave function with itself is always complete, whereas the overlap of two distinct wave functions is often only partial. In this case, with this hypothetical infinitely deep quantum well, the overlap of the wave functions for the intersubband case is higher than the usual situation.

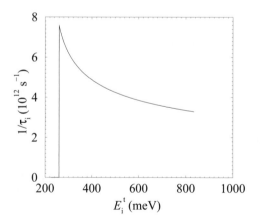

Figure 10.11   The intrasubband scattering rate via LO phonon emission for the same case as that shown in Fig. 10.8

10.6   AVERAGING OVER CARRIER DISTRIBUTIONS

The formula shown in equation (10.151) gives the lifetime of a carrier in a particular subband with a definite in-plane wave vector k_i with respect to scattering with an LO phonon into another subband. In real situations, there isn't just one carrier in the initial subband, plus an empty final subband; in fact, there are generally Fermi–Dirac distributions in both of the subbands (see Section 2.4). It is then more useful to know the *mean* scattering rate (or lifetime) of a carrier.

A simple weighted mean over a distribution of carriers in the initial subband might look like the following:

$$\text{mean}\left(\frac{1}{\tau_i}\right) = \frac{\int \frac{1}{\tau_i} f_i^{\text{FD}}(E) \; dE}{\int f_i^{\text{FD}}(E) \; dE} \tag{10.152}$$

where the subscript 'i' on the distribution functions indicates the subband, i.e. to be evaluated with the 'quasi' Fermi energy of that subband. However, this still disregards the distribution in the final subband; in this case, filled states could prevent carriers from scattering into them, thus reducing the probability of an event. This effect of *final-state blocking* can be incorporated into the above to give:

$$\frac{1}{\tau_{if}} = \frac{\int \frac{1}{\tau_i} f_i^{\text{FD}}(E)(1 - f_f^{\text{FD}}(E - \hbar\omega)) \; dE}{\int f_i^{\text{FD}}(E) \; dE} \tag{10.153}$$

where the double subscript 'if' is used to indicate that this scattering rate is an average over the subband populations in the initial and final states. The integrals are evaluated from the subband minimum of the initial state up to some defined maximum. In the calculations that follow, this maximum has been chosen to be the energy of the highest point in the potential profile, which means physically that any carriers above the barrier are assumed to ionize rapidly. An alternative to this may be to choose the Fermi energy plus $10kT$, or similar. Equation (10.153) may be simplified slightly, as the denominator is simply the number of carriers in the subband divided by the density of states (see equation (2.48)).

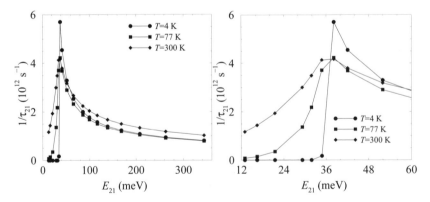

Figure 10.12   The mean scattering rate averaged over distributions in both the initial and final subbands, as a function of the subband separation, shown for three different temperatures

Fig. 10.12 shows this calculated mean for the same system as before, i.e. a GaAs infinitely deep quantum well, for the case of scattering via the emission of a LO phonon from the second subband to the ground subband. In this series of calculations the width of the well was varied in order to scan the difference between the energy band minima (labelled here as $E_{21} = E_2 - E_1$) through the LO phonon energy, and the electron density in each subband was assumed to be 10^{10} cm^{-2}. The *electron temperature*, i.e. the temperature inserted into the Fermi–Dirac distribution function was taken equal to the lattice temperature—in systems under excitation where non-equilibrium distributions will be present, this assumption is almost certainly not true. Again, the infinitely deep quantum well is a good illustrative example, as the overlap of the wave functions, which is contained within the form factor G_{if}, does not change, i.e. the effect on the scattering rate is due entirely to the energy separation.

In the long-range scan, Fig. 10.12 (left), it can be seen that, as the subband separation decreases, the scattering rate increases up to almost a 'resonance' point and then decreases rapidly. At energies above the resonance, the scattering rate has only a weak temperature dependence, but below it, the dependence is stronger. Fig. 10.12 (right) illustrates this resonance effect more clearly, for a smaller range of energies. As may be expected for this fixed phonon energy, which in GaAs is 36 meV, the peak in the scattering rate occurs when the subband separation is equal to the phonon energy. The right-hand figure highlights well the strong temperature dependence of the scattering rate for subband separations below the LO phonon energy. At very low temperatures, the 'cut-off' in the scattering is almost as complete as that shown by the single-carrier case in the previous section. However, as the temperature increases, the carrier distributions broaden, so although the subband separation remains below the phonon energy, the carriers in the upper level spread up the subband, with a proportion having enough kinetic energy to be able to emit a LO phonon and scatter to the lower level. As the temperature increases, this proportion increases and hence the mean scattering rate also increases.

10.7   RATIO OF EMISSION TO ABSORPTION

In the previous section, the importance of the carrier energy for emission has been demonstrated. Converse to this, for the case of carrier scattering by absorption of a phonon, it is not the energy of the carrier that is the important issue but rather the number of phonons available—the more phonons, then the more likely an absorption process. The phonon density, given in equation (10.2), increases as the temperature rises, thus increasing the probability of an absorption.

Fig. 10.13 displays the results of calculations of the ratio of the emission to the electron–LO phonon absorption rate, for the same series of quantum wells as in the previous section. It can be seen that the emission rate is always larger than the absorption rate; thus given a carrier population in an excited subband, then when left to reach equilibrium the carriers will always emit more phonons than they absorb and hence scatter down to the ground state.

The linearities of the graphs shown in Fig. 10.13 hints at a relationship of the form:

$$\frac{\tau_{12}}{\tau_{21}} \propto \exp\left(\text{constant} \times E_{21}\right) \tag{10.154}$$

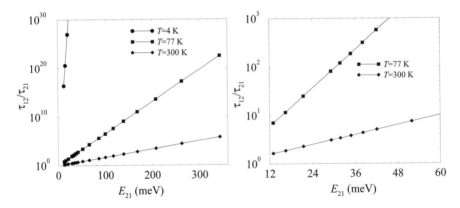

Figure 10.13   The ratio of the mean electron–LO phonon emission $(1/\tau_{21})$ to absorption $(1/\tau_{12})$ rate, as a function of the subband separation (left) and a more detailed view around the LO phonon energy (right)

and in fact, a numerical analysis of the data shows that the 'constant' is equal to $1/kT$, i.e.

$$\frac{\tau_{12}}{\tau_{21}} \propto \exp\left(\frac{E_{21}}{kT}\right) \tag{10.155}$$

This simple relationship between the ratio of the emission to the absorption scattering rates, the subband separation and the temperature is helpful in summarizing the data presented in Fig. 10.13. For a fixed temperature, this ratio increases as the energy separation between the two levels increases, while for a given subband separation, increasing temperature leads to a decrease in the ratio.

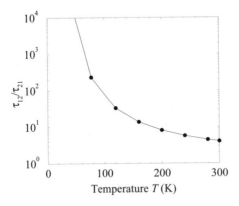

Figure 10.14   The ratio of the mean electron–LO phonon emission $(1/\tau_{21})$ to absorption $(1/\tau_{12})$ rate, as a function of temperature T for a fixed subband separation E_{21} equal to the LO phonon energy of 36 meV

These conclusions are not that obvious *a priori*, and go beyond the simple intuitive picture that the ratio of the scattering rates is controlled by the ratio of the phonon densities, i.e. $(N_0 + 1)/N_0$.

For a more detailed investigation into the temperature dependency of the ratio of the emission to absorption scattering rates, consider the data shown in Fig. 10.14, which corresponds to the fixed subband separation of E_{21} equal to the LO phonon energy, which in this case is 36 meV. At room temperature, the ratio is equal to the well-known result of 4; however, as the temperature decreases, the ratio of these mean scattering rates increases very rapidly and by 77 K emission is more than two orders of magnitude more likely than absorption.

10.8 SCREENING OF THE LO PHONON INTERACTION

The longitundinal optical phonon interaction is a polar interaction, thus it can be influenced by the presence of other charges. In particular, in a doped semiconductor there can be many charges which are free to move in an electromagnetic field and, like Lenz's law in electromagnetic induction, they move to oppose any change. This idea is known as 'screening' and its effect is to reduce the scattering rate due to LO phonons.

The screening model of Park *et al.* [215] can be implemented in a simple way in the formalism here by making the substitution:

$$K_z^2 \longrightarrow K_z^2 \left(1 + \frac{\lambda_s^2}{K_z^2}\right)^2 \tag{10.156}$$

in equation (10.151) except in the form factor $G_{\mathrm{if}}(K_z)$. The quantity λ_s is known as the 'inverse screening length' and, for systems with a majority carrier type, this would simplify to:

$$\lambda_s^2 = \frac{e^2}{\pi\hbar^2\epsilon_s} \sum_j \left\{ \frac{\sqrt{2m^*E_j}\, m^* f^{\mathrm{FD}}(E_j)}{\pi\hbar} \right\} \tag{10.157}$$

where the index 'j' includes all occupied subbands. The unusual parameters employed in Figs. 10.15 and 10.16 were chosen to give direct comparisons with Figs. 1 and 2 of Park *et al.* and the effects of introducing screening are very similar.

The peaks in the curves in Fig. 10.15 occur when the energy separation between the subbands is in resonance with the LO phonon energy, in this case 36 meV. Around this energy, screening at this carrier density can reduce the scattering rate by a factor of 2 or 3, though further from resonance screening has a reduced effect. Fig. 10.16 shows the effect of the temperature of the electron distributions in the subbands on the lifetime for two different carrier densities. Such elevated electron temperatures occur commonly in electronic and optoelectronic devices, which are all driven by energy some of which is absorbed directly by the electron gas.

10.9 ACOUSTIC DEFORMATION POTENTIAL SCATTERING

Combining Lundstrom's [211] equations (2.56) and (2.59), summing over all bulk phonon wave functions, which are of the form:

$$\frac{e^{-i\mathbf{K}_{xy}\bullet\mathbf{r}_{xy}}}{A^{\frac{1}{2}}} \frac{e^{-iK_z\bullet z}}{L^{\frac{1}{2}}} \tag{10.158}$$

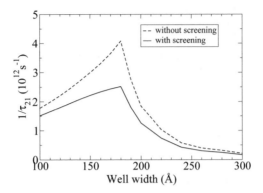

Figure 10.15 The electron–LO phonon scattering rate from the first excited to the ground subband as a function of well width for a GaAs quantum well surrounded by 60 Å $Ga_{0.7}Al_{0.3}As$ barriers. The lattice temperature was taken as 15 K, the electron temperature as 100 K and the carrier density as 10^{11} cm^{-2}.

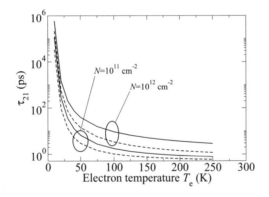

Figure 10.16 The electron–LO phonon scattering rate from the first excited to the ground subband as a function of the temperature of the electron distributions (known as the 'electron temperature') for a 220 Å GaAs quantum well surrounded by 60 Å $Ga_{0.7}Al_{0.3}As$ barriers. The lattice temperature was taken as 15 K.

and taking a carrier wave function in a heterostructure as before:

$$\psi = \psi(z)\frac{e^{-i\mathbf{k}\bullet\mathbf{r}_{xy}}}{A^{\frac{1}{2}}} \tag{10.159}$$

gives on insertion into Fermi's Golden Rule:

$$\frac{1}{\tau_i} = \frac{2\pi}{\hbar}\sum_{\mathbf{k}_f}\frac{D_A^2\hbar}{2\rho\omega_s}\left(N_0 + \frac{1}{2} \mp \frac{1}{2}\right)$$

$$\times \left| \int \int \psi_f^*(z) \frac{e^{i\mathbf{k_f}\bullet\mathbf{r}_{xy}}}{A^{\frac{1}{2}}} \sum_{\mathbf{K}_{xy}} \sum_{K_z} K \frac{e^{-i\mathbf{K}_{xy}\bullet\mathbf{r}_{xy}}}{A^{\frac{1}{2}}} \frac{e^{-iK_z z}}{L^{\frac{1}{2}}} \psi_i(z) \frac{e^{-i\mathbf{k_i}\bullet\mathbf{r}_{xy}}}{A^{\frac{1}{2}}} dz \ d\mathbf{r}_{xy} \right|^2$$

$$\times \delta(E_f^t - E_i^t) \tag{10.160}$$

which gives:

$$\frac{1}{\tau_i} = \frac{2\pi}{\hbar} \sum_{\mathbf{k}_f} \frac{D_A^2 \hbar}{2\rho\omega_s} \left(N_0 + \frac{1}{2} \mp \frac{1}{2} \right)$$

$$\times \left| \sum_{\mathbf{K}_{xy}} \sum_{K_z} \int \psi_f^*(z) \frac{e^{-iK_z z}}{L^{\frac{1}{2}}} \psi_i(z) \ dzK \int \frac{e^{i\mathbf{k_f}\bullet\mathbf{r}_{xy}}}{A^{\frac{1}{2}}} \frac{e^{-i\mathbf{K}_{xy}\bullet\mathbf{r}_{xy}}}{A^{\frac{1}{2}}} \frac{e^{-i\mathbf{k_i}\bullet\mathbf{r}_{xy}}}{A^{\frac{1}{2}}} \ d\mathbf{r}_{xy} \right|^2$$

$$\times \delta(E_f^t - E_i^t) \tag{10.161}$$

Now the integral over z is just the form factor (see equation (10.84)), hence:

$$\frac{1}{\tau_i} = \frac{2\pi}{\hbar} \sum_{\mathbf{k}_f} \frac{D_A^2 \hbar}{2\rho\omega_s} \left(N_0 + \frac{1}{2} \mp \frac{1}{2} \right)$$

$$\times \left| \sum_{\mathbf{K}_{xy}} \sum_{K_z} \frac{G_{if}(K_z)}{L^{\frac{1}{2}}} K \int \frac{e^{-i(\mathbf{k_i}-\mathbf{k_f}+\mathbf{K}_{xy})\bullet\mathbf{r}_{xy}}}{A^{\frac{3}{2}}} d\mathbf{r}_{xy} \right|^2 \delta(E_f^t - E_i^t) \tag{10.162}$$

Converting the integral over the x–y plane (denoted by $d\mathbf{r}_{xy}$) into a δ-function gives a factor of 2π per dimension, i.e.

$$\frac{1}{\tau_i} = \frac{2\pi}{\hbar} \sum_{\mathbf{k}_f} \frac{D_A^2 \hbar}{2\rho\omega_s L A^3} \left(N_0 + \frac{1}{2} \mp \frac{1}{2} \right)$$

$$\times \left| \sum_{\mathbf{K}_{xy}} \sum_{K_z} G_{if}(K_z) K (2\pi)^2 \delta(\mathbf{k_i} - \mathbf{k_f} + \mathbf{K}_{xy}) \right|^2 \delta(E_f^t - E_i^t) \tag{10.163}$$

Now converting the summations over K_z and $\mathbf{K}_{xy}$ into integrals introduces factors of $L/(2\pi)$ and $A/(2\pi)^2$ respectively, giving:

$$\frac{1}{\tau_i} = \frac{2\pi}{\hbar} \sum_{\mathbf{k}_f} \frac{D_A^2 \hbar}{2\rho\omega_s L A^3} \left(N_0 + \frac{1}{2} \mp \frac{1}{2} \right)$$

$$\times \left| \int \int \frac{L}{2\pi} G_{if}(K_z) K \frac{A}{(2\pi)^2} (2\pi)^2 \delta(\mathbf{k_i} - \mathbf{k_f} + \mathbf{K}_{xy}) \ d\mathbf{K}_{xy} \ dK_z \right|^2 \delta(E_f^t - E_i^t) \tag{10.164}$$

Just simplifying the constants, then:

$$\frac{1}{\tau_i} = \sum_{\mathbf{k}_f} \frac{D_A^2}{2\rho\omega_s A} \frac{L}{2\pi} \left(N_0 + \frac{1}{2} \mp \frac{1}{2} \right)$$

$$\times \left| \int \int G_{if}(K_z) K \delta\left(\mathbf{k}_i - \mathbf{k}_f + \mathbf{K}_{xy}\right) \ \mathrm{d}\mathbf{K}_{xy} \ \mathrm{d}K_z \right|^2 \delta(E_f^t - E_i^t) \tag{10.165}$$

Approximating the acoustic branch of the phonon dispersion curve by a linear function then the angular frequency of a sound wave is given by:

$$\omega_s = v_s K \tag{10.166}$$

hence as ω_s is dependent upon the phonon wave vector, it needs to be included within the integrals:

$$\frac{1}{\tau_i} = \sum_{\mathbf{k}_f} \frac{D_A^2}{2\rho v_s A} \frac{L}{2\pi} \left(N_0 + \frac{1}{2} \mp \frac{1}{2}\right)$$

$$\times \left| \int \int G_{if}(K_z) \sqrt{K} \delta\left(\mathbf{k}_i - \mathbf{k}_f + \mathbf{K}_{xy}\right) \ \mathrm{d}\mathbf{K}_{xy} \ \mathrm{d}K_z \right|^2 \delta(E_f^t - E_i^t) \tag{10.167}$$

Converting the sum over final carrier momentum states $\mathbf{k}_f$ into an integral introduces a factor of $A/(2\pi)^2$, thus:

$$\frac{1}{\tau_i} = \frac{D_A^2}{2\rho v_s A} \frac{L}{2\pi} \frac{A}{(2\pi)^2} \left(N_0 + \frac{1}{2} \mp \frac{1}{2}\right)$$

$$\times \int \left| \int \int G_{if}(K_z) \sqrt{K} \delta\left(\mathbf{k}_i - \mathbf{k}_f + \mathbf{K}_{xy}\right) \ \mathrm{d}\mathbf{K}_{xy} \ \mathrm{d}K_z \right|^2 \mathrm{d}\mathbf{k}_f \ \delta(E_f^t - E_i^t) \tag{10.168}$$

Expanding the square:

$$\frac{1}{\tau_i} = \frac{D_A^2}{2\rho v_s (2\pi)^2} \frac{L}{2\pi} \left(N_0 + \frac{1}{2} \mp \frac{1}{2}\right)$$

$$\times \int \left(\int \int G_{if}(K_z) \sqrt{(K_{xy}^2 + K_z^2)^{\frac{1}{2}}} \delta\left(\mathbf{k}_i - \mathbf{k}_f + \mathbf{K}_{xy}\right) \ \mathrm{d}\mathbf{K}_{xy} \ \mathrm{d}K_z \right)$$

$$\times \left(\int \int G_{if}(K_z') \sqrt{(K_{xy}'^2 + K_z'^2)^{\frac{1}{2}}} \delta\left(\mathbf{k}_i - \mathbf{k}_f + \mathbf{K}_{xy}'\right) \ \mathrm{d}\mathbf{K}_{xy}' \ \mathrm{d}K_z' \right)$$

$$\mathrm{d}\mathbf{k}_f \ \delta(E_f^t - E_i^t) \tag{10.169}$$

Performing the integration over $\mathbf{K}_{xy}'$ first, then the second δ-function implies that there is only a contribution when $\mathbf{K}_{xy}' = -\mathbf{k}_i + \mathbf{k}_f$:

$$\frac{1}{\tau_i} = \frac{D_A^2}{2\rho v_s (2\pi)^2} \frac{L}{2\pi} \left(N_0 + \frac{1}{2} \mp \frac{1}{2}\right)$$

$$\times \int \left(\int \int G_{if}(K_z) \sqrt{(K_{xy}^2 + K_z^2)^{\frac{1}{2}}} \delta\left(\mathbf{k}_i - \mathbf{k}_f + \mathbf{K}_{xy}\right) \ \mathrm{d}\mathbf{K}_{xy} \ \mathrm{d}K_z \right)$$

$$\times \left(\int G_{if}(K_z') \sqrt{(| - \mathbf{k}_i + \mathbf{k}_f|^2 + K_z'^2)^{\frac{1}{2}}} \ \mathrm{d}K_z' \right)$$

$$\mathrm{d}\mathbf{k}_f \ \delta(E_f^t - E_i^t) \tag{10.170}$$

Now performing the integral over the final carrier momemtum $\mathbf{k}_f$, the first δ-function limits the integrals to contributions when $\mathbf{k}_f = \mathbf{k}_i + \mathbf{K}_{xy}$, i.e.

$$\frac{1}{\tau_i} = \frac{D_A^2}{2\rho v_s (2\pi)^2} \frac{L}{2\pi} \left(N_0 + \frac{1}{2} \mp \frac{1}{2} \right)$$

$$\times \left(\int \int G_{if}(K_z) \sqrt{(K_{xy}^2 + K_z^2)^{\frac{1}{2}}} \, d\mathbf{K}_{xy} \, dK_z \right)$$

$$\times \left(\int G_{if}(K_z') \sqrt{(| -\mathbf{k}_i + \mathbf{k}_i + \mathbf{K}_{xy}|^2 + K_z'^2)^{\frac{1}{2}}} \, dK_z' \right)$$

$$\delta(E_f^t - E_i^t) \tag{10.171}$$

and therefore:

$$\frac{1}{\tau_i} = \frac{D_A^2}{2\rho v_s (2\pi)^2} \frac{L}{2\pi} \left(N_0 + \frac{1}{2} \mp \frac{1}{2} \right)$$

$$\times \int \int \int G_{if}(K_z) G_{if}(K_z') \sqrt{(K_{xy}^2 + K_z^2)^{\frac{1}{2}}} \sqrt{(K_{xy}^2 + K_z'^2)^{\frac{1}{2}}} \, dK_z \, dK_z' \, d\mathbf{K}_{xy}$$

$$\delta(E_f^t - E_i^t) \tag{10.172}$$

Utilizing the 'questionable' mathematical trick as before (see Section 10.4) allows the mathematics to be forced through to yield expressions that agree with those quoted, but again it is worth reiterating the statement that maybe the scattering rate should be a 'square of an integral' rather than an 'integral of a square'. The procedure equates K_z' to K_z and introduces a factor of $2\pi/L$ thus giving:

$$\frac{1}{\tau_i} = \frac{D_A^2}{2\rho v_s (2\pi)^2} \left(N_0 + \frac{1}{2} \mp \frac{1}{2} \right)$$

$$\times \int \int (G_{if}(K_z))^2 (K_{xy}^2 + K_z^2)^{\frac{1}{2}} \, dK_z \, d\mathbf{K}_{xy} \, \delta(E_f^t - E_i^t) \tag{10.173}$$

Now, as before, the initial and final total energies of the carrier–phonon system are given by:

$$E_i^t = E_i + \frac{\hbar^2 \mathbf{k}_i^2}{2m^*} \pm \hbar\omega_s \tag{10.174}$$

$$E_f^t = E_f + \frac{\hbar^2 \mathbf{k}_f^2}{2m^*} \tag{10.175}$$

Hence substituting into equation (10.173):

$$\frac{1}{\tau_i} = \frac{D_A^2}{2\rho v_s (2\pi)^2} \left(N_0 + \frac{1}{2} \mp \frac{1}{2} \right)$$

$$\times \int \int (G_{if}(K_z))^2 (K_{xy}^2 + K_z^2)^{\frac{1}{2}} \delta \left(E_f + \frac{\hbar^2 \mathbf{k}_f^2}{2m^*} - E_i - \frac{\hbar^2 \mathbf{k}_i^2}{2m^*} \mp \hbar\omega_s \right) \, dK_z \, d\mathbf{K}_{xy} \tag{10.176}$$

Following the prescription in Fig. 10.4 and defining the angle between the initial and final carrier momentum states as ϕ, and applying the cosine rule gives:

$$\mathbf{k}_f^2 = \mathbf{k}_i^2 + \mathbf{K}_{xy}^2 - 2k_i K_{xy} \cos(\pi - \phi) \tag{10.177}$$

$$\therefore \mathbf{k}_f^2 = \mathbf{k}_i^2 + \mathbf{K}_{xy}^2 + 2k_i K_{xy} \cos\phi \tag{10.178}$$

Hence substituting into equation (10.176), then:

$$\frac{1}{\tau_i} = \frac{D_A^2}{2\rho v_s (2\pi)^2} \left(N_0 + \frac{1}{2} \mp \frac{1}{2}\right) \int \int (G_{if}(K_z))^2 (K_{xy}^2 + K_z^2)^{\frac{1}{2}}$$

$$\times \delta\left(E_f + \frac{\hbar^2}{2m^*}\left(k_i^2 + K_{xy}^2 + 2k_i K_{xy} \cos\phi\right) - E_i - \frac{\hbar^2 k_i^2}{2m^*} \mp \hbar\omega_s\right) dK_z \, d\mathbf{K}_{xy} \tag{10.179}$$

Now the energies of phonons from the acoustic branch are generally very small, i.e. a few meV, which is small when compared to typical intersubband separations $\Delta E = E_f - E_i$, which range from several tens to several hundred meV, hence in this formalism the phonon energy $\hbar\omega_s$ will be approximated as zero. The physical implication of this is that, *from the viewpoint of the calculation of scattering rates, the acoustic deformation potential scattering is assumed elastic.*

Implementing this and taking the factor $\hbar^2/2m$ out of the δ-function then:

$$\frac{1}{\tau_i} = \frac{D_A^2 m^*}{\rho v_s (2\pi)^2 \hbar^2} \left(N_0 + \frac{1}{2} \mp \frac{1}{2}\right) \int \int (G_{if}(K_z))^2 (K_{xy}^2 + K_z^2)^{\frac{1}{2}}$$

$$\times \delta\left(K_{xy}^2 + 2k_i K_{xy} \cos\phi + \frac{2m^* \Delta E}{\hbar^2}\right) dK_z \, d\mathbf{K}_{xy} \tag{10.180}$$

Transforming the integral over the in-plane phonon wave vector $\mathbf{K}_{xy}$ into polar coordinates gives:

$$\frac{1}{\tau_i} = \frac{D_A^2 m^*}{\rho v_s (2\pi)^2 \hbar^2} \left(N_0 + \frac{1}{2} \mp \frac{1}{2}\right) \int \int_0^{2\pi} \int (G_{if}(K_z))^2 (K_{xy}^2 + K_z^2)^{\frac{1}{2}}$$

$$\times \delta\left(K_{xy}^2 + 2k_i K_{xy} \cos\phi + \frac{2m^* \Delta E}{\hbar^2}\right) K_{xy} \, dK_{xy} \, d\phi \, dK_z \tag{10.181}$$

The argument within the δ-function can be factorized—as K_{xy} is a magnitude then solutions for it must be positive, hence the argument can be factorized as $(K_{xy} - \alpha_1)(K_{xy} - \alpha_2)$, where the roots α_1 and α_2 are given by:

$$\alpha_{1,2} = -k_i \cos\phi \pm \sqrt{k_i^2 \cos^2\phi - \frac{2m^* \Delta E}{\hbar^2}} \tag{10.182}$$

where clearly $\alpha_1 > \alpha_2$. Equation (10.181) can therefore be written as:

$$\frac{1}{\tau_i} = \frac{D_A^2 m^*}{\rho v_s (2\pi)^2 \hbar^2} \left(N_0 + \frac{1}{2} \mp \frac{1}{2}\right) \int_0^\infty \int_0^{2\pi} \int_0^\infty (G_{if}(K_z))^2 (K_{xy}^2 + K_z^2)^{\frac{1}{2}}$$

$$\times \delta\left((K_{xy} - \alpha_1)(K_{xy} - \alpha_2)\right) K_{xy} \, dK_{xy} \, d\phi \, dK_z \tag{10.183}$$

Consider now the integration over K_{xy}, there are only two contributions to the integral, one when K_{xy} is around α_1 and the other when K_{xy} is around α_2. Thus the δ-function can be split into two components:

$$\frac{1}{\tau_\mathrm{i}} = \frac{D_A^2 m^*}{\rho v_s (2\pi)^2 \hbar^2} \left(N_0 + \frac{1}{2} \mp \frac{1}{2} \right) \int_0^\infty \int_0^{2\pi} \int_0^\infty \left(G_\mathrm{if}(K_z) \right)^2 \left(K_{xy}^2 + K_z^2 \right)^{\frac{1}{2}}$$

$$\times \left[\frac{1}{|K_{xy} - \alpha_2|} \delta \left(K_{xy} - \alpha_1 \right) + \frac{1}{|K_{xy} - \alpha_1|} \delta \left(K_{xy} - \alpha_2 \right) \right] K_{xy} \, \mathrm{d}K_{xy} \, \mathrm{d}\phi \, \mathrm{d}K_z \quad (10.184)$$

and now actually performing the integration over K_{xy} gives:

$$\frac{1}{\tau_\mathrm{i}} = \frac{D_A^2 m^*}{\rho v_s (2\pi)^2 \hbar^2} \left(N_0 + \frac{1}{2} \mp \frac{1}{2} \right) \int_0^\infty \int_0^{2\pi} \left(G_\mathrm{if}(K_z) \right)^2$$

$$\times \left(\frac{\Theta \left(\alpha_1 \right) \alpha_1 \sqrt{\alpha_1^2 + K_z^2}}{|\alpha_1 - \alpha_2|} + \frac{\Theta \left(\alpha_2 \right) \alpha_2 \sqrt{\alpha_2^2 + K_z^2}}{|\alpha_2 - \alpha_1|} \right) \, \mathrm{d}\phi \, \mathrm{d}K_z \quad (10.185)$$

where the Heaviside functions ensure there are only contributions for positive α_1 and α_2. Recalling $\alpha_1 > \alpha_2$ then finally:

$$\frac{1}{\tau_\mathrm{i}} = \frac{D_A^2 m^*}{\rho v_s (2\pi)^2 \hbar^2} \left(N_0 + \frac{1}{2} \mp \frac{1}{2} \right) \int_0^\infty \int_0^{2\pi} \left(G_\mathrm{if}(K_z) \right)^2$$

$$\times \left(\frac{\Theta(\alpha_1)\alpha_1 \sqrt{\alpha_1^2 + K_z^2} + \Theta(\alpha_2)\alpha_2 \sqrt{\alpha_2^2 + K_z^2}}{\alpha_1 - \alpha_2} \right) \, \mathrm{d}\phi \, \mathrm{d}K_z \quad (10.186)$$

which is the same result as Piorek [214].

10.10   APPLICATION TO CONDUCTION SUBBANDS

Fig. 10.17(a) shows the effect of the energy separation on the intersubband scattering rate due to acoustic phonon emission. It can be seen that, for these thermalized distributions, the scattering rate is almost linear. A comparison with the equivalent electron–electron scattering rate, as shown in Figs. 10.25 and 10.26, shows that acoustic deformation potential scattering is much slower (an order of magnitude or more) than electron–electron scattering for small (less than 20 meV) intersubband energy separations at the same carrier density. However, it can be comparable or faster at larger (greater than 100 meV) energy separations. Fig. 10.17(b) compares the emission rate given in (a) with the absorption rate (shown in dashed lines). It can be seen that, for the same pair of initial (2) and final (1) states, the absorption rate remains a little below the emission rate due to the ratio of $N_0 + 1$ to N_0. In contrast to this the emission and absorption rates for carriers to scatter from the ground state (1) to the first excited state (2), which are visible in the lower section of Fig. 10.17(b) reach a maximum and then decrease as the separation between the initial and final states increases. This is to be expected as the acoustic phonon can only supply a few meV of energy and as the carriers in the initial state are thermalized most of them are concentrated near the subband minimum and simply cannot gain enough energy to transfer into the higher state.

(a)   (b)

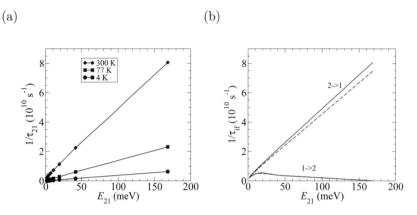

Figure 10.17 The intersubband acoustic deformation potential scattering rates as a function of the subband energy separation $E_{21} = E_2 - E_1$ for an infinitely deep quantum well with electron densities of 10×10^{10} cm^{-2} in each of the lowest two subbands, (a) shows the effect of temperature on the scattering rate from the second ($n = 2$) to the first ($n = 1$) subband due to the emission of acoustic phonons and (b) compares the scattering rates from the second to the first and the first to the second subbands due to emission (solid lines) and absorption (dashed lines) of acoustic phonons at the fixed temperature of 300 K.

Fig. 10.18(a) shows the microscopic effect of acoustic deformation potential scattering on carriers *within* a subband. In particular, it can be seen that the rate for emission

(a)   (b)

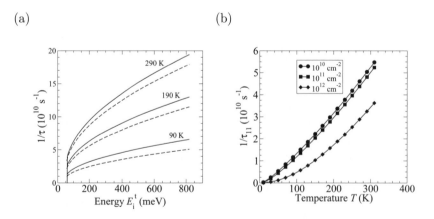

Figure 10.18 (a) The intrasubband acoustic deformation potential scattering rates due to phonon emission (solid lines) and phonon absorption (dashed lines) as a function of the total energy E_i^t of the carrier in the initial state (in this case the subband index $i = 1$) in a 100 Å GaAs infinitely deep quantum well with a carrier density of 10×10^{10} cm^{-2}, and (b) the thermally averaged intrasubband scattering rates due to acoustic phonon emission again in the ground state of a 100 Å GaAs infinitely deep quantum well as a function of the temperature and for several different carrier densities.

of acoustic phonons is greater than that for absorption for all initial carrier energies and temperatures and this is why carriers which have a lot of kinetic energy within a subband lose this energy to the lattice and eventually form a thermalized distribution[†]. Note the presence of the subband minimum produces the 'cut-off' in the data at energies just larger than 50 meV.

Fig. 10.18(b) illustrates the results of calculations of the effect of temperature and carrier density on the *thermally averaged* acoustic deformation potential scattering rates. Increasing temperature implies the presence of more phonons and hence the scattering rate increases, while the variations due to the carrier density changes are much smaller and are due to final state blocking. It is interesting to note that, for (b), the thermally averaged phonon absorption rates (which are not shown) are almost identical to the emission rates under this assumption of thermalized (Fermi–Dirac) carrier distributions, this is a reflection of the idea of an equilibrium.

10.11   OPTICAL DEFORMATION POTENTIAL SCATTERING

Taking the optical deformation potential as a sum over all phonon wave vectors with the interaction term of Lundstrom [211] (equation (2.67)) then in analogy with equation (10.160) for acoustic deformation potential scattering, Fermi's Golden Rule gives:

$$\frac{1}{\tau_i} = \frac{2\pi}{\hbar} \sum_{\mathbf{k}_f} \frac{D_o^2 \hbar}{2\rho\omega_o} \left(N_0 + \frac{1}{2} \mp \frac{1}{2} \right)$$

$$\times \left| \int \int \psi_f^*(z) \frac{e^{i\mathbf{k}_f \cdot \mathbf{r}_{xy}}}{A^{\frac{1}{2}}} \sum_{\mathbf{K}_{xy}} \sum_{K_z} \frac{e^{-i\mathbf{K}_{xy} \cdot \mathbf{r}_{xy}}}{A^{\frac{1}{2}}} \frac{e^{-iK_z z}}{L^{\frac{1}{2}}} \psi_i(z) \frac{e^{-i\mathbf{k}_i \cdot \mathbf{r}_{xy}}}{A^{\frac{1}{2}}} dz\, d\mathbf{r}_{xy} \right|^2$$

$$\times \delta(E_f^t - E_i^t) \qquad (10.187)$$

The derivation can be taken forward in exactly the same manner as for acoustic deformation potential scattering, except without the substitution for ω_s as given in equation (10.166), also assuming parabolic subbands then in analogy to equation (10.179) in Section 10.9:

$$\frac{1}{\tau_i} = \frac{D_o^2}{2\rho\omega_o(2\pi)^2} \left(N_0 + \frac{1}{2} \mp \frac{1}{2} \right) \int \int (G_{\mathrm{if}}(K_z))^2$$

$$\times \delta \left(E_f + \frac{\hbar^2}{2m^*} \left(\mathbf{k}_i^2 + \mathbf{K}_{xy}^2 + 2k_i K_{xy} \cos\phi \right) - E_i - \frac{\hbar^2 \mathbf{k}_i^2}{2m^*} \mp \hbar\omega_o \right) dK_z\, dK_{xy} \quad (10.188)$$

Now the energies of phonons from the optical branch are not small, hence the phonon energy cannot be approximated to zero and the collision considered elastic, rather the term $\mp\hbar\omega_o$ has to be retained. Labelling $E_f - E_i \mp \hbar\omega_o$ as Δ, then

$$\frac{1}{\tau_i} = \frac{D_o^2}{2\rho\omega_o(2\pi)^2} \left(N_0 + \frac{1}{2} \mp \frac{1}{2} \right) \int \int (G_{\mathrm{if}}(K_z))^2$$

[†]This result helps justify the assumption which gives the thermal averaging of carriers, used so often in this chapter, some physical meaning.

$$\times \delta \left(\frac{\hbar^2 K_{xy}^2}{2m^*} + \frac{\hbar^2 k_i K_{xy} \cos \phi}{m^*} + \Delta \right) \, dK_z \, d\mathbf{K}_{xy} \tag{10.189}$$

and taking the factor $\hbar^2/(2m^*)$ out of the δ-function then:

$$\frac{1}{\tau_i} = \frac{D_o^2 m^*}{\rho \omega_o (2\pi)^2 \hbar^2} \left(N_0 + \frac{1}{2} \mp \frac{1}{2} \right) \int \int (G_{if}(K_z))^2$$

$$\times \delta \left(K_{xy}^2 + 2k_i K_{xy} \cos \phi + \frac{2m^* \Delta}{\hbar^2} \right) \, dK_z \, d\mathbf{K}_{xy} \tag{10.190}$$

Transforming the integral over the in-plane phonon wave vector $\mathbf{K}_{xy}$ into plane polar coordinates then:

$$\frac{1}{\tau_i} = \frac{D_o^2 m^*}{\rho \omega_o (2\pi)^2 \hbar^2} \left(N_0 + \frac{1}{2} \mp \frac{1}{2} \right) \int \int_0^{2\pi} \int (G_{if}(K_z))^2$$

$$\times \delta \left(K_{xy}^2 + 2k_i K_{xy} \cos \phi + \frac{2m^* \Delta}{\hbar^2} \right) K_{xy} \, dK_{xy} \, d\phi \, dK_z \tag{10.191}$$

Again, following previous procedures, consider the integral over K_{xy}. There is only a contribution to this integral when the argument within the δ-function is zero. Now clearly solutions for K_{xy} must be positive (since K_{xy} is the *length* of the in-plane phonon wave vector), hence the argument of the δ-function can be factorized as $(K_{xy} - \alpha_1)(K_{xy} - \alpha_2)$, where again the roots are given by:

$$\alpha_{1,2} = -k_i \cos \phi \pm \sqrt{k_i^2 \cos^2 \phi - \frac{2m^* \Delta}{\hbar^2}} \tag{10.192}$$

and again $\alpha_1 > \alpha_2$. Equation (10.191) can therefore be written:

$$\frac{1}{\tau_i} = \frac{D_o^2 m^*}{\rho \omega_o (2\pi)^2 \hbar^2} \left(N_0 + \frac{1}{2} \mp \frac{1}{2} \right) \int \int_0^{2\pi} \int (G_{if}(K_z))^2$$

$$\times \delta \left((K_{xy} - \alpha_1)(K_{xy} - \alpha_2) \right) K_{xy} \, dK_{xy} \, d\phi \, dK_z \tag{10.193}$$

Again following the same procedure as for acoustic deformation potential scattering then:

$$\frac{1}{\tau_i} = \frac{D_o^2 m^*}{\rho \omega_o (2\pi)^2 \hbar^2} \left(N_0 + \frac{1}{2} \mp \frac{1}{2} \right) \int_0^\infty \int_0^{2\pi} \int_0^\infty (G_{if}(K_z))^2$$

$$\times \left[\frac{1}{|K_{xy} - \alpha_2|} \delta (K_{xy} - \alpha_1) + \frac{1}{|K_{xy} - \alpha_1|} \delta (K_{xy} - \alpha_2) \right] K_{xy} \, dK_{xy} \, d\phi \, dK_z \tag{10.194}$$

and now performing the integration over K_{xy} then:

$$\frac{1}{\tau_i} = \frac{D_o^2 m^*}{\rho \omega_o (2\pi)^2 \hbar^2} \left(N_0 + \frac{1}{2} \mp \frac{1}{2} \right) \int_0^\infty \int_0^{2\pi} (G_{if}(K_z))^2$$

$$\times \left[\frac{\alpha_1}{|\alpha_1 - \alpha_2|} + \frac{\alpha_2}{|\alpha_2 - \alpha_1|} \right] \, d\phi \, dK_z \tag{10.195}$$

and with $\alpha_1 > \alpha_2$ then:

$$\frac{1}{\tau_i} = \frac{D_o^2 m^*}{\rho \omega_o (2\pi)^2 \hbar^2} \left(N_0 + \frac{1}{2} \mp \frac{1}{2} \right) \int_0^\infty \int_0^{2\pi} (G_{if}(K_z))^2 \frac{\alpha_1 + \alpha_2}{\alpha_1 - \alpha_2} \, d\phi \, dK_z \qquad (10.196)$$

But:

$$\alpha_1 + \alpha_2 = -2k_i \cos\phi \quad \text{and} \quad \alpha_1 - \alpha_2 = 2\sqrt{k_i^2 \cos^2\phi - \frac{2m^*\Delta}{\hbar^2}} \qquad (10.197)$$

Therefore:

$$\frac{1}{\tau_i} = -\frac{D_o^2 m^*}{\rho \omega_o (2\pi)^2 \hbar^2} \left(N_0 + \frac{1}{2} \mp \frac{1}{2} \right)$$

$$\times \int_0^\infty \int_0^{2\pi} (G_{if}(K_z))^2 \frac{k_i \cos\phi}{\sqrt{k_i^2 \cos^2\phi - \frac{2m^*\Delta}{\hbar^2}}} \, d\phi \, dK_z \qquad (10.198)$$

and as these integrals are independent of each other then finally the optical deformation scattering rate for a carrier with an initial wave vector k_i into all final states of a given subband is given by:

$$\frac{1}{\tau_i} = -\frac{D_o^2 m^*}{\rho \omega_o (2\pi)^2 \hbar^2} \left(N_0 + \frac{1}{2} \mp \frac{1}{2} \right)$$

$$\times \int_0^\infty (G_{if}(K_z))^2 \, dK_z \int_0^{2\pi} \frac{k_i \cos\phi}{\sqrt{k_i^2 \cos^2\phi - \frac{2m^*\Delta}{\hbar^2}}} \, d\phi \qquad (10.199)$$

10.12 CONFINED AND INTERFACE PHONON MODES

Forming quantum wells or superlattices clearly changes the electronic energy levels of a crystal from what they are in an infinite bulk crystal, which is the main subject of this book. In fact, all of the crystal properties are changed to a greater or lesser extent. Perhaps of secondary importance to the effect on the electronic energy levels, and the subsequent changes that this induces in scattering rates, exciton energies, impurity energies, etc. are the fundamental changes introduced to the *phonon modes*. It can be appreciated that such changes are likely as the bulk LO phonon in GaAs has an energy of around 36 meV while in AlAs it is closer to 50 meV. Hence, forming a superlattice with alternating layers of GaAs and AlAs is going to have some effect on the phonon energies.

Various models have been put forward to account for this change in symmetry. At one end of the scale, some models consider each semiconductor layer as a continuum of material with macroscopic-like properties, namely the *Dielectric Continuum model*dielectric continuum model, (see, for example [212]), or the *Hydrodynamic model* (see, for example [216]). Alternative approaches have considered the allowed vibrational modes calculated directly from the viewpoint of individual atomic potentials (see, for example, [217]). See Adachi [14], p. 70, for an introduction.

Such improved models for phonons in heterostructures lead to modes which are confined to the individual semiconductor layers—*confined modes*—while some propagate along boundaries between the layers—the so-called *interface modes*. Recent work has shown that, while the electron–phonon scattering rates from these individual modes are quite different, the *total* rate from all of the modes collectively is quite similar to that from bulk phonons (see Kinsler *et al.* [218]). However, this is still a very active area of research and future developments will need to be monitored.

10.13 CARRIER–CARRIER SCATTERING

Fermi's Golden Rule describes the lifetime of a particle in a particular state with respect to scattering by a *time-varying potential*. For phonon scattering, this harmonic potential is derived from the phonon wave function, which is itself a travelling wave. For the case of one carrier scattering against another due to the Coulomb potential, there appears to be no time dependency. The *Born approximation* is often cited in the literature when discussing carrier–carrier scattering; this is just a way of working scattering from a constant potential into Fermi's Golden Rule. This is achieved by considering that the perturbing potential is 'switched on' only when the particle reaches the same proximity. For an excellent introduction to the Born approximation, see Liboff ([219], p. 621).

Therefore, the perturbating potential appearing in Fermi's Golden Rule for the interaction of two isolated carriers *is* the Coulombic interaction, i.e.

$$\tilde{\mathcal{H}} = \frac{e^2}{4\pi\epsilon r} \tag{10.200}$$

where $\epsilon = \epsilon_r \epsilon_0$ is the permitivity of the material and r is the separation of the electrons. Now the initial and final states, $|i\rangle$ and $|f\rangle$, respectively, of the *system* both consist of two electron (or hole) wave functions, as carrier–carrier scattering is a two-body problem, and thus there is a much greater variety of scattering mechanisms possible than in the essentially one-body problem encountered in phonon scattering. Fig. 10.19 illustrates all of the possible mechanisms in a two-level system, where at least one of the carriers changes its subband, these are usually referred to as *intersubband* transitions. However, the distinction now is not quite so clear (see below). The central diagram in Fig. 10.19 illustrates the symmetric intersubband event, '22→11', which moves two carriers down a level. The left and right figures show *Auger-type* intersubband scattering, where one carrier relaxes down to a lower subband, giving its excess energy to another carrier which remains within its original subband

In addition to the above, there are also scattering events where the number of carries in each subband does not change. Some of these are illustrated schematically in Fig. 10.20. Clearly the first of these, '22→22' is an intrasubband event; however, the second and third events are more difficult to categorize precisely because, although the number of carriers in each subband remain the same, the interaction itself is between carriers in different subbands. Pauli exclusion prevents carriers with the same spin occupying the same region of space, which therefore lowers their probability of scattering; in this work attention will be focused on collisions between particles with anti-parallel spins. Such considerations of spin-dependent scattering are often referred to by the term *exchange* [220]. Given that there are four possible carrier states involved, then in a N-level system there are 4^N different scattering events. In this two-level system these are as follows: 11–11, 11–12, 11–21, 11–22, 12–11, 12–12, 12–21, 12–22, 21–11, 21–12, 21–21, 21–22, 22–11, 22–12, 22–21, and 22–22. Note that completely different events of the type 'ij-fj'

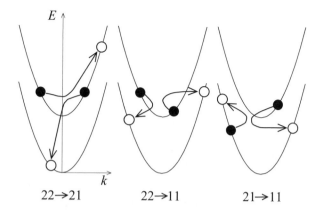

Figure 10.19   Illustration of various intersubband carrier–carrier scattering mechanisms in a two-level system

are possible in quantum wells with three or more subbands, and interactions of this type have been shown to be important in optically pumped intersubband lasers [220].

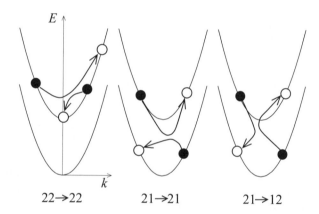

Figure 10.20   Illustration of various 'intrasubband' carrier–carrier scattering mechanisms in a two-level system

Therefore, taking a heterostructure wave function, of the form shown in equation (10.80), then the matrix element in Fermi's Golden Rule (equation (10.1)) becomes:

$$\langle f | \tilde{\mathcal{H}} | i \rangle =$$

$$\left\langle \psi_f(z) \frac{e^{-i\mathbf{k}_f \cdot \mathbf{r}_{xy}}}{\sqrt{A}} \psi_g(z') \frac{e^{-i\mathbf{k}_g \cdot \mathbf{r}'_{xy}}}{\sqrt{A}} \left| \frac{e^2}{4\pi\epsilon r} \right| \psi_i(z) \frac{e^{-i\mathbf{k}_i \cdot \mathbf{r}_{xy}}}{\sqrt{A}} \psi_j(z') \frac{e^{-i\mathbf{k}_j \cdot \mathbf{r}'_{xy}}}{\sqrt{A}} \right\rangle \qquad (10.201)$$

where the subband indices of the initial states are labelled 'i' and 'j' and those of the final states 'f' and 'g'. The decoupled form of the wave functions, with a component of the motion confined along the z-axis and an in-plane (x–y) travelling wave, suggests that the integrals should be evaluated across the plane and along the growth axis, and that the separation of the carriers be expressed as:

$$r = \sqrt{\left|\mathbf{r}_{xy} - \mathbf{r}'_{xy}\right|^2 + (z - z')^2} \tag{10.202}$$

Therefore obtain:

$$\langle f| \tilde{\mathcal{H}} |i \rangle = \frac{e^2}{4\pi\epsilon A^2} \int_{-\infty}^{+\infty} \int_{-\infty}^{+\infty} \int \int \psi_f^*(z)\psi_g^*(z')\psi_i(z)\psi_j(z')$$

$$\times \frac{e^{-i(\mathbf{k}_i \bullet \mathbf{r}_{xy} + \mathbf{k}_j \bullet \mathbf{r}'_{xy})} e^{i(\mathbf{k}_f \bullet \mathbf{r}_{xy} + \mathbf{k}_g \bullet \mathbf{r}'_{xy})}}{\sqrt{\left|\mathbf{r}_{xy} - \mathbf{r}'_{xy}\right|^2 + (z - z')^2}} \, d\mathbf{r}'_{xy} \, d\mathbf{r}_{xy} \, dz' \, dz \tag{10.203}$$

Goodnick and Lugli [221] (later re-iterated by Smet *et al.* [222]), followed earlier methods for carrier–carrier scattering in bulk (see, for example, Ziman [223], p. 170 and Takenaka *et al.* [224]), and took the two-dimensional Fourier Transform of the Coulombic potential to give

$$\langle f| \tilde{\mathcal{H}} |i \rangle = \frac{2\pi e^2}{4\pi\epsilon A q_{xy}} A_{ijfg}(q_{xy}) \, \delta(\mathbf{k}_f + \mathbf{k}_g - \mathbf{k}_i - \mathbf{k}_j) \tag{10.204}$$

where $q_{xy} = |\mathbf{k}_i - \mathbf{k}_f|$, and A_{ijfg} is a form factor, i.e.

$$A_{ijfg} = \int_{-\infty}^{+\infty} \int_{-\infty}^{+\infty} \psi_i(z)\psi_j(z')\psi_f^*(z)\psi_g^*(z') \, e^{-q_{xy}|z-z'|} \, dz' \, dz \tag{10.205}$$

Using the form for the matrix element shown in equation (10.204) and then substituting directly into Fermi's Golden Rule (equation (10.1)) gives the lifetime of a carrier in subband 'i' as follows:

$$\frac{1}{\tau_i} = \frac{2\pi}{\hbar} \sum_{f,g} \left| \frac{2\pi e^2}{4\pi\epsilon A q_{xy}} A_{ijfg}(q_{xy}) \right|^2 \delta(\mathbf{k}_f + \mathbf{k}_g - \mathbf{k}_i - \mathbf{k}_j) \, \delta(E_f^t + E_g^t - E_i^t - E_j^t) \tag{10.206}$$

Converting the summations over both final-state wave vectors into integrals introduces a factor of $L/(2\pi)$ per dimension, thus giving a factor of $A^2/(2\pi)^4$ in total (where the general area $A = L^2$) (see Section 2.3), therefore:

$$\frac{1}{\tau_i} = \frac{2\pi}{\hbar} \frac{A^2}{(2\pi)^4} \int \int \left| \frac{2\pi e^2}{4\pi\epsilon A q_{xy}} A_{ijfg}(q_{xy}) \right|^2 \delta(\mathbf{k}_f + \mathbf{k}_g - \mathbf{k}_i - \mathbf{k}_j)$$

$$\times \, \delta(E_f^t + E_g^t - E_i^t - E_j^t) \, d\mathbf{k}_g \, d\mathbf{k}_f \tag{10.207}$$

and thus:

$$\frac{1}{\tau_i} = \frac{e^4}{2\pi\hbar(4\pi\epsilon)^2} \int \int \frac{|A_{ijfg}(q_{xy})|^2}{q_{xy}^2} \delta(\mathbf{k}_f + \mathbf{k}_g - \mathbf{k}_i - \mathbf{k}_j)$$

$$\times \, \delta(E_f^t + E_g^t - E_i^t - E_j^t) \, d\mathbf{k}_g \, d\mathbf{k}_f \tag{10.208}$$

Integrating over all of the states of the second carrier (given by $\mathbf{k}_j$) and introducing Fermi–Dirac distribution functions to account for state occupancy, then obtain:

$$\frac{1}{\tau_i} = \frac{e^4}{2\pi\hbar(4\pi\epsilon)^2} \int \int \int \frac{|A_{ijfg}(q_{xy})|^2}{q_{xy}^2} f_j^{FD}(\mathbf{k}_j) \left[1 - f_f^{FD}(\mathbf{k}_f)\right] \left[1 - f_g^{FD}(\mathbf{k}_g)\right]$$

$$\times \ \delta(\mathbf{k_f} + \mathbf{k_g} - \mathbf{k_i} - \mathbf{k_j}) \ \delta(E_f^t + E_g^t - E_i^t - E_j^t) \ d\mathbf{k_g} \ d\mathbf{k_f} \ d\mathbf{k_j} \qquad (10.209)$$

which is equation (49) in Smet *et al.* [222]. Following their notation, collect the distribution functions together and label them as $P_{j,f,g}(\mathbf{k_j}, \mathbf{k_f}, \mathbf{k_g})$. The first δ-function summarizes in-plane momentum conservation and limits the integral over $\mathbf{k_g}$ to a contribution when $\mathbf{k_g} = \mathbf{k_i} + \mathbf{k_j} - \mathbf{k_f}$. In addition, the total energy of the carriers, E_i^t, etc. are equal to the energy of the relevant subband minima, E_i say, plus the in-plane kinetic energy; thus:

$$\frac{1}{\tau_i} = \frac{e^4}{2\pi\hbar(4\pi\epsilon)^2} \int \int \frac{|A_{ijfg}(q_{xy})|^2}{q_{xy}^2} \ P_{j,f,g}(\mathbf{k_j}, \mathbf{k_f}, \mathbf{k_g})$$

$$\times \ \delta \left(E_f + \frac{\hbar^2 \mathbf{k_f^2}}{2m^*} + E_g + \frac{\hbar^2 \mathbf{k_g^2}}{2m^*} - E_i - \frac{\hbar^2 \mathbf{k_i^2}}{2m^*} - E_j - \frac{\hbar^2 \mathbf{k_j^2}}{2m^*} \right) \ d\mathbf{k_f} \ d\mathbf{k_j} \qquad (10.210)$$

and therefore:

$$\frac{1}{\tau_i} = \frac{m^* e^4}{\pi\hbar^3 (4\pi\epsilon)^2} \int \int \frac{|A_{ijfg}(q_{xy})|^2}{q_{xy}^2} \ P_{j,f,g}(\mathbf{k_j}, \mathbf{k_f}, \mathbf{k_g})$$

$$\times \ \delta \left(\mathbf{k_f^2} + \mathbf{k_g^2} - \mathbf{k_i^2} - \mathbf{k_j^2} + \frac{2m^*}{\hbar^2}(E_f + E_g - E_i - E_j) \right) \ d\mathbf{k_f} \ d\mathbf{k_j} \qquad (10.211)$$

where $\mathbf{k_g}$ is known in terms of the other three wave vectors. It is the assumption of parabolic subbands in this last step that will be the limiting factor for the application of this method to hole–hole scattering—a point mentioned earlier in the context of the carrier–LO phonon scattering rate derivation. Now equation (10.211) represents the scattering rate of a carrier at a particular wave vector $\mathbf{k_i}$ averaged over all of the other initial particle states $\mathbf{k_j}$, and hence the only unknown in this remaining δ-function is the wave vector $\mathbf{k_f}$. Contributions to the integral over $\mathbf{k_f}$ occur when the argument of the δ-function is zero, and indeed given the form for this argument, it is clear that the solutions for $\mathbf{k_f}$ map out an ellipse.

The standard procedure [221, 222] is then to introduce relative wave vectors:

$$\mathbf{k_{ij}} = \mathbf{k_j} - \mathbf{k_i} \qquad (10.212)$$
$$\mathbf{k_{fg}} = \mathbf{k_g} - \mathbf{k_f} \qquad (10.213)$$

and replace the integration over $\mathbf{k_f}$ in equation (10.211) with an integration over $\mathbf{k_{fg}}$. Since:

$$\mathbf{k_f} = \mathbf{k_i} + \mathbf{k_j} - \mathbf{k_g} = \mathbf{k_i} + \mathbf{k_j} - \mathbf{k_{fg}} - \mathbf{k_f} \qquad (10.214)$$

i.e.

$$\mathbf{k_f} = \frac{1}{2}(\mathbf{k_i} + \mathbf{k_j} - \mathbf{k_{fg}}) \qquad (10.215)$$

It follows that:

$$d(\mathbf{k_f})_x = -\frac{1}{2}d(\mathbf{k_{fg}})_x \quad \text{and} \quad d(\mathbf{k_f})_y = -\frac{1}{2}d(\mathbf{k_{fg}})_y \qquad (10.216)$$

so:

$$d\mathbf{k_f} = \frac{1}{4}d\mathbf{k_{fg}} \qquad (10.217)$$

as pointed out by Moško [225], which should then be substituted into equation (10.211). In order to perform the integration over $\mathbf{k}_{fg}$ it is converted to plane polar coordinates with $d\mathbf{k}_{fg} = k_{fg}dk_{fg}d\theta$, where θ is an angle measured from k_{ij} and the trajectory in the k_{fg}-θ plane is deduced from the condition that the argument of the δ-function in equation (10.211) must be zero.

The conservation of momentum diagram therefore looks like Fig. 10.21, where:

$$\mathbf{k}_{sum} = \mathbf{k}_i + \mathbf{k}_j = \mathbf{k}_f + \mathbf{k}_g \tag{10.218}$$

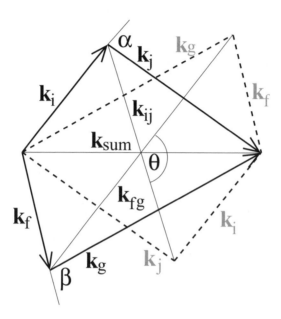

Figure 10.21   The conservation of momentum diagram for the two-body electron–electron scattering event

In this work, the occupancy of the final states will be assumed to be small, such that the distribution functions dependent upon $\mathbf{k}_f$ and $\mathbf{k}_g$ can be ignored. The effect of putting $f_f^{FD}(\mathbf{k}_f)$ and $f_g^{FD}(\mathbf{k}_g)$ to zero can be seen to place an upper limit on the scattering rate. The physical interpretation of this approximation is that *final-state blocking* is ignored, i.e. the process by which a scattering event is prevented because the required final state is already occupied. This is a common simplification [226] which allows the particular values of the final-state wave vectors to remain unknown. The approximation that this introduces is 'good' for the typical carrier densities encountered in devices based on quantum well heterostructures [227]. Recently, a generalization of this approach has been put forward that allows the specific values of the final carrier wave vectors to be calculated, thus avoiding this approximation; this method is also an example of making a different choice for the angle θ [220].

Using the notation $k = |\mathbf{k}|$, and applying the cosine rule to Fig. 10.21, then:

$$k_{sum}^2 = k_i^2 + k_j^2 + 2k_ik_j\cos\alpha \tag{10.219}$$

$$k_{sum}^2 = k_f^2 + k_g^2 + 2k_fk_g\cos\beta \tag{10.220}$$

and also:

$$k_{ij}^2 = k_i^2 + k_j^2 - 2k_i k_j \cos \alpha \qquad (10.221)$$

$$k_{fg}^2 = k_f^2 + k_g^2 - 2k_f k_g \cos \beta \qquad (10.222)$$

Summing equation (10.219) with equations (10.221) and (10.220) with 10.222 gives:

$$k_{sum}^2 + k_{ij}^2 = 2k_i^2 + 2k_j^2 \qquad (10.223)$$

$$k_{sum}^2 + k_{fg}^2 = 2k_f^2 + 2k_g^2 \qquad (10.224)$$

and by eliminating k_{sum}^2 gives:

$$k_{fg}^2 = k_{ij}^2 + 2\left(k_f^2 + k_g^2 - k_i^2 - k_j^2\right) \qquad (10.225)$$

Energy conservation has already been considered earlier within the δ-function, i.e.

$$E_f + \frac{\hbar^2 \mathbf{k}_f^2}{2m^*} + E_g + \frac{\hbar^2 \mathbf{k}_g^2}{2m^*} = E_i + \frac{\hbar^2 \mathbf{k}_i^2}{2m^*} + E_j + \frac{\hbar^2 \mathbf{k}_j^2}{2m^*} \qquad (10.226)$$

thus giving:

$$k_f^2 + k_g^2 - k_i^2 - k_j^2 = \frac{2m^*}{\hbar^2}\left(E_i + E_j - E_f - E_g\right) \qquad (10.227)$$

Substituting into equation (10.225) then gives:

$$k_{fg}^2 = k_{ij}^2 + \frac{4m^*}{\hbar^2}\left(E_i + E_j - E_f - E_g\right) = k_{ij}^2 + \Delta k_0^2, \quad \text{say} \qquad (10.228)$$

where the notation of Smet et al. has been employed.

Consider the entity $\mathbf{q}_{xy}$, for which q_{xy} is the magnitude. This vector $\mathbf{q}_{xy} = \mathbf{k}_i - \mathbf{k}_f$ is illustrated in the equivalent diagram (Fig. 10.22). In addition, this figure emphasizes the difference between the relative wave vectors $\mathbf{k}_{ij}$ and $\mathbf{k}_{fg}$, which turns out to be equal to $2\mathbf{q}_{xy}$, and therefore:

$$\mathbf{q}_{xy} = \frac{\mathbf{k}_{ij} - \mathbf{k}_{fg}}{2} \qquad (10.229)$$

With an angle of θ between them, the cosine rule then gives:

$$(2q_{xy})^2 = k_{ij}^2 + k_{fg}^2 - 2k_{ij}k_{fg}\cos\theta \qquad (10.230)$$

As the aim is to eliminate the final-state wave vectors from the definition of q_{xy}, then substitute for k_{fg} from equation (10.228), thus giving:

$$(2q_{xy})^2 = 2k_{ij}^2 + \Delta k_0^2 - 2k_{ij}\sqrt{k_{ij}^2 + \Delta k_0^2}\cos\theta \qquad (10.231)$$

Summarizing then, the expression for the scattering rate of a carrier of a particular wave vector $\mathbf{k}_i$ with another carrier, is given by:

$$\frac{1}{\tau_i} = \frac{m^* e^4}{4\pi\hbar^3 (4\pi\epsilon)^2} \int \int_0^{2\pi} \frac{|A_{ijfg}(q_{xy})|^2}{q_{xy}^2} P_{j,f,g}(\mathbf{k}_j, \mathbf{k}_f, \mathbf{k}_g) \, d\theta \, d\mathbf{k}_j \qquad (10.232)$$

which is equation (51) in Smet et al. [222], but with an additional factor of 4 in the denominator [225]. In this present treatise the factor $P_{jfg}(\mathbf{k}_j, \mathbf{k}_f, \mathbf{k}_g)$ has been assumed to

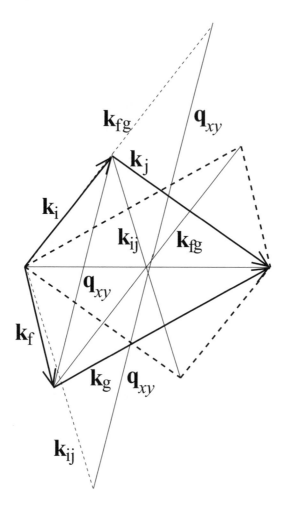

Figure 10.22   The conservation of momentum diagram for the two-body electron–electron scattering event

be dependent upon the initial-state distribution only, i.e. $P_{\mathrm{jfg}}(\mathbf{k}_{\mathrm{j}}, \mathbf{k}_{\mathrm{f}}, \mathbf{k}_{\mathrm{g}}) = f_{\mathrm{j}}^{\mathrm{FD}}(\mathbf{k}_{\mathrm{j}})$, and q_{xy} is given by equation (10.231). The integration over the vector $\mathbf{k}_{\mathrm{j}}$ can be performed by effectively switching to plane polar coordinates, and integrating along the length k_{ij} and around the angle α between $\mathbf{k}_{\mathrm{i}}$ and $\mathbf{k}_{\mathrm{j}}$.

With this change in the integral, equation (10.232) then becomes:

$$\frac{1}{\tau_{\mathrm{i}}} = \frac{m^* e^4}{4\pi\hbar^3 (4\pi\epsilon)^2} \int \int_0^{2\pi} \int_0^{2\pi} \frac{|A_{\mathrm{ijfg}}(q_{xy})|^2}{q_{xy}^2} P_{\mathrm{j}}(\mathbf{k}_{\mathrm{j}}) \ \mathrm{d}\theta \ \mathrm{d}\alpha \ k_{\mathrm{j}} \ \mathrm{d}k_{\mathrm{j}} \tag{10.233}$$

where q_{xy} is given by equation (10.231), Δk_0^2 is given by equation (10.228), and from Fig. 10.23:

$$k_{\mathrm{ij}}^2 = k_{\mathrm{i}}^2 + k_{\mathrm{j}}^2 - 2k_{\mathrm{i}}k_{\mathrm{j}} \cos\alpha \tag{10.234}$$

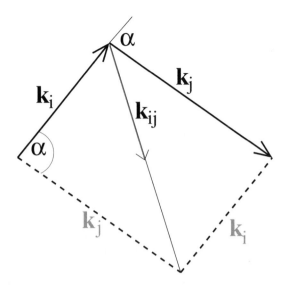

Figure 10.23   The relationship between the two initial carrier states

As in the case of phonon scattering, the limits of integration of the wave vector are taken from zero (the subband minima) to that corresponding to the barrier height of the heterostructure in question.

10.14   ADDITION OF SCREENING

So far, the theoretical treatise has only considered carrier–carrier scattering of two carriers in an empty environment. In semiconductor heterostructures, carrier–carrier scattering occurs because the system is doped, and therefore by definition there will be many carriers. In such instances, the force between any two carriers is not just the bare Coulombic repulsion, as the other mobile carriers are able to respond to any change in the electrostatic field, with the result being a reduction in the probability of scattering; the other carriers are said to *screen* the interaction.

One of the simplest models for screening [222] considers only the carriers within the same subband as the initial carrier state; it then proceeds by replacing the dielectric constant $\epsilon = \epsilon_r \epsilon_0$ with one which is dependent upon the relative wave vector q_{xy}, i.e. $\epsilon = \epsilon_r \epsilon_0 \epsilon_{sc}$, where:

$$\epsilon_{sc} = 1 + \frac{2\pi e^2}{(4\pi\epsilon)q_{xy}} \Pi_{ii}(q_{xy}) A_{iiii}(q_{xy}) \tag{10.235}$$

One consequence of this replacement is that the $q_{xy} = 0$ pole is removed from the scattering rate. Combining the results of Ando *et al.* [228] and using a Heaviside unit step function for convenience, the polarization factor at absolute zero is then given by:

$$\Pi_{ii}(q_{xy}) = \frac{m^*}{\pi\hbar^2}\left[1 - \Theta(q_{xy} - 2k_F)\sqrt{1 - \left(\frac{2k_F}{q_{xy}}\right)^2}\right] \tag{10.236}$$

where k_F is the two-dimensional Fermi wave vector [228] applicable to the initial state 'i' and is defined only at absolute zero, i.e.

$$k_F = \sqrt{\frac{2\pi N_i}{g}} \tag{10.237}$$

(see [1], p. 36, for example, for the three-dimensional equivalent). The factor g accounts for any degeneracy which can lead to multiple subband valleys; in GaAs, this is just 1. Maldague [229] showed that, under certain approximations, this polarizability can be generalized to any temperature with:

$$\Pi_{ii}(q_{xy}, T) = \int_0^\infty \frac{\Pi_{ii}(q_{xy}, T=0)}{4kT \cosh^2\left((E_F - E)/(2kT)\right)} \, dE \tag{10.238}$$

where E_F is the quasi-Fermi energy and E the minimum (called E_i here) of subband 'i' (see Section 2.4). This integral can actually be performed analytically[‡]:

$$\Pi_{ii}(q_{xy}, T) = \frac{\Pi_{ii}(q_{xy}, 0)}{4kT} \int_0^\infty \frac{1}{\cosh^2\left((E_F - E)/(2kT)\right)} \, dE \tag{10.239}$$

using the standard result $\int 1/\cosh^2\theta \, d\theta = \tanh\theta$ then:

$$\Pi_{ii}(q_{xy}, T) = \frac{\Pi_{ii}(q_{xy}, 0)}{4kT} \left[\tanh\left(\frac{E_F - E}{2kT}\right)\right]_0^\infty \tag{10.240}$$

then:

$$\Pi_{ii}(q_{xy}, T) = \frac{\Pi_{ii}(q_{xy}, 0)}{2} \left\{1 + \tanh\left(\frac{E_F}{2kT}\right)\right\} \tag{10.241}$$

Studying this equation, it can be seen that as $T \to 0$, $\Pi_{ii}(q_{xy}, T)$ recovers the low temperature limit. In the limit of high temperature it is exactly half this value.

Fig. 10.24 shows the results of a calculation of the 22–11 electron–electron scattering rate as a function of the electron energy of the initial state 'i', in an infinitely deep 400 Å wide GaAs quantum well. The figure compares the rates with and without the screening term at a temperature of 77 K and for carrier densities of 1×10^{10} cm^{-2} (left) followed by 100×10^{10} cm^{-2} (right) carriers in each level.

The first conclusion to be drawn is that the scattering rate increases as the energy of the initial electron in state 'i' decreases towards the subband minimum. In addition, the scattering rates themselves are nearly two orders of magnitude higher in the higher-carrier-density case illustrated on the right-hand side of Fig. 10.24 than in the lower-density case illustrated on the left. Furthermore, it is apparent that screening reduces the scattering rate, and the higher the carrier density, then the larger the effect of screening, as might be expected.

10.15   AVERAGING OVER AN INITIAL STATE POPULATION

The method documented so far gives the carrier–carrier scattering rate for a particular carrier energy averaged over another initial state distribution; as before for the carrier–

[‡]Thanks to Jim McTavish for pointing this out.

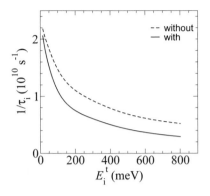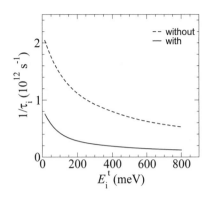

Figure 10.24   The electron–electron scattering rate as a function of the initial electron energy, with and without screening, for a carrier density of 1×10^{10} cm^{-2} (left) and 100×10^{10} cm^{-2} (right)

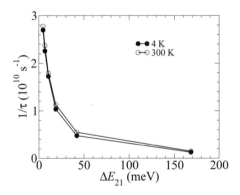

Figure 10.25   The mean 22–11 electron–electron scattering rate as a function of the subband separation, including screening and with a carrier density of 1×10^{10} cm^{-2} in each level

LO phonon case, it is usually more instructive to average this quantity over the Fermi–Dirac distribution of carriers in the initial state. This could be achieved by using the following mean:

$$\frac{1}{\tau} = \frac{\int \frac{1}{\tau_i} f_i^{\mathrm{FD}}(E_i^k) \ \mathrm{d}E_i^k}{\int f_i^{\mathrm{FD}}(E_i^k) \ \mathrm{d}E_i^k} \tag{10.242}$$

where the superscript k has been introduced to indicated that this is the in-plane kinetic energy associated with the wave vector $\mathbf{k}_i$. By using the result of equation (2.50) in

Section 2.4, the denominator is then given by:

$$\int f_i^{\mathrm{FD}}(E_i^k) \ dE_i^k = N_i \frac{\pi \hbar^2}{m^*} \tag{10.243}$$

As the carrier's in-plane energy is expressed in terms of a wave vector and:

$$E_i^k = \frac{\hbar^2 k_i^2}{2m^*}, \quad \text{then} \quad dE_i^k = \frac{\hbar^2 k_i}{m^*} \ dk_i \tag{10.244}$$

Hence the above mean can be expressed as:

$$\frac{1}{\tau} = \frac{\int \frac{1}{\tau_i} f_i^{\mathrm{FD}}(k_i) \ k_i \ dk_i}{\pi N_i} \tag{10.245}$$

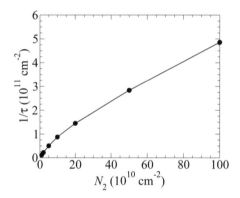

Figure 10.26   The mean 22–11 electron–electron scattering rate as a function of the carrier density in level $|2\rangle$, for a 300 Å infinitely deep GaAs quantum well at 300 K

Fig. 10.25 shows the electron–electron scattering rate averaged over the initial carrier distribution, as given by equation (10.245), for the 22–11 mechanism in a infinitely deep quantum well. The subband separation, $\Delta E_{21} = E_2 - E_1$, was varied by adjusting the quantum well width. It can be seen that the electron–electron scattering rate increases as the subband separation decreases and that in this particular approach, with the low carrier density of 1×10^{10} cm^{-2}, the rate is only a weak function of temperature.

The effect of increasing carrier density is shown in Fig. 10.26. The range of carrier densities was chosen to span roughly those found in intersubband devices. It can be seen that, with this model, the intersubband scattering is almost proportional to the carrier density, although it must be noted that in this work any effects due to Pauli exclusion are being ignored—this took the form of assuming that the final-state populations were small, as discussed earlier. Inclusion of significant populations in a final-state could lead to final-state blocking being significant, and hence interfere with the proportionality, but at this stage of gaining an intuitive feel for carrier–carrier scattering, such investigations necessitate the introduction of yet more variables and unnecessary complication. As in

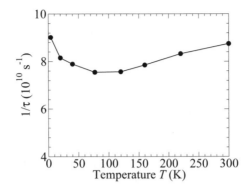

Figure 10.27   The mean 22–11 electron–electron scattering rate as a function of temperature, for a 300 Å infinitely deep GaAs quantum well, with a carrier density of 10×10^{10} cm^{-2}

the previous example, it was found that these intersubband scattering rates were quite insensitive to the temperature. This is substantiated further in Fig. 10.27, which shows the temperature dependence of the 22–11 intersubband scattering rate for 10×10^{10} cm^{-2} electrons in the same 300 Å infinitely deep quantum well. The rate dips to around 10% less than its low temperature value before increasing towards room temperature. The temperature dependency within the model comes from the screening term as well as both of the initial-state distribution functions.

10.16   INTRASUBBAND VERSUS INTERSUBBAND

So far, only the results of calculations of intersubband carrier–carrier scattering rates have been presented, but as mentioned before intrasubband events are also possible. The latter are characterized by no change in the numbers of carriers in each subband. Considering only the most simplistic mechanisms of the type 'ii–ii', then Fig. 10.28 plots the carrier density dependency and Fig. 10.29 the temperature dependency of the 22–22 electron–electron scattering rate, again for a typical 300 Å infinitely deep GaAs quantum well. These figures are the intrasubband equivalents of the intersubband cases of Figs. 10.26 and 10.27, respectively, in the last section. Note that, as all of the initial and final states are within the same subband, no dependence on an intersubband energy separation is required.

Therefore, comparing Fig. 10.28 with Fig. 10.26, it can be seen that the dependence of the scattering rate on the carrier density is quite different in this the intrasubband case. The 'almost' proportionality is evident at the two lowest temperatures, but for 300 K there is a clear non-monotonic behaviour.

Fig. 10.29 shows the temperature dependence of this 22–22 rate; this is stronger than in the intersubband case of Fig. 10.27 dipping by 10% again before increasing markedly to its room temperature value.

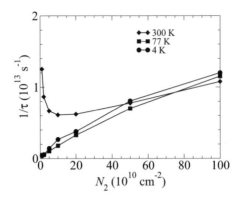

Figure 10.28   The mean 22–22 electron–electron scattering rate as a function of the carrier density, for a 300 Å infinitely deep GaAs quantum well

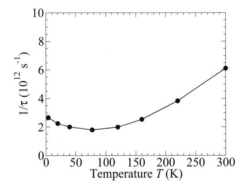

Figure 10.29   The mean 22–22 electron–electron scattering rate as a function of temperature, for a 300 Å infinitely deep GaAs quantum well, with a carrier density of 10×10^{10} cm^{-2}

Overall, the intrasubband scattering rates are around one to two orders of magnitude higher than the closest intersubband equivalent. This is a general result which is observed in a range of semiconductor heterostructures. In fact, in this idealistic infinitely deep quantum well, the intersubband rates represent something of a maximum as the overlap of the carrier wave functions is complete, and hence the ratio of the intrasubband to intersubband rates deduced here is probably conservative.

10.17 THERMALIZED DISTRIBUTIONS

This last result, i.e. that the intrasubband scattering rates are much larger than their typical intersubband equivalents, is a very important result. Its implication is that, given a quantum well or a quantum wire (basically any system which has carrier dispersion) with a number of subbands, then the carriers within each subband scatter much more rapidly than they do between the subbands. Thus, if the system is stimulated somehow, either by optical excitation or electrical injection of carriers, then the distributions of carriers within each subband will, in the first instance, reorganize themselves independently of each other, before intersubband scattering begins to redistribute the carriers between the levels. This justifies the initial assumption that the carrier populations in each subband can be represented by Fermi–Dirac distributions, although perhaps with independent quasi-Fermi energies (see Section 2.4)—in this situation, the carriers are said to be *independently thermalized.* This is a common assumption in intersubband devices (see, for example [230]). It is also possible that the temperature required to describe these distributions is not equal to the lattice temperature; such a scenario will either require a thorough analysis of the kinetics of the system or else a Monte Carlo simulation of the complete subband structure.

Should the means of excitation, i.e. the source of input energy, be removed from the quantum well system, then the carriers will relax down to a state where the populations can all be described by a single Fermi energy—this would represent thermal equilibrium. Thus, the previous case of an excited subband structure is often given the rather grand descriptive title of a *non-equilibrium carrier distribution.*

10.18 AUGER-TYPE INTERSUBBAND PROCESSES

The various carrier–carrier scattering rates were discussed earlier, and illustrated in Figs. 10.19 and 10.20. Until now, the calculations themselves have only considered the 'symmetric' 22–11 process, and only within an infinitely deep quantum well. This particular heterostructure was just an illustrative example, and provided a way of reducing the number of material parameters that had to be specified. In order to calculate the scattering rates due to the 'asymmetric' 22–21 and 21–11 processes, it is necessary to base the calculations in a different quantum well system; for this purpose, a single 200 Å GaAs quantum well surrounded by $Ga_{0.8}Al_{0.2}As$ barriers was chosen. This is also an opportunity to reiterate and demonstrate that the carrier–carrier scattering rate derived is valid for any one-particle eigenstates, as calculated with the methods outlined in Chapters 2 and 3.

The asymmetric intersubband processes are often referred to as *Auger-type processes* because they resemble traditional interband Auger scattering, where one carrier is able to give up its potential energy to another carrier and hence relax down a level. These processes are forbidden in symmetric potentials, and the simplest way to break the symmetry of the above finite quantum well is to apply an electric field. Fig. 10.30 illustrates the potential profile and the wave functions of the ground $|1\rangle$ and first excited $|2\rangle$ states of the quantum well with the maximum applied electric field employed.

Fig. 10.31 shows the results of calculations of the mean scattering rates for all three possible intersubband mechanisms in this two-level system. The selection rules arise quite naturally in the formalism; hence, at zero field when the potential and hence the

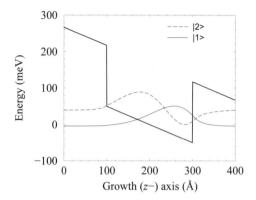

Figure 10.30 The potential profile and two lowest-energy eigenstates of a 200 Å GaAs quantum well surrounded by $Ga_{0.8}Al_{0.2}As$ barriers, at an applied electric field of 50 kVcm^{-1}

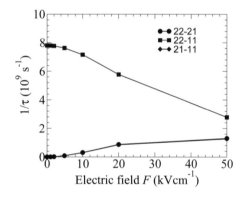

Figure 10.31 The mean intersubband electron–electron scattering rate as a function of the applied electric field for the quantum well in Fig. 10.30 with 1×10^{10} electrons per cm^2 in each level; note that the scattering rates for the 22–21 and 21–11 processes are almost identical

wave functions are symmetric (and anti-symmetric), only the symmetric 22–11 channel is allowed. However, as the electric field is increased and the symmetry of the potential broken, the asymmetric rates increase and the symmetric rate decreases. At the maximum value of the electric field employed, the Auger-type rates are a significant fraction of the total scattering rate, thus illustrating that their inclusion is essential in modelling such systems. Indeed, quantum well systems exist in which the Auger-type scattering rates dominate (see, for example [231]). Note that the 22–11 rate moves two electrons down a level, whereas the two Auger-type rates only move one electron down at a time; this fact must be accounted for when solving subband population rate equations.

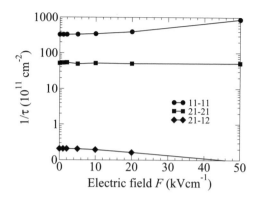

Figure 10.32   The mean intrasubband electron–electron scattering rates as a function of the applied electric field for the quantum well in Fig. 10.30 with 1×10^{10} electrons per cm^2 in each level

10.19   ASYMMETRIC INTRASUBBAND PROCESSES

For completeness, Fig. 10.32 shows the scattering rates of possible intrasubband mechanisms for the same finite quantum well as in the previous section. With equal carrier densities employed in each level, the 'asymmetric' 21–21 and 21–12 intrasubband events are significant too; particularly the former. The implication of these results is that there will be a substantial interaction between the carrier distributions in the different subbands. Therefore, even disregarding the transfer of carriers between levels with intersubband scattering, the distributions are coupled in some way and will not be independently thermalized, as often conjectured. The high scattering rate of the 21–21 channel implies that there will be a redistribution of kinetic energy between the subbands without a redistribution of carriers, and thus it seems likely that, although the electron temperature will not be equal to that of the lattice, the same electron temperature may be applicable to the entire multi-subband carrier distribution.

10.20   EMPIRICAL RELATIONSHIPS

Simple empirical relationships giving the dependency of some computationally complex quantity, such as a carrier scattering rate, on some simple material parameter, are often useful for device designers and experimentalists who need an intuitive feel for the underlying physics without having recourse to lengthy calculations. Earlier, a relationship was derived for the ratio of emission to absorption rates for carrier–LO phonon scattering (see, equation (10.155)). As hinted at then, the subband separation and temperature dependencies of the 22–11 carrier–carrier scattering rate imply simple relationships (see Section 10.15). In particular, the hyperbolic energy dependence suggests that:

$$\frac{1}{\tau} \propto \frac{1}{\Delta E_{21}}$$

(10.246)

and the linear carrier density dependence:

$$\frac{1}{\tau} \propto N \tag{10.247}$$

where in this simple example of a symmetric 'ii–ff' rate, the N refers to the carrier density in the initial state 'i'.

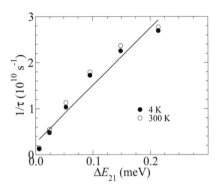
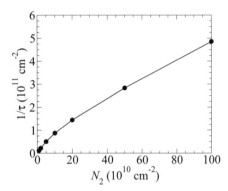

Figure 10.33   The mean 22–11 intersubband electron–electron scattering rate plotted against the reciprocal of the subband separation (left) and the carrier density (right)

The diagrams shown in Fig. 10.33 substantiate these postulated relationships, i.e. the graphs are quite linear and close to the origin. Thus, given a quantum well system with a particular doping profile, and hence a specific carrier density, it is possible to answer questions such as 'What happens to the intersubband electron–electron scattering rate if we increase the well width and hence halve the subband separation?'. The answer is of course, that the rate will double. Similar questions can be asked about variations in doping profile and hence carrier densities for fixed layer thicknesses.

Both results *can* be combined as follows:

$$\frac{1}{\tau} \propto \frac{N}{\Delta E_{21}} \tag{10.248}$$

10.21   CARRIER–PHOTON SCATTERING

The emission (or absorption) of light by a charge carrier, whether an electron or a hole, is essentially a scattering event between an initial state 'i' and a final state 'f'. The electromagnetic field is the time-dependent perturbation $\tilde{\mathcal{H}}$ which induces this event [219]. The transition rate from the initial electronic state $|i\rangle$ to the final state $|f\rangle$ is given by Fermi's Golden Rule (see [232] for its application to light):

$$\frac{1}{\tau_i} = \frac{2\pi}{\hbar} \sum_{f} \left| \langle f| \tilde{\mathcal{H}} |i\rangle \right|^2 \delta(E_f^c - E_i^c \mp \hbar\omega) \tag{10.249}$$

where the superscript on the energies has been introduced to indicate that these are the total carrier energies, which contain both kinetic and potential energy components. The δ-function now explicitly contains the photon energy, with the upper sign of the $\mp$ representing absorption and the lower emission. For example, for absorption the total initial energy is the sum of the carrier energy and the photon energy, i.e.

$$E_i^c + \hbar\omega = E_f^c, \quad \text{and hence} \quad E_f^c - E_i^c - \hbar\omega = 0 \tag{10.250}$$

and for emission:

$$E_i^c = E_f^c + \hbar\omega, \quad \text{and hence} \quad E_f^c - E_i^c + \hbar\omega = 0 \tag{10.251}$$

The photon momentum is assumed to be zero, and hence momentum conservation does not have to be considered with an additional δ-function, with the physical implication of this being that transitions on the band diagrams are always vertical whether they are between bands (interband) or within the same band (intraband or intersubband) (see Fig. 10.34).

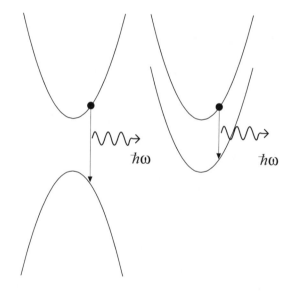

Figure 10.34   Carrier–photon scattering events result in vertical transitions; interband (left) and intraband (intersubband) (right)

Although a great deal has been achieved within the envelope function approximation, it is necessary now to reintroduce the rapidly varying component of the wave function, writing this as the product $u\psi$ (see Section 1.7). Then, as reiterated by Adachi ([14], p. 476) and Basu ([11], p. 299), Coon and Karunasiri [233] showed that the matrix element in equation 10.249 becomes:

$$\langle f| \tilde{\mathcal{H}} |i\rangle = \langle u_f| \tilde{\mathcal{H}} |u_i\rangle_{\text{cell}} \langle \psi_f|\psi_i\rangle + \langle u_f|u_i\rangle_{\text{cell}} \langle \psi_f| \tilde{\mathcal{H}} |\psi_i\rangle \tag{10.252}$$

In the case of interband transitions between the conduction and valence band, the second term gives zero since the Bloch functions u_f and u_i at the same point in the Brillouin zone, in two different bands, are orthogonal, i.e.

$$\langle u_{\mathrm{f}} | u_{\mathrm{i}} \rangle_{\mathrm{cell}} = \int_{\mathrm{cell}} u_{\mathrm{f}}(\mathbf{r}) u_{\mathrm{i}}(\mathbf{r}) \ \mathrm{d}\mathbf{r} = 0 \tag{10.253}$$

and therefore:

$$\langle \mathrm{f} | \tilde{\mathcal{H}} | \mathrm{i} \rangle = \langle u_{\mathrm{f}} | \tilde{\mathcal{H}} | u_{\mathrm{i}} \rangle_{\mathrm{cell}} \langle \psi_{\mathrm{f}} | \psi_{\mathrm{i}} \rangle \tag{10.254}$$

Hence the envelope function overlap integral, $\langle \psi_{\mathrm{f}} | \psi_{\mathrm{i}} \rangle$, determines which transitions are allowed and which are forbidden, which was the result quoted earlier in Section 6.8 during the discussion of excitons.

Radiative interband transitions are, of course, very important in both light sources (see, for example [232]) and detectors (see, for example, works such as [234–236]). This mechanism gives access to a frequency range from the near-infrared [237] to the blue, depending on the material system employed. For a review of wide-bandgap III–V emitters and lasers, see Nakamura [13], while for II–VI see Gunshor and Nurmikko [238]. Much discussion has already centred around interband transitions in the calculation of energy levels (see Chapters 2 and 3), exciton binding energies, and oscillator strengths (see Chapter 6), for further information, see Ivchenko and Pikus ([239], p. 162), or Bastard ([18], p. 237).

For intersubband transitions the first term on the right-hand side of equation (10.252) is zero, and since the subband envelope functions ψ_{f} and ψ_{i} are both eigenfunctions of the same Hermitian operator (the conduction or valence band Hamiltonian), they are therefore orthogonal, i.e.

$$\langle \psi_{\mathrm{f}} | \psi_{\mathrm{i}} \rangle = \int_{\mathrm{all \ space}} \psi_{\mathrm{f}}(\mathbf{r}) \psi_{\mathrm{i}}(\mathbf{r}) \ \mathrm{d}\mathbf{r} = 0 \tag{10.255}$$

Therefore, the matrix element becomes:

$$\langle \mathrm{f} | \tilde{\mathcal{H}} | \mathrm{i} \rangle = \langle u_{\mathrm{f}} | u_{\mathrm{i}} \rangle_{\mathrm{cell}} \langle \psi_{\mathrm{f}} | \tilde{\mathcal{H}} | \psi_{\mathrm{i}} \rangle \tag{10.256}$$

where the time-dependent perturbing potential $\tilde{\mathcal{H}}$ is given by:

$$\tilde{\mathcal{H}} = \frac{e}{m^*} \mathbf{A} \cdot \mathbf{p} \tag{10.257}$$

(see [222], [14], p. 476 and, most appropriately [240], p. 184). The vector $\mathbf{A}$ is the vector potential of the electromagnetic field and hence consists of a magnitude and a direction, with the latter being represented by the unit polarization vector $\hat{\mathbf{e}}$. The linear momentum operator $\mathbf{p} = -i\hbar\nabla$, and therefore:

$$\langle \mathrm{f} | \tilde{\mathcal{H}} | \mathrm{i} \rangle \propto \langle \psi_{\mathrm{f}} | \hat{\mathbf{e}} \cdot \nabla | \psi_{\mathrm{i}} \rangle \tag{10.258}$$

In this case of intersubband scattering, the envelope wave functions ψ_{i} and ψ_{f} are functions of the displacement along the growth (z-) axis only. Therefore, the gradient operator will also only have a z- component, and hence:

$$\langle \mathrm{f} | \tilde{\mathcal{H}} | \mathrm{i} \rangle \propto \hat{e}_z \left\langle \psi_{\mathrm{f}} \left| \frac{\partial \psi_{\mathrm{i}}}{\partial z} \right. \right\rangle \tag{10.259}$$

This implies that transitions are only allowed when there is a component of the polarization vector $\hat{\mathbf{e}}$ along the growth (z–)axis [241, 242], which means that no intersubband absorption occurs for normal (i.e. along the growth (z–)axis) incident light. This is illustrated in the top diagram of Fig. 10.35. For the case of normal-incidence light, there is no component of the polarization vector $\hat{\mathbf{e}}$ along the growth (z-) axis of the heterostructure, and hence no intersubband absorption occurs. When the incident light is at an angle, the z-component $\hat{\mathbf{e}}_z$ is non-zero, and some absorption is allowed. This is a major difference between intersubband and interband transitions.

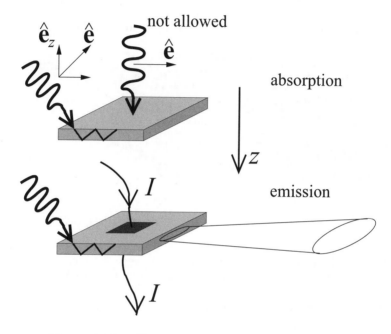

Figure 10.35 Natural intersubband device geometries

In addition, any laser made from intersubband transitions, whether optically pumped or electrically injected, as illustrated schematically in the bottom diagram of Fig. 10.35, will be an edge-emitter [243, 244].

These conclusions about natural device geometries are valid for the simple single parabolic band wave functions considered in the first instance. However, there is a wealth of literature available about possible band structure effects, including non-parabolicity, anisotropy, and degeneracy which could break these selection rules (for experimental work, see, for example [245], and for a detailed theoretical analysis, see Batty and Shore [246], and references therein).

The actual lifetime for intersubband spontaneous radiative emission is obtained by summing equation (10.249) over all photon modes; this requires some assumption to the cavity, i.e. the region of dielectric that the modes occupy. Conveniently, Smet *et al.* [222] quote the results as:

$$\frac{1}{\tau_i} = \frac{e^2 n \omega^2}{6\pi \epsilon m^* c^3} O_{\text{if}} \qquad (10.260)$$

where n is the refractive index at the emission wavelength, for a three-dimensional distribution of photon modes, and:

$$\frac{1}{\tau_i} = \frac{e^2\omega}{4\epsilon m^*c^2 W_z}O_{if} \tag{10.261}$$

for a two-dimensional photonic density of states. The latter is a common scenario in a quantum well system which is often surrounded by barriers and/or a substrate of different dielectric material, which can act as a microcavity producing confinement of the optical modes along the growth (z-) axis, as illustrated in Fig. 10.36.

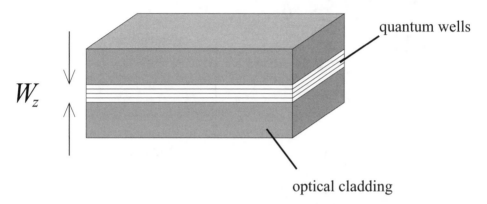

Figure 10.36   Schematic representation of a typical semiconductor quantum well system surrounded by dielectric material, thus producing confinement of optical modes

The oscillator strength O_{if} is dependent upon the dipole matrix element [243], i.e.

$$O_{if} = \frac{2m^*\omega}{\hbar}|\langle\psi_i|z|\psi_f\rangle|^2 \tag{10.262}$$

By using the approach of Burt [247] this can be shown to be equivalent to:

$$O_{if} = \frac{2\hbar}{m^*\omega}\left|\left\langle\psi_i\left|\frac{\partial\psi_f}{\partial z}\right\rangle\right|\right|^2 \tag{10.263}$$

Given that the matrix element is squared, then it is equivalent (see [12], p. 26), to that used earlier to derive the selection rules.

Fig. 10.37 compares the results of calculations of the three-dimensional and two-dimensional spontaneous radiative emission lifetimes, between the two lowest conduction band eigenstates of a GaAs infinitely deep quantum well; the emission wavelength has been varied by adjusting the quantum well width. This figure shows that for short-wavelength emission in the mid-infrared (<10 μm), the radiative lifetime calculated by using the 3D distribution of photon modes is shorter than that calculated with the 2D distribution, with W_z taken as 3 μm. However, as the wavelength increases, and the emission energy decreases, shorter lifetimes for radiative emission are obtained with the 2D equation. Fig. 10.37 basically confirms the work of Smet et al. [222]. The refractive index n was calculated by using the first-order Sellmeier equation [248] and the data of Seraphin and Bennett [249].

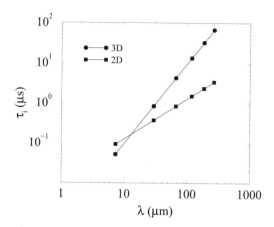

Figure 10.37   Comparison of the three-dimensional (3D) and two-dimensional (2D) spontaneous radiative emission lifetimes

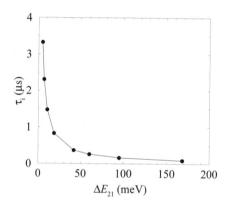

Figure 10.38   The 2D spontaneous lifetime as a function of subband separation

Fig. 10.38 plots the 2D spontaneous radiative lifetime as in the previous figure, but this time as a function of the subband separation. It can be seen that the lifetime increases as the energy separation decreases, with what appears to be a hyperbolic dependence. This is the reason why intersubband emitters and lasers are thought to be more difficult to fabricate at longer wavelengths (see, for example [250]). Fig. 10.39 confirms the hyperbolic dependence of the lifetime on the subband separation by plotting its inverse, i.e. the scattering rate against $\Delta E_{21} = E_2 - E_1$.
Notice the striking linearity implying that

$$\frac{1}{\tau_i} \propto \Delta E_{21} \tag{10.264}$$

This is not as obvious as it seems, for in the expression for this scattering rate, the ω in the numerator cancels with the ω in the denominator. Hence, the functional dependencies are controlled by the matrix element.

The independence of the spontaneous radiative lifetime on temperature *in this model* should also be noted. In fact, this is mainly because for simplicity the material depen-

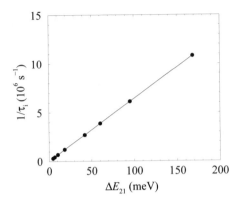

Figure 10.39   The 2D spontaneous emission rate as a function of subband separation

dencies on temperature have been ignored, in order to deduce as general results as are possible. In reality, the energy gaps do change slightly and hence the emission wavelength also changes (see Adachi [14] for further information). While in these models the emission frequency doesn't change, the intensity, i.e. the number of photons emitted, does change, in fact, the latter decreases. This is because of the increased competition with the non-radiative channels.

10.22   CARRIER SCATTERING IN QUANTUM WIRES AND DOTS

Carrier scattering in lower-dimensional systems, wires and dots, will be an important issue as they are introduced into future generations of opto-electronic devices. For the theoretical understanding of the carrier dynamics, scattering rate expressions will be needed.

Although the scattering rates derived here are not applicable to such systems, the techniques employed to obtain them do have some relevance. Quantum wires resemble quantum wells more closely than dots, as they still retain carrier dispersion. Therefore, developing a model of carrier scattering with *bulk* LO phonons will require expressing the carrier wave function as a product of a two-dimensional envelope function with a one-dimensional plane wave, as opposed to the other way around. The remainder of the derivation should follow in a similar fashion. Similar arguments for the route to carrier–carrier scattering in quantum wires may also be applicable.

However, for quantum dots the situation is quite different as the carriers never possess dispersion; hence, for scattering with phonons of fixed energies, it would appear *a priori* that this will only occur for resonance conditions, i.e. when the sublevel separation is equal to a phonon energy. Carrier–carrier scattering between the sublevels of quantum dots resembles the Coulomb interaction in multi-electron atoms, with the latter being an area where much work has been done.

Theoretical, and hence computational studies of carrier scattering in quantum wires and dots are still quite rare in the literature, although see, for example [251–256]. For an introduction to some aspects of the particulars of optical processes (carrier–photon scattering) in wires and dots, see Basu ([11], p. 343).

CHAPTER 11

ELECTRON TRANSPORT

11.1 INTRODUCTION

Since their invention [257], the development of quantum cascade lasers (QCLs) has been rapid, with improving temperature dependency [258] and increasing [227] or decreasing wavelength [259, 260]. Mid-infrared GaAs/AlGaAs QCLs devices have achieved pulsed room temperature operation, for example the triple quantum well QCL [261] emitting at 9.3 μm, bound to continuum QCLs at 11 μm [262] and 13.5 μm [263] and the superlattice QCL [264] at 12.6 μm. In addition, continuous wave operation up to 150 K has also been reported [265]. Since the first far-infrared (>30 μm) or terahertz (1–10×10^{12} Hz) QCL laser action [266], recent reports have given wavelengths as long as 161μm (1.9 THz) [267] in a one-well injector structure, based on electron-LO phonon depopulation and at 150 μm (2.0 THz) in a superlattice bound-to-continuum structure [268]. Furthermore, very long wavelength QCLs operating between 0.68 and 3.3 THz including a 1 THz laser at 215 K and 3 THz at 255 K in strong magnetic fields have been demonstrated [269]. The numerous quantum wells separated by often thin barriers and the single carrier type (these are unipolar devices) make QCLs an ideal testbed for studies of quantum mechanical transport and this is the focus of this chapter.

Quantum Wells, Wires and Dots, Third Edition. P. Harrison
©2009 John Wiley & Sons, Ltd.

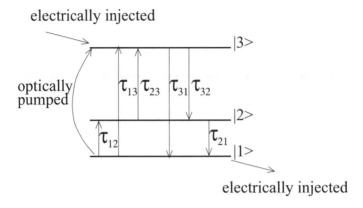

Figure 11.1   The scattering processes in a three-level laser

11.2   MID-INFRARED QUANTUM CASCADE LASERS

This mid-infrared (4–14 μm) device is rather esoteric and specialized, and inclusion of a discussion of its operation would seem out of place in this present work. However, this is included for one particular reason: the quantum cascade laser is a *textbook* example of electron scattering. This laser is a unipolar laser in that there is only one carrier type, as opposed to the vast majority of laser diodes and light-emitting diodes, which are bipolar and generate light through interband recombination of electrons with holes.

The simplest active region of any solid state laser would incorporate three energy levels, which in this case would be quantum well subbands, as illustrated schematically in Fig. 11.1. Each level contains a certain number of electrons, say n_1, n_2 and n_3, which scatter between the subbands (as indicated by the arrows). Energy can be input to the system, either by optically exciting electrons from the ground state $|1\rangle$ to the second excited state $|3\rangle$, or by injecting them directly into the uppermost state. The quantum cascade laser is an example of the latter, with electrons being removed from the lowest state $|1\rangle$ to be recycled, with these constituting the injected current in an additional stage. Thus, one electron can produce many photons.

Consider the rate equation for the number of electrons (population) of each level, i.e.

$$\frac{dn_3}{dt} = I_{in} + \frac{n_1}{\tau_{13}} + \frac{n_2}{\tau_{23}} - \frac{n_3}{\tau_{32}} - \frac{n_3}{\tau_{31}} \qquad (11.1)$$

$$\frac{dn_2}{dt} = \frac{n_1}{\tau_{12}} + \frac{n_3}{\tau_{32}} - \frac{n_2}{\tau_{21}} - \frac{n_2}{\tau_{23}} \qquad (11.2)$$

$$\frac{dn_1}{dt} = \frac{n_3}{\tau_{31}} + \frac{n_2}{\tau_{21}} - \frac{n_1}{\tau_{12}} - \frac{n_1}{\tau_{13}} - I_{out} \qquad (11.3)$$

where I_{in} represents the injection rate (the number of electrons per unit time), which in the steady-state is equal to I_{out}. This is a *semi-classical* description as, although the electron energies and wave functions have arisen from solutions of Schrödinger's quantum mechanical wave equation and the scattering rates have been derived from Fermi's Golden Rule, the interpretation now is one of a flow of particles from one state to another at a calculable rate.

Accepting that this is a model for describing the movement of electrons between quantized energy states, or in this case two-dimensional subbands, implies mathematically that the functions describing the distributions of electron energies within the subbands are all of the same form so that they can be integrated out of the Boltzmann transport equation. It is further assumed that the electron distributions in each subband can be described by a Fermi–Dirac distribution function as in Section 2.4, characterized by some, as yet, unknown temperature. There are good arguments for this assumption: firstly, calculations in Chapter 10 showed that intrasubband scattering is much, much faster than intersubband scattering and so any events that cause electrons to be scattered into a different subband will quickly cause intrasubband scattering, which will redistribute any energy changes across the subband—driving the subband towards an equilibrium Fermi–Dirac distribution, possibly with a different temperature. This is described as 'thermalization' and the electron distributions are said to be 'thermalized'. Furthermore, intrasubband scattering mechanisms occur, which couple subbands together: electron–electron scattering of the form 21-21* are as fast as any other intrasubband events and so any subband with electrons with a higher average kinetic energy, i.e. a higher temperature, will be driven through these 'bi-intrasubband scattering' events towards thermal equilibrium with the other subbands. So, it can also be expected that the subbands will have the same characteristic (electron) temperature [270]; further justification for this is provided at the end of the chapter.

Returning to the rate equations: consider the population of the second level, where in the steady-state, the net rate of change is zero. Also assume that the temperature is relatively low, and hence the absorption rates can be ignored, then:

$$\frac{n_3}{\tau_{32}} = \frac{n_2}{\tau_{21}} \tag{11.4}$$

Furthermore, if:

$$\frac{1}{\tau_{21}} > \frac{1}{\tau_{32}} \tag{11.5}$$

then $n_3 > n_2$, i.e. a population inversion will exist between levels $|3\rangle$ and $|2\rangle$, thus fulfilling a necessary condition for stimulated emission. The ratio n_3/n_2 is known as the *population ratio*, which in this analysis would be given by:

$$\frac{n_3}{n_2} = \frac{\tau_{32}}{\tau_{21}} \tag{11.6}$$

Perhaps the simplest way to realize such a three-level system is within a triple quantum well structure, with an energy level in each well—the subband minima can then be altered (almost) independently merely by adjusting the quantum well widths.

Equation (11.5) suggests that considering ways of enhancing the scattering rate from the second level to the first may be a productive way of engineering a population inversion. With this in mind, consider a GaAs triple quantum well surrounded by $Ga_{0.8}Al_{0.2}As$ barriers, with well widths of 56.5, 96.1 and 84.8 Å, respectively (integral numbers of monolayers), separated by barriers of width 56.5 and 28.25 Å. The central well has been chosen to be the widest, such that at zero applied electric field it contains the ground state. As the field is increased, an anti-crossing with the state in the

---

*This is the shorthand notation of Chapter 10 and means that electrons in initial states '2' and '1' scatter and end up in final states '2' and '1', respectively.

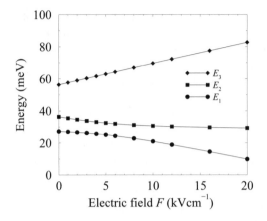

Figure 11.2   The electric field dependence of the lowest three subband minima

right-hand well (84.8 Å) will be inevitable, thus hopefully leading to an increase in the scattering rate, which depopulates level $|2\rangle$.

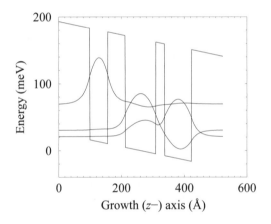

Figure 11.3   A quantum cascade laser active region at $F=10$ kVcm^{-1}

This behaviour can be clearly seen in Fig. 11.2, where the latter plots the lowest three subband minima as a function of applied electric field. The wave functions themselves are illustrated in Fig. 11.3, just beyond the anti-crossing, at $F=10$ kVcm^{-1}. The large overlap between $|2\rangle$ and $|1\rangle$ is apparent, which will hopefully lead to a strong depopulation of the lower laser ground state, i.e. a high $1/\tau_{21}$. In comparison, the overlap between $|3\rangle$ and $|2\rangle$ is smaller, thus implying a longer carrier lifetime in the upper laser level, i.e. a small $1/\tau_{32}$.

To confirm whether there is indeed a population inversion requires a calculation of the scattering rates themselves. Using the methods outlined in the previous chapter, Fig. 11.4 displays the electron–LO phonon and Fig. 11.5 the electron–electron scattering rates, as a function of the applied electric field at 77 K and with a carrier density of 10×10^{10} cm^{-2} in each level.

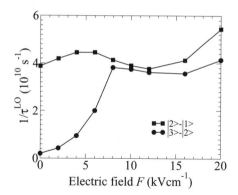

Figure 11.4   The electron–LO phonon scattering rates from level $|3\rangle$ to $|2\rangle$ and from level $|2\rangle$ to $|1\rangle$

The nature of the anti-crossing is evident from the data in Fig. 11.4, as the $|3\rangle$ to $|2\rangle$ scattering rate changes rapidly at the anti-crossing, as level $|2\rangle$ moves from being confined in the right-hand well to the central well. This increases the overlap with state $|3\rangle$ confined in the left-hand well and hence the scattering rate also increases. The depopulation of the lower lasing level, i.e. the rate $|2\rangle$ to $|1\rangle$, mirrors the subband separation between these levels, and indicates that there is mixing between these states for a considerable range of electric fields.

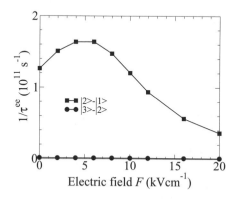

Figure 11.5   The electron–electron scattering rates from level $|3\rangle$ to $|2\rangle$ and from levels $|2\rangle$ to $|1\rangle$

Fig. 11.5 shows the corresponding electron–electron scattering rates; note these are total rates and include *all* contributions, i.e. the $|3\rangle$ to $|2\rangle$ rate includes 33–32, 33–22, and 32–22, while the $|2\rangle$ to $|1\rangle$ rate includes 22–21, 22–11, and 21–11. It is immediately apparent that this electron–electron depopulation rate of level $|2\rangle$, i.e. the rate at which carriers are removed from $|2\rangle$ to $|1\rangle$, is very much faster than the repopulation rate, i.e. the rate at which carriers scatter into $|2\rangle$. This is due to the close proximity in both energy and real space of levels $|1\rangle$ and $|2\rangle$. A more general observation, from comparing Figs 11.4 with 11.5, shows that the electron–electron scattering rate out of level $|2\rangle$ is very much faster than that due to scattering with LO phonons. In addition, it is also apparent that, in both cases, the depopulation rate is larger then the repopulation rate, and hence a population inversion is likely.

This is confirmed by Fig. 11.6, which plots the ratio of the lifetimes—a simple quantity which has been shown to approximate the population ratio between levels $|3\rangle$ and $|2\rangle$. Thus, in conclusion, using the methods outlined, this active-layer design has been shown to exhibit a population inversion between levels $|3\rangle$ and $|2\rangle$ at 77 K. The corresponding emission energy, given by the subband separation $E_3 - E_2$, varies between 33 and 53 meV, which, in turn, corresponds to 37 and 23 μm, respectively.

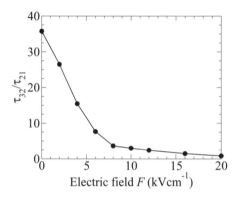

Figure 11.6   The ratio of the lifetimes τ_{32}/τ_{21} of electrons in the triple quantum well active region of the quantum cascade laser

11.3   REALISTIC QUANTUM CASCADE LASER

The above is a simplified model of a quantum cascade laser to give some insight into the internal workings of those devices. A more realistic model would treat the injection and extraction currents specifically and actually try to calculate them. In order to do this, it would be necessary to model more than one period so that the scattering rates from one period to the next can be calculated. The quantum cascade laser of Sirtori *et al.* [271] has become a textbook example because it was the first GaAs-based quantum cascade laser and the material parameters are more familiar to everybody. It has the classic design incorporating a simple active region with three quantum wells providing the ground state, the lower and upper laser levels and this device has a well-defined injector/extractor region consisting of five quantum wells. This is illustrated in Fig. 11.7.

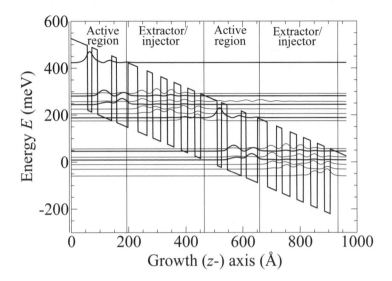

Figure 11.7 A band diagram showing the squares of the wave functions of the electron conduction band states at a field of 52 kVcm^{-1} for the GaAs/Ga$_{0.67}$Al$_{0.33}$As quantum cascade laser of Sirtori *et al.* [271].

Starting at the left-hand edge of the structure then, the device has three quantum wells in the active region. The highest energy level is mainly localized in the first quantum well and is the upper laser level, labelled as '16' in Fig. 11.8. It is quite separated in both energy and space from the next highest energy level and therefore the lifetime of carriers in the upper level is long (a few picoseconds!) and hence a high density of carriers can build up there. The carriers are eventually scattered out of this level, mainly into the lower laser level, labelled as '14' in Fig. 11.8 through both non-radiative (phonon) scattering and stimulated emission within the laser. The third and lowest energy level within the active region, labelled as '12' in Fig. 11.8, is usually called the ground state and this is generally designed to have a large spatial overlap with the lower laser level and to have an energy separation of the order of the LO phonon energy (36 meV in GaAs) to encourage electrons to scatter out of the lower laser level and into it. As described above in Section 11.2, it is this emptying of the lower laser level that produces the population inversion, i.e. more carriers in the higher energy level (16) than the lower level (14).

In contrast to a traditional diode laser, the electron is still in the conduction band after traversing the active region and generating a photon (or a phonon!), and so the principle of the *cascade* laser is to push this electron into another active region and hence try to encourage it to create another photon, and then another, and so on. This is done from a device design point of view by putting several more states close in energy with reasonable spatial overlap to increase the scattering rate and hence reduce the lifetime of the electrons in the active region ground state. In the example device here, a five quantum well (non-uniform) superlattice is used as illustrated in Fig. 11.7. These 'extractor' states are labelled as 9, 10, 11, 13 and 15 in Fig. 11.8. As discussed in Chapter 10, providing multiple states for the electrons to scatter into increases the probability that they will scatter and that is one of the roles of the extractor. Another role is to provide multiple energy levels close together, certainly separated by less than an LO phonon energy so that the electrons must make diagonally inward transitions, as in Fig. 10.10, which forces them to lose in-plane kinetic energy ($\hbar^2 k_{xy}^2/2m^*$) and settle towards the bottom of the subband: this idea of cooling the carriers is thought to be beneficial for the operation of quantum cascade lasers, the performance of which does decrease as the temperature increases.

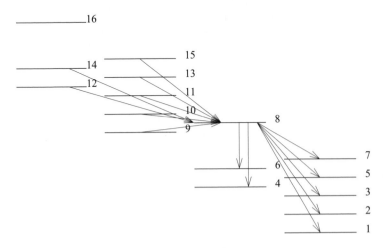

Figure 11.8   A schematic diagram of the energy levels within the two adjacent periods of the quantum cascade laser of the previous figure. Note the energy levels are numbered over increasing energy within a period. The arrows show all the possible scattering rates that are included in the computation for the example of the upper laser level of the lower energy (right-hand) period.

11.4   RATE EQUATIONS

The electron densities in each subband of the quantum cascade laser can again be represented by rate equations as in Section 11.2 however, with many more levels the rate equations involve more terms. For example, choosing a level near the middle of the structure, i.e. the upper laser level '8' as in Fig. 11.8 and assuming that electrons scatter no further than a complete period, then the rate equation describing the population density (neglecting stimulated emission) is:

$$\frac{dn_8}{dt} = \frac{n_{15}}{\tau_{15,8}} + \frac{n_{14}}{\tau_{14,8}} + \frac{n_{13}}{\tau_{13,8}} + \frac{n_{12}}{\tau_{12,8}} + \frac{n_{11}}{\tau_{11,8}} + \frac{n_{10}}{\tau_{10,8}} + \frac{n_9}{\tau_{9,8}} + \frac{n_7}{\tau_{7,8}} + \frac{n_6}{\tau_{6,8}} + \frac{n_5}{\tau_{5,8}} +$$

$$\frac{n_4}{\tau_{4,8}} + \frac{n_3}{\tau_{3,8}} + \frac{n_2}{\tau_{2,8}} + \frac{n_1}{\tau_{1,8}} - \frac{n_8}{\tau_{8,15}} - \frac{n_8}{\tau_{8,14}} - \frac{n_8}{\tau_{8,13}} - \frac{n_8}{\tau_{8,12}} - \frac{n_8}{\tau_{8,11}} - \frac{n_8}{\tau_{8,10}} - \frac{n_8}{\tau_{8,9}} - \frac{n_8}{\tau_{8,7}}$$

$$-\frac{n_8}{\tau_{8,6}} - \frac{n_8}{\tau_{8,5}} - \frac{n_8}{\tau_{8,4}} + \frac{n_8}{\tau_{8,3}} + \frac{n_8}{\tau_{8,2}} + \frac{n_8}{\tau_{8,1}} = 0 \qquad (11.7)$$

which is equated to zero as it is the *steady state* that is of interest. This can be put more succinctly as:

$$\frac{dn_8}{dt} = \sum_{j=1}^{15} \left(\frac{n_j}{\tau_{j,8}} - \frac{n_8}{\tau_{8,j}} \right) = 0 \tag{11.8}$$

where in this case there is no need to manually exclude the $j = 8$ term as this will cancel. This is a specific example for state '8' of Fig. 11.8, which now needs generalizing to any state in the double period that must be solved. If there are N_{sp} electron energy 'states per period', then equation (11.8) can be written for any state 'i' as:

$$\frac{dn_i}{dt} = \sum_{j=i-(N_{sp}-1)}^{i+(N_{sp}-1)} \left(\frac{n_j}{\tau_{j,i}} - \frac{n_i}{\tau_{i,j}} \right) = 0 \tag{11.9}$$

where in this example there are eight states per period (three in the active region and five in the extractor/injector regions) and so, with $N_{sp} = 8$, then the summation over 'j' states goes over seven states of *lower* energy and seven states of *higher* energy than the state in question. In this example the index 'i' varies over the 16 states as indicated in Figs 11.7 and 11.8. Equation 11.9 thus gives one rate equation for each level in the system, but they are not linearly independent and so it is also necessary to have an additional piece of information in order to derive a unique solution and so the total sheet carrier density per period is introduced and taken as equal to the total sheet doping density:

$$\sum_{i=1}^{N_{sp}} n_i = \int_{period} N_d(z) \ dz \tag{11.10}$$

where $N_d(z)$ is the density of dopant atoms per unit volume, which is usually modulated, hence it is written as a function of position z. Thus the integral of $N_d(z)$ gives the sheet doping density per period. It is usually assumed in quantum cascade laser modelling that all the donor atoms are ionized.

If the index of the state of interest 'i' is less than N_{sp}, then the summation in equation (11.9) will contain carrier lifetimes τ_{ij} and τ_{ji}, which refer to states that are outside of the 2-period domain in Figs 11.7 and 11.8. In fact, with this approach this happens for all the states in the lower energy or right-hand period and the converse happens in the higher energy or left-hand period, as often τ_{ij} and τ_{ji} refer to lifetimes for which i or j are greater than the number of states in the 2-period unit cell. To overcome this, periodicity is invoked since:

$$\frac{1}{\tau_{i,j}} = \frac{1}{\tau_{i+N_{sp},j+N_{sp}}} = \frac{1}{\tau_{i-N_{sp},j-N_{sp}}} \tag{11.11}$$

and the indices are shifted by a period to the left or the right until the scattering channel lies completely within the 2-period domain. Thus the reason for choosing *two* quantum cascade laser periods in the model is now clear: it is the *minimum* number that can be used that includes all intra- (within the same) and inter- (between two neighbouring) period scattering channels within the device.

11.5   SELF-CONSISTENT SOLUTION OF THE RATE EQUATIONS

As discussed in Chapter 10, the lifetimes $\tau_{i,j}$ and $\tau_{j,i}$ in equation (11.9) depend on the number of electrons in the initial and final states, i.e. in this case n_i and n_j. So equation (11.9) should be written more correctly as:

$$\frac{\mathrm{d}n_i}{\mathrm{d}t} = \sum_{j=i-(N_{\mathrm{sp}}-1)}^{i+(N_{\mathrm{sp}}-1)} \left(\frac{n_j}{\tau_{j,i}(n_j, n_i)} - \frac{n_i}{\tau_{i,j}(n_i, n_j)} \right) = 0 \qquad (11.12)$$

This can be rearranged to give:

$$\sum_{j=i-(N_{\mathrm{sp}}-1)}^{i+(N_{\mathrm{sp}}-1)} \frac{n_j}{\tau_{j,i}(n_j, n_i)} = \sum_{j=i-(N_{\mathrm{sp}}-1)}^{i+(N_{\mathrm{sp}}-1)} \frac{n_i}{\tau_{i,j}(n_i, n_j)} \qquad (11.13)$$

As n_i is independent of the summation on the right-hand side of the above equation then:

$$\sum_{j=i-(N_{\mathrm{sp}}-1)}^{i+(N_{\mathrm{sp}}-1)} \frac{n_j}{\tau_{j,i}(n_j, n_i)} = n_i \sum_{j=i-(N_{\mathrm{sp}}-1)}^{i+(N_{\mathrm{sp}}-1)} \frac{1}{\tau_{i,j}(n_i, n_j)} \qquad (11.14)$$

hence:

$$n_i = \sum_{j=i-(N_{\mathrm{sp}}-1)}^{i+(N_{\mathrm{sp}}-1)} \frac{n_j}{\tau_{j,i}(n_j, n_i)} \bigg/ \sum_{j=i-(N_{\mathrm{sp}}-1)}^{i+(N_{\mathrm{sp}}-1)} \frac{1}{\tau_{i,j}(n_i, n_j)} \qquad (11.15)$$

Of course, as the lifetimes $\tau_{j,i}$ and $\tau_{i,j}$ are complicated functions of the carrier densities n_i and n_j, then it is not possible to completely separate n_i in equation (11.15) and therefore it must be solved self-consistently. One way to do this is to start with initial approximations for the carrier densities n_i and n_j (perhaps take the carrier densities to be equal across all the subbands) and then use the right-hand side of equation (11.15) to generate a new value for n_i, i.e. interpret equation (11.15) as one of a set of $2N_{\mathrm{sp}}$ interative equations:

$$n_i^{(n+1)} = \sum_{j=i-(N_{\mathrm{sp}}-1)}^{i+(N_{\mathrm{sp}}-1)} \frac{n_j^{(n)}}{\tau_{j,i}(n_j^{(n)}, n_i^{(n)})} \bigg/ \sum_{j=i-(N_{\mathrm{sp}}-1)}^{i+(N_{\mathrm{sp}}-1)} \frac{1}{\tau_{i,j}(n_i^{(n)}, n_j^{(n)})} \qquad (11.16)$$

where $n_i^{(n)}$ is the nth approximation to n_i. These are solved in sequence for increasing values over the set of subbands i, which then give improved approximations for each of the carrier densities n_i. After each new set of n_i are generated, the necessary additional information of the total carrier density in each period, from equation (11.10), is input by renormalizing according to this, i.e.

$$n_i^{(n+1)} \longrightarrow \frac{n_i^{(n+1)}}{\sum_{j=1}^{N_{\mathrm{sp}}} n_j^{(n+1)}} \qquad (11.17)$$

This new set of carrier densities are then used to calculate a new set of lifetimes $\tau_{i,j}(n_i, n_j)$, which then allows for the calculation of a further improved set of n_i using equation (11.16) and the process is repeated until convergence is acheived, i.e. until

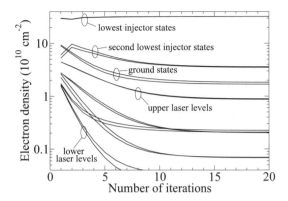

Figure 11.9   The electron densities in each subband through the self-consistent iteration for the mid-infrared quantum cascade laser of interest at a temperature of 77 K and a bias field equivalent to Fig. 11.7 of 52 kVcm^{-1}

the new set of carrier densities $n_i^{(n+1)}$ is equal (within some predefined accuracy) to the previous set $n_i^{(n)}$.

Fig. 11.9 shows the results of such a self-consistent iteration for the solution of the electron densities in each subband for the quantum cascade laser in Fig. 11.7. It can be seen from the figure that the electron densities in each level do converge as the number of iterations increases above about 10. Also, the method naturally reproduces the periodicity of the electron populations in equivalent subbands a period apart, which is reassuring and lends weight to the validity of the technique.

The calculations have shown, for this particular device at this temperature and bias, that the subband with the highest population is the lowest state in the extraction/injector region. The second highest populations are to be found in the second lowest injector state, with the third highest populations being in the lowest energy state (the so-called ground state) found in the active region (indicated by one of the thicker lines in Fig. 11.7. Note that the total carrier density of 39×10^{10} cm^{-2} in each period of the quantum cascade laser is calculated from the doping density and assumes all the electrons are ionized from the donors. The position of the donors is not included in this calculation as the self-consistent solution of Poisson's equation, as described in Section 3.17, is not included in these calculations. The effect of this on the band edge potential at this medium carrier density is still quite low, see, for example [272].

Fig. 11.9 shows quite clearly when self-consistency is reached that there are approximately two orders of magnitude more carriers in the upper laser level than the lower level laser, which is good because it indicates that there is a strong population inversion as observed in experiment (the device does lase!).

11.6   CALCULATION OF THE CURRENT DENSITY

Once the set of equations (11.16) has been iterated to consistency, then the carrier densities n_i, together with the lifetimes of the carriers in each level in relation to scattering into every other level $\tau_{i,j}$, are also known: this is a complete description of the electronic structure of the laser or whatever other device has been studied and, from this, physical observables such as the current density and gain can be calculated.

The terms like $n_i/\tau_{i,j}$ represent the number of carriers (in this case electrons) per unit area that make a transition from the state i to the state j every $\tau_{i,j}$ seconds. So, it really just represents a current density, i.e. a charge per unit time. A simple model for the current density in the device as a whole is therefore just to take the sum of all such terms which move electrons across some reference plane in the device, as illustrated in Fig. 11.10.

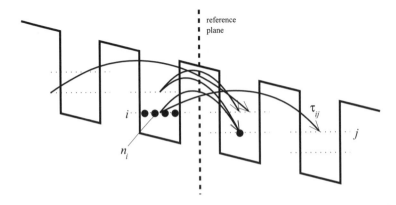

Figure 11.10   Calculating the current density by summing the contributions over all the transitions across a reference plane.

For convenience in this case, the reference plane for the current density calculation can be taken as the boundary between the two distinct quantum cascade laser periods, as illustrated in Figs 11.7 and 11.8. Hence, in that case the current density would follow as:

$$J = e \left(\sum_{i=N_{\text{sp}}+1}^{2N_{\text{sp}}} \sum_{j=1}^{N_{\text{sp}}} \frac{n_i}{\tau_{i,j}} - \frac{n_j}{\tau_{j,i}} \right) \tag{11.18}$$

where the second term accounts for back-scattering.

A more detailed model [273] might not assume that the carriers jump the same distance, i.e. a period, in each scattering event and could include the relative separations between the expectation values of the position of the wave functions, but this simple model is fine here.

11.7   PHONON AND CARRIER–CARRIER SCATTERING TRANSPORT

Fig. 11.11 shows the results of solving for the current density for different bias fields at 77 K for the quantum cascade laser of interest here, when including LO phonons only, and then adding in acoustic phonons and finally electron–electron scattering. The calculations were performed in this sequence in order to gain an appreciation of their relative contributions to the physical observable of the current density.

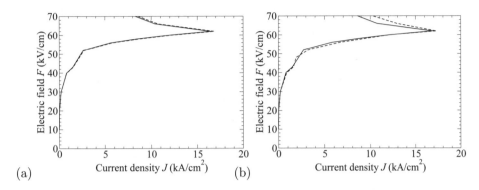

(a)   (b)

Figure 11.11   The electric field F versus current density J for the quantum cascade laser as above. In (a) the solid line represents the results of calculations including just LO phonon scattering whereas the dashed line also includes acoustic phonon scattering, in (b) the solid line represents the results of calculations including LO and acoustic phonon scattering, whereas the dashed line also includes electron–electron scattering.

Fig. 11.11(a) shows the effect of including just phonon scattering mechanisms. For interest, following the procedures discussed above gave 200 different phonon scattering channels, which had to be evaluated for both LO and acoustic phonon scattering. It can be seen from (a) that LO phonon scattering dominates the transport in this mid-infrared quantum cascade laser, with acoustic phonon scattering making an almost negligible contribution. The same can also be said for the 552 electron–electron scattering rates that were evaluated: (b) shows that these also have a small contribution to the current density. These are fairly general conclusions that have been observed in calculations of other devices [274–276] and the reason is because the intersubband energy separations are typically of the order of a few tens of meV and the temperature was mid-range. At very low temperatures, LO phonon scattering at energy spacings less than 10–20 meV becomes slow and low energy scattering such as acoustic and electron–electron will become more important. This is also the case in the longer wavelength terahertz or far-infrared quantum cascade lasers where the energy gaps and operating temperatures are both usually lower [277–279].

11.8   ELECTRON TEMPERATURE

As stated at the beginning of the chapter, this rate equation method assumes that the electron distribution in each subband can be described by a Fermi–Dirac distribution function and, the temperature characterizing each distribution is the same. In the calculations that followed it was assumed that this temperature was equal to the lattice temperature. In fact, it is now possible to discard that last assumption and calculate an *average* electron temperature based on an energy balance method. To be more specific, the argument is that, given that any intersubband device will always reach a steady state under continuous operating conditions, then the rate at which the electron distributions gain *kinetic* energy (relative to the particular subband minimum) through scattering will balance with the rate at which they lose kinetic energy to the lattice[†].

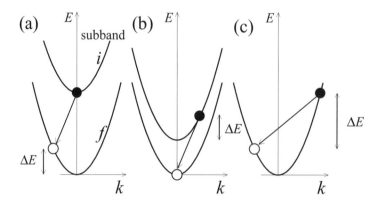

Figure 11.12   Schematic diagram illustrating inelastic (longitudinal optic phonon) intersubband scattering processes for subband separations (a) greater than and (b) less than the LO phonon energy; (c) illustrates an inelastic intrasubband scattering process.

For example, consider the two instances of intersubband scattering as in Fig. 11.12(a) and (b). In Fig. 11.12(a) the subband separation $E_i - E_f$ between the initial electron state $|i\rangle$ and the final state $|f\rangle$ is greater than the dominant phonon energy (in III-Vs the polar optic (LO) phonon), hence any phonon emission event produces a carrier in the lower subband *with more kinetic energy in relation to the subband minimum*. This energy adds to the *total* kinetic energy of the electron distribution but, because of the presence of very rapid sub-picosecond intrasubband carrier–carrier scattering events [270], this additional energy is quickly shared out amongst the electrons in that subband: the consequence is that the electron distribution within that subband can still be described by a Fermi–Dirac distribution function, albeit with a different temperature. This is a process known as 'rethermalization'.

Now the electrons in this distribution have more kinetic energy in total than previously and hence the distribution as a whole has a higher temperature than the other subband populations: thermodynamics then drives the system to redistribute this energy. This occurs through 'bi-intrasubband carrier–carrier scattering' (of the form $|j\rangle |g\rangle \rightarrow |j\rangle |g\rangle$) which redistribute this energy throughout the set of subbands within the quantum device to equalize the temperatures of the distributions [270].

In Fig. 11.12(b) the subband separation $E_i - E_f$ is less than the phonon energy, hence a scattering event from the upper to the lower subband *reduces* the total kinetic energy of the electrons.

In both cases (a) and (b), the change in the total kinetic energy of the electron distributions can be written as: $\Delta E = E_i - E_f - E_{\mathrm{LO}}$, where $\Delta E > 0$ should be interpreted as an increase and $\Delta E < 0$ a decrease in this total energy. If there are n_i

---

[†]The lattice temperature itself will also reach a steady-state value, and this is calculable too using electro-thermal modelling [280, 281], but this lies beyond the scope of this work and so it will continue to be assumed that the lattice temperature is constant across the physical dimensions of the device and is equal to the known temperature of some heatsink.

carriers in the initial state and the LO phonon transitions have associated scattering rates of $1/\tau_{if}^{\text{em.}}$ and $1/\tau_{if}^{\text{abs.}}$ for emission and absorption processes, respectively, then the net kinetic energy generation rate from *intersubband* scattering is:

$$\sum_{f\neq i}\sum_{i}\left[\frac{n_i}{\tau_{if}^{\text{em.}}}(E_i-E_f-E_{\text{LO}})+\frac{n_i}{\tau_{if}^{\text{abs.}}}(E_i-E_f+E_{\text{LO}})\right]$$

where the indices on the summations imply over all initial and final states in the quantum system. The case when $f=i$ will be discussed below.

One of the main features of quantum cascade lasers is the bridging (injector) regions between the active layers. These regions serve to remove the electrons from one active region and supply them, at the correct energy, to the next active region. There is, however, another important role that they provide—the opportunity for hot (high kinetic energy) carriers to cool. The injector–collector regions can be optimized to do this by designing the separation of the subbands to be less than the LO phonon energy. Hence, they encourage the carriers to scatter as in Fig. 11.12(b), thus losing kinetic energy and cooling the electron distribution (of course, phonons are generated in this process, which heat the lattice, but as stated above it is assumed here that the device is thermally well connected to a large heat sink, which keeps the lattice temperature constant). Another mechanism that contributes to this cooling is illustrated in Fig. 11.12(c)—intrasubband phonon emission. Such transitions lead to a *decrease* in the energy by an amount E_{LO}; again if this has a scattering rate of $1/\tau_{ii}^{\text{em.}}$, then the corresponding kinetic energy loss rate from *intrasubband* scattering is:

$$\sum_{i}\left[\frac{n_i}{\tau_{ii}^{\text{em.}}}E_{\text{LO}}+\frac{n_i}{\tau_{ii}^{\text{abs.}}}(-E_{\text{LO}})\right]$$

where $1/\tau_{ii}^{\text{abs.}}$ accounts for intrasubband phonon re-absorption, which reduces the above energy loss rate by $-E_{\text{LO}}$. This is just the case when $f=i$ as in the previous expression; thus the two will be able to be combined succinctly (see below).

Carrier–carrier (electron–electron or hole–hole) scattering events are described as *elastic*, which means that the total energy of the particles before the event is the same as that after. However, intersubband electron–electron transitions do convert potential energy into kinetic energy (or vice versa), which from the viewpoint of this work would lead to an increase (decrease) in the total kinetic energy of a subband population. Note, the potential energy as defined here includes the quantized component of the kinetic energy, i.e. the energy of the subband minimum.

The expressions for the intersubband and intrasubband cases above for both phonon and carrier–carrier scattering can thus be combined into one expression and, as it has been argued that in the steady-state the net kinetic energy generation rate is zero, then:

$$\Delta=\sum_{\text{em.,abs.,c–c}}\sum_{f}\sum_{i}\frac{n_i}{\tau_{if}}(E_i-E_f+\delta E)=0 \tag{11.19}$$

where the change in energy δE is equal to $-E_{\text{LO}}$ for phonon emission (em.), $+E_{\text{LO}}$ for absorption (abs.) and zero for carrier–carrier (c–c) scattering.

Equation (11.19) is a kinetic energy balance equation: the scattering rates in equation (11.19) are functions of the subband populations n_i and the electron temperatures:

if the electron temperature is too low, the number of intrasubband scattering events, as in Fig. 11.12(c), is too small and the kinetic energy equation cannot be balanced. The computational procedure therefore is to vary the electron temperature (which, as argued above, is assumed to be the same for all subbands) until the kinetic energy balance equation is satisfied self-consistently.

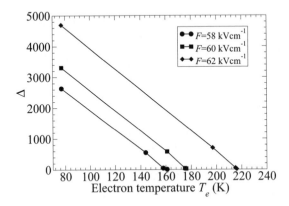

Figure 11.13   The calculated value of the kinetic energy generation rate Δ as a function of the electron temperature and for several different electric fields.

Fig. 11.13 illustrates this procedure for deducing the electron temperature. Following the standard method for calculating the carrier densities and lifetimes across all levels of the device the self-consistent iteration is extended to include the effect of the electron temperature. The data in Fig. 11.13 shows the variation of the net kinetic energy generation rate Δ with the electron temperature T_e for the cascade laser of interest here. The solutions for the electron temperatures occur when $\Delta = 0$. Detailed study of the contributions to the electron temperature highlight two interesting points: firstly that phonon absorption mechanisms are important even at low (77 K) lattice temperatures, and secondly that, for lattice temperatures above 50 K, electron–electron scattering is not an important electron heating mechanism (accounting for less than 1% of the total heat generation rate) in this mid-infrared laser.

Fig. 11.14 shows the results of including the electron temperature in the self-consistent iteration. It can be seen that although the current density doesn't change very much from the case when the electron temperature is fixed at the lattice temperature; when allowed to vary, the electron temperature is considerably higher. As might be expected the electron temperature increases as the current density through the device increases; which is merely a reflection on the increased energy being forced into the device. Fig. 11.15 summarizes this and shows that indeed there is a linear dependence of the electron temperature on the power density in the device: an effect that has been observed in experiment [282, 283] thus helping to substantiate the assumptions and approach of the model.

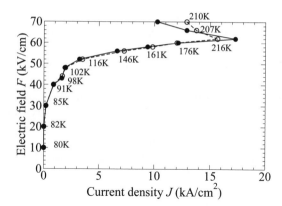

Figure 11.14 The effect of elevated electron temperatures on the electric field F versus current density J for the quantum cascade laser as above. The closed symbols represent the data for the electron temperature fixed at the lattice temperature of 77 K and the open symbols show the equivalent points where the electron temperature is allowed to vary and is calculated self-consistently. The actual calculated values of T_e are shown for each electric field point.

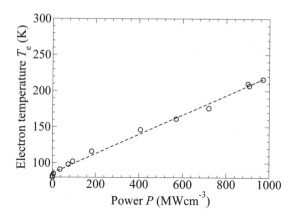

Figure 11.15 The electron temperature T_e plotted as a function of the input power density $F \times J$.

11.9 CALCULATION OF THE GAIN

As mentioned in Section 11.6, the self-consistent iteration yields the density of electrons in each quantum well subband and the lifetimes for transitions to every other subband: it is a fundamental description, which allows the physical observable of the current density to be calculated as well as the gain of the laser. Convergence of the iteration gives the steady-state behaviour of the laser and the modal gain for an optical transition between levels i and j is given by [274]:

$$G_M = \frac{4\pi e^2}{\epsilon_0 n} \frac{|\langle \psi_i | z | \psi_j \rangle|^2}{2\gamma_{ij} L_p \lambda} \Gamma(n_i - n_j) \equiv g\Gamma J \tag{11.20}$$

where λ is the free-space wavelength of the emission, n is the mode refractive index, $2\gamma_{ij}$ is the full-width at half-maxima of the electroluminesence spectrum of the device below lasing threshold (usually taken from experimental measurements but it can be calculated, see the discussion and references in [273]), L_p is the length of one quantum cascade laser period and Γ is the modal overlap factor, see Chapter 13, particularly equation (13.70). A positive modal gain therefore requires $n_i > n_j$, i.e. a population inversion between the two levels generating the photon emission.

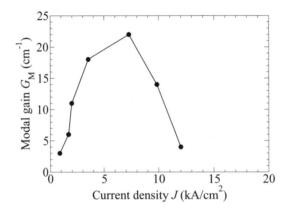

Figure 11.16   The modal gain G_M as a function of the current density J.

Fig. 11.16 illustrates results of the calculations of the modal gain as a function of the current density for the quantum cascade laser of interest in this chapter. It can be seen from the figure that the modal gain increases as the amount of current through the device increases, reaches a peak and then decreases. The latter happens even though the current density is still increasing and is due to a decrease in the overlap (the matrix element $\langle \psi_i | z | \psi_j \rangle$) between the upper and lower laser levels as the bias across the device increases. When the current is increased to such an extent that the modal gain is larger than the optical losses of the cavity or waveguide, lasing can occur. This current is known as the 'threshold current', see Chapter 13 and Section 13.4.2, in particular.

11.10   QCLS, QWIPS, QDIPS AND OTHER METHODS

This rate equation method has been developed further to include self-consistent itera-tion with Poisson's equation (see Section 3.17) to form a 'self-self-consistent' solution to transport in semiconductor heterostructures. This has been applied successfully to the problem of the effect of doping density on the transport characteristics, namely the current–voltage curves, to explain the experimentally observed saturation of the maxi-mum current density which occurs for increased doping density [284–287]. In addition, meaningful predictions have been made relating to the amount of current that escapes from the quantum well system to form a leakage current. This phenomenon known as 'thermionic emission' has been reported on [288].

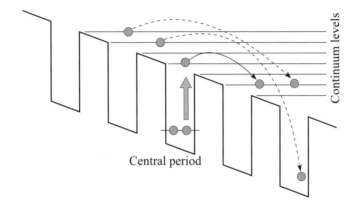

Figure 11.17   Schematic diagram of the electron energy states in a quantum well infrared photodetector or QWIP. The photo-ionization process, electron recapture and continuum state currents are illustrated.

The versatility of the method is also demonstrated by the application to quantum well infrared photodetectors (QWIPs). In the first instance QWIPs seem to have a simpler band structure, see Fig. 11.17; however, the calculations are more complicated as much of the current propagates through the continuum states above quantum well potentials. To calculate the current accurately, the scattering rates must be summed over the continuum states until convergence is achieved. However, with this method, quantitative predictions of the internal (unobservable) physics such as capture probability and photoconductive gain can be made, as well as physical observables such as the current–voltage curve and the responsivity [289].

Although it needed quite some modification, this general method of rate equations with calculated scattering rates to represent electron densities and the time to scatter between levels has also been applied to quantum dot infrared photodetectors (QDIPs) [290, 291] and its applicability has been confirmed through successful explanation of experimental data [292–294].

So summarizing this semi-classical method for calculating electron transport properties in semiconductor heterostructures has been applied and validated through comparison with experiment to mid-infrared and far-infrared quantum cascade lasers, quantum well infrared photodetectors and quantum dot infrared photodetectors. There is another way of effectively solving for the subband population densities *without* making the assumption that the electron distributions in each subband are thermalized and this is achieved by tracking the momenta of an ensemble of electrons in a stochastic or Monte Carlo simulation. For a simple introduction to stochastic simulations of carrier scattering see Ref. [171]. Recent simulations of heterostructured devices in similar conditions to those here showed that the carrier distributions are indeed very close to Fermi–Dirac distributions [295, 296] and that the electron temperatures of the subbands are similar [297], both conclusions lending weight to the approach introduced in this chapter. Other deeper quantum approaches to similar devices do exist, for example, for a density matrix approach to quantum cascade lasers, see Ref. [298] and references therein, but these methods lie beyond the scope of this introduction.

CHAPTER 12

OPTICAL PROPERTIES OF QUANTUM WELLS

contributed by D. Indjin

12.1 INTERSUBBAND ABSORPTION IN QUANTUM WELLS

It is well established that semiconductor quantum wells (QWs) allow for intersubband, along with intraband, optical transitions, which is due to the non-bulk-like energy spectrum [299]. Depending on the quantization properties of the states involved, intersubband transitions may be bound–bound, bound–free or free–free. The latter can be neglected in this analysis because it makes a very small contribution to the total intersubband absorption in realistic QW structures. Intersubband transitions are quite interesting and have been studied extensively [300] as they have actual or potential applications in quantum well infrared photodetectors and unipolar (quantum cascade) lasers in the infrared regime, because typical energies in realistic structures are in the tens or hundreds of meV range.

A thorough understanding of the intersubband absorption is of crucial importance for evaluating the performance of semiconductor intersubband devices and hereafter will be derived in detail. In calculating the intersubband absorption, one can start from Fermi's Golden Rule:

$$w_{if} = \frac{2\pi}{\hbar}|\tilde{\mathcal{H}}_{if}|^2\delta(E_f - E_i \pm \hbar\omega) \tag{12.1}$$

Quantum Wells, Wires and Dots, Third Edition. P. Harrison
©2009 John Wiley & Sons, Ltd.

which represents the rate at which electrons of energy E_i scatter from an initial $|i\rangle$ into a final $|f\rangle$ state of energy E_f owing to the interaction with photons of energy $\hbar\omega$ (i.e. due to a time-dependent perturbation $\tilde{\mathcal{H}}$). Based on perturbation theory, the total Hamiltonian of the system can be represented as a sum of the non-perturbed Hamiltonian and the perturbation due to the interaction with the electromagnetic field as:

$$\mathcal{H} = \mathcal{H}_0 + \tilde{\mathcal{H}}(t) \tag{12.2}$$

where $\mathcal{H}_0 = \mathcal{P}^2/2m_0 + U(\mathbf{r})$ represents the 'microscopic' Hamiltonian for an electron in the crystal potential, corresponding to complete wave functions. The potential $U(\mathbf{r})$ can be presented as a sum of a rapidly varying potential and a slowly varying term i.e. $U(\mathbf{r}) = U_f(\mathbf{r}) + U_s(\mathbf{r})$, and m_0 is free electron mass.

The Hamiltonian for a particle in an electromagnetic field can be found by substituting the momentum operator in Schrödinger equation with $\mathcal{P} \to (\mathcal{P} + e\mathbf{A}(\mathbf{r}, t))$, yielding the total Hamiltonian as:

$$\mathcal{H}(\mathbf{r}, t) = \frac{1}{2m_0} [\mathcal{P} + e\mathbf{A}(\mathbf{r}, t)]^2 - eV(\mathbf{r}, t) + U(\mathbf{r}) \tag{12.3}$$

where $\mathbf{A}(\mathbf{r}, t)$ and $V(\mathbf{r}, t)$ are the vector and scalar potentials, respectively. In order to determine a unique representation of the electromagnetic wave, the Coulomb gauge is introduced as:

$$\operatorname{div}\mathbf{A}(\mathbf{r}, t) = 0 \quad \text{and} \quad V(\mathbf{r}, t) = 0$$

Furthermore, equation (12.3) can be simplified by recalling that the $\mathcal{P}$ and $\mathbf{A}(\mathbf{r}, t)$ commute i.e.

$$[\mathcal{P}, \mathbf{A}(\mathbf{r}, t)] = -i\hbar \operatorname{div}\mathbf{A} = 0$$

Therefore, the Hamiltonian can be written as [301]:

$$\tilde{\mathcal{H}} = \frac{e}{m_0}\mathbf{A}_\bullet\mathcal{P} + \frac{e^2}{2m_0}\mathbf{A}^2 \tag{12.4}$$

The magnitude of the photon wave vector for infrared radiation is much smaller than the wave vector of electrons in semiconductor materials, i.e. the corresponding photon wavelength is much larger than the QW domain in which the electron wave function is confined. Therefore, the spatial dependence of the magnetic vector potential can be neglected: $\mathbf{A} = A(t)_\bullet\mathbf{e}$, where $\mathbf{e}$ is the polarization vector of unity intensity, $|\mathbf{e}| = 1$. The matrix element in equation (12.1) has only the first term in the equation (12.4) while the contribution from the second term vanishes because of orthogonality of the wave functions, yielding the final form of the perturbation Hamiltonian as:

$$\tilde{\mathcal{H}}(\mathbf{r}, t) = \frac{e}{m_0}\mathbf{A}(\mathbf{r}, t)_\bullet\mathcal{P}. \tag{12.5}$$

Now, the time-dependent magnitude of the magnetic vector potential is, without the loss of generality, simply given as:

$$A(t) = A_0 \cos(\omega t) \tag{12.6}$$

where ω is the angular frequency of the electromagnetic radiation (photon energy is $\hbar\omega$). The matrix element in Fermi's Golden Rule can be then written as:

$$\tilde{\mathcal{H}}_{if} = \langle \Psi_i | \tilde{\mathcal{H}}(\mathbf{r}, t) | \Psi_f \rangle = \frac{e}{m_0} P_{if} A(t) \qquad (12.7)$$

where $P_{if} = \langle \Psi_i | \mathbf{e} \cdot \mathcal{P} | \Psi_f \rangle$ is the momentum matrix element and Ψ_i are the full electron wave functions in the crystal potential. Substituting equation (12.7) into (12.1) gives:

$$w_{if} = \frac{2\pi}{\hbar} \left(\frac{e}{m_0} \right)^2 |P_{if}|^2 A(t)^2 \delta(E_f - E_i \pm \hbar\omega) \qquad (12.8)$$

To calculate the effective absorption coefficient, the scattering rate w_{if} should be averaged over the period $T_A = 2\pi/\omega$:

$$\overline{w_{if}} = \frac{2\pi}{\hbar} \left(\frac{e}{m_0} \right)^2 |P_{if}|^2 \frac{A_0^2}{2} \delta(E_f - E_i \pm \hbar\omega) \qquad (12.9)$$

where the time-dependent term $A(t)$ (equation 12.6) is averaged as:

$$\overline{A(t)^2} = \frac{1}{T_A} \int_0^{T_A} A_0^2 \cos^2(\omega t) dt = \frac{A_0^2}{2}$$

It is clear that, for calculation of the intersubband absorption, the most important factor is to correctly determine the corresponding momentum matrix element.

For a particular case of a semiconductor QW, both the slow varying potential and the effective mass are functions only of the z coordinate and the electron state is described within the envelope function approximation, hence the electron envelope function can be written as a product of an exponential part representing the free motion in the x-y plane and a z-dependent function as:

$$\psi(\mathbf{r}) = \psi(z) e^{i\mathbf{k}_{xy}\mathbf{R}} \qquad (12.10)$$

where $\psi(\mathbf{r})$ is the envelope wave function, $\mathbf{k}_{xy} = \{k_x, k_y\}$ is the in-plane wave vector and $\mathbf{R} = \{x, y\}$.

The *net* scattering rate can be found by summing over all possible interaction events of photons being absorbed or emitted in electron transitions between the initial state $\langle i, \mathbf{k}_i |$ and the final state $\langle f, \mathbf{k}_f |$, i.e.

$$\overline{W_{if}} = \sum_{k_x} \sum_{k_y} \overline{w_{if}} [\underbrace{f^{FD}(E_i)(1 - f^{FD}(E_f))}_{\text{absorption term}} - \underbrace{f^{FD}(E_f)(1 - f^{FD}(E_i))}_{\text{emission term}}] =$$

$$= \sum_{k_x} \sum_{k_y} \overline{w_{if}} [f^{FD}(E_i) - f^{FD}(E_f)] \qquad (12.11)$$

The average intensity of incident radiation is by definition given as:

$$\overline{I_{po}} = \overline{w_{em}} v_\Phi = \overline{w_{em}} \frac{c}{\bar{n}} \qquad (12.12)$$

where v_Φ is the velocity of light in media, $\bar{n}$ is the the average refractive index and c the velocity of light in vacuum. $\overline{w_{em}}$ is the density of electromagnetic energy, averaged over the time period, given as the sum of the electrostatic and magnetic energy, i.e.

$$\overline{w_{em}} = \overline{w_e} + \overline{w_m} = 2\overline{w_e} = 2 \times \frac{1}{2} \varepsilon_r \varepsilon_0 \overline{|\mathbf{E}|^2} \qquad (12.13)$$

where **E** is the electric field vector calculated as:

$$\mathbf{E} = -\frac{\partial \mathbf{A}}{\partial t} = -\frac{\partial (A(t)\bullet\mathbf{e})}{\partial t} = \omega A_0 \sin(\omega t)\bullet\mathbf{e} \tag{12.14}$$

Substituting equation (12.14) into equation (12.13) gives:

$$\overline{w_{\text{em}}} = \frac{1}{2}\varepsilon_r\varepsilon_0\omega^2 A_0^2 \tag{12.15}$$

Thus, equation (12.12), together with $\varepsilon_r = \bar{n}^2$, gives an expression for the average incident intensity of radiation as:

$$\overline{I_{\text{po}}} = \frac{1}{2}\varepsilon_0\bar{n}c\omega^2 A_0^2 \tag{12.16}$$

On the other hand, the intensity of light at a depth z in the semiconductor structure is given by:

$$\overline{I_{\text{p}}(z)} = \overline{I_{\text{po}}}\exp(-\alpha_{\text{if}}z) \tag{12.17}$$

which assuming a small structure and/or small absorption ($\alpha_{\text{if}}z \ll 1$) reads:

$$\overline{I_{\text{p}}(z)} = \overline{I_{\text{po}}}(1 - \alpha_{\text{if}}z) \tag{12.18}$$

The above equation defines the absorption coefficient in the form:

$$\alpha_{\text{if}} = \frac{\frac{\overline{I_{\text{po}}}-\overline{I_{\text{p}}(z)}}{z}}{\overline{I_{\text{po}}}} = \frac{\overline{W_{\text{if}}}\hbar\omega}{V_z\overline{I_{\text{po}}}} \tag{12.19}$$

where $(\overline{I_{\text{po}}} - \overline{I_{\text{p}}(z)})/z = \overline{W_{\text{if}}}\hbar\omega/V_z$ represents the density of electromagnetic energy absorbed in the volume $V_z = zL_xL_y$ where z is the thickness of the QW structure along the growth-axis and L_x and L_y are the dimensions of the structure in the x and y directions, respectively.

Considering the whole QW structure shown in Fig. 2.4 of Chapter 2, with dimensions L_x, L_y and $z \to L_z$, the absorption coefficient reads:

$$\alpha_{\text{if}} = \frac{2\pi e^2}{\bar{n}\varepsilon_0\omega m_0^2 cL_xL_yL_z}\sum_{k_x}\sum_{k_y}|P_{\text{if}}|^2 F_{\text{if}}\delta(E_{\text{f}} - E_{\text{i}} - \hbar\omega)\frac{\Delta k_x}{\Delta k_x}\frac{\Delta k_y}{\Delta k_y} \tag{12.20}$$

where $F_{\text{if}} = f^{\text{FD}}(E_{\text{i}}) - f^{\text{FD}}(E_{\text{f}})$ represents the difference in the populations of states $|i\rangle$ and $|f\rangle$ (the difference of Fermi–Dirac functions for the initial and and final states). The factor 2 was introduced to account for electrons of both spin orientations. In the case of an in-plane finite structure, the quasi-discrete steps Δk_x and Δk_y can be introduced in the derivation, representing the 'distance' of subsequent values of the **k** vector components k_x and k_y in the lateral dimensions. Based on the 'box' boundary conditions applied across the structure (vanishing of the wave functions at the structure edges) $\Delta k_x = \pi/L_x$ and $\Delta k_y = \pi/L_y$ and assuming that the structure is close to being infinite ($L_x, L_y \to \infty$), $\Delta k_x \to 0$ and $\Delta k_y \to 0$ can be substituted with dk_x and dk_y and hence, equation (12.20) becomes an integral as:

$$\alpha_{\text{if}} = \frac{2e^2}{\bar{n}\varepsilon_0\omega m_0^2 c\pi L_z}\int_{k_x}\int_{k_y}|P_{\text{if}}|^2 F_{\text{if}}\delta(E_{\text{f}} - E_{\text{i}} - \hbar\omega)\ dk_xdk_y \tag{12.21}$$

Because of the fact that, if $L_z \to +\infty$, then also $\alpha_{\mathrm{if}} \to +\infty$, it is more convenient to represent the absorption properties of the QW structure in terms of the fractional absorption defined as:

$$A_{\mathrm{if}} = \alpha_{\mathrm{if}} L_z \qquad (12.22)$$

The fractional absorption is a dimensionless quantity without units and is commonly given as a percentage, defining the proportion of radiation absorbed by the QW structure in one pass.

As the states' dispersion is axially isotropic, introducing an in-plane wave vector $k_{xy}^2 = k_x^2 + k_y^2$, transforms the integral $\int_0^\infty \int_0^\infty (...) \, \mathrm{d}k_x \mathrm{d}k_y = (\pi/4) \int_0^\infty (...) \, \mathrm{d}k_{xy}^2$, based on which equation (12.21) reads:

$$A_{\mathrm{if}} = \frac{e^2}{2\bar{n}\varepsilon_0 \omega m_0^2 c} \int_0^\infty |P_{\mathrm{if}}|^2 F_{\mathrm{if}} \delta(E_{\mathrm{f}} - E_{\mathrm{i}} - \hbar\omega) \, \mathrm{d}k_{xy}^2 =$$

$$= \frac{e^2}{2\bar{n}\varepsilon_0 \omega m_0^2 c} \left[|P_{\mathrm{if}}|^2 \, F_{\mathrm{if}} \left| \frac{\mathrm{d}(k_{xy}^2)}{\mathrm{d}(E_{\mathrm{f}} - E_{\mathrm{i}})} \right| \right]_{k_{xy}^2 = k_{xy0}^2} \qquad (12.23)$$

where the property of the δ-function $\int_{-\infty}^{+\infty} f(x)\delta(g(x)) = \sum_i \frac{f(x_i)}{|g'(x_i)|}$, $g(x_i) = 0$ has been used. k_{xy0} is obtained by solving:

$$E_{\mathrm{f}}(k_{xy0}^2) - E_{\mathrm{i}}(k_{xy0}^2) = \hbar\omega \qquad (12.24)$$

Now, attention is turned to calculating the momentum matrix element P_{if}, as a crucial step in evaluating the fractional absorption. It can be shown that the envelope function and the Hamiltonian contain all the necessary information about the full wave function required for estimating the momentum matrix element. Therefore, the full crystal Hamiltonian and corresponding wave functions, in the matrix element definition, can be substituted with the effective mass Hamiltonian and the associated wave functions in the envelope function approximation derived in Chapter 2. Based on the well-known commutation relation between the non-perturbated Hamiltonian and the position operator:

$$\mathcal{P} = \frac{m_0}{i\hbar} [\mathcal{H}_0, \mathbf{r}] \qquad (12.25)$$

the momentum matrix element can be written as [299]:

$$P_{\mathrm{if}} = \frac{1}{A} \langle \psi_{\mathrm{i}}(z) e^{i\mathbf{k_i r}} | \mathbf{e} \cdot \mathcal{P} | \psi_{\mathrm{f}}(z) e^{i\mathbf{k_f r}} \rangle. \qquad (12.26)$$

As the matrix elements of $\mathcal{P}_x$ and $\mathcal{P}_y$ are equal to zero due to the orthogonality of the envelope functions, only the z-polarization can induce intersubband transition in QWs. Hence, the P_{if} reads:

$$P_{\mathrm{if}} = \frac{1}{A} \int_V \psi_{\mathrm{f}}^*(z) e^{-i\mathbf{k} \cdot \mathbf{r}} \mathcal{P}_z \psi_{\mathrm{i}}(z) e^{i\mathbf{k} \cdot \mathbf{r}} \, \mathrm{d}z \mathrm{d}\mathbf{r} \qquad (12.27)$$

Therefore:

$$P_{if} = \frac{1}{A} \int_z \psi_f^*(z) \mathcal{P}_z \psi_i(z) dz \int_A e^{i(\mathbf{k}_\bullet \mathbf{r} - \mathbf{k}_\bullet \mathbf{r})} \, d\mathbf{r} =$$

$$= \int_z \psi_f^*(z) \mathcal{P}_z \psi_i(z) \, dz = \frac{m_0}{i\hbar} \int_z \psi_f^*(z)(\mathcal{H}_0 z - z\mathcal{H}_0)\psi_i(z) \, dz =$$

$$= -i\frac{m_0}{\hbar}(E_f - E_i) \int_z \psi_f^*(z) z \psi_i(z) \, dz = -i m_0 \omega d_{if} \qquad (12.28)$$

Thus the momentum matrix element is reduced to the dipole matrix element $d_{if} = \langle \psi_i | z | \psi_f \rangle$, which is considerably easier to calculate. In the derivation the **k**-selection rule (in-plane wave vector conservation) was implicitly used, i.e. ($\mathbf{k}_i = \mathbf{k}_f$). Generally speaking, in a semiconductor QW the effective mass and slowly varying potential are z-dependent ($m = m(z)$ and $U_s(\mathbf{r}) = U_s(z) \equiv U(z)$), therefore the Shrödinger equation in the envelope function approximation [301]:

$$-\frac{\hbar^2}{2} \frac{d}{dz} \left\{ \frac{1}{m(z)} \frac{d\psi(z)}{dz} \right\} + \left[U(z) + \frac{\hbar^2 k_{xy}^2}{2m(z)} \right] \psi(z) = E\psi(z) \qquad (12.29)$$

should be solved for each in-plane wave vector k_{xy}. In order to simplify the procedure, it is a reasonable approximation, based on time-independent perturbation theory (see, for example Ref. [302]), to introduce an *in-plane effective mass* in which case the Shrödinger equation can be solved only for $k_{xy} = 0$ and then for each state $|j\rangle$ (where j is either the initial or final state i.e. $j = i, f$) calculate in-plane effective mass. Therefore, the eigenenergy corresponding to an arbitrary in-plane wave vector k_{xy} is given as a small perturbation of the energy obtained solving equation (12.29) for $k_{xy} = 0$, i.e.

$$E_j(k_{xy}^2) = E_j(k_{xy}^2 = 0) + \int \psi_{j_0}^*(z) \frac{\hbar^2 k_{xy}^2}{2m(z)} \psi_{j_0}(z) \, dz = E_j(0) + \frac{\hbar^2 k_{xy}^2}{2m_{tj}} \qquad (12.30)$$

hence, the in-plane effective mass is calculated as:

$$\frac{1}{m_{tj}} = \langle j | 1/m(z) | j \rangle = \int \frac{1}{m(z)} |\psi_{j_0}(z)|^2 \, dz \qquad (12.31)$$

where the wave function $\psi_{j_0}(z)$ is also obtained as a solution of equation (12.29) for $k_{xy} = 0$. Obviously, introduction of the in-plane effective mass is an approximation analogous to the parabolic band approximation.

12.2   BOUND–BOUND TRANSITIONS

Adopting the in-plane effective mass approximation, we can write:

$$E_f(k_{xy0}^2) - E_i(k_{xy0}^2) = \hbar\omega_0 - \frac{\hbar^2 k_{xy0}^2}{2m_{if}} \qquad (12.32)$$

where $\hbar\omega_0 = E_f(0) - E_i(0)$ is the transition energy for $k_{xy}^2 = 0$ and $m_{if}^{-1} = m_{ti}^{-1} - m_{tf}^{-1}$ is the difference of the inverse in-plane effective masses of the initial and final states.

Usually, in the real structures, $m_{\rm tf} > m_{\rm ti}$ and equation (12.24) has a solution only for $\hbar\omega < \hbar\omega_0$. Solution of equation (12.23) in the case of *equal* in-plane effective masses will yield the fractional absorption with the shape of a Dirac impulse, in order to satisfy the energy conservation law. However, in addition to band dispersion, numerous mechanisms contribute to intersubband absorption line broadening such as non-parabolic band dispersion, scattering processes, etc. Therefore, a more appropriate and experimentally justified lineshape is the Lorentzian given as [299]:

$$L(\hbar\omega, \hbar\omega_0) = \frac{\Gamma/2\pi}{(\hbar\omega - \hbar\omega_0)^2 + (\Gamma/2)^2}, \tag{12.33}$$

where Γ is the full-width at half maximum (FWHM) usually taken from experiment.

In order to incorporate the absorption line broadening, the δ-function in equation (12.23) has to be substituted with a normalized Lorentzian form $L(\hbar\omega, \hbar\omega_0)$, therefore the normalizing condition $\int_{-\infty}^{\infty} L\, {\rm d}(\hbar\omega) = 1$ is satisfied. Hence, the expression for the intersubband absorption in QWs reads:

$$A_{\rm if}(\omega) = \frac{e^2\omega}{2\overline{n}\varepsilon_0 c}\, |d_{\rm if}|^2 \int_0^{\infty} L(\hbar\omega, \hbar\omega_0) F_{\rm if}(k_{xy}^2)\ {\rm d}k_{xy}^2 \tag{12.34}$$

In equation (12.30) the transition energy $\hbar\omega_0$ and the corresponding Lorentzian are independent of the in-plane wave vector (k_{xy}^2). Equation (12.34) therefore takes the form:

$$A_{\rm if}(\omega) = \frac{e^2\omega}{2\overline{n}\varepsilon_0 c}\, |d_{\rm if}|^2\, L(\hbar\omega, \hbar\omega_0) \int_0^{\infty} \left(f^{\rm FD}(E_{\rm i}, k_{xy}^2) - f^{\rm FD}(E_{\rm f}, k_{xy}^2) \right)\ {\rm d}k_{xy}^2 \tag{12.35}$$

Integrating the Fermi–Dirac distribution analytically over the subband dispersion i.e. $\int f^{\rm FD}(E_{\rm j}, k_{xy}^2)\ {\rm d}k_{xy}^2 = 2\pi N_{\rm S_j}$, where $N_{\rm S_j}$ is the sheet carrier concentration (the electron density per unit well surface) of state j, the final formula for fractional absorption on *bound–bound transitions* reads:

$$A_{\rm if}(\omega) = \frac{e^2\omega\pi}{\overline{n}\varepsilon_0 c}\, |d_{\rm if}|^2\, L(\hbar\omega, \hbar\omega_0)\, (N_{\rm S_i} - N_{\rm S_f}) \tag{12.36}$$

Here, the sheet electron density is:

$$N_{\rm S_j} = \frac{1}{2\pi} \int_0^{\infty} f^{\rm FD}(E_{\rm F}, E_{\rm j})\ {\rm d}k_{xy}^2 = \frac{kT m_{\rm tj}}{\pi\hbar^2} \ln\left[1 + e^{(E_{\rm F} - E_{\rm j}(0))/(kT)}\right] \tag{12.37}$$

where again $m_{\rm tj}$ is the in-plane effective mass of state j, $(j = i, f)$ and ω_0 and $E_{\rm j}(0)$ are the transition frequency and bound state energy at $k_{xy}^2 = 0$.

12.3 BOUND–FREE TRANSITIONS

In calculating the absorption on bound(discrete)–free(continuum) transitions one has to account for the two-fold degeneracy in the continuum, and choose the two degenerate wave functions to be orthogonal, in order to avoid overcounting or undercounting the continuum states. Therefore, consideration should be given to transitions from an initial bound state to both free final states corresponding to an energy in the continuum spectrum. Bound–free transitions have practical relevance for QW infrared photodetectors. Introduce the notation that the index i of an initial state becomes b (bound-state) and the index f of a final state becomes c (continuum-state). Assuming that continuum spectrum is initially discretized, equation (12.23) reads:

$$A_{if} = \frac{4\pi\beta\omega|m_{bc}|}{\bar{n}\hbar}[|d_{bc}|^2 F_{bc}]_{k_{xy}^2=k_{xy0}^2} \tag{12.38}$$

where $\beta = e^2/(4\pi\varepsilon_0\hbar c)$, $1/m_{bc} = 1/m_b + 1/m_c$ and $|d_{bc}|^2 = |d_{bc_1}|^2 + |d_{bc_2}|^2$ is the contribution of the two free states to the dipole matrix element (note that in symmetric structures because of wave function parity, one of the two terms is zero). The energy of the free state is given as:

$$E_c(k_{xy}^2) = \frac{\hbar^2}{2m_c}(k_z^2 + k_{xy}^2) \tag{12.39}$$

while the energy of the bound state is given with equation (12.30) with index j becoming b. The in-plane wave vector k_{xy0} is obtained by solving equation (12.24) as:

$$k_{xy0}^2 = \frac{m_b}{m_c - m_b}\left[k_z^2 - \frac{2m_c}{\hbar^2}(E_b(0) + \hbar\omega)\right] \tag{12.40}$$

The factor in square brackets should not be smaller than zero which, in turn, determines the minimal value of k_z:

$$k_{z_{min}} = \begin{cases} 0, & \hbar\omega + E_b(0) \le 0 \\ \sqrt{\frac{2m_c}{\hbar^2}[\hbar\omega + E_b(0)]}, & \hbar\omega + E_b(0) > 0 \end{cases}$$

The total bound–continuum fractional absorption is found by a summation over discretized continuum spectrum:

$$A_{bc} = \sum_{k_z} \frac{4\pi\beta\omega|m_{bc}|}{\bar{n}\hbar} F_{bc} d_{bc}^2 \frac{\Delta k_z}{\Delta k_z} \tag{12.41}$$

where the normalization constant of a continuum wave function is proportional to $1/\sqrt{L_z}$, i.e. the dipole matrix element reads:

$$d_{bc} = \frac{1}{\sqrt{L_z}}d_{bc}^* = \frac{1}{\sqrt{L_z}}\langle\psi_c|z|\psi_b\rangle \tag{12.42}$$

where d_{bc}^* is the dipole matrix element calculated with the free state wave function without the normalization factor $1/\sqrt{L_z}$. Again, the assumption can be made that the structure is close to being infinite ($L_z \to \infty$), and box-boundary conditions can be applied in the z-direction (vanishing of wave functions at structure edges) to obtain $\Delta k_z = \pi/L_z$. $\Delta k_z \to 0$ can be substituted with dk_z and hence, equation (12.42) becomes an integral and the total bound–continuum fractional absorption is found by integration over the continuum states [303]:

$$A_{bc} = \frac{4\beta\omega m_{bc}}{\bar{n}\hbar}\int_{k_{z_{min}}}^{+\infty} [F_{bc}d_{bc}^*]_{k_{xy}^2=k_{xy0}^2} \, dk_z \tag{12.43}$$

where $F_{\rm bc}$ denotes the difference of Fermi–Dirac functions for bound and free states. Obviously, in the case of a symmetric structure, taking into account wave function parity, only transitions from even bound states to odd free states or from odd bound to even free states will exist.

It should be noted that Lorentzian broadening was not included in the above equations because in bound–free transitions the 'band structure' linewidth is much larger than the Lorentzian linewidth, therefore, in contrast to bound–bound transitions the latter effect may be neglected in realistic structures. In addition, thanks to the another integration in the bound–free absorption calculation, the experimentally non-realistic feature of the δ-function peak is lost, hence inclusion of the Lorentzian is not necessary.

12.4 FERMI LEVEL

12.4.1 Bulk doping

In general, a single semiconductor QW is a structure where a thin layer of one material (for example, GaAs) is sandwiched between much wider (bulk) layers of another material/alloy (for example, AlGaAs). In the case that thick-bulk layers are uniformly doped, assuming that all donors of volume density N_D are ionized and neglecting the hole concentration, the electron concentration in these layers is $n = N_D$. On the other hand, the position of the Fermi energy in the structure is dictated by the electron concentration in the bulk material, i.e.

$$n = \frac{2}{\sqrt{\pi}} \left(\frac{m^* kT}{2\pi \hbar^2} \right)^{3/2} \int_0^\infty \frac{x^{1/2} dx}{\exp(x - \eta) + 1} = \frac{2N_{\rm CB}}{\sqrt{\pi}} F_{1/2}(\eta) \qquad (12.44)$$

where m^* is effective mass of bulk material, for example $m^* = m^*_{\rm AlGaAs}$ and $N_{\rm CB} = \left(\frac{m^* kT}{2\pi \hbar^2} \right)^{3/2}$ is the effective density of states in the conduction band. $F_{1/2}(\eta)$ is the Fermi–Dirac integral of order of $1/2$:

$$F_{1/2}(\eta) = \int_0^\infty \frac{x^{1/2} dx}{\exp(x - \eta) + 1} \qquad (12.45)$$

where $x = E/kT$ is the carrier energy in units of kT and $\eta = E_F/kT$ is the Fermi energy in units of kT.

It is known that the Fermi–Dirac integral cannot be integrated analytically; however, it is shown (see, for example [304]), that the function:

$$F_{1/2}(\eta) \approx \frac{2\sqrt{\pi}}{3\sqrt{\pi}a^{-3/8} + 4\exp(-\eta)} \qquad (12.46)$$

approximates the Fermi–Dirac integral of order of $1/2$ with relative error not exceeding 0.4% for value of η included in the range from $-\infty$ to $+\infty$ where:

$$a = \eta^4 + 33.6\eta\{1 - 0.68\exp[-0.17(\eta + 1)^2]\} + 50. \qquad (12.47)$$

For a given electron concentration, equations (12.44) and (12.46) still require numerical solution for the Fermi energy. In semiconductors, however, the Fermi energy is below the conduction band minimum ($\eta < 0$) and the Fermi–Dirac integral of order of $1/2$ can be reasonably well approximated as:

$$F_{1/2}(\eta) \approx e^{\eta} - 0.25e^{2\eta}, \tag{12.48}$$

resulting in a more explicit relation for the Fermi energy:

$$E_{\mathrm{F}} = kT \ln \left[2 \left(1 - \sqrt{1 - \frac{\eta\sqrt{\pi}}{2N_{\mathrm{CB}}}} \right) \right] \tag{12.49}$$

12.4.2   Quantum well doping

A common situation is when the thin layer (GaAs for example) of the QW is uniformly doped with the donor volume density N_{D}. If the width of the well material is l_w, then the total sheet electron density is given by:

$$N_{\mathrm{S}} = N_{\mathrm{D}}l_w \tag{12.50}$$

where, neglecting the electron concentration on the continuum states of the QW, the total sheet density is, on the other hand, given as a sum over all subbands: $N_{\mathrm{S}} = \sum N_{\mathrm{S}_j}$.

Following the procedure described in the Section 2.4, knowing that the subband minima of the jth subband is actually obtained for $k_{xy} = 0$ and assuming the in-plane effective mass approximation, the sheet density of jth subband reads:

$$N_{\mathrm{S}_j} = \frac{1}{2\pi} \int_0^{\infty} f^{\mathrm{FD}}(E_{\mathrm{F}}, E_j) \ \mathrm{d}(k_{xy}^2) = \int_{E_j(0)}^{+\infty} \frac{1}{1 + e^{\frac{E_{\mathrm{F}} - E_j(0)}{kT}}} \frac{m_{\mathrm{tj}}}{\pi\hbar^2} \ \mathrm{d}E \tag{12.51}$$

Using $\int \frac{dx}{1+e^x} = -\ln(1 + e^{-x})$ it is straightforward to obtain:

$$N_{\mathrm{S}_j} = \frac{kTm_{\mathrm{tj}}}{\pi\hbar^2} \ln \left[1 + e^{(E_{\mathrm{F}} - E_j(0))/(kT)} \right] \tag{12.52}$$

After summation over all the QW subbands for a given doping density/electron concentration, the Fermi energy should be obtained numerically from:

$$N_{\mathrm{D}}l_w = \sum_j \frac{kTm_{\mathrm{tj}}}{\pi\hbar^2} \ln \left[1 + e^{(E_{\mathrm{F}} - E_j(0))/(kT)} \right] \tag{12.53}$$

The above condition can be simplified in an explicit form, assuming that only the ground (bound) level (j=1) of the QW is populated, $N_{\mathrm{S}} = N_{\mathrm{S}_1}$, which is a reasonably good approximation at lower temperatures, i.e. (quantum limit case)

$$E_{\mathrm{F}} = E_1(0) + kT \left[\exp \left(\frac{N_{\mathrm{S}_1}\pi\hbar^2}{m_{\mathrm{t}1}kT} \right) - 1 \right] \tag{12.54}$$

12.5   RECTANGULAR QUANTUM WELL

Consider a symmetric square (rectangular) quantum well with the finite barrier height V_0 based on the GaAs/Al$_x$Ga$_{1-x}$As system (see Fig. 12.1). Because of simplicity, the conduction band edge in bulk AlGaAs is chosen as reference energy level. The potential and effective mass z-axis dependence of the well reads:

$$U(z) = \begin{cases} -V_0, \ |z| < \frac{l_w}{2} \\ 0, \ |z| > \frac{l_w}{2} \end{cases}$$

$$m(z) = \begin{cases} m_{\mathrm{GaAs}} = m_{\mathrm{W}}, \ |z| < \frac{l_w}{2} \\ m_{\mathrm{AlGaAs}} = m_{\mathrm{B}}, \ |z| > \frac{l_w}{2} \end{cases} \tag{12.55}$$

where l_w is the width of the well. Therefore, the bound-state spectrum belongs to the energy range $-V_0 < E < 0$, while the free (continuum) state spectrum is for $E > 0$. As the potential is symmetric, then the wave functions will have definite parity. As already discussed in Chapter 2, for the even wave functions of the nth bound-state, the solutions of equation (12.29) for $k_{xy} = 0$ are of the form:

$$\psi_n^{\mathrm{E}}(z) = \begin{cases} A_n^{\mathrm{E}} \cos (k_n z), \ |z| < \frac{l_w}{2} \\ B_n^{\mathrm{E}} e^{-\kappa_n z}, \ z > \frac{l_w}{2} \\ B_n^{\mathrm{E}} e^{\kappa_n z}, \ z < -\frac{l_w}{2} \end{cases} \tag{12.56}$$

where $k_n = \sqrt{2m_{\mathrm{W}}(V_0 + E_n)/\hbar^2}$ and $\kappa_n = \sqrt{-2m_{\mathrm{B}}E_n/\hbar^2}$. Constants were obtained from the boundary condition of wave function continuity at the interface between the well and barrier material ($z = l_w/2$) and after normalization of bound-state wave function $\int_{-\infty}^{+\infty} |\psi_n^{\mathrm{E}}(z)|^2 \, dz = 1$

$$A_n^{\mathrm{E}} = \sqrt{2} \left[l_w + \frac{1}{\kappa_n} + \frac{1}{k_n} \sin(k_n l_w) + \frac{1}{\kappa_n} \cos(k_n l_w) \right]^{-1/2},$$

$$\tag{12.57}$$

$$B_n^{\mathrm{E}} = A_n^{\mathrm{E}} \cos(k_n \tfrac{l_w}{2}) e^{\frac{\kappa_n l_w}{2}}.$$

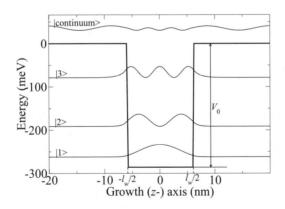

Figure 12.1 A rectangular AlGaAs quantum well with three bound states. Eigenfunctions squared are also presented as well as a representative of the continuum spectrum wave functions

Again, the boundary conditions in which the wave function and its first derivative divided by the effective mass $(1/m)(d\psi/dz)$ are continuous at the interface between the barrier and the well materials, after eliminating integration constants, determines the quantization condition from which the even parity solution eigenenergies can be found (Section 2.6):

$$\kappa_n = \frac{m_B k_n}{m_W} \tan\left(k_n \frac{l_w}{2}\right) \tag{12.58}$$

Substitution of equation (12.58) into equation (12.57) gives an alternative form of the normalization constant A_n^E [302]:

$$A_n^E = \sqrt{2}\left[l_w + \frac{2}{\kappa_n}\left(\cos^2(k_n \frac{l_w}{2}) + \frac{m_B}{m_W}\sin^2(k_n \frac{l_w}{2})\right)\right]^{-1/2} \tag{12.59}$$

Similarly, for odd parity solutions:

$$\psi_n^O(z) = \begin{cases} A_n^O \sin(k_n z), \; |z| < \frac{l_w}{2} \\ B_n^O e^{-\kappa_n z}, \; z > \frac{l_w}{2} \\ -B_n^O e^{\kappa_n z}, \; z < -\frac{l_w}{2} \end{cases} \tag{12.60}$$

with normalization constants given as:

$$A_n^O = \sqrt{2}\left[l_w + \frac{1}{\kappa_n} - \frac{1}{k_n}\sin(k_n l_w) - \frac{1}{\kappa_n}\cos(k_n l_w)\right]^{-1/2}, \tag{12.61}$$

$$B_n^O = A_n^O \sin(k_n \frac{l_w}{2})e^{\frac{\kappa_n l_w}{2}}. $$

Thus, energies of the odd parity states are given by:

$$\kappa_n = -\frac{m_B k_n}{m_W}\cot\left(k_n \frac{l_w}{2}\right) \tag{12.62}$$

which also gives an alternative form of the normalization constant A_n^O [302]:

$$A_n^O = \sqrt{2}\left[l_w + \frac{2}{\kappa_n}\left(\sin^2(k_n \frac{l_w}{2}) + \frac{m_B}{m_W}\cos^2(k_n \frac{l_w}{2})\right)\right]^{-1/2} \tag{12.63}$$

Once the eigenenergies and wave functions are known, the bound–bound dipole matrix element d_{if} in equation (12.36) can be calculated straightforwardly.

Fig. 12.2 shows the calculated bound–bound absorption spectra as a function of the photon energy $\hbar\omega$, for a rectangular GaAs quantum well embedded in $Al_{0.33}Ga_{0.67}As$ bulk at a temperature $T = 77$ K. The thickness of the well is $l_w = 80$ Å, which supports two bound states. The particular value of the Al mole fraction of $x = 0.33$ gives the depth of the potential well of $V_0 = 286$ meV and the effective mass of the barrier material $m_B = 0.094m_0$. The calculated energies of the ground state and the first excited state with respect to the conduction band edge in bulk are $E_1 = -243$ meV and $E_2 = -116$ meV, respectively. The doping level was chosen to provide a value of the Fermi energy $E_F = -230$ meV at a given cryogenic temperature and the Lorentzian broadening (full width at half maximum) was taken to be $\Gamma = 5$ meV [305]. Fig. 12.2 illustrates the characteriztic bell-shape intersubband absorption spectra, which has a calculated magnitude of around 4% with the position of the peak corresponding to a difference of energies of the first excited and ground (bound) state $E_2 - E_1 = 127$ meV.

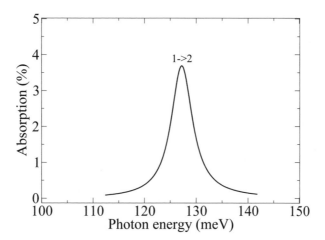

Figure 12.2   Calculated bound–bound absorption as a function of the photon energy $\hbar\omega$ for a 8 nm thick GaAs quantum well in $\mathrm{Al_{0.33}Ga_{0.67}As}$ bulk at $T = 77$ K. The value of the Fermi energy is $E_F = -230$ meV. Position of the absorption peak corresponds to a difference of energies of the two bound levels in the quantum well.

In the second example, a somewhat larger GaAs-$\mathrm{Al_{0.33}Ga_{0.67}As}$ QW, 160 Å thick, with four bound levels is considered. The calculated energies of the bound states are $E_1 = -271$ meV, $E_2 = -226$ meV, $E_3 = -153$ meV, and $E_4 = -54$ meV and the Fermi energy is again set to $E_F = -230$ meV at $T = 77$ K. The calculated bound–bound absorption spectra is depicted in Fig. 12.3. As already mentioned, due to the symmetry of the structure, only transitions between opposite parity states are allowed, with the position of the main peak corresponding to the difference between the ground and the first excited state energy $E_2 - E_1 = 45$ meV. The satellite peaks at $E_3 - E_2 = 73$ meV and at $E_4 - E_1 = 217$ meV are the result of the $2 \rightarrow 3$ and $1 \rightarrow 4$ transitions respectively. Although allowed, absorption from the level 3 into the level 4 at $E_4 - E_3 = 99$ meV is very weak, because of the fact that the level 3 is almost unpopulated at low temperatures.

To find the wave functions associated with the continuum spectrum $(E > 0)$ in the symmetric rectangular QW one can take into account the double degeneracy of the free states through the fact that, for each energy in equation (12.39) corresponding even and odd parity solutions of the Schrödinger equation can be constructed [303]. Therefore, an even free state wave function takes the form:

$$\psi_c^E(z) = \begin{cases} \cos(k_z^w z), & |z| < \frac{l_w}{2} \\ \cos(k_z z + \varphi_E), & z > \frac{l_w}{2} \\ \cos(k_z z - \varphi_E), & z < -\frac{l_w}{2} \end{cases} \tag{12.64}$$

where $k_z^w = \sqrt{\frac{m_W}{m_B} k_z^2 + \frac{2m_W V_0}{\hbar^2}}$ and φ_E is obtained from boundary conditions as $\varphi_E = \arctan\left(\frac{k_z^w m_B}{k_z m_W} \tan(k_z^w \frac{l_w}{2})\right) - k_z \frac{l_w}{2}$.

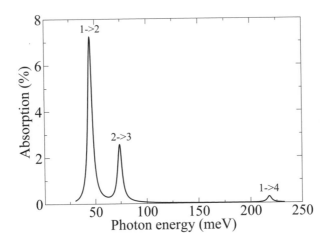

Figure 12.3 Calculated bound–bound absorption as a function of the photon energy $\hbar\omega$ for a 16 nm thick GaAs quantum well in Al$_{0.33}$Ga$_{0.67}$As bulk supporting four bound states at $T = 77$ K. The value of the Fermi energy was set to $E_F = -230$ meV. The positions of the absorption peaks correspond to the difference of energies between bound levels in the quantum well. Due to the symmetry, only parity changing, odd–even or even–odd, transitions are allowed, with a very small $3 \rightarrow 4$ absorption because of the low population of level 3.

Similarly, an odd free state wave function reads:

$$\psi_c^O(z) = \begin{cases} \sin(k_z^w z), \, |z| < \frac{l_w}{2} \\ \sin(k_z z + \varphi_O), z > \frac{l_w}{2} \\ \sin(k_z z - \varphi_O), z < -\frac{l_w}{2} \end{cases} \tag{12.65}$$

where $\varphi_O = \arctan\left(\frac{k_z m_W}{k_z^w m_B} \tan(k_z^w \frac{l_w}{2})\right) - k_z \frac{l_w}{2}$ is obtained from the boundary conditions.

The wave functions of the free states in equations (12.64) and (12.65) are given in non-normalized form as required for the dipole matrix element d_{bc}^* calculation in equation (12.43). The in-plane effective mass associated with the even state can now be calculated:

$$\frac{1}{m_c} = \frac{\left\langle \psi_c^E \left| 1/m(z) \right| \psi_c^E \right\rangle}{\left\langle \psi_c^E \left| \psi_c^E \right\rangle}=$$

$$= \frac{\left[\frac{1}{m_W} \left(\frac{l_w}{2} + \frac{\cos(k_z^w l_w)}{2k_z^w} \right) + \frac{1}{m_B} \left(\frac{L_z}{2} - \frac{l_w}{2} + \cos\left(\frac{2(k_z z + \varphi_E)}{2k_z} \right) \right)\Big|_{z=l_w/2}^{L_z/2} \right]}{\left[\frac{\cos(k_z^w l_w)}{2k_z^w} + \frac{L_z}{2} + \cos\left(\frac{2(k_z z + \varphi_E)}{2k_z} \right)\Big|_{z=l_w/2}^{L_z/2} \right]} \tag{12.66}$$

which, in the case of a long structure, gives $\lim\limits_{L_z \to \infty} m_c = m_B = m_{\text{AlGaAs}}$. This is an expected result as the free states are not confined in the thin GaAs well layer. The same conclusion applies for the odd parity wave function and the corresponding in-plane effective mass.

Finally, the matrix element for bound–free transitions should be calculated. Because of the inversion symmetry, only parity changing transitions between even bound and odd free state or between odd bound and even free states are allowed and have a non-zero dipole matrix element. For the $(n$th$)$ even bound–odd free transition:

$$d^*_{bc} = A^E_n I^{E \to O}_1 + B^E_n (I^S_2 \cos \varphi_O + I^C_2 \sin \varphi_O) \tag{12.67}$$

and for the odd bound–even free transition:

$$d^*_{bc} = A^O_n I^{O \to E}_1 + B^O_n (I^C_2 \cos \varphi_E - I^S_2 \sin \varphi_E) \tag{12.68}$$

where:

$$
\begin{aligned}
I^{E \to O}_1 &= \left(-\frac{l_w}{2} \right) \left(\frac{\cos \left((k^w_z - k_n) \frac{l_w}{2} \right)}{k^w_z - k_n} + \frac{\cos \left((k^w_z + k_n) \frac{l_w}{2} \right)}{k^w_z + k_n} \right) \\
&+ \left(\frac{\sin \left((k^w_z - k_n) \frac{l_w}{2} \right)}{(k^w_z - k_n)^2} + \frac{\sin \left((k^w_z + k_n) \frac{l_w}{2} \right)}{(k^w_z + k_n)^2} \right)
\end{aligned}
$$

$$
\begin{aligned}
I^{O \to E}_1 &= \left(-\frac{l_w}{2} \right) \left(\frac{\cos \left((k_n - k^w_z) \frac{l_w}{2} \right)}{k_n - k^w_z} + \frac{\cos \left((k^w_z + k_n) \frac{l_w}{2} \right)}{k^w_z + k_n} \right) \\
&+ \left(\frac{\sin \left((k_n - k^w_z) \frac{l_w}{2} \right)}{(k_n - k^w_z)^2} + \frac{\sin \left((k^w_z + k_n) \frac{l_w}{2} \right)}{(k^w_z + k_n)^2} \right)
\end{aligned}
$$

$$
\begin{aligned}
I^S_2 &= \frac{2 e^{-\kappa_n \frac{l_w}{2}}}{\kappa^2_n + k^2_z} \left[\frac{l_w}{2} \left(-\kappa_n \sin \left(k_z \frac{l_w}{2} \right) + k_z \cos \left(k_z \frac{l_w}{2} \right) \right) \right] \\
&- \frac{2 e^{-\kappa_n \frac{l_w}{2}}}{\kappa^2_n + k^2_z} \left[-\frac{(\kappa^2_n - k^2_z) \cos \left(k_z \frac{l_w}{2} \right) + 2 \kappa_n k_z \cos \left(k_z \frac{l_w}{2} \right)}{\kappa^2_n + k^2_z} \right]
\end{aligned}
$$

$$
\begin{aligned}
I^C_2 &= \frac{2 e^{-\kappa_n \frac{l_w}{2}}}{\kappa^2_n + k^2_z} \left[\frac{l_w}{2} \left(-\kappa_n \cos \left(k_z \frac{l_w}{2} \right) + k_z \sin \left(k_z \frac{l_w}{2} \right) \right) \right] \\
&- \frac{2 e^{-\kappa_n \frac{l_w}{2}}}{\kappa^2_n + k^2_z} \left[\frac{(\kappa^2_n - k^2_z) \cos \left(k_z \frac{l_w}{2} \right) - 2 \kappa_n k_z \sin \left(k_z \frac{l_w}{2} \right)}{\kappa^2_n + k^2_z} \right]
\end{aligned}
$$

The wave functions of free states are delocalized (see Fig. 12.1) and consequently the dipole matrix elements between the confined bound states and the continuum states are much smaller than for bound–bound transitions. Hence, the values of the bound–free absorptions are clearly way below the bound–bound absorptions in QWs where both type of intersubband transitions exists. However, of particular interest are QWs that support only one bound state. In such structures, which have practical relevance in infrared photodetectors, intersubband bound–free absorption become an important mechanism.

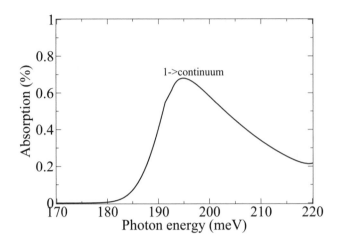

Figure 12.4   Bound–free absorption as a function of the photon energy $\hbar\omega$ for a 4 nm thick GaAs quantum well in $Al_{0.33}Ga_{0.67}As$ bulk, calculated at $T = 77$ K. The quantum well supports only one bound state and all the absorption comes from the bound $\rightarrow$ to continuum transitions. The value of Fermi energy is set to $E_F = -160$ meV.

Fig. 12.4 displays the calculated bound–free absorption spectra, at $T = 77$ K, as a function of the photon energy $\hbar\omega$, again for the rectangular GaAs-$Al_{0.33}Ga_{0.67}As$ narrow quantum well with a 40 Å thick GaAs layer. The structure has only one bound state with an eigenenergy of $E_b = -183$ meV. The doping level is chosen to set the Fermi energy at $E_F = -160$ meV. The absorption line shape yields a step rise at the ionization threshold $\hbar\omega = |E_b|$ and a more slowly decaying tail on the high-energy side. The asymmetric bell-like absorption feature arises from the existence of a virtual-resonance state in the continuum. This state corresponds to the transmission resonance, when the free state wave function has maximal amplitude in the region above the well. Therefore, if the energy of the photon $\hbar\omega$ approximately coincides with the difference between the energy of the virtual level and the bound state energy, the bound–continuum dipole matrix elements have large values and hence strong absorption peaks.

To summarize, in semiconductor QW structures the subband energy levels and the resulting intersubband absorption properties can be specifically tailored with great design freedom. Such structures can, in some sense, be regarded as artificial, man-made atoms or molecules, although they contain two-dimensional subbands and not really discrete energy levels such as, for instance, quantum dots. But because the joint density of states for intersubband transitions is essentially a δ-function, quantum wells behave like atoms as far as their intersubband absorption properties are concerned. This can be exploited for novel, efficient devices like harmonic generators, or infrared detectors and lasers.

12.6 INTERSUBBAND OPTICAL NON-LINEARITIES

It is well known that quantum confined structures, like semiconductor quantum wells or superlattices exhibit a variety of enhanced optical properties compared with bulk materials. An electron in a confining potential is accelerated in this potential under incident laser radiation. The accelerated electron oscillates and radiates according to its motion in the potential. This motion is the origin of the observed optical properties of the material, such as the refractive index. If the potential is asymmetric and the driving electromagnetic field is large enough to force the electron into the noticeably non-harmonic portion of the potential, then the re-radiated light will contain noticeable fractions of higher harmonics. The actual shape of the effective confining quantum-well potential determines the relative strength of the ensuing harmonics, such as in second harmonic generation (SHG) or third harmonic generation (THG). In analogy, an electron in this anharmonic potential driven by two different incident electric fields will lead to sum-frequency and/or difference-frequency generation (SFG and/or DFG) of the two incident fields.

Due to high degree of flexibility in design, with an appropriate choice of materials and structural parameters for optimization, it is possible to achieve remarkably large values of the second- and third-order non-linear susceptibilities in semiconductor QWs. Optical non-linearities can be efficiently employed in performing a number of opto-electronic functions such as modulation, limiting and thresholding, switching, etc. maintained over a long interaction length.

Following the seminal work of Gurnick and DeTemple [306] and their prediction that 10–100 times larger second-order optical non-linearities compared to the lattice value for a specially shaped, graded, asymmetric potential $Al_xGa_{1-x}As$ heterostructure are possible, non-linear optical effects based on intersubband transitions in semiconductor QW structures have been extensively studied, both experimentally and theoretically [307–313]. These structures possess quite remarkable linear and non-linear optical properties due to considerable values of the dipole transition matrix elements. A particular non-linear effect may be grossly enhanced by achieving the resonance conditions [314], when the spacing between relevant bound states coincides with the incident photon energy. Having obtained a resonance in a quantum well, however, the property of interest still depends on the combination of the matrix elements, i.e. via the wave functions, on the potential shape, which may be varied to optimize the structure [315].

Although considerable successes have been demonstrated for intersubband optical non-linearities, there are several drawbacks worth noting. The non-linearities are only strongly enhanced for light travelling in the plane of the QWs, as significant optical dipole matrix elements only exist for an electric field polarized normal to the QW plane, i.e. the plane of electron confinement. As a result, light needs to be efficiently coupled into the in-plane configuration, which is inherently more difficult than normal incidence coupling. Another issue is the resonant absorption accompanying most intersubband optical non-linearities. The strength of the optical non-linearity is approximately proportional to the density of electrons in the asymmetric QW potential, assuming only one energy level, usually the lowest one, is occupied. Energy levels show resonant enhancement of both the non-linearity and the absorption, with the latter outweighing the former under most conditions. Finally, a challenge is related to phase matching. As semiconductor materials have noticeable non-zero material dispersion across the wide wavelength spans of non-linear optics, the co-propagation of linear and non-linear light generally leads to destructive interference of the non-linearly generated light along the propagation direction of the primary pump beam.

Full analysis of all non-linear effects is beyond the scope of this chapter; however, the theoretical framework used to calculate the strength of the optical non-linearities and the expected non-linear optical output power will be briefly presented for perhaps the most common mechanism—resonant second harmonic generation, which enables the generation of coherent radiation at twice the frequency of available lasers. A full theoretical model employs the density matrix formalism and the reader can see, for example [316] for a detail derivation. A general theoretical analysis of SHG can then be followed by a calculation of the energy levels and the wave functions from solving Schrödinger's equation in the envelope function approximation for a rectangular asymmetric QW structure, in order to optimize effective QW potential shapes. The electromagnetic fields are treated classically with Maxwell's equations.

12.7   ELECTRIC POLARIZATION

The electromagnetic field of a light wave propagating through a medium exerts a force on loosely bound, valence electrons. In normal circumstances this force is weak and in a linear isotropic medium the resulting polarization is parallel with, and directly proportional to, the applied field. In other words, a light beam travelling through the material induces motion of the charged particles that constitute the given medium. In a dielectric medium, the charges are bound together and will start to oscillate in the applied electric field, i.e. they form oscillating electric dipoles. The contributions from the magnetic field part of the light and from electric quadrupoles are much weaker and are usually neglected. This is called the electric dipole approximation. The oscillating dipoles add up to form a macroscopic polarization P, which is used to describe the material response to the incident field. For low light intensities, i.e. small amplitudes of the electric field E, the charges can follow the field almost exactly and the relationship between P and E is essentially linear, i.e.

$$P = \varepsilon_0 \chi E \tag{12.69}$$

where χ is a dimensionless constant known as the dielectric susceptibility. With the invention of lasers, strong fields have become accessible and they are comparable with fields holding together the atoms in a crystal lattice. For large amplitudes of the field, particle motion will be distorted and equation (12.69) cannot be applied because the non-linear terms become important. Provided that these new terms are still small compared with the linear one, the polarization can be expanded as a series in E:

$$P = \varepsilon_0 (\chi_1 E + \chi_2 E^2 + \chi_3 E^3 + ...) \tag{12.70}$$

where χ_1 is the linear susceptibility of the medium while $\chi_2, \chi_3, ...$, describe the non-linear optical properties.

Given the harmonic electromagnetic field of the input radiation as $E(t) = E_0 \sin(\omega t)$, the QW polarization can be written as:

$$
\begin{aligned}
P(t) &= \varepsilon_0 [\chi_1 E_0 \sin(\omega t) + \chi_2 E_0^2 \sin^2(\omega t) + \chi_3 E_0^2 \sin^2(\omega t) + ...) \qquad (12.71)\\
&= \varepsilon_0 \chi_1 E_0 \sin(\omega t) + \frac{1}{2} \varepsilon_0 \chi_2 E_0^2 [1 - \cos(2\omega t)]\\
&+ \frac{1}{4} \varepsilon_0 \chi_3 E_0^3 [3 \sin(\omega t) - \sin(3\omega t)] + ...
\end{aligned}
$$

where the quadratic polarization term contains a constant static term and a term oscillating at double the incident frequency. The static polarization produces a DC-electric field in the medium, thus producing an optical rectification. The polarization oscillating at twice the applied frequency ($2\omega t$) represents second harmonic generation characterized with the second-order susceptibility χ_2. Similarly, the third-order susceptibility χ_3 characterizes third harmonic generation.

12.8   INTERSUBBAND SECOND HARMONIC GENERATION

Consider the general circumstance in which the incident radiation consists of two distinct frequency components ω_0 and ω_1. The strength of total electric field can be represented in the form:

$$
E(t) = E_0 e^{-i\omega_0 t} + E_1 e^{-i\omega_1 t} \qquad (12.72)
$$

then assuming that the second-order contribution to the non-linear polarization is of the form $P_2(t) = \chi_2 E^2$, the non-linear polarization is given by:

$$
\begin{aligned}
P_2(t) &= \chi_2 \varepsilon_0 [E_0^2 e^{-2i\omega_0 t} + E_1^2 e^{-2i\omega_1 t} + 2E_0 E_1 e^{-i(\omega_0 + \omega_1)t}\\
&+ 2E_0 E_1^* e^{-i(\omega_0 - \omega_1)t}] + 2\chi_2 \varepsilon_0 [E_0 E_0^* + E_1 E_1^*] \qquad (12.73)
\end{aligned}
$$

The complex amplitudes of the various frequency components of the non-linear polarization are therefore given by:

$$
\begin{aligned}
P(2\omega_0) &= \chi_2 E_0^2\\
P(2\omega_1) &= \chi_2 E_1^2\\
P(\omega_0 + \omega_1) &= 2\chi_2 E_0 E_1 \qquad (12.74)\\
P(\omega_0 - \omega_1) &= 2\chi_2 E_0 E_1^*\\
P(0) &= 2\chi_2 (E_0 E_0^* + E_1 E_1^*)
\end{aligned}
$$

In the above expressions $P(2\omega_0)$ and $P(2\omega_1)$ represent the second harmonic generation (SHG). $P(\omega_0 + \omega_1)$ represents the sum frequency generation (SFG), $P(\omega_0 - \omega_1)$ the difference frequency generation (DFG) and $P(0)$ the optical rectification. Out of four non-zero frequency components, only one will be present in appreciable intensity in typical radiation. The reason for this behaviour is that the non-linear polarization can efficiently produce an output signal only if a certain phase matching condition is satisfied, which is usually the case for only one frequency component. In practice, one chooses which frequency component will be radiated by selecting the polarization of the input radiation and orientation of non-linear crystal.

We consider an n-doped structure based on direct bandgap semiconductors and take the bandgap throughout it to be large enough that interband transitions may be neglected. The polarization response of the structure to the pump field with photon energy

$\hbar\omega$ is then mainly governed by intersubband transitions between quantized conduction band states. We will assume that in-plane dispersion is the same for all subbands, i.e. possible differences can be described by the linewidth of the transition. Analogously to the intersubband absorption, intersubband non-linear susceptibility has a substantial value only for z-polarized radiation.

Taking a realistic n-doped QW with its quantized states well separated from each other, one may take only a few most important states (the well-populated ground and and a few excited ones), approximately resonant with the incident photons energies $\hbar\omega_0$ and $\hbar\omega_1$ for further consideration. From the density matrix method [316] applied *to a three-level system*, one finds the expression for χ_2 at frequency $\omega_2 = \omega_0 + \omega_1$:

$$\chi_2 = \frac{e^3}{\hbar^2 L_z \varepsilon_0} \left(\frac{1}{\omega_2 + \omega_{13} - i\Gamma_{13}} \right) \left[\frac{(n_2 - n_1)d_{12}d_{23}d_{31}}{(\omega_{12} + \omega_0 - i\Gamma_{12})} - \frac{(n_3 - n_2)d_{12}d_{23}d_{31}}{(\omega_{23} + \omega_1 - i\Gamma_{23})} \right] \quad (12.75)$$

where d_{ij} are the transition dipole matrix elements, L_z as earlier, the length of the structure, the n_i electron sheet densities (population) in the ground $(i = 1)$, first excited $(i = 2)$ and second excited state $(i = 3)$, ω_{ij} transition frequency $(\omega_{ij} < 0$ for $i < j)$, and Γ_{ij} the off-diagonal relaxation rates $(i \to j$ transition line widths$)$.

Now, if we have a monochromatic (laser) pump i.e. *only one* incident radiation frequency ω exists, then we have $\omega_0 = \omega_1 = \omega$ and $\omega_2 = \omega_0 + \omega_1 = 2\omega$, in which case equation (12.75) transforms into:

$$\chi_2 = \frac{e^3}{\hbar^2 L_z \varepsilon_0} \frac{d_{12}d_{23}d_{31}}{2\omega + \omega_{13} - i\Gamma_{13}} \left[\frac{n_2 - n_1}{(\omega_{12} + \omega_0 - i\Gamma_{12})} - \frac{n_3 - n_2}{(\omega_{23} + \omega_1 - i\Gamma_{23})} \right]$$
$$(12.76)$$

In order to maximize non-linear optical susceptibility and perform an efficient second harmonic generation in QWs, the resonant conditions should be fulfilled, i.e. the difference between subsequent bound state energies should be equal, and at the same time be equal to the incident radiation photon energy: $E_3 - E_2 = E_2 - E_1 = \hbar\omega$ (also called *double-resonant condition*).

For realistic n-doped QWs in thermal equilibrium, only the ground level is well populated, hence the difference between the electron populations between the second and third excited state, $n_3 - n_2$, in equation (12.76) can be neglected. Therefore, at exact resonance $\omega_{32} = \omega_{21} = \omega$, and when the off-diagonal relaxation rates are taken to be equal $\Gamma_{12} = \Gamma_{23} \equiv \Gamma$ (though this is not essential), using $\omega_{ij} = -\omega_{ji}$ equation (12.76) reads:

$$\chi_2^{resonant} = \frac{e^3(n_1 - n_2)}{L_z \varepsilon_0} \frac{d_{12}d_{23}d_{21}}{(\hbar\Gamma)^2} \quad (12.77)$$

Furthermore, we can also assume that population of the first excited state is much lower than population of the ground state $(n_2 \ll n_1)$ in which case the total sheet electron density N_S (dictated by the doping level) equals to the population of the ground state, i.e. $n_1 = N_S$. Hence, equation (12.77) becomes:

$$\chi_2^{resonant} = \frac{e^3 N_S}{L_z \varepsilon_0} \frac{d_{12}d_{23}d_{21}}{(\hbar\Gamma)^2} \quad (12.78)$$

In conclusion, the resonant second harmonic generation requires three equidistant energy levels in a quantum well system, with the second-order optical susceptibility proportional to the product of the cyclic dipole matrix elements. In order to further maximize χ_2, one should clearly maximize the corresponding products of the dipole matrix elements in the numerator of equation (12.78) by appropriate tailoring of the QW profile (and hence the wave functions) while preserving the level spacing. The presence of the d_{31} matrix element rules out symmetric QWs, like in a parabolic quantum well, because of the definite parity of the wave functions when $d_{31} = 0$, so one should consider asymmetric structures only.

12.9   MAXIMIZATION OF RESONANT SUSCEPTIBILITY

To find the best potential shape, which will maximize $\chi_2^{\mathrm{resonant}}$, i.e. the product of matrix elements $\Pi^{(2)} = d_{12}d_{23}d_{21}$, the QW potential should be varied, and hence (related to it in ternary alloys) the effective mass, subjected to the constraint that state spacing should be as desired. Various options are possible such as digitally grading the QW, using a continually graded QW, multiple QWs in an external electric field, step graded quantum well, coupled double quantum well, etc. Here concentration will be focused on the last two cases as they can provide a semi-analytical textbook example of intersubband SHG optimization.

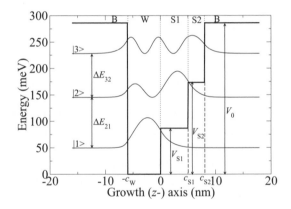

Figure 12.5   The potential (conduction band edge) in step QWs. The structure design parameters, used in the text, are all denoted

Consider an asymmetric step QW with a stepwise constant potential and effective mass, as shown in Fig. 12.5. This flexible double-step potential shape can be easily transformed into the simple single-step QW setting $c_{S1} = c_{S2} \equiv c_S$, $V_{S1} = V_{S2} = V_S$ or into an asymmetric double-QW choosing $V_{S1} = V_0$, $V_{S2} = 0$, $c_W = c_{W1}$, $c_{S1} = c_B$ and $c_{S2} = c_{W2}$, see Fig. 12.6, which are frequently used in resonant SHG [311,314,315]. As in Chapter 2, the Schrödinger equation should be solved for the QW in Fig. 12.5. The conventional exponential/hyperbolic or plane-wave type of solutions in separate layers (B, W, S1 and S2) of the structure, for the range of energies below the first step,

$E < V_{S1}$, reads:

$$
\psi(z) = \begin{cases}
Ae^{k_B z}, z < -c_W \\
C\cos(k_W z) + D\sin(k_W z), -c_W \le z < 0 \\
F\cosh(\kappa_{S1} z) + G\sinh(\kappa_{S1} z), 0 \le z < c_{S1} \\
M\cosh(\kappa_{S2} z) + N\sinh(\kappa_{S2} z), c_{S1} \le z < c_{S2} \\
He^{-k_B z}, c_{S2} \le z
\end{cases}
\tag{12.79}
$$

where $\kappa_B = [2m_B(V_0 - E)/\hbar^2]^{1/2}$, $k_W = [2m_W E/\hbar^2]^{1/2}$, $\kappa_{S1} = [2m_{S1}(V_{S1} - E)/\hbar^2]^{1/2}$, and $\kappa_{S2} = [2m_{S2}(V_{S2} - E)/\hbar^2]^{1/2}$. In the present analysis, depending on the incident photons energies, bound state energies can be some way above the conduction band minima, and inclusion of the band non-parabolicity, as pointed out in Chapter 3 could be useful. Therefore, a convenient energy-dependent effective mass model can be used, i.e. $m_W(E) = m_W(0)[1 + E/E_{gB}]$, $m_{S1}(E) = m_{S1}(0)[1 + (E - V_{S1})/E_{gS1}]$, $m_{S2}(E) = m_{S2}(0)[1 + (E - V_{S2})/E_{gS2}]$, and $m_B(E) = m_B(0)[1 + (E - V_0)/E_{gB}]$ (see Section 3.16, where, in this case, because of simplicity $\alpha \approx 1$ is used).

Applying the boundary conditions, i.e. continuity of $\psi(z)$ and $(1/m(z))(d\psi(z)/dz)$ at $z = -c_W$, $z = 0$, $z = c_{S1}$ and $z = c_{S2}$, in the energy range $E < V_{S1}$, a system of eight homogeneous equations is obtained, the non-trivial solution of which requires that:

$$
\begin{aligned}
\Phi(E) = {} & \left[\frac{k_W}{m_W}\cos(k_W c_W) + \frac{k_B}{m_B}\sin(k_W c_W) \right] \\
& \times \left\{ \sinh[\kappa_{S2}(c_{S2} - c_{S1})] \left[\frac{\kappa_{S2}^2}{m_{S2}^2}\cosh(\kappa_{S1} c_{S2}) + \frac{\kappa_{S1}}{m_{S1}}\frac{\kappa_B}{m_B}\sinh(k_{S1} c_{S1}) \right] \right. \\
& \left. + \cosh[\kappa_{S2}(c_{S2} - c_{S1})] \left[\frac{\kappa_{S2}}{m_{S2}}\frac{\kappa_B}{m_B}\cosh(\kappa_{S1} c_{S2}) + \frac{\kappa_{S1}}{m_{S1}}\frac{\kappa_{S2}}{m_{S2}}\sinh(\kappa_{S1} c_{S1}) \right] \right\} \\
& + \frac{k_W}{m_W}\frac{m_{S1}}{\kappa_{S1}} \left[\frac{\kappa_B}{m_B}\cos(k_W c_W) - \frac{k_W}{m_W}\sin(k_W c_W) \right] \\
& \times \left\{ \sinh[\kappa_{S2}(c_{S2} - c_{S1})] \left[\frac{\kappa_{S2}^2}{m_{S2}^2}\sinh(\kappa_{S1} c_{S1}) + \frac{\kappa_{S1}}{m_{S1}}\frac{\kappa_B}{m_B}\cosh(\kappa_{S1} c_{S1}) \right] \right. \\
& \left. + \cosh[\kappa_{S2}(c_{S2} - c_{S1})] \left[\frac{\kappa_{S2}}{m_{S2}}\frac{\kappa_B}{m_B}\sinh(\kappa_{S1} c_{S1}) + \frac{\kappa_{S1}}{m_{S1}}\frac{\kappa_{S2}}{m_{S2}}\cosh(\kappa_{S1} c_{S1}) \right] \right\} = 0
\end{aligned}
\tag{12.80}
$$

The integration constants in equation (12.79) are then simply derived from the boundary conditions and the normalization condition $\int_{-\infty}^{+\infty} \psi^2(z)\,dz = 1$. In the energy range above the first step and below the second step ($V_{S1} < E < V_{S2}$), define $k_{S1} = [2m_{S1}(E - V_{S1})/\hbar^2]^{1/2}$ and equation (12.80) is modified by substitutions: $\kappa_{S1} \to ik_{S1}$, $\sinh(\kappa_{S1} c_{S1}) \to i\sin(k_{S1} c_{S1})$ and $\cosh(\kappa_{S1} c_{S1}) \to \cos(k_{S1} c_{S1})$. An analogous substitution $\kappa_{S2} \to ik_{S2}$ applies in the energy range above the second step ($V_{S2} < E < V_0$) where $k_{S2} = [2m_{S2}(E - V_{S2})/\hbar^2]^{1/2}$. This completely defines the function $\Phi(E)$, the zeros of which are the energies of quantized states in an asymmetric double-step QW, with band non-parabolicity included.

Having chosen the alloy system to work with (e.g. $Al_x Ga_{1-x} As$), it is reasonable to take the well layer to comprise a pure well-type semiconductor (GaAs in this instance), because, with dipole matrix elements roughly scaling as $m^{-1/2}$, there is no benefit from allowing the well layer to be made of the alloy. Thus, m_W is defined from the start, and in the step and barrier layers, which are made of the alloy, with suitable compositions x_{S1}, x_{S2} and x_B, the effective mass and potential are uniquely related to each other, i.e. $m_{S1,S2,B} = m_{S1,S2,B}(x_{S1,S2,B})$ and $V_{S1,S2,B} = V_{S1,S2,B}(x_{S1,S2,B})$. Therefore, $\Phi(E)$ is a non-linear function of six independent parameters, say the widths (coordinates) c_W, c_{S1} and c_{S2} and potentials V_{S1}, V_{S2} and V_0. All possible potential shapes for the QW in Fig. 12.5 , i.e. the values of the six parameters, which result in the double-resonant condition $\Delta E_{32} = \Delta E_{21} \equiv \Delta E = \hbar\omega$ (three states spaced by ΔE), may be obtained from the system of three non-linear equations [315]:

$$\Phi(c_W, c_{S1}, c_{S2}, V_0, V_{S1}, V_{S2}, E_1) = 0$$
$$\Phi(c_W, c_{S1}, c_{S2}, V_0, V_{S1}, V_{S2}, E_1 + \Delta E) = 0 \qquad (12.81)$$
$$\Phi(c_W, c_{S1}, c_{S2}, V_0, V_{S1}, V_{S2}, E_1 + 2\Delta E) = 0$$

where E_1, the ground state energy measured from the well bottom, is an additional free parameter. The system of equations (12.81) may then be solved for three parameters out of seven, the remaining four being input parameters to be used for the QW shape variation, with values of all the seven parameters subject either to obvious physical constraints or to limitations imposed by the chosen alloy system or by the technological feasibility of the structure. By evaluation of the matrix elements (that can be done analytically, though via rather cumbersome expressions) for each individual solution, it is quite straightforward to search the entire free-parameter space and find the best of all QWs, which maximizes the product of the dipole matrix elements $\Pi^{(2)} = d_{12}d_{23}d_{21}$ in equation (12.78).

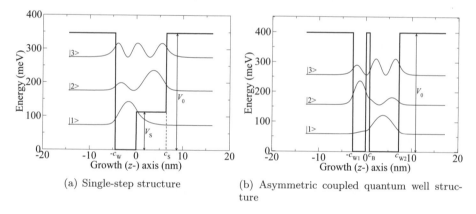

(a) Single-step structure

(b) Asymmetric coupled quantum well structure

Figure 12.6   The potential (conduction band edge) of $Al_x Ga_{1-x} As$ QWs being the target of optimization of the resonant second harmonic generation

Now move from the general QW structure shown in Fig. 12.5 to, technologically more realistic and significantly less demanding to grow, structures based on single-step QW or asymmetric coupled QWs as depicted in Fig.12.6. The single-step QW in Fig. 12.6(a) can be straightforwardly analysed, modifying the original structure shown in Fig. 12.5 by elimination of the layer S2, i.e. $c_{S1} = c_{S2} \equiv c_S$, $V_{S1} = V_S$, $m_{S1} = m_S$, and $\kappa_{S1} = \kappa_S$. Equation (12.80) then simplifies to:

$$\Phi(E) = \sin(k_W c_W) \left[\sinh(\kappa_S c_S) \frac{\kappa_B}{m_B} \frac{m_S}{\kappa_S} \left(\frac{k_W^2}{m_W^2} - \frac{\kappa_S^2}{m_S^2} \right) + \cosh(\kappa_S c_S) \left(\frac{k_W^2}{m_W^2} - \frac{\kappa_B^2}{m_B^2} \right) \right]$$
$$- \cos(k_W c_W) \frac{k_W}{m_W} \left[\frac{m_S}{\kappa_S} \sinh(\kappa_S c_S) \left(\frac{\kappa_B^2}{m_B^2} + \frac{\kappa_S^2}{m_S^2} \right) + 2 \cosh(\kappa_S c_S) \frac{\kappa_B}{m_B} \right] = 0$$
$$(12.82)$$

for the range of energies below the step ($E < V_{S1}$). Again, in the energy range above the step, ($V_S < E < V_0$), after $k_S = [2m_S(E - V_S)/\hbar^2]^{1/2}$, equation (12.82) is modified by the substitutions: $\kappa_S \to ik_S$, $\sinh(\kappa_S c_S) \to i\sin(k_S c_S)$ and $\cosh(\kappa_S c_S) \to \cos(k_S c_S)$.

Hence, the function $\Phi(E)$, is defined for the full range of bound energies. Now, the system of non-linear equations (12.81) is reduced and can be solved numerically for three parameters out of five, the remaining two being input parameters to be used for the QW shape optimization in order to maximize the resonant second harmonic generation of the desired incident photon radiation $\hbar\omega = \Delta E$.

Fig. 12.7 illustrates the numerical optimization of $Al_x Ga_{1-x} As$ single-step QW in Fig. 12.6(a). Incident photon radiation is chosen to be $\hbar\omega = 100$ meV which, for example, corresponds to $\lambda \approx 12.4$ μm GaAs-based quantum-cascade laser operating at room temperature [317,318]. The width of the well layer c_W and barrier potential height V_0 were used as free parameters. Other parameters (c_S, V_S, E_1) were coming out from the solution of system (12.81). The barrier potential V_0 was kept at $V_0 = 340$ meV which corresponds to the common value of $x = 0.4$ for the aluminium mole fraction in the barrier material.

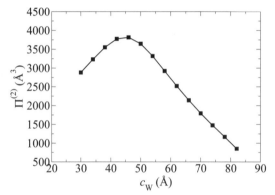

(a) The product of matrix elements $\Pi^{(2)}$ as it depends on the choice of well width

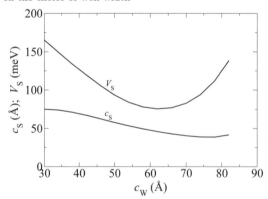

(b) Step width and step potential as a function of well width

Figure 12.7   Optimization of an $Al_x Ga_{1-x} As$ single-step QW under the double resonance condition, $\hbar\omega = 100$ meV. The barrier height was kept constant at $V_0 = 340$ meV

The largest value of the dipole matrix element product of around 3800 Å^3 was obtained for a well width of $c_W = 45$ Å, as shown in Fig. 12.7(a). Corresponding values of the step width $c_S = 64$ Å and the step potential of $V_S = 110$ meV can be found in Fig. 12.7(b).

The results of another example of QW optimization for the same incident photon energy of $\hbar\omega = 100$ meV are presented in Fig. 12.8. This time an asymmetric coupled QW structure shown in Fig. 12.6(b), was analysed. This structure can be obtained by modifying the original structure in Fig. 12.5 using $V_{S1} = V_0$, $V_{S2} = 0$, $c_W = c_{W1}$, $c_{S1} = c_B$ and $c_{S2} = c_{W2}$.

Choosing the well width of the first well c_{W1} as a free parameter and keeping another free parameter, the barrier potential, constant at $V_0 = 400$ meV (aluminium mole fraction in barrier $x = 0.46$) and solving equation (12.81), the maximal product of the matrix elements $\Pi^{(2)} \approx 3900$ Å^3 was calculated for $c_{W1} = 28$ Å, as seen in Fig. 12.8(a). The optimal thicknesses of other layers of coupled QWs i.e. thin barrier thickness $c_B = 8.5$ Å and the second well width $c_{W2} - c_B = 64$ Å were also obtained automatically during the non-linear optimization, see Fig. 12.8(b).

The scheme presented for QW shape optimization is quite general. It can be used for other intersubband non-linear processes, which may not require equispaced states (off-resonant harmonic generation, parametric down-conversion, etc.), hence QWs intended for these processes can also be optimized in the same fashion. On the practical side, it may be implemented with reasonable effort and computation time only for structures comprising not more than a few layers of different widths and compositions. Yet it is exactly such simple structures that are of the largest technical importance at present.

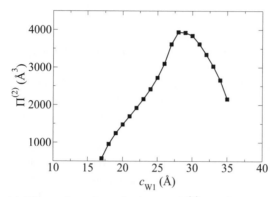

(a) The product of matrix elements $\Pi^{(2)}$ as a functiion the first well width

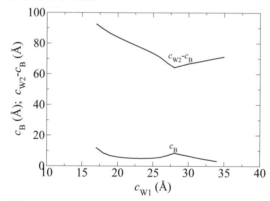

(b) The barrier thickness and the second well width as a function of the first well width

Figure 12.8 Optimization of an $Al_xGa_{1-x}As/GaAs$ asymmetric coupled QWs under the double resonance condition, $\hbar\omega = 100$ meV. The barrier height was kept constant at $V_0 = 400$ meV

CHAPTER 13

OPTICAL WAVEGUIDES

contributed by C. A. Evans

13.1 INTRODUCTION TO OPTICAL WAVEGUIDES

Optical waveguides are an integral aspect of optoelectronic devices. Gaining the ability to direct the flow of light, just as copper wires and co-axial cable direct the flow of electrons, enabled a 'quantum leap' to be made in areas such as photonics and communications. Optoelectronic devices such as amplifiers, optical fibres, power splitters, reflectors, directional couplers, polarizers, modulators, converters and frequency shifters rely on the ability to be able to guide light to where it is needed. Optical waveguides also play a large role in the operation of semiconductor laser diodes. The gain medium is placed within a cavity, which is terminated at either end with partially reflecting mirrors. This configuration allows photons to travel back and forth along the cavity, causing more and more stimulated emission until the optical gain overcomes the absorption loss and lasing commences. This behaviour gives rises to standing wave patterns between the mirrors, which are known as *longitudinal* waveguide modes*. The rate at which stimulated emission occurs from an excited state in a laser is directly proportional to the density of photons at the emission frequency in the laser cavity. To this end, it is

*For more information on longitudinal modes see [319].

beneficial to increase the density of photons and this is done by confining the photons within the cavity using an optical waveguide. An optical waveguide therefore provides transverse light confinement, preventing beam divergence as it propagates along the cavity helping maintain a high optical intensity and hence gain along its entire length. A schematic representation of a typical laser waveguide cavity is shown in Fig. 13.1.

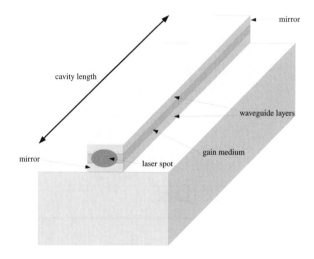

Figure 13.1   Schematic diagram of a laser waveguide cavity.

Although the major focus of this chapter is the application of optical waveguides to semiconductor lasers, much of the theory is equally applicable to other types of waveguide structures. The basic building block that will be focused upon is the *planar waveguide*. The planar waveguide is spatially inhomogeneous in one direction and assumed to stretch to infinity in the other two. Light confinement is therefore one-dimensional and the light is trapped in the region of highest refractive index. Examples of common types of planar waveguides are shown in Fig. 13.2. For the example on the left, light is confined to the

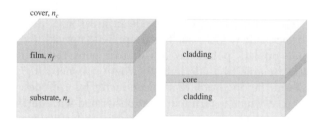

Figure 13.2   Examples of types of planar waveguide. Left: Light is confined in the film. Right: Light is confined to the core. Darker shading corresponds to higher refractive index.

thin film on top of the substrate. The high-index film can be deposited on the low-index substrate through either physical or chemical means. The cover region is generally formed by the surrounding air. An alternative configuration is shown on the right in

Fig. 13.2 in which the high-index core region is surrounded by lower-index cladding. This type of planar waveguide is typical of laser diode waveguides.

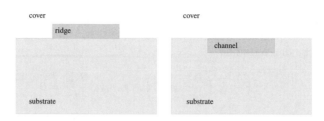

Figure 13.3   Examples of types of two-dimensional waveguides. Left: Light is confined in the high-index ridge. Right: Light is confined to the high-index channel.

It is also possible to construct waveguides that confine light in two dimensions. These types of waveguides are commonly called *ridge* or *channel* waveguides and examples of these are shown in Fig. 13.3. Light is generally confined within the ridge or channel region with vertical confinement provided through the same means as for planar waveguides, i.e. by the lower-index substrate and cover regions. Additionally, horizontal confinement is also provided (e.g. in the case of a ridge waveguide) by the lower index cover region which surrounds the sides of the ridge. This horizontal confinement is important for laser diodes as it prevents the light beam suffering from divergence as it propagates along the laser cavity. It would therefore seem necessary to study the optical properties of these two-dimensional structures in relation to laser diode waveguides. However, if the width of the ridge or channel is large compared with the wavelength of light ($>10\lambda$) in the region, then these types of waveguides can be considered planar waveguides, which simplifies the analysis considerably.

13.2   OPTICAL WAVEGUIDE ANALYSIS

This section will discuss the propagation of light in a planar multi-layer waveguide. Due to its simplicity, studying the planar waveguide is a good starting point for understanding more complicated waveguide structures. The theory presented in this section is discussed in more depth in several texts, for instance [319], [320] and [321] give a good introduction to optics and photonics aimed both at undergraduates and post-graduates.

Starting from Maxwell's equations, this section derives the equations that are used to describe light propagation in planar waveguides. Once these governing equations have been obtained, a method for solving them is presented, together with an example calculation.

13.2.1   The wave equation

For light that is propagating in an isotropic, non-magnetic (permeability $\mu = \mu_0$), perfect dielectric (conductivity $\sigma = 0$) medium, Maxwell's equations reduce to:

$$\nabla \times \mathbf{E} = -\mu_0 \frac{\partial \mathbf{H}}{\partial t} \tag{13.1}$$

$$\nabla \times \mathbf{H} = \epsilon_0 n^2 \frac{\partial \mathbf{E}}{\partial t} \qquad (13.2)$$

where $\mathbf{E}$ and $\mathbf{H}$ are the electric and magnetic fields, ϵ_0 is the permittivity of free space and n is the index of refraction of the material in which the light propagates and $n^2 = \epsilon_r$ where ϵ_r is the permittivity of the medium. Now, if the medium is inhomogeneous and its refractive index is position dependent, $n = n(\mathbf{r})$, then the following wave equations for $\mathbf{E}$ and $\mathbf{H}$ can be derived [320]:

$$\nabla^2 \mathbf{E} + \nabla \left(\frac{1}{n^2} \nabla n^2 \mathbf{E} \right) - \epsilon_0 \mu_0 n^2 \frac{\partial^2 \mathbf{E}}{\partial t^2} = 0 \qquad (13.3)$$

$$\nabla^2 \mathbf{H} + \frac{1}{n^2} \nabla n^2 \times (\nabla \times \mathbf{H}) - \epsilon_0 \mu_0 n^2 \frac{\partial^2 \mathbf{H}}{\partial t^2} = 0 \qquad (13.4)$$

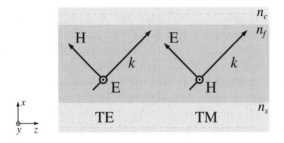

Figure 13.4 Relevant field directions for TE and TM propagation in a planar waveguide.

In the case of the planar waveguide depicted in Fig 13.2, the refractive index is dependent only upon a single Cartesian coordinate (x) and, assuming that propagation is in the z-direction, the fields do not depend on the y-direction and the solutions to equations (13.3) and (13.4) take the form:

$$\mathbf{E}(\mathbf{r}, t) = \mathbf{E_0}(x) e^{i(\omega t - \beta z)} \qquad (13.5)$$

$$\mathbf{H}(\mathbf{r}, t) = \mathbf{H_0}(x) e^{i(\omega t - \beta z)} \qquad (13.6)$$

where $\mathbf{E_0}$ and $\mathbf{H_0}$ are the complex field amplitudes, ω is the angular frequency and β the propagation constant. These two expressions determine the electromagnetic (EM) field for a propagating mode, characterized through its propagation constant. Therefore, for a given planar waveguide (defined by its refractive index profile $n(x)$), the solutions required are the complex field amplitudes and the propagation constants. Since the complex field amplitudes are determined by the propagation constants, the task of solving a waveguide problem becomes one of finding the propagation constants. A method to calculate the propagation constants of a planar waveguide is presented in Section 13.2.2.

As has just been mentioned, the fields do not depend on the y-direction. Setting $\partial/\partial y = 0$ in equations (13.1) and (13.2), Maxwell's equations reduce to:

$$\epsilon_0 n^2 \frac{\partial E_x}{\partial t} = -\frac{\partial H_y}{\partial z} \qquad (13.7)$$

$$\epsilon_0 n^2 \frac{\partial E_y}{\partial t} = \frac{\partial H_x}{\partial z} - \frac{\partial H_z}{\partial x} \qquad (13.8)$$

$$\epsilon_0 n^2 \frac{\partial E_z}{\partial t} = \frac{\partial H_y}{\partial x} \tag{13.9}$$

and

$$-\mu_0 \frac{\partial H_x}{\partial t} = -\frac{\partial E_y}{\partial z} \tag{13.10}$$

$$-\mu_0 \frac{\partial H_y}{\partial t} = \frac{\partial E_x}{\partial z} - \frac{\partial E_z}{\partial x} \tag{13.11}$$

$$-\mu_0 \frac{\partial H_z}{\partial t} = -\frac{\partial E_y}{\partial x}. \tag{13.12}$$

This group of equations contains two sets of solutions. Equations (13.8), (13.10) and (13.12) contain terms only in E_y, H_x and H_z, while equations (13.7), (13.9) and (13.11) contain only H_y, E_x and E_z terms. The first set form solutions in which the electric field has no components in the direction of propagation (i.e. only transverse components). These solutions are known as *transverse electric* (TE) modes. Similarly, the second set form solutions in which the magnetic field has no components in the direction of propagation and are known as *transverse magnetic* (TM) modes.

Transverse electric modes By substituting the equations for the electric and magnetic fields from equations (13.5) and (13.6) into the above set of solutions for TE modes, the following coupled equations are obtained:

$$H_x = -\frac{\beta}{\omega \mu_0} E_y \tag{13.13}$$

$$H_z = \frac{i}{\omega \mu_0} \frac{\partial E_y}{\partial x} \tag{13.14}$$

$$i\beta H_x + \frac{\partial H_z}{\partial x} = -i\omega \epsilon_0 n(x)^2 E_y \tag{13.15}$$

which link the field components E_y, H_x and H_z. By substituting equations (13.13) and (13.14) into equation (13.15), defining $k_0 = 2\pi/\lambda_0$, where λ_0 is the free-space wavelength, then $\omega = 2\pi c/\lambda_0 = k_0 c$ with c being the speed of light and equal to $(\epsilon_0 \mu_0)^{-1/2}$, the following wave equation containing only the E_y component is obtained:

$$\frac{\mathrm{d}^2 E_y(x)}{\mathrm{d}x^2} - \kappa^2 E_y(x) = 0 \tag{13.16}$$

where

$$\kappa = \sqrt{\beta^2 - k_0^2 n^2} \tag{13.17}$$

is the transverse wave vector. This can be solved by applying appropriate boundary conditions at the interfaces of the planar waveguide. From Maxwell's equations it can be shown that, at the interface between two dielectric media, the boundary conditions are such that the tangential components of the electric and magnetic fields must be continuous across the interface. In the case of the geometry in Fig. 13.4, the interfaces are in the y-z plane and so E_y and H_z must be continuous across the interface. From equation (13.14) it can be seen that H_z is proportional to the first derivative of E_y and hence the TE boundary conditions imply that E_y and its first derivative must be continuous across the interfaces.

Transverse magnetic modes   Following the same procedure as carried out in the previous section for TE modes but utilizing the set of solutions for TM modes, the following coupled equations are obtained linking H_y, E_x and E_z:

$$E_x = \frac{\beta}{\omega\epsilon_0 n(x)^2} H_y \tag{13.18}$$

$$E_z = \frac{-i}{\omega\epsilon_0 n(x)^2} \frac{\partial H_y}{\partial x} \tag{13.19}$$

$$i\beta E_x + \frac{\partial E_z}{\partial x} = i\omega\mu_0 H_y. \tag{13.20}$$

Substituting equations (13.18) and (13.19) into equation (13.20) gives the wave equation for TM modes:

$$\frac{\mathrm{d}^2 H_y(x)}{\mathrm{d}x^2} - \frac{1}{n(x)^2}\frac{\mathrm{d}n(x)^2}{\mathrm{d}x}\frac{\mathrm{d}H_y(x)}{\mathrm{d}x} - \kappa^2 H_y(x) = 0 \tag{13.21}$$

As also is the case for TE modes, appropriate boundary conditions are necessary and for TM modes, the requirement for continuous tangential electric and magnetic field components across an interface means that, for the geometry in Fig. 13.4, E_z and H_y must be continuous. From equation (13.19), it can be seen that E_z is proportional to the first derivative of H_y multiplied by $(1/n^2)$ and hence the TM boundary conditions require that H_y and $(1/n^2)\mathrm{d}H_y/\mathrm{d}x$ must be continuous across the interfaces.

13.2.2   The transfer matrix method

We now have wave equations for TE and TM modes in planar waveguides and, in this section, a transfer matrix method for solving the wave equations will be presented. For simple three-layer planar waveguides, relatively straight forward equations can be derived analytically, in closed form, and solved numerically to give the propagation constant of the modes supported by the waveguide (see, for example [319]). However, this method is not tractable for more complicated planar waveguide structures and a transfer matrix method becomes more suitable. The transfer matrix method is a problem solving technique that is used frequently in physics and mathematics due to its simplicity and is commonly used in optical waveguide problems [322–324]. The geometry of a multi-layer planar waveguide is shown in Fig. 13.5.

The wave equation for TE modes is given in equation (13.16) and it can be shown to have a general solution of the form:

$$E_i(x) = A_i \exp[-\kappa_i(x - x_{i-1})] + B_i \exp[\kappa_i(x - x_{i-1})] \tag{13.22}$$

which can be thought of a superposition of forward and backward propagating electric fields. $E_i(x)$ is the y-component of the electric field amplitude of the wave propagating in the ith layer, A_i and B_i are the complex field coefficients, $\kappa_i = \sqrt{\beta^2 - k_0^2 n_i^2}$ are the transverse wave vectors (κ_i can be either real or imaginary) and x_{i-1} is the location of the interface between the ith and $(i-1)$th layer.

The transfer matrix is derived by applying the appropriate boundary conditions at the interface between the layers in the waveguide. As discussed in the previous section,

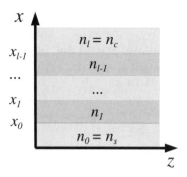

Figure 13.5   Geometry of a multi-layer planar waveguide.

for TE modes, the boundary conditions imply that E_y and its first derivative (dE_y/dx) are continuous across the interfaces. Therefore in the ith layer:

$$E_i(x) = A_i \exp[-\kappa_i(x - x_{i-1})] + B_i \exp[\kappa_i(x - x_{i-1})] \tag{13.23}$$

$$\frac{dE_i(x)}{dx} = \kappa_i \left(-A_i \exp[-\kappa_i(x - x_{i-1})] + B_i \exp[\kappa_i(x - x_{i-1})]\right) \tag{13.24}$$

and in the $(i + 1)$th layer:

$$E_{i+1}(x) = A_{i+1} \exp[-\kappa_{i+1}(x - x_i)] + B_{i+1} \exp[\kappa_{i+1}(x - x_i)] \tag{13.25}$$

$$\frac{dE_{i+1}(x)}{dx} = \kappa_{i+1} \left(-A_{i+1} \exp[-\kappa_{i+1}(x - x_i)] + B_{i+1} \exp[\kappa_{i+1}(x - x_i)]\right) \tag{13.26}$$

At the interface between the ith and $(i + 1)$th layers, i.e. at $x = x_i$, the values of equations (13.23) and (13.25) are equal and give the following:

$$A_i \exp[-\kappa_i d_i] + B_i \exp[\kappa_i d_i] = A_{i+1} + B_{i+1} \tag{13.27}$$

where $d_i = x_i - x_{i-1}$ is the thickness of the ith layer. Similarly, the values of equations (13.24) and (13.26) are also equal at $x = x_i$ and give the following:

$$\kappa_i \left(-A_i \exp[-\kappa_i d_i] + B_i \exp[\kappa_i d_i]\right) = \kappa_{i+1} \left(-A_{i+1} + B_{i+1}\right) \tag{13.28}$$

The above equations can be manipulated to give expressions for A_{i+1} and B_{i+1}. Equation (13.27) gives:

$$A_{i+1} = A_i \exp[-\kappa_i d_i] + B_i \exp[\kappa_i d_i] - B_{i+1} \tag{13.29}$$

and equation (13.28) gives:

$$B_{i+1} = A_{i+1} - \frac{\kappa_i}{\kappa_{i+1}} A_i \exp[-\kappa_i d_i] + \frac{\kappa_i}{\kappa_{i+1}} B_i \exp[\kappa_i d_i] \tag{13.30}$$

Inserting these into one another, the following expressions are obtained:

$$A_{i+1} = \frac{1}{2} \left[A_i \left(1 + \frac{\kappa_i}{\kappa_{i+1}} \exp[-\kappa_i d_i]\right) + B_i \left(1 - \frac{\kappa_i}{\kappa_{i+1}} \exp[\kappa_i d_i]\right)\right] \tag{13.31}$$

$$B_{i+1} = \frac{1}{2}\left[A_i\left(1 - \frac{\kappa_i}{\kappa_{i+1}}\exp[-\kappa_i d_i]\right) + B_i\left(1 + \frac{\kappa_i}{\kappa_{i+1}}\exp[\kappa_i d_i]\right)\right] \qquad (13.32)$$

The same procedure is can also be carried out for TM modes. In each individual layer, the refractive index is constant and so $dn(x)^2/dx = 0$ and the wave equation for TM modes (equation (13.21)) reduces to:

$$\frac{d^2 H_y(x)}{dx^2} - \kappa^2 H_y(x) = 0 \qquad (13.33)$$

which is equivalent to that for TE modes and has a general solution of the form:

$$H_i(x) = A_i\exp[-\kappa_i(x - x_{i-1})] + B_i\exp[\kappa_i(x - x_{i-1})] \qquad (13.34)$$

where $H_i(x)$ is the y-component of the magnetic field amplitude of the wave propagating in the ith layer. It should be noted that the complex coefficients A_i and B_i are *not* the same as for TE modes.

TM boundary conditions require that H_y and $(1/n^2)dH_y/dx$ are continuous across the interfaces between the layers. Therefore in the ith layer:

$$H_i(x) = A_i\exp[-\kappa_i(x - x_{i-1})] + B_i\exp[\kappa_i(x - x_{i-1})] \qquad (13.35)$$

$$\frac{1}{n^2}\frac{dH_i(x)}{dx} = \frac{1}{n_i^2}\left[\kappa_i\left(-A_i\exp[-\kappa_i(x - x_{i-1})] + B_i\exp[\kappa_i(x - x_{i-1})]\right)\right] \qquad (13.36)$$

and in the $(i + 1)$th layer:

$$H_{i+1}(x) = A_{i+1}\exp[-\kappa_{i+1}(x - x_i)] + B_{i+1}\exp[\kappa_{i+1}(x - x_i)] \qquad (13.37)$$

$$\frac{1}{n^2}\frac{dH_{i+1}(x)}{dx} = \frac{1}{n_{i+1}^2}\left[\kappa_{i+1}\left(-A_{i+1}\exp[-\kappa_{i+1}(x - x_i)] + B_{i+1}\exp[\kappa_{i+1}(x - x_i)]\right)\right] \qquad (13.38)$$

The same procedure is then followed as for TE modes to obtain the following expressions:

$$A_{i+1} = \frac{1}{2}\left[A_i\left(1 + \frac{n_{i+1}^2}{n_i^2}\frac{\kappa_i}{\kappa_{i+1}}\exp[-\kappa_i d_i]\right) + B_i\left(1 - \frac{n_{i+1}^2}{n_i^2}\frac{\kappa_i}{\kappa_{i+1}}\exp[\kappa_i d_i]\right)\right] \qquad (13.39)$$

$$B_{i+1} = \frac{1}{2}\left[A_i\left(1 - \frac{n_{i+1}^2}{n_i^2}\frac{\kappa_i}{\kappa_{i+1}}\exp[-\kappa_i d_i]\right) + B_i\left(1 + \frac{n_{i+1}^2}{n_i^2}\frac{\kappa_i}{\kappa_{i+1}}\exp[\kappa_i d_i]\right)\right] \qquad (13.40)$$

Noting the similarity between the above two equations for A_{i+1} and B_{i+1} and equations (13.31) and (13.32), these can be put into matrix form:

$$\begin{pmatrix} A_{i+1} \\ B_{i+1} \end{pmatrix} = \mathcal{Q}_i \begin{pmatrix} A_i \\ B_i \end{pmatrix}$$

where the transfer matrix of the ith layer, $\mathcal{Q}_i$, is given by:

$$\mathcal{Q}_i = \frac{1}{2}\begin{pmatrix} \left[1 + f_i\frac{\kappa_i}{\kappa_{i+1}}\right]\exp[-\kappa_i d_i] & \left[1 - f_i\frac{\kappa_i}{\kappa_{i+1}}\right]\exp[\kappa_i d_i] \\ \left[1 - f_i\frac{\kappa_i}{\kappa_{i+1}}\right]\exp[-\kappa_i d_i] & \left[1 + f_i\frac{\kappa_i}{\kappa_{i+1}}\right]\exp[\kappa_i d_i] \end{pmatrix}$$

For the case of TE modes, $f_i=1$ and for TM modes, $f_i = n_{i+1}^2/n_i^2$. The transfer matrix Q_i relates the complex field coefficients at the interface between the ith and $(i-1)$th layer and for a multi-layer waveguide with l layers, the transfer matrix of the whole waveguide is given by:

$$Q_{\text{wg}} = \prod_{i=l-1}^{0} Q_i \qquad (13.41)$$

and relates the field coefficients in the cover and the substrate layer, i.e.

$$\begin{pmatrix} A_{\text{c}} \\ B_{\text{c}} \end{pmatrix} = Q_{\text{wg}} \begin{pmatrix} A_{\text{s}} \\ B_{\text{s}} \end{pmatrix}$$

where the elements of the transfer matrix are denoted as:

$$Q_{\text{wg}} = \begin{pmatrix} q_{11} & q_{12} \\ q_{21} & q_{22} \end{pmatrix}$$

13.2.3   Guided modes in multi-layer waveguides

Now that the transfer matrix that describes the waveguide has been obtained, the propagation constants of the modes supported by the waveguide can be calculated. For the case of *guided* modes, which are generally the most interesting, the light field must decay in the substrate and cover regions, i.e. for the case of the multi-layer waveguide geometry in Fig. 13.5, only the outwards decaying real exponentials are permitted in the two outermost semi-infinite layers. From equation (13.17), it can be seen that this is only possible for the case in which the propagation constants of the modes fulfil the condition:

$$(k_0 n_{\text{s}}, k_0 n_{\text{c}}) < \beta \qquad (13.42)$$

In the opposite case, no linear combination of oscillatory solutions can have the outwards decaying property. The guided mode condition can be re-written in terms of the *effective mode index* N, which is equal to β/k_0 and is therefore dimensionless:

$$n_{\text{s}}, n_{\text{c}} < N \qquad (13.43)$$

The effective mode index represents the average refractive index that is 'seen' by that particular mode. Other modes also exist, in particular *radiation* or *leaky* modes. These modes are those in which the light is not confined to the waveguide and leaks into the surrounding cover and substrate regions causing losses as the light wave propagates along the waveguide. This situation occurs when $n_{\text{c}} < N < n_{\text{s}}$ or $N < n_{\text{c}}$.

According to equations (13.22) and (13.34), the guided mode condition requires that $A_{\text{s}} = B_{\text{c}} = 0$. If A_{s} and B_{s} are set to zero and unity, respectively, B_{c} will equal zero when $q_{22} = 0$. This leads to the dispersion relation for the waveguide $q_{22}(\beta) = 0$, which is solved numerically (for example, using a Newton–Raphson technique) in order to obtain the propagation constants of the waveguide modes.

The only parameters that are left to find are the complex field coefficients A_i and B_i. These are found from the relation:

$$\begin{pmatrix} A_i \\ B_i \end{pmatrix} = \prod_{j=i-1}^{0} Q_j \begin{pmatrix} 0 \\ 1 \end{pmatrix}$$

Once A_i and B_i are known, the field profile can be plotted and the waveguide is described completely except for its amplitude. The amplitude of the field is determined by A_i and B_i, but these values are only determined relative to B_s (hence set to unity). Waveguide modes are yet another example of the linear eigenvalue problem, where the absolute field amplitude is determined from only an external condition—in this case the power of a particular mode.

Modes in a planar waveguide   We will now consider an example multi-layer planar waveguide. The substrate, film and cover have refractive index values of $n_s = 3.4$, $n_f = 3.5$ and $n_c = 1$, respectively, and the thickness of the film is 5 μm. The modes have been calculated for a free space wavelength of 1.55 μm. There exist five guided TE modes in the structure, which can clearly be see from Fig. 13.6.

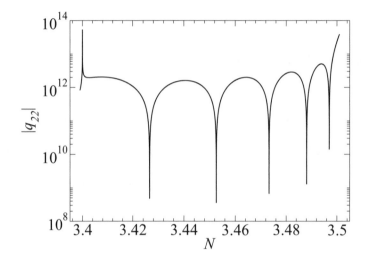

Figure 13.6   Absolute value of q_{22} as a function of effective mode index.   Each minima represents a TE mode.

Since q_{22} is a complex quantity, its absolute value is plotted as a function of effective mode index N. Each minimum in the plot represents a mode, the fact that $|q_{22}|$ does not equal zero at each mode is entirely down to numerical accuracy of the zero-finding algorithm. A finer step can be used in the search for zeroes, which will obtain smaller values of $|q_{22}|$ but with negligible difference to the final value of N. Modes are generally labelled according to their index $m = 0, 1, 2, ...$, etc. starting with the mode with the highest effective mode index. The $m = 0$ mode is known as the fundamental mode. The effective mode indices of both the TE and TM modes supported by the waveguide are listed in Table 13.1.

A plot of the normalized $E_y(x)$ for each TE mode is shown in Fig. 13.7. It can be seen from the figure that the electric field and its first derivative are indeed continuous across the interfaces and the field does decay in the cover and substrate regions. Similar results are obtained for the TM modes except, there exists a discontinuity in the derivative of the magnetic field at the interfaces. The boundary condition for TM modes states that $(1/n^2)\mathrm{d}H_y/\mathrm{d}x$ must be continuous across an interface and therefore, if two adjacent

Table 13.1   Effective mode indexes for the TE and TM modes supported by the waveguide.

Mode	N	Mode	N
TE_0	3.497026	TM_0	3.496930
TE_1	3.488109	TM_1	3.487731
TE_2	3.473278	TM_2	3.472446
TE_3	3.452630	TM_3	3.451207
TE_4	3.426512	TM_4	3.424463

regions have different values of n, then dH_y/dx cannot be continuous across the interface in order for the boundary condition to hold.

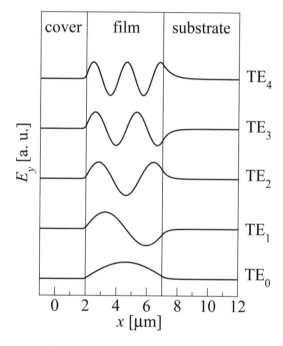

Figure 13.7   Normalized electric field profiles for the TE modes supported in the waveguide. The profiles have been shifted vertically for clarity.

13.3 OPTICAL PROPERTIES OF MATERIALS

In general, the dielectric function of a semiconductor is dispersive and complex:

$$\epsilon(\omega) = \epsilon_r(\omega) + i\epsilon_i(\omega) \tag{13.44}$$

and therefore the refractive index is also complex and frequency dependent:

$$n'(\omega) = n(\omega) + ik(\omega) = \sqrt{\epsilon(\omega)} \tag{13.45}$$

where n is the ordinary (real) refractive index and k is the imaginary part that is also known as the extinction coefficient. The real and imaginary parts of the dielectric constant can be described in terms of the real and imaginary parts of the refractive index:

$$\epsilon_r = n^2 - k^2 \tag{13.46}$$

and

$$\epsilon_i = 2nk \tag{13.47}$$

Similarly, the real and imaginary parts of the refractive index can be described in terms of the real and imaginary parts of the dielectric constant:

$$n = \sqrt{\frac{\sqrt{\epsilon_r^2 + \epsilon_i^2} + \epsilon_r}{2}} \tag{13.48}$$

and

$$k = \sqrt{\frac{\sqrt{\epsilon_r^2 + \epsilon_i^2} - \epsilon_r}{2}} \tag{13.49}$$

In general, a material with a purely real refractive index will have zero loss, while a non-zero extinction coefficient will lead to an absorption coefficient:

$$\alpha_k == -2k_0 k = -\frac{4\pi}{\lambda}k \tag{13.50}$$

It should be noted that, in the notation employed in this chapter, a negative value of k denotes losses, while a positive value denotes gain (e.g. a laser active medium). When α_k is positive, it could also be referred to as a gain coefficient.

13.3.1 Semiconductors

When modelling waveguides, knowledge of the wavelength dependence of both n and k is crucial. Tabulated values of n and k are available over a wide wavelength range in a host of semiconductors (e.g. [325]). For wavelengths close to and above the band gap, strong interband absorption occurs, which is characterized by large values of k. For longer wavelengths (i.e. energies below the band edge) these effects are negligible and non-zero values of k are caused by the interaction of photons with phonons. These effects can be modelled very well via a damped harmonic oscillator model [326]:

$$\epsilon(\omega) = \epsilon_\infty \left(1 + \frac{\omega_{LO}^2 - \omega_{TO}^2}{\omega_{TO}^2 - \omega^2 - i\omega\gamma_{ph}} \right) \tag{13.51}$$

where ϵ_∞ is the high frequency dielectric constant, ω_{LO} and ω_{TO} are the long-wavelength longitudinal-optical (LO) and transverse-optical (TO) phonon frequencies, respectively and γ_{ph} is the phonon damping constant. Table 13.2 lists the values used in equation (13.51) for the important III-V semiconductor binaries [325, 326].

The real and imaginary parts of the dielectric function of GaAs are plotted in Fig. 13.8. The dispersion that can be seen in the figure is caused by the EM field at these frequencies (called the Reststrahlen region) interacting with fundamental lattice vibrations. This results in the absorption or emission of EM waves due to lattice vibrations being

Material	ϵ_∞	ω_{LO} (meV)	ω_{TO} (meV)	γ_{ph} (meV)
GaAs	10.89	36.22	33.32	0.30
AlAs	8.48	49.78	44.86	0.99
InAs	12.25	29.76	27.03	0.50
InP	9.61	42.78	37.65	0.43

Table 13.2   Binary material parameters used in the damped single harmonic oscillator model

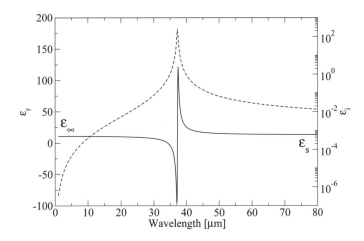

Figure 13.8   The real (solid line) and imaginary (dashed line) parts of the dielectric function of GaAs.

annihilated or created. It is important to see the difference in the real part of the dielectric constant above and below the Reststrahlen region. On the long wavelength side of the Reststrahlen region (lower frequency), the refractive index is higher and tends towards the root of the static dielectric constant (ϵ_s), while on the shorter wavelength side (higher frequency) the refractive index is lower and tends towards the root of the high frequency dielectric constant (ϵ_∞). The difference in the dielectric constants is due to different polarization mechanisms that come into play at different frequencies. At frequencies below the Reststrahlen band (far-infrared), the phase velocity is determined by both electronic and ionic polarization; however, above the Reststrahlen band at optical frequencies, ionic polarization is too slow to respond to the EM field and hence the phase velocity is determined by only electronic polarization and is therefore larger. Given that the refractive index can be defined as $n = \sqrt{\epsilon} = c/\nu$ (where ν is the phase velocity), a larger phase velocity results in a lower refractive index. The real part of the dielectric constant is equal to zero when $\omega = \omega_{TO}$.

From equation (13.47) it is apparent that k (and hence α_k) is related directly to ϵ_i and, therefore for wavelengths around the Reststrahlen region extremely strong absorption occurs preventing laser emission in this wavelength range.

13.3.2   Influence of free-carriers

When a semiconductor is doped, equation (13.51) must be modified to include contributions from the free-carriers. This contribution is commonly modelled using the classical Drude free-electron model. In this instance the dielectric function (ignoring contributions from phonons) is given by:

$$\epsilon(\omega) = \epsilon_\infty \left(1 - \frac{\omega_p^2}{\omega^2 + i\omega\gamma_{\text{pl}}} \right) \tag{13.52}$$

where the plasma frequency is defined as:

$$\omega_p = \sqrt{\frac{N_d q^2}{\epsilon_0 \epsilon_\infty m^* m_0}} \tag{13.53}$$

where N_d is the doping (i.e. free carrier) density and m^* the effective mass (see Table 13.3).

Table 13.3   Semiconductor effective masses

Material	m^*
GaAs	0.067
AlAs	0.150
InAs	0.023
InP	0.080

The plasma damping frequency is given by:

$$\gamma_{\text{pl}} = \frac{q}{m^* m_0 \mu} \tag{13.54}$$

where μ is the carrier mobility. From this equation it is apparent that the plasma damping frequency is actually the inverse of the electron relaxation time τ. Equation (13.52) can be split into real and imaginary parts:

$$\text{Re}[\epsilon(\omega)] = \epsilon_\infty \left(1 - \frac{\omega_p^2}{\omega^2 + \gamma_{\text{pl}}^2} \right) \tag{13.55}$$

and

$$\text{Im}[\epsilon(\omega)] = -\frac{\epsilon_\infty \gamma_{\text{pl}} \omega_p^2}{\omega(\omega^2 + \gamma_{\text{pl}}^2)} \tag{13.56}$$

For the situation when $\omega \gg \gamma_{\text{pl}}$:

$$\text{Re}[\epsilon(\omega)] = \epsilon_\infty \left(1 - \frac{\omega_p^2}{\omega^2} \right) \tag{13.57}$$

and, if $\omega = \omega_p$, then $\mathrm{Re}[\epsilon(\omega)] = 0$ and the material is defined as a resonant plasma. When $\mathrm{Re}[\epsilon(\omega)] < 0$, the electron gas in the plasma exhibits metallic behaviour while, for situations in which $\mathrm{Re}[\epsilon(\omega)] > 0$, the electron gas exhibits dielectric behaviour. Therefore, by tuning the plasma frequency of a semiconductor via suitable choice of doping density, one can modify the dielectric constant and hence the refractive index. This effect is highlighted in Fig. 13.9, which plots the real part of the refractive index of GaAs as a function of doping density at a wavelength of 9.4 μm.

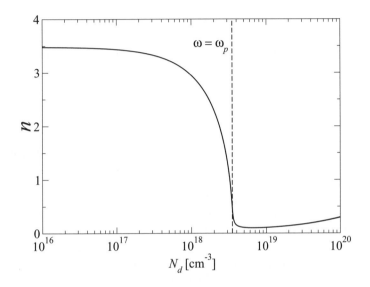

Figure 13.9   Real part of the refractive index of GaAs at various doping densities at a wavelength of 9.4 μm. The dashed vertical line indicates the doping density at which the corresponding plasma frequency equals the photon frequency.

From the figure it is apparent that, in the limit of low doping, the refractive index approaches the root of ϵ_∞. As the doping is increased and ω_p approaches ω, the refractive index decreases and experiences a strong reduction when $\omega_p = \omega$.

The Drude model can also be used to calculate the dielectric constant of metals. The plasma frequencies and plasma damping frequencies have been extracted from measurements for a wide range of metals. Some of these are listed in Table 13.4 and ϵ_∞ has been taken to be unity [327].

Table 13.4   Drude parameters for various metals

Metal	ω_p [eV]	γ_{pl} [meV]
Au	9.02	26.67
Ag	9.01	17.98
Cu	7.39	9.08
Al	14.76	81.83

For frequencies above the plasma frequency, the Drude scattering time is generally long enough such that $\omega \gg \gamma_{pl}$ and equation (13.56) reduces to:

$$\text{Im}[\epsilon(\omega)] = -\frac{\epsilon_\infty \gamma_{pl} \omega_p^2}{\omega^3} \tag{13.58}$$

Combining this with equations (13.47) and (13.50) one arrives at:

$$\alpha_k = \frac{k_0 \epsilon_i}{n} = \frac{\epsilon_\infty \gamma_{pl} \omega_p^2}{4\pi^2 c^3} \lambda^2 \tag{13.59}$$

and in the first instance it is apparent that the loss due to free-carriers is proportional to λ^2 which is an important aspect when it comes to long-wavelength waveguides as discussed in Section 13.4.2.

13.3.3   Carrier mobility model

From equation (13.54) it can be seen that the plasma damping frequency is related to the carrier mobility. Since the free-carrier effects discussed in the last section only come into play at longer wavelengths and in this chapter are only covered in Section 13.4.2 when discussing unipolar lasers, only the electron mobility is required and not the hole mobility. The situation is complicated slightly since the electron mobility itself is a function of doping density due to the influence of ionized impurity scattering. It is therefore useful to have a simple model that accurately describes the influence of doping density on the carrier mobility in a variety of semiconductors. One such model is the empirical Caughey–Thomas mobility model [328, 329]. In this model, the lattice contribution to the electron mobility is given by:

$$\mu^L = \mu_{300}^L \left(\frac{T}{300}\right)^{\gamma_0} \tag{13.60}$$

The total mobility including the reduction caused by ionized impurity scattering is given by:

$$\mu = \mu^{\text{min}} + \frac{\mu^L - \mu^{\text{min}}}{1 + \sqrt{\frac{N_d}{N^{\text{ref}}}}} \tag{13.61}$$

where

$$\mu^{\text{min}} = \mu_{300}^{\text{min}} \left(\frac{T}{300}\right)^{\gamma_1} \tag{13.62}$$

and

$$N^{\text{ref}} = N_{300}^{\text{ref}} \left(\frac{T}{300}\right)^{\gamma_2} \tag{13.63}$$

The model parameters are summarised in Table 13.5 for a variety of III-V semiconductors [329] and the resulting mobilities as a function of doping density are plotted in Fig. 13.10. The unknown values can be set to zero in order to remove the temperature dependence. As in the rest of this chapter, T is assumed to be 300 K in all cases. The figure clearly shows the decrease in mobility as the doping density increases due to the increasing influence of ionized impurity scattering.

Table 13.5   Parameters used in the Caughey–Thomas mobility model.

Material	μ_{300}^{L} [cm^2/Vs]	μ_{300}^{min} [cm^2/Vs]	N_{300}^{ref} [cm^{-3}]	γ_0	γ_1	γ_2
GaAs	8500	800	1×10^{17}	-2.2	-0.9	6.2
AlAs	410	10	1×10^{17}	-2.1	-	-
InAs	32500	11700	4.4×10^{16}	-1.7	-0.33	3.6
InP	5300	1520	6.4×10^{16}	-1.9	2	3.7

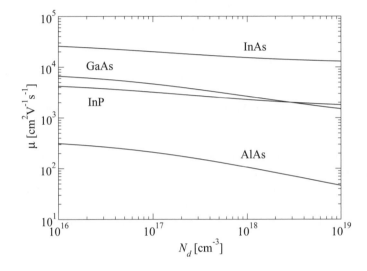

Figure 13.10   Electron mobility of III-V binaries as a function of doping density as calculated by the Caughey–Thomas mobility model.

13.3.4   Influence of doping

The previous sections have introduced models that can be used to calculate the refractive index of various semiconductors as a function of wavelength and doping density. For a given semiconductor at a given wavelength, the phonon contribution to the dielectric constant is calculated using equation (13.51). If the semiconductor is doped, the dielectric constant is modified by the presence of free-carriers and is described via the Drude model given in equation (13.52) with the plasma frequency determined by equation (13.53) and the plasma damping frequency by equation (13.54). Both of these parameters depend on the doping density, the plasma frequency directly and the damping frequency indirectly via the mobility, which can be calculated using the model described in Section 13.3.3. It is therefore interesting to study the behaviour of the optical properties of a semiconductor as a function of wavelength and doping density. Fig. 13.11 shows a plot of the real part of the refractive index of GaAs as a function of wavelength for different values of doping density.

The figure shows that, irrespective of the doping density, there is dispersion around the Reststrahlen region caused by the interaction of the light wave with phonons in the

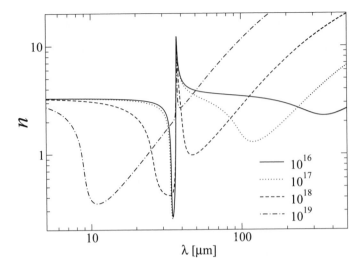

Figure 13.11   Refractive index of GaAs as a function of wavelength for various doping densities.

material. Each curve also shows a strong reduction in the real part of the refractive index when the wavelength corresponds to the plasma wavelength of the material. The location of this reduction depends on the doping density and occurs at shorter wavelengths (i.e. higher frequencies) for larger doping densities. Fig. 13.12 shows a plot of the free-carrier loss of GaAs as a function of wavelength for different values of doping density.

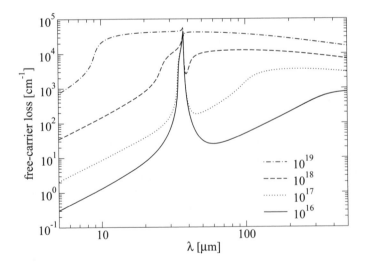

Figure 13.12   Free-carrier loss in GaAs as a function of wavelength for various doping densities.

It can be seen from that figure that the highest value of free-carrier loss occurs in the Reststrahlen region and is practically independent of the doping for densities less than $10^{19}\,\mathrm{cm}^{-3}$. If the contribution from the Reststrahlen region is ignored, the approximation that free-carrier loss is proportional to λ^2 is apparent only until the plasma frequency when the loss actually begins to fall with increasing wavelength.

13.4 APPLICATION TO WAVEGUIDES OF LASER DEVICES

This section discusses the application of the previously presented waveguide theory and material optical properties to actual laser devices. An example is given of a typical double heterostructure (DH) laser optical waveguide, together with examples of waveguides that are used for lasers emitting at much longer wavelengths.

A laser waveguide is typically characterized by its confinement factor Γ, the fraction of the light field that overlaps with the active medium and its loss $\alpha_{\rm wg}$ that must be overcome by the gain of the active medium before lasing can commence. Up to now, only planar waveguides with real refractive indices have been considered. From equation (13.50) it can be seen that a material with a complex refractive index will have a corresponding absorption loss, which is the main contribution to the waveguide loss $\alpha_{\rm wg}$[†]. For waveguides containing complex refractive index materials, the propagation constant β becomes a complex quantity and is given by:

$$\gamma = \beta - i\alpha \tag{13.64}$$

where β has the same meaning as previously and α is the attenuation constant. The method presented in Section 13.2 is still valid, except that β is replaced by γ and, in order to find the solution of the dispersion relation $q_{22}(\gamma)$, a multi-dimension root-finding algorithm (such as a steepest descent method) must be used.

The light intensity I is related to the field through $I \sim |\mathbf{E}|^2$ and for the field described in equation (13.5):

$$I = |E_0 e^{i(\omega t - \gamma z)}|^2 = |E_0 e^{i\omega t} e^{-i\beta z} e^{-\alpha z}|^2 = |E_0|^2 e^{-2\alpha z} \tag{13.65}$$

and therefore the loss suffered by the light intensity propagating in the z-direction in the waveguide cavity is given by $\alpha_{\rm wg} = -2\alpha$.

For a device with a gain coefficient g, loss $\alpha_{\rm wg}$ and cavity length L, the threshold condition is such that:

$$R_1 R_2 \exp[2(g - \alpha_{\rm wg})L] = 1 \tag{13.66}$$

where R_1 and R_2 are the reflectivities of the mirrors at the ends of the laser cavity. If the mirrors are defined by cleaving the facets, the reflectivity is just the Fresnel reflection coefficient, which for a cavity surrounded by air is given by:

$$R_1 = R_2 = \left(\frac{n-1}{n+1}\right)^2 \tag{13.67}$$

where n is the refractive index of the cavity material. The above threshold condition has the meaning that, under steady-state operation, the round trip gain equals unity and can be re-arranged to give an expression for the gain at threshold:

$$g = \alpha_{\rm wg} + \frac{1}{2L} \ln\left(\frac{1}{R_1 R_2}\right) \tag{13.68}$$

[†]The waveguide loss can be non-zero for purely real refractive index profiles if leaky modes exist in the structure, but this is beyond the scope of this chapter.

in which the second term on the right-hand side represents the loss suffered at the mirrors distributed over the entire cavity length L and is commonly referred to as the mirror loss α_{m}. Now, the gain in the above equation is the gain experienced by the laser mode field and, to get it, the actual material gain of the active medium must be multiplied by the confinement factor Γ of the waveguide, as the light field of the laser will only experience gain in the fraction of it that overlaps with the active medium. We can therefore get an expression for the threshold gain required by the device in order for lasing to occur:

$$g_{\mathrm{th}} = \frac{\alpha_{\mathrm{wg}} + \alpha_{\mathrm{m}}}{\Gamma} \tag{13.69}$$

where the confinement factor Γ is commonly defined as the fraction of the light intensity in the waveguide that overlaps with the gain medium:

$$\Gamma = \frac{\int_{\mathrm{active}} |E|^2 \mathrm{d}x}{\int_{-\infty}^{\infty} |E|^2 \mathrm{d}x} \tag{13.70}$$

The confinement factor is a useful quantity that is often used to compare different waveguides and the behaviour of different modes in a single waveguide.

13.4.1   Double heterostructure laser waveguide

For the simple case of an $\mathrm{Al}_x\mathrm{Ga}_{1-x}\mathrm{As}$-based DH laser, the active region is sandwiched between two $\mathrm{Al}_x\mathrm{Ga}_{1-x}\mathrm{As}$ layers, one doped n-type and the other p-type[‡]. These layers have a larger aluminium fraction (i.e. larger x) and therefore a wider band gap. Since the band gap and refractive index are inversely related to one another in the vast majority of III-V semiconductors, the wider band gap layers have a lower refractive index than the laser core and therefore, rather fortuitously, simultaneously act as waveguide cladding layers. This type of waveguide is depicted schematically in Fig. 13.2.

Consider an example DH laser with an active region aluminium content of $x = 0.198$ and cladding layers consisting of $x = 0.491$. The photon energy is typically of the order of 40 meV above the band gap and, for this particular active region composition, corresponds to an operating wavelength of $\sim$730 nm (1.7 eV) [331]. At wavelengths close to the band gap, strong interband absorption occurs under no-pumping conditions causing large values of extinction coefficient in the active region material. In the wider band gap cladding layers, the absorption is much weaker at the emission wavelength and can generally be ignored. For the AlGaAs compositions in this particular example, the active region has a complex refractive index of $n' = 3.635 - i0.002$, corresponding to an absorption coefficient of 344.3 cm^{-1} at 730 nm, while the cladding has a purely real refractive index of 3.368 [331]. It has been assumed that the refractive indices do not depend on the doping level or type of electrical conductivity and that free-carrier absorption is negligible, which is reasonable considering the short emission wavelength.

For this type of symmetric three-layer waveguide, the confinement factor of the waveguide depends on the active region thickness d and the refractive index contrast between the core and cladding layers. For given material compositions, the refractive index contrast is fixed and so Γ can be varied by changing d. In the limit of zero thickness $\Gamma = 0$ and increases with increasing d until it saturates at 100 per cent (see Fig. 13.13). One

---

[‡]For a more in-depth discussion of DH lasers, see Refs. [319, 330].

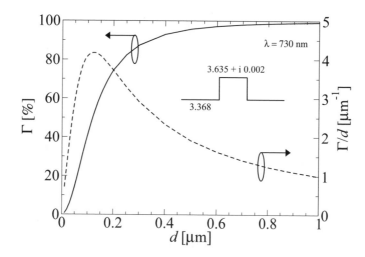

Figure 13.13   Confinement factor (solid line) and ratio of confinement factor to thickness (dashed line) as a function of active region thickness.

could be expected to think that optimum performance is obtained when $\Gamma = 100\%$; however, it has been been shown that the modal gain in a DH laser is proportional to the ratio of Γ/d [332], as a thicker active region requires a larger current density to achieve transparency. For a given current density, the performance of a DH laser with too narrow an active region will be dominated by its small confinement factor, while a DH laser with a wider active region will be limited due to its small gain. Fig. 13.13 shows a plot of Γ and the ratio Γ/d as a function of d.

From the figure it can be seen that the optimum thickness of the active region is $d = 0.125\,\mu$m. Even though the confinement factor Γ is only equal to 52.3%, the ratio of Γ to d is at its peak. At this optimum thickness, the propagation constant of the TE$_0$ mode is $\gamma = 2.97516 \times 10^6 - i9.46859 \times 10^3\,\text{m}^{-1}$, giving rise to an effective mode index of $N = 3.45663$ and a propagation loss of $\alpha = 189.37\,\text{cm}^{-1}$. For a cavity of length $L = 0.5\,$mm and a reflectivity of 0.304, the mirror loss is calculated to be $23.82\,\text{cm}^{-1}$. In an interband laser the absorption and gain are inextricably linked so, as the gain increases (by increasing the current density for instance), the interband absorption concomitantly decreases. The threshold gain is therefore simply the mirror loss divided by the confinement factor ($g_{\text{th}} = 45.55\,\text{cm}^{-1}$) since, at the point of lasing, the interband absorption and gain are balanced. The electric field distribution (E_y) of the TE$_0$ mode and the real part of the refractive index of the DH laser is plotted in Fig. 13.14.

13.4.2   Quantum cascade laser waveguides

Quantum cascade lasers (QCLs) [333] are unipolar lasers that operate on a fundamentally different principle to the DH lasers that were discussed in the previous section. The emission wavelengths of DH lasers are essentially fixed by the band gap of the active medium, while QCLs allow the emission wavelength to be tuned over a wide range from the mid-infrared to the far-infrared (THz) region through suitable design of the quantum well and barrier thicknesses. As the emission wavelength is increased, a waveguide

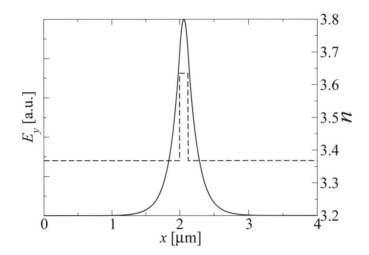

Figure 13.14   Electric field distribution of the TE$_0$ mode (solid line) and the real part of the refractive index profile (dashed line).

that provides optical confinement through a simple refractive index mismatch becomes impractical since the required thicknesses of the cladding layers become simply too thick to be fabricated. Also, free-carrier absorption (equation (13.59)) becomes the dominant loss mechanism at longer wavelengths and can have large values causing serious implications for device performance. Therefore, as QCL active region designs have evolved to longer and longer wavelengths, so too have the optical waveguides. This section presents typical waveguides that are used for mid- and far-infrared GaAs-based QCLs [334, 335].

Mid-infrared waveguide   As can be seen in Fig. 13.9, by tailoring the plasma frequency of GaAs (and indeed any semiconductor) so that it is close to the emission frequency of the laser through suitable choice of the doping density, one can achieve a significant reduction in the real part of the refractive index. The resulting refractive index contrast that would occur with the laser active region suggests that these highly doped, low-index layers could be used as waveguide cladding layers to provide optical confinement. This is true but, as can be seen from Fig. 13.12, the resulting free-carrier absorption at these high-doping levels is extremely large and would kill any possibility of laser action. Therefore, in order to circumvent these issues, Sirtori *et al.* placed low-doped layers (3.5 μm thick) between the active region (1.6 μm) and the highly doped cladding layers (1 μm) in order to separate the optical mode from the lossy highly doped regions [336]. Since the optical confinement is provided by the anomalous dispersion around the plasma frequency, these waveguides are known as double plasmon-enhanced waveguides. Fig. 13.15 shows the real part of the refractive index and the free-carrier loss of the waveguide layers calculated at a QCL emission frequency of $\lambda = 9.4\,\mu$m using the methods set out in Section 13.3.

Using the field directions depicted in Fig. 13.4, for the case of QCLs, the quantum wells and barriers that form the active region are grown in the x direction and following the intersubband selection rules, the electric field must be oriented parallel to the growth direction for intersubband transitions to occur. From Fig. 13.4 it can be seen that there is no E_x field component in TE modes and therefore only TM modes are of

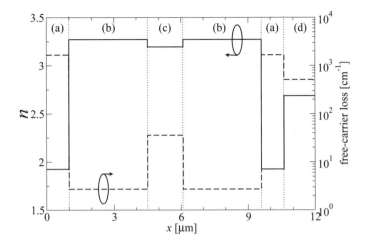

Figure 13.15   Real part of the refractive index profile (solid line) and free-carrier loss (dashed line). Layer (a) highly doped plasmon layer, $N_d = 6 \times 10^{18}$ cm^{-3}, (b) low-doped spacer layer, $N_d = 4 \times 10^{16}$ cm^{-3}, (c) 1.6 μm thick active region, (d) substrate $N_d = 3 \times 10^{18}$ cm^{-3}.

interest when it comes to QCL waveguides. When searching for TM mode solutions to the wave equation, H_y is calculated and therefore it seems natural to plot this value. In this case, since generally only E_x is of interest, which is related to H_y via equation (13.20), the mode intensity ($\sim |E_x|^2$) will be plotted in the rest of this section, as this is the quantity that is used in equation (13.70) to compute the confinement factor Γ.

The complex propagation constant of the TM$_0$ mode of the double plasmon-enhanced waveguide at 9.4 μm is calculated to be $\gamma = 2.14414 \times 10^6 - i8.7040 \times 10^2$ m^{-1}, corresponding to an effective mode index of $N = 3.20775$ and a waveguide loss of $\alpha_{\mathrm{wg}} = 17.461$ cm^{-1}. The mode intensity of the TM$_0$ mode is plotted in Fig. 13.16. The discon-

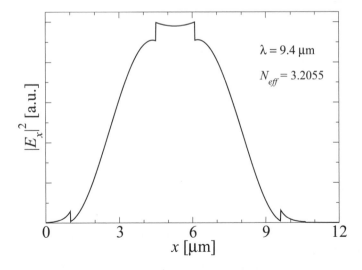

Figure 13.16   Mode intensity of the TM$_0$ mode in a double plasmon-enhanced QCL waveguide.

tinuities that are observed at the interfaces between layers arise directly from equation (13.20), H_y is set by Maxwell's equations to be continuous across the interfaces but the different refractive indices of the layers cause the discontinuity in the E_x field.

It has been shown that equation (13.70) is only strictly valid when calculating the confinement factor of TE modes and can lead to inaccuracies when dealing with TM modes [337]. An alternative method to find Γ can be used, as discussed next. This amounts to incorporating the material gain directly into the waveguide calculation, in order to find the threshold gain g_{th}. From equation (13.47) it can be seen that the imaginary part of the permittivity is given by $\epsilon_i = 2nk$. From equation (13.50) it is also known that the extinction coefficient k can be related to the gain/loss of a material via the relation $\alpha_k = -2k_0k$. Therefore, there is a relation that links the imaginary part of the permittivity to the material gain (defined for $\alpha_k < 0$) of the medium:

$$\epsilon_i = \frac{n}{k_0} g_{mat} \tag{13.71}$$

The previous examples have been carried out in the limit of zero gain, i.e. the absence of lasing. Indeed, the imaginary part of the permittivity of the active region (and hence the refractive index) is a function of the material gain and, at the onset of lasing, when the gain balances the losses, the medium is transparent. Fig. 13.17 shows the waveguide loss α_{wg} as a function of the material gain g_{mat} of the active region. Note that, in the convention used in this chapter, a negative value of waveguide 'loss' denotes gain while a positive value indicates loss.

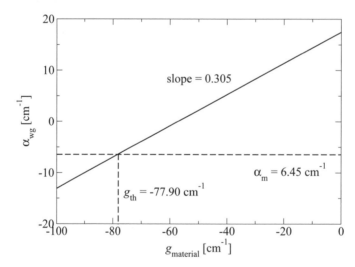

Figure 13.17   Waveguide loss/gain as a function of material gain. The mirror loss α_m is indicated by the dashed horizontal line. The intercept of the mirror loss and gain curve gives the threshold gain g_{th}.

It can be seen from the figure that, for a material gain of $g_{mat} = -57.2\,\mathrm{cm}^{-1}$, α_{wg} is zero and the medium appears transparent. In order to extract the threshold gain g_{th}, there must be enough excess gain in order to balance the mirror loss. For a cavity length of $L = 2\,\mathrm{mm}$ and a reflectivity of $R = 0.275$, the mirror loss has a value of $6.45\,\mathrm{cm}^{-1}$. From the figure, it can be seen that, for a material gain of $g_{mat} = -77.90\,\mathrm{cm}^{-1}$, the waveguide has a gain of $-6.45\,\mathrm{cm}^{-1}$, which balances the mirror loss and therefore $g_{\mathrm{th}} = -77.90\,\mathrm{cm}^{-1}$. Since the permittivity itself is a function of the material gain, so must be the waveguide loss and field distributions. In this essence, the problem is non-linear but, as can be seen from the above figure, for this particular waveguide and range of material gain, the relationship is linear to a very good approximation. In this limit, the slope gives an effective confinement factor and in this case is $\Gamma = 30.5\%$, in very good agreement with the value of $\Gamma = 31.2\%$ that is calculated using equation (13.70). Therefore, for this type of waveguide, equation (13.70) is equally valid for both TE and TM modes.

Far-infrared waveguide   For QCLs with longer emission wavelengths, the double plasmon-enhanced waveguide cannot be used because, as equation (13.59) shows, the free-carrier loss increases according to λ^2 and therefore at these longer wavelengths, even the low-doped spacer layers would have extremely unfavourable values of loss, making laser emission difficult. Therefore, alternative waveguide designs based upon surface-plasmons [338,339] have been developed for use in the far-infrared. A plasmon can be thought of as a quantization of the plasma oscillations in a free electron gas. A surface-plasmon is an electromagnetic wave that is confined to the interface between materials of opposite signs of dielectric constant, i.e. a semiconductor and metal and exists if $|\epsilon_m| > \epsilon_s$, $\epsilon_m < 0$ and $\epsilon_s > 0$. The loss suffered by a surface-plasmon as it propagates along the interface is given by the following [338]:

$$\alpha_{\mathrm{sp}} = \frac{4\pi}{\lambda} \frac{n_{\mathrm{m}} n_{\mathrm{s}}^3}{|k_{\mathrm{m}}^3|} \tag{13.72}$$

where n_{m} and n_{s} are the real parts of the refractive index of the metal and semiconductor, respectively and k_{m} is the extinction coefficient of the metal. The $1/\lambda$ dependence of the loss makes waveguides based upon surface-plasmons extremely favourable for long wavelength devices. This is especially true, since n_{m} and k_{m} themselves depend on λ and so the loss in fact drops faster than $1/\lambda$. The world's first terahertz QCL [335] utilized a so-called semi-insulating surface-plasmon waveguide. In this waveguide, a surface-plasmon exists on top of the $\sim11\,\mu$m-thick laser active region due to a 200 nm thick GaAs layer doped to $5 \times 10^{18}\,\mathrm{cm}^{-3}$ followed by a metallic (gold) layer. A highly doped layer ($2 \times 10^{18}\,\mathrm{cm}^{-3}$, 800 nm) is placed under the active region on top of a semi-insulating GaAs substrate. A semi-insulating substrate is used in order to minimise the free-carrier losses at these long wavelengths. At a wavelength of $68\,\mu$m, the highly doped layer has a complex permittivity of $\epsilon = -94.86 + i41.35$ meaning a surface-plasmon mode is supported at the interface between the highly doped layer and the active region. The overall mode profile of the whole waveguide (Fig. 13.18) can be thought of as a hybridization of this and the metal-semiconductor plasmon mode.

The propagation constant of the TM_0 mode is calculated to be $\gamma = 3.46740 \times 10^5 + i1.27905 \times 10^3\,\mathrm{m}^{-1}$, corresponding to an effective mode index of $N = 3.76003$ and a waveguide loss of $\alpha_{\mathrm{wg}} = 25.581\,\mathrm{cm}^{-1}$. For a 3 mm long cavity, the mirror loss is calculated as $3.63\,\mathrm{cm}^{-1}$. The threshold gain and effective confinement factor can be calculated using the same approach as for the double plasmon-enhanced waveguide. A plot of the waveguide loss/gain as a function of material gain is given in Fig. 13.19.

From the intercept of the mirror loss and gain curves, a threshold gain of $g_{\mathrm{th}} = -56.34\,\mathrm{cm}^{-1}$ is estimated for this waveguide. From the slope of the gain curve an

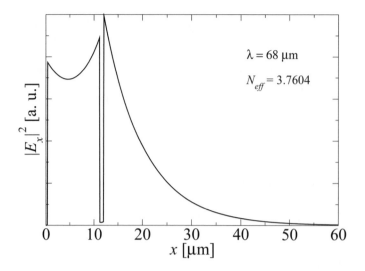

Figure 13.18   Mode intensity of the TM_0 mode in a semi-insulating surface-plasmon QCL waveguide.

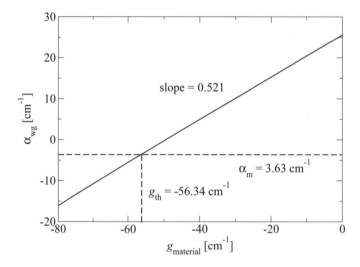

Figure 13.19   Waveguide loss/gain as a function of material gain. The mirror loss α_m is indicated by the dashed horizontal line.

effective confinement factor of $\Gamma = 52.1\%$ is calculated, in good agreement with the value of 49.8% given by equation (13.70).

Acknowledgments

The author is extremely grateful to Zoran Ikonić for useful discussions.

CHAPTER 14

MULTIBAND ENVELOPE FUNCTION (K.P) METHOD

contributed by Z. Ikonić

14.1 SYMMETRY, BASIS STATES AND BAND STRUCTURE

The band structures of common semiconductors such as GaAs, InP, Si, Ge, etc., which have the diamond/zinc blende crystal structure, are more complicated in the valence band than they are in the conduction band. This is related to the fact that, looking microscopically, at the level of the crystalline unit cell, the conduction band states are mostly s-like in character (i.e. the microscopic 'Bloch' wave function has the symmetry of an atomic s orbital), while the valence band states are mostly p-like in character. There are three degenerate p-type atomic orbitals, the symmetry of which is denoted as x, y, and z, which stems from the direction along which the orbitals are aligned. It is therefore natural that all three of them will take part in valence band state wave functions.

Detailed microscopic calculations, e.g. using the empirical pseudopotential method, see Chapter 15, show that there are two valence bands degenerate at the centre of the Brillouin zone (the Γ-point), and close by (a few tens to a few hundreds of meV below) there is a third band. The first two are called heavy-hole (HH) and light-hole (LH) bands, and they cease to be degenerate for finite values of the wave vector $\mathbf{k} = k_x, k_y, k_z$: the energy of the former descends at a slower rate as the wave vector moves away from the

---

Quantum Wells, Wires and Dots, Third Edition. P. Harrison
©2009 John Wiley & Sons, Ltd.

Γ-point, which corresponds to a larger effective mass, hence the name. The third band is called the spin–orbit split-off (SO) band. An example of the valence band dispersion is given in Fig. 14.1.

Any one of the bands is an energy eigenstate of the bulk material, and 'pure' (single-band) states with definite energy may therefore exist in bulk. However, in the case of position-dependent potentials this will no longer be true: states in quantum wells, for instance, will be 'mixtures', with *all* the bulk bands contributing to their wave functions. The contribution of a particular bulk band to a quantized state generally depends on its energy spacing from that band: the smaller the energy, the larger the relative contribution will be (this follows from quantum mechanical perturbation theory). When two or more bands are degenerate, or almost degenerate, states in their vicinity are likely to have similar contributions from these bands.

In a quantum well type of structure, for energies which are not far from the conduction band edges of the constituent materials, a quantized state wave function will have contributions mostly from the conduction bands of these materials, each of which has s-like character, and there will be a single envelope wave function (solution of the effective mass Schrödinger equation), which represents the amplitude of these, s-like Bloch functions. On the other hand, a quantized state near the valence band edges of any of the constituent materials is expected to comprise two or three of the bulk valence bands with comparable contributions, each having its own envelope function. Depending on the accuracy required in the calculation of the quantized states, and on the positions of the bands in the bulk materials, different number of bands may be included in the calculation. It may sometimes suffice to explicitly include just HH and LH bands, or, on other occasions, also the SO band in the description of the system.

Generally, the Hamiltonians that describe states in such situations are matrices, or systems of coupled Schrödinger equations, which will deliver the possible energies and wave functions expressed as a set of envelope functions (which vary slowly over a crystalline unit cell), themselves representing the amplitudes of the corresponding basis states (usually the bulk bands). For this reason, the method is known as the 'multiband envelope function', or 'multiband effective mass method' and, because the interaction of bulk bands is described via the **k.p** perturbation, it is also known as the **k.p** method.

Clearly, the concept of bulk band mixing in forming quantized states of a system applies to more remote bands as well. The conduction band quantized states will thus include contributions from bulk valence band states, and vice versa, and there exist extended versions of the **k.p** method that explicitly include the HH, LH, SO, and the conduction band, or still wider variants including even more remote bands. However, in this work attention will be focused on 4- and 6-band Hamiltonians, which explicitly include the valence band states. The number of bands the Hamiltonian is named after is the number of bands that are explicitly included, i.e. their envelope wave functions are explicitly evaluated. However, such Hamiltonians do *implicitly* account for the existence of other, more remote bands, and their influence is incorporated via the values of the material parameters. In this context, the conventional (conduction band) effective mass Schrödinger equation is just a special case of the multiband envelope function model, where the existence of bands other than the conduction band is accounted for by using the effective, rather than the free electron mass, and only the conduction band envelope wave function is calculated explicitly. The 4- and 6-band Hamiltonians for the valence envelope wave functions were derived by Luttinger and Kohn [340] using **k.p** perturbation theory, while the 8-band model (that includes the conduction band) was developed by Pidgeon and Brown [341].

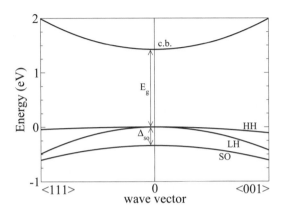

Figure 14.1   The valence and conduction bands of a group IV or III–V semiconductor near the Γ-point of the Brillouin zone.

14.2 VALENCE BAND STRUCTURE AND THE 6 × 6 HAMILTONIAN

It is beyond the scope of this book to derive the Hamiltonian that describes the valence band states, so its form will be stated and it will be employed in the calculation of quantized states within the valence band of nanostructures in order to illustrate how it is used. In many cases good accuracy can be obtained by using the so-called 6×6 Hamiltonian, although its shortened version, the 4×4 Hamiltonian is frequently just as good—it all depends on the energy range of interest. As mentioned above, a valid basis set can be the atomic p-like orbitals with x-, y-, and z-like spatial symmetry, here denoted as $|X\rangle$, $|Y\rangle$, $|Z\rangle$, each also having the spin projection along the z-axis equal to either $+1/2$ or $-1/2$, denoted as ↑ or ↓, respectively. However, common practice is to change from the $|X\uparrow\rangle$, $|Y\uparrow\rangle$, $|Z\uparrow\rangle$, $|X\downarrow\rangle$, $|Y\downarrow\rangle$, $|Z\downarrow\rangle$ basis into another one, such that its member functions are simultaneously the eigenstates of the angular momentum operator (with eigenvalues J equal to $3\hbar/2$ or $\hbar/2$), and of its projection along the z-axis m_J (with eigenvalues equal to $\pm 3/2$ or $\pm 1/2$). This is achieved by making appropriate linear combinations of the atomic basis states. The list of this new set of $|J, m_J\rangle$ basis states is given below in equation (14.1).

The precise form of the Hamiltonian depends on the pre-factors in equation (14.1), for example, the presence of the imaginary number i or -1 does not change the state properties, and it also depends on how these states are ordered in the list. There is no unique choice that is universally accepted in the literature; however, one of the frequently used possibilities, see, for example [342], which will be adopted here, reads:

$$
\begin{aligned}
|3/2, 3/2\rangle &= (1/\sqrt{2})\,|(X + iY)\uparrow\rangle \\
|3/2, -3/2\rangle &= (1/\sqrt{2})\,|(X - iY)\downarrow\rangle \\
|3/2, 1/2\rangle &= (1/\sqrt{6})\,|(X + iY)\downarrow\rangle - (\sqrt{2}/\sqrt{3})|Z\uparrow\rangle \\
|3/2, -1/2\rangle &= -(1/\sqrt{6})\,|(X - iY)\uparrow\rangle - (\sqrt{2}/\sqrt{3})\,|Z\downarrow\rangle \\
|1/2, 1/2\rangle &= (1/\sqrt{3})\,|(X + iY)\downarrow\rangle + (1/\sqrt{3})|Z\uparrow\rangle \\
|1/2, -1/2\rangle &= -(1/\sqrt{3})\,|(X - iY)\uparrow\rangle + (1/\sqrt{3})\,|Z\downarrow\rangle
\end{aligned}
\tag{14.1}
$$

Just as with the original ($|X\uparrow\rangle$, $|Y\uparrow\rangle$, $|Z\uparrow\rangle$, $|X\downarrow\rangle$, $|Y\downarrow\rangle$, $|Z\downarrow\rangle$) states, the new basis states are all orthogonal to each other. In this $|J, m_J\rangle$ basis the 6×6 Hamiltonian that describes the HH, LH and SO bands for the bulk reads:

$$
\mathbf{H} =
\begin{bmatrix}
P+Q & 0 & -S & R & (1/\sqrt{2})S & \sqrt{2}R \\
0 & P+Q & -R^\dagger & -S^\dagger & -\sqrt{2}R^\dagger & (1/\sqrt{2})S^\dagger \\
-S^\dagger & -R & P-Q & 0 & \sqrt{2}Q & \sqrt{3/2}S \\
R^\dagger & -S & 0 & P-Q & -\sqrt{3/2}S^\dagger & \sqrt{2}Q \\
(1/\sqrt{2})S^\dagger & -\sqrt{2}R & \sqrt{2}Q & -\sqrt{3/2}S & P+\Delta_{SO} & 0 \\
\sqrt{2}R^\dagger & (1/\sqrt{2})S & \sqrt{3/2}S^\dagger & \sqrt{2}Q & 0 & P+\Delta_{SO}
\end{bmatrix}
\tag{14.2}
$$

where

$$
\begin{aligned}
P &= \left(\frac{\hbar^2}{2m_0}\right)\gamma_1(k_x^2 + k_y^2 + k_z^2) \\
Q &= \left(\frac{\hbar^2}{2m_0}\right)\gamma_2(k_x^2 + k_y^2 - 2k_z^2) \\
R &= \left(\frac{\hbar^2}{2m_0}\right)\sqrt{3}\left[-\gamma_2(k_x^2 - k_y^2) + 2i\gamma_3 k_x k_y\right] \\
S &= \left(\frac{\hbar^2}{2m_0}\right)2\sqrt{3}\gamma_3 k_z k_-
\end{aligned}
\tag{14.3}
$$

where $k_- = k_x - ik_y$, $\gamma_{1,2,3}$ are the Luttinger parameters, and Δ_{SO} is the spin–orbit splitting—the spacing between the HH (or LH) band and the SO band at the ($k = 0$) Γ-point. The values of $\gamma_{1,2,3}$ and Δ_{SO} in some common semiconductors are given in Table 14.1.

In writing equation (14.2) the convention has been used that the hole energy is measured from the top of the valence band downwards (the inverted energy picture). This is because it is usually easier to look at the hole band structure in the same manner as that for electrons, and this is possible if only holes are considered. For the true energy picture (as in Fig. 14.1), all the terms in the Hamiltonian should be multiplied by -1. Furthermore, it is important to note that the coordinate system in which this Hamiltonian is written is not orientated arbitrarily—the axes x, y, and z are aligned along the edges of the crystalline cubic unit cell.

The Schrödinger equation corresponding to this Hamiltonian may still be written as

$$
\mathcal{H}\psi = E\psi
\tag{14.4}
$$

but $\psi = \psi(\mathbf{r})$ is a *vector* (a six-component array). Note that the Hamiltonian is a Hermitian matrix with simple scalars as its elements (which depend on the material parameters) and wave vector components k_x, k_y, and k_z. If the eigenenergies and the corresponding eigenvectors are found, the latter will obviously be lists of (possibly complex-valued) scalar constants. What would these mean? It is implicitly assumed that the wave function has a plane wave form, i.e. all the six components of ψ have the common, plane wave type of spatial behaviour, i.e. $\sim \exp(i\mathbf{k}.\mathbf{r})$. This makes them the components of the envelope wave function and, if interested in a more detailed form, each of them multiplies a corresponding basis state from the list in equation (14.1) and can be added together to construct the 'true' microscopic wave function of a state with energy E. The eigenvectors are generally 'full', i.e. their entries usually have non-zero values, which means that a plane-wave state has all the six basis states admixed. This is in contrast to the conduction band, where one usually deals with 'pure' spin-up or spin-down states.

Table 14.1 Parameters relevant for the valence band structure in a few semiconductors [343, 344].

	Si	Ge	GaAs	AlAs	InAs
γ_1	4.22	13.4	6.98	3.76	20.0
γ_2	0.39	4.25	2.06	0.82	8.5
γ_3	1.44	5.69	2.93	1.42	9.2
Δ_{SO} (eV)	0.044	0.290	0.341	0.28	0.39
C_{11} (Mbar)	1.675	1.315	1.221	1.250	0.833
C_{12} (Mbar)	0.650	0.494	0.566	0.534	0.453
a_{latt} (Å)	5.431	5.657	5.653	5.661	6.058
a_v (eV)	2.46	1.24	−1.116	−2.47	−1.00
b (eV)	−2.10	−2.86	−2.0	−2.3	−1.8
VBO (eV)			−0.80	−1.33	−0.59

To have the usual meaning of a wave function, an eigenvector ψ has to be normalized to unity, i.e. all the components of a vector may have to be multiplied by a suitable constant so that:

$$\psi^\dagger \psi = 1 \tag{14.5}$$

is satisfied. The '$\dagger$' symbol means the Hermitian conjugate, i.e. the transpose (a column-vector becomes a row-vector) followed by the complex conjugate (take the complex conjugate of all the elements).

In order to find eigenenergies for a specified $\mathbf{k}$, solutions can be sought to $\det|H - E| = 0$, which delivers a sixth-order polynomial in the energy E. This could only be solved numerically, unless it is noted that it can be factored into two identical third-order polynomials thus allowing for analytic, though lengthy, solutions. However, the situation is quite simple at the zone centre, i.e. if we set $k_x = k_y = k_z = 0$ the Hamiltonian becomes:

$$\mathcal{H}(\mathbf{k}=0) = \begin{bmatrix} 0 & 0 & 0 & 0 & 0 & 0 \\ 0 & 0 & 0 & 0 & 0 & 0 \\ 0 & 0 & 0 & 0 & 0 & 0 \\ 0 & 0 & 0 & 0 & 0 & 0 \\ 0 & 0 & 0 & 0 & \Delta_{SO} & 0 \\ 0 & 0 & 0 & 0 & 0 & \Delta_{SO} \end{bmatrix} \qquad (14.6)$$

This has a four-fold degenerate eigenvalue $E = 0$, and a two-fold degenerate eigenvalue $E = \Delta_{SO}$, consistent with the concept that there are HH and LH branches degenerate at $\mathbf{k} = 0$, and an SO branch displaced by Δ_{SO} from them. As for the eigenvectors: for each of the four degenerate values $E = 0$, just one of the four amplitudes of $|3/2, 3/2\rangle$, $|3/2, -3/2\rangle$, $|3/2, 1/2\rangle$, $|3/2, -1/2\rangle$ has unity value, the others being equal to zero (though, due to degeneracy, any other linear combination of these states would do just as well). Similarly, the eigenvectors corresponding to the two degenerate eigenvalues $E = \Delta_{SO}$ may be taken to have non-zero amplitude of the $|1/2, 1/2\rangle$, and of the $|1/2, -1/2\rangle$ states. For finite values of $\mathbf{k}$, the four-fold degeneracy splits into two pairs of doubly degenerate states (HH and LH have different energies), so all three branches remain doubly degenerate—in analogy to the spin-degeneracy of conduction band states.

14.3 4 × 4 VALENCE BAND HAMILTONIAN

For non-zero $\mathbf{k}$, the expressions for the eigenenergies would be quite lengthy, so consider in more detail a 'shortened' version, i.e. a 4×4 Hamiltonian that includes only the HH and LH states in its basis. This is obtained by excluding the $|1/2, \pm 1/2\rangle$ states, i.e. by removing the fifth and sixth rows and columns from the Hamiltonian matrix in equation (14.2):

$$\mathcal{H} = \begin{bmatrix} P+Q & 0 & -S & R \\ 0 & P+Q & -R^\dagger & -S^\dagger \\ -S^\dagger & -R & P-Q & 0 \\ R^\dagger & -S & 0 & P-Q \end{bmatrix} \qquad (14.7)$$

Finding $\det|\mathbf{H} - E|$ gives a fourth-order polynomial in E, but it is easy to see that it is a square of a quadratic polynomial:

$$(P+Q-E)(P-Q-E) - |R|^2 - |S|^2 = 0 \qquad (14.8)$$

Using the expressions in equation (14.3), it can be found that the two solutions read:

$$E_\pm(k_x, k_y, k_z) = \gamma_1 k^2 \pm 2\sqrt{\gamma_2^2 k^4 - 3(\gamma_2^2 - \gamma_3^2)(k_x^2 k_y^2 + k_x^2 k_z^2 + k_y^2 k_z^2)} \qquad (14.9)$$

where $k^2 = k_x^2 + k_y^2 + k_z^2$.

It is immediately clear from equation (14.9) that both solutions, also called dispersion branches, are non-parabolic (E does not depend only on k^2) and anisotropic (E depends on $k_{x,y,z}$ in a manner different from just that of its modulus k, i.e. the energy varies differently in different directions of the wave vector space). This latter property is also called 'band warping'.

Consider now the special case when two components of the wave vector are zero, e.g. $k_x = k_y = 0$, while $k_z \neq 0$. From the definitions in equation (14.3), $R = S = 0$, hence:

$$E_+ = k_z^2 \hbar^2 (\gamma_1 + 2\gamma_2)/2m_0 \quad \text{and} \quad E_- = k_z^2 \hbar^2 (\gamma_1 - 2\gamma_2)/2m_0 \qquad (14.10)$$

Clearly, the E_- branch is characterized by an effective mass in the z-direction equal to $m_0/(\gamma_1 - 2\gamma_2)$, and this is larger than the effective mass of the E_+ branch, which amounts to $m_0/(\gamma_1 + 2\gamma_2)$. For this reason, the E_- branch is called the 'heavy hole' (HH) branch, and its energy is denoted as E_{HH}, while the E_+ branch is referred to as the 'light hole' (LH) branch, and its energy denoted as E_{LH}. Under these conditions, where in fact $E_{HH} = P + Q$, the first two equations in $(\mathcal{H} - E_{HH})\psi = 0$ have all zero entries, while the third and fourth equations do not. From this it can be concluded that elements 3 and 4 in the vector ψ are zero, while elements 1 and 2 are not. Therefore, $[1\ 0\ 0\ 0]^T$ can be chosen for one HH state vector, and $[0\ 1\ 0\ 0]^T$ for the other (remember that E_{HH} is doubly degenerate), though any pair of linear combinations of these two vectors would do just as well. However, if choosing to make such linear combinations, care should be taken to make them mutually orthogonal in order to avoid complications in any further use of the wave functions. In any case, for $k_z \neq 0$, the HH branch has only $|3/2, \pm 3/2\rangle$ basis states as its constituents. Similar reasoning for the E_{LH} branch shows that, for $k_z \neq 0$, the LH branch includes only $|3/2, \pm 1/2\rangle$ states.

For finite k_x and/or k_y, however, the system of linear equations is generally full, which indicates that a hole state has a finite contribution from all the basis states. This may seem puzzling, in view of the fact that there is essentially no difference between the x and z directions in the bulk crystal. However, this difference arises because of the choice of basis states: each of them has a definite projection of momentum and spin along the z-axis, and therefore (as is known from quantum mechanics) does not have the same property along the x-axis.

14.4 COMPLEX BAND STRUCTURE

While equation (14.8) and its solutions in equation (14.9) give the possible values of hole energy for a particular wave vector $\mathbf{k}$, it is interesting to consider the reverse problem: what values of k_z may holes have if their energy E and the other two components of the wave vector (k_x and k_y) are specified? For the sake of simplicity, k_y is set to zero, while k_x and E are generally non-zero and are real. Equation (14.8) in its expanded form then reads:

$$(\gamma_1^2 - \gamma_2^2)k_z^4 + [(2\gamma_1^2 + 8\gamma_2^2 - 12\gamma_3^2)k_x^2 - 2E\gamma_1]\,k_z^2$$

$$+[E - (\gamma_1 + 2\gamma_2)k_x^2][E - (\gamma_1 - 2\gamma_2)k_x^2] \equiv Ak_z^4 + Bk_z^2 + C = 0 \qquad (14.11)$$

which is, of course, quadratic in k_z^2, with the two solutions:

$$(k_z^2)_{1,2} = \frac{-B \pm \sqrt{B^2 - 4AC}}{2A} \qquad (14.12)$$

each of which delivers two k_zs as a positive and a negative root of k_z^2. In real semiconductors the values of the Luttinger parameters γ_1 and γ_2 are such that A is always positive, while B and C may be of either sign, depending on E and k_x, as well as the material parameters.

Consider the case of $4AC < 0$, i.e. $C < 0$. From the form used to write C, it is clear that this will happen when k_x and E are such that $(\gamma_1 - \gamma_2)k_x^2 < E < (\gamma_1 + 2\gamma_2)k_x^2$ is satisfied. The $\sqrt{B^2 - 4AC}$ is then larger than B, and regardless of the sign of B one of the roots $(k_z^2)_{1,2}$ will be positive and the other one negative. Hence there will be a pair of real and a pair of purely imaginary k_z values.

Consider now the case of $4AC > B^2$: equation (14.12) then contains the square root of a negative number, which will imply that both roots will be fully complex numbers (with both the real and imaginary parts non-zero). Finally, for $0 < 4AC < B^2$, both values of $(k_z^2)_{1,2}$ are either positive or negative, depending on the sign of B, i.e. all roots are either real or imaginary.

Real-valued solutions for k_z imply conventional plane-wave envelope wave functions, which is an allowed state in an infinite bulk. Complex k_z implies an 'evanescent' wave, which decays in one direction and increases in the opposite direction, and which may simultaneously oscillate (if the real part of $k_z \neq 0$). Because of the infinite length in the z-direction, and the inability to normalize the wave function, such states are not allowed in an infinite bulk crystal, but are perfectly allowed in finite regions, just as is the case with electrons, and can therefore appear in structures of finite extent.

The above considerations were presented for the 4×4 model; however, the general conclusions about the possible types of evanescent waves remain for more elaborate models as well. The list of all possible states at a particular energy E, that behave exponentially is called the 'complex band structure'. The wave functions vary along a particular direction, say z, as $\exp(ik_z z)$, where k_z is a real or complex wave vector.

The case of electrons in the conduction band is quite different as they have a scalar effective mass and really a very simple complex band structure: their wave vector can be only real or imaginary: $k_z = \pm\sqrt{2m^*E/\hbar^2 - k_x^2 - k_y^2}$, but never fully complex.

For holes, this situation only occurs at $k_x = k_y = 0$, when $(k_z^2)_1 = E/(\gamma_1 - 2\gamma_2)$ and $(k_z^2)_2 = E/(\gamma_1 + 2\gamma_2)$, so all k_zs are real for $E > 0$ and imaginary for $E < 0$.

14.5 BLOCK-DIAGONALIZATION OF THE HAMILTONIAN

There is an elegant method, which allows one to simplify the 6×6 Hamiltonian, by recasting it into a block-diagonal form: it then has two 3×3 matrices as its diagonal elements and the remaining two off-diagonal elements are zero-matrices. This is achieved by a suitable change of basis, i.e. by creating a new basis from linear combinations of the existing basis states. Essentially, this means that there are two completely independent sets of states, which are eigenstates of either one or the other 3×3 block. Since this block-diagonalization reduces the size of the system to be considered at any time, it is useful when attempting to obtain a result in analytical form. Block-diagonalization can also be useful in cases that are more complicated than straightforward bulk material, for example, strained bulk or two-dimensional quantum-well systems.

The derivation of the transformation of the Hamiltonian into a block-diagonal form will not be reproduced here, the result will merely be stated, i.e. upon introducing the new basis:

$$|F_1\rangle = \alpha\,|3/2,-3/2\rangle - \alpha^*\,|3/2,3/2\rangle$$
$$|F_2\rangle = \beta\,|3/2,1/2\rangle + \beta^*\,|3/2,-1/2\rangle$$
$$|F_3\rangle = \beta\,|1/2,1/2\rangle + \beta^*\,|1/2,-1/2\rangle$$
$$|F_4\rangle = \alpha\,|3/2,-3/2\rangle + \alpha^*\,|3/2,3/2\rangle$$
$$|F_5\rangle = \beta\,|3/2,1/2\rangle - \beta^*\,|3/2,-1/2\rangle$$
$$|F_6\rangle = \beta\,|1/2,1/2\rangle - \beta^*\,|1/2,-1/2\rangle \tag{14.13}$$

where:

$$\alpha = \frac{1}{\sqrt{2}}\exp[i(\phi/2+\eta+\pi/4)], \quad \beta = \frac{1}{\sqrt{2}}\exp[i(\phi/2-\eta-3\pi/4)]$$

$$\phi = \arctan(k_y/k_x), \quad \eta = \frac{1}{2}\arctan[(\gamma_3/\gamma_2)\tan(2\phi)] \tag{14.14}$$

the Hamiltonian takes a block-diagonal form:

$$\mathbf{H} = \begin{bmatrix} \mathbf{H}_+ & 0 \\ 0 & \mathbf{H}_- \end{bmatrix} \tag{14.15}$$

The two 3×3 blocks read:

$$\mathbf{H}_\pm = \begin{bmatrix} P+Q & R\mp iS & \sqrt{2}R\pm iS/\sqrt{2} \\ R\pm iS^\dagger & P-Q & \sqrt{2}Q\mp i\sqrt{3/2}S \\ \sqrt{2}R\mp iS^\dagger/\sqrt{2} & \sqrt{2}Q\pm i\sqrt{3/2}S^\dagger & P+\Delta_{\mathrm{SO}} \end{bmatrix} \tag{14.16}$$

where P and Q are the same as above, while R and S now read:

$$R = -\sqrt{3}\left(\frac{\hbar^2}{2m_0}\right)\gamma_\phi k_\|^2, \quad S = 2\sqrt{3}\left(\frac{\hbar^2}{2m_0}\right)\gamma_3 k_\| k_z$$

$$\gamma_\phi = \sqrt{\bar{\gamma}^2+\mu^2-2\bar{\gamma}\mu\cos\phi}, \quad \bar{\gamma} = \frac{1}{2}(\gamma_2+\gamma_3), \quad \mu = \frac{1}{2}(\gamma_3-\gamma_2) \tag{14.17}$$

where $k_\|^2 = k_x^2 + k_y^2$ is the in-plane wave vector.

Without going into detail, it should be noted that, in order to make this transformation possible, the element R from equation (14.2) had to be approximated as given in equation (14.17). A simpler approximation, with $\gamma_\phi = (\gamma_2+\gamma_3)/2$ has previously been in use, but the one given above is much better. The important point to note is that the transformation in equation (14.13) does not depend on k_z, and when this becomes an operator $k_z \to -i\partial/\partial z$ (for application to quantum wells) there are no difficulties in changing from one basis to another.

To find the band energies and wave functions, the block-diagonal form of the Hamiltonian may then be used. Solving one 3×3 block at a time, the wave function is now a vector of length 3, i.e.

$$\mathbf{H}_\pm\psi_\pm = E\psi_\pm$$

$$\psi_+(z) = \begin{bmatrix} F_1 \\ F_2 \\ F_3 \end{bmatrix}, \quad \psi_-(z) = \begin{bmatrix} F_4 \\ F_5 \\ F_6 \end{bmatrix} \tag{14.18}$$

Block-diagonalization may also be applied to the 4×4 Hamiltonian of equation (14.7), using the basis states F_1, F_2, F_4, and F_5 defined in equation (14.13), which transforms the full 4×4 matrix into one which has two 2×2 blocks. Alternatively, one can simply cut out the third row and third column from the two blocks defined in equation (14.16) to obtain the same result. Taking, for example, the upper 2×2 block:

$$\begin{bmatrix} P + Q - E & R - iS \\ R + iS & P - Q - E \end{bmatrix} \cdot \begin{bmatrix} F_1 \\ F_2 \end{bmatrix} = 0 \tag{14.19}$$

and evaluating $\det |H_+ - E| = 0$ gives:

$$E^2 - 2PE + P^2 - Q^2 - |R - iS|^2 = 0 \tag{14.20}$$

which has two non-degenerate solutions:

$$E_\pm = P \pm \sqrt{Q^2 + |R - iS|^2} \tag{14.21}$$

in which it can immediately be recognized that $E_- = E_{HH}$ and $E_+ = E_{LH}$ from the previous section. The same values would be obtained from the lower 2×2 block.

Putting these solutions (one at a time!) back into either of the two equations of the homogeneous linear system in equation (14.19), as convenient, and solving it, gives the eigenvectors for the two eigenenergies as:

$$\begin{bmatrix} F_1 \\ F_2 \end{bmatrix}_{E_-} = \begin{bmatrix} P - Q - E_- \\ -(R + iS) \end{bmatrix} \qquad \begin{bmatrix} F_1 \\ F_2 \end{bmatrix}_{E_+} = \begin{bmatrix} (R - iS) \\ E_+ - P - Q \end{bmatrix} \tag{14.22}$$

These eigenvectors are not yet normalized to unity, which can be done simply by dividing each by the square root of the sum of its component moduli squared. In the special case when $k_x = k_y = 0$ and $k_z \neq 0$, the HH state vector is simply $\begin{bmatrix} 1 & 0 \end{bmatrix}^T$ (i.e. only a particular linear combination of $|3/2, \pm 3/2\rangle$ states is present), while the LH state vector is $\begin{bmatrix} 0 & 1 \end{bmatrix}^T$ (only $|3/2, \pm 1/2\rangle$ states are present).

14.6   THE VALENCE BAND IN STRAINED CUBIC SEMICONDUCTORS

In this section the influence of strain on the valence band structure will be considered. Strain may be created in several different ways. One way is to subject a semiconductor to hydrostatic pressure, in which case the cubic unit cell would become compressed equally in all three directions. Another is to apply a uniaxial pressure along some axis (e.g. z), while leaving the surfaces perpendicular to the other two axes free, which would result in compressive strain along z and tensile strain along x and y. Yet another way of straining a semiconductor is to grow an epitaxial lattice-mismatched layer of a material on top of a substrate (i.e grow a layer of a material with a different lattice constant from that of the substrate). The thin (epitaxial) layer is forced to acquire the in-plane lattice constant of the substrate, and the perpendicular lattice constant will then also change.

Strain is described by six components: ϵ_{xx}, ϵ_{yy}, ϵ_{zz}, ϵ_{xy}, ϵ_{xz}, and ϵ_{yz}. It modifies various terms in the Hamiltonian (see equation (14.2)), which then acquire strain-dependent contributions, see, for example [345]:

$$P \to P + P_\epsilon \quad , \qquad P_\epsilon = -a_v(\epsilon_{xx} + \epsilon_{yy} + \epsilon_{zz})$$

$$Q \to Q + Q_\epsilon \quad , \qquad Q_\epsilon = -\frac{b}{2}(\epsilon_{xx} + \epsilon_{yy} - 2\epsilon_{zz})$$

$$R \to R + R_\epsilon \quad , \qquad R_\epsilon = \frac{\sqrt{3}}{2}b(\epsilon_{xx} - \epsilon_{yy}) - id\epsilon_{xy}$$

$$S \to S + S_\epsilon \quad , \qquad S_\epsilon = -d(\epsilon_{zx} - i\epsilon_{yz}) \tag{14.23}$$

where a_v and b, and d are the Pikus–Bir deformation potentials, describing the influence of hydrostatic, uniaxial and shear strain, respectively.

In the (practically most important) case of biaxial strain generated by lattice-mismatched growth of a semiconductor layer on a [001] oriented substrate, the components of strain become:

$$\epsilon_{xx} = \epsilon_{yy} \neq \epsilon_{zz},$$

$$\epsilon_{xy} = \epsilon_{yz} = \epsilon_{zx} = 0 \tag{14.24}$$

where the non-zero strain components can be evaluated from:

$$\epsilon_{xx} = \epsilon_{yy} = \frac{a_0 - a_{\text{latt}}}{a_{\text{latt}}}$$

$$\epsilon_{zz} = -\frac{2C_{12}}{C_{11}}\epsilon_{xx} \tag{14.25}$$

and where C_{11} and C_{12} are the stiffness constants, a_0 is the lattice constant of the substrate (which is taken to be unstrained, because it is much thicker than the layer grown on top of it), and a_{latt} is the lattice constant of the unstrained epitaxial material.

Looking again at the special case of $\mathbf{k} = 0$, the Hamiltonian takes the form:

$$\mathbf{H} = \begin{bmatrix} P_\epsilon + Q_\epsilon & 0 & 0 & 0 & 0 & 0 \\ 0 & P_\epsilon + Q_\epsilon & 0 & 0 & 0 & 0 \\ 0 & 0 & P_\epsilon - Q_\epsilon & 0 & \sqrt{2}Q_\epsilon & 0 \\ 0 & 0 & 0 & P_\epsilon - Q_\epsilon & 0 & \sqrt{2}Q_\epsilon \\ 0 & 0 & \sqrt{2}Q_\epsilon & 0 & P_\epsilon + \Delta_{\text{SO}} & 0 \\ 0 & 0 & 0 & \sqrt{2}Q_\epsilon & 0 & P_\epsilon + \Delta_{\text{SO}} \end{bmatrix} \tag{14.26}$$

This is a bit more complicated than the unstrained case. For example, the presence of non-zero off-diagonal matrix elements on positions that link LH and SO states implies that these two bands are coupled even at $\mathbf{k} = 0$; however, the HH band remains independent. Away from the zone centre, however, all the three bands are coupled. If the size of the system is reduced to a 4×4 Hamiltonian, by removing the $|1/2, \pm 1/2\rangle$ states (the fifth and sixth rows and columns), a simple diagonal matrix is obtained and its eigen energies can readily be written, each of which is two fold degenerate:

$$E_{\text{HH}}(0) = P_\epsilon + Q_\epsilon$$

$$E_{\text{LH}}(0) = P_\epsilon - Q_\epsilon \tag{14.27}$$

Consider the case when a_{latt} is larger than a_0, so that the thin epitaxial layer experiences a compressive in-plane strain (i.e. $\epsilon_{xx} < 0$). Therefore, according to equation (14.25), the result of this accommodation of the in-plane lattice constant to the value of the substrate is that ϵ_{zz} must be greater than zero. The result is a finite value of P_ϵ which, as it only occurs along the diagonal terms in equations (14.2) or (14.15), has the trivial effect in bulk of a rigid energy shift of all the bands by that amount, with no other physical consequences.

The term Q_ϵ, however, has non-trivial effects. The value of b in semiconductors is negative, therefore $Q_\epsilon < 0$, and the HH band edge will therefore decrease, while the LH band edge will increase (with the inverted energy picture), hence the HH and LH bands cease to be degenerate, with the light-holes at the zone centre having an energy greater than that of the heavy-holes. In the opposite case with a_{latt} smaller than a_0, the thin layer acquires a tensile in-plane strain, and the above effects are reversed.

Returning to the 6×6 Hamiltonian, the eigenvalue equation is clearly a sixth-order polynomial in the energy E. It can be written as a square of a third-order polynomial, which can be solved analytically, but this is not an easy thing to do. Again, looking at the special case of $\mathbf{k} = 0$ summarized in equation (14.26), notice that the diagonal elements $P_\epsilon + Q_\epsilon$ stand alone in both their rows and columns, which means that this is a doubly degenerate eigenvalue and can be factored out. The remaining problem fourth-order polynomial is then easier to solve and, in fact, it can be seen that it is the square of a quadratic polynomial. The eigen energies, which are the edges of the three valence bands in a strained semiconductor, follow as:

$$
\begin{aligned}
E_{HH}(0) &= P_\epsilon + Q_\epsilon \\
E_{LH}(0) &= P_\epsilon - \frac{1}{2}\left(Q_\epsilon - \Delta_{SO} + \sqrt{\Delta_{SO}^2 + 2\Delta_{SO}Q_\epsilon + 9Q_\epsilon^2}\right) \\
E_{SO}(0) &= P_\epsilon - \frac{1}{2}\left(Q_\epsilon - \Delta_{SO} - \sqrt{\Delta_{SO}^2 + 2\Delta_{SO}Q_\epsilon + 9Q_\epsilon^2}\right)
\end{aligned}
\tag{14.28}
$$

It is important to note, for the purpose of a later discussion on the band offset at an interface between different materials, that the average ('centre of mass') of these three valence band edges shifts only because of the hydrostatic component P_ϵ, from its unstrained value of $\Delta_{SO}/3$. In the case of a large spin–orbit splitting ($\Delta_{SO} >> Q_\epsilon$), by taking the first two terms in Taylor series expansions of the roots in equation (14.28), the results in equation (14.27) are recovered, while the SO band in this limit appears relatively insensitive to strain.

Values of the parameters necessary for strain calculations in a few common semiconductors are given in Table 14.1.

14.7   HOLE SUBBANDS IN HETEROSTRUCTURES

In a heterostructure, where the material composition is modulated, and hence all the material parameters, as well as the valence band edge energies, become position dependent, the Hamiltonian in equation (14.2) has to be modified appropriately. For some time it has been accepted that equation (14.2) is modified in analogy with the effective mass Schrödinger equation: in a 2D system, for example, k_x and k_y remain as they are, because only a plane-wave type of solution is allowed in the non-quantizing* directions, along which the potential is constant; however, the wave vector k_z along the quantizing (growth) direction is substituted for its quantum mechanical operator $-i\partial/\partial z$, and

---

*In the plane of the quantum well.

then, in order to preserve the hermiticity of the Hamiltonian, the terms of the type γk_z^2 become $k_z \gamma k_z$, while terms of the type γk_z become $(\gamma k_z + k_z \gamma)/2$ (note that the γs are position dependent in a heterostructure).

Furthermore, the diagonal elements of the Hamiltonian are amended with the potential V, which could have contributions from the valence band offset in the particular material, the potential from an external electrostatic field, or the self-consistent space-charge electrostatic potential. Following this, it was shown by Foreman [342] that further modifications are necessary, which improve the accuracy of the method, and under these developments the Hamiltonian now reads:

$$\mathbf{H} =$$

$$
\begin{bmatrix}
P+Q+V & 0 & -S_- & R & (1/\sqrt{2})S_- & \sqrt{2}R \\
0 & P+Q+V & -R^\dagger & -S_+ & -\sqrt{2}R^\dagger & (1/\sqrt{2})S_+ \\
-S_-^\dagger & -R & P-Q+V & C & \sqrt{2}Q & \sqrt{3/2}\Sigma_- \\
R^\dagger & -S_+^\dagger & C^\dagger & P-Q+V & -\sqrt{3/2}\Sigma_+ & \sqrt{2}Q \\
(1/\sqrt{2})S_-^\dagger & -\sqrt{2}R & \sqrt{2}Q & -\sqrt{3/2}\Sigma_+^\dagger & P+\Delta_{\mathrm{SO}}+V & -C \\
\sqrt{2}R^\dagger & (1/\sqrt{2})S_+^\dagger & \sqrt{3/2}\Sigma_-^\dagger & \sqrt{2}Q & -C^\dagger & P+\Delta_{\mathrm{SO}}+V
\end{bmatrix}
\quad (14.29)
$$

where:

$$P = \left(\frac{\hbar^2}{2m_0}\right)\gamma_1(k_x^2 + k_y^2 + k_z^2)$$

$$Q = \left(\frac{\hbar^2}{2m_0}\right)\gamma_2(k_x^2 + k_y^2 - 2k_z^2)$$

$$R = \sqrt{3}\left(\frac{\hbar^2}{2m_0}\right)(-\bar{\gamma}k_-^2 + \mu k_+^2)$$

$$S_\pm = 2\sqrt{3}\left(\frac{\hbar^2}{2m_0}\right)k_\pm\left[(\sigma - \delta)k_z + k_z\pi\right]$$

$$\Sigma_\pm = 2\sqrt{3}\left(\frac{\hbar^2}{2m_0}\right)k_\pm\left\{\left[\frac{1}{3}(\sigma - \delta) + \frac{2}{3}\pi\right]k_z + k_z\left[\frac{2}{3}(\sigma - \delta) + \frac{1}{3}\pi\right]\right\}$$

$$C = 2\left(\frac{\hbar^2}{2m_0}\right)k_-\left[k_z(\sigma - \delta - \pi) - (\sigma - \delta - \pi)k_z\right]$$

$$k_\pm = k_x \pm ik_y, \qquad \bar{\gamma} = \frac{1}{2}(\gamma_2 + \gamma_3), \qquad \mu = \frac{1}{2}(\gamma_3 - \gamma_2)$$

$$\sigma = \bar{\gamma} - \frac{1}{2}\delta, \qquad \pi = \mu + \frac{3}{2}\delta, \qquad \delta = \frac{1}{9}(1 + \gamma_1 + \gamma_2 - 3\gamma_3) \quad (14.30)$$

It is possible to show that equation (14.29) reduces to equation (14.2) if the Luttinger parameters (γ_1, γ_2 and γ_3) are set as constants.

As for the case of the bulk Hamiltonian, equation (14.29) can also be block-diagonalized into two 3×3 blocks, which read:

$$
\mathbf{H}_\pm =
\begin{bmatrix}
P+Q+V & R \mp iS & \sqrt{2}R \pm iS/\sqrt{2} \\
R \pm iS^\dagger & P-Q \mp iC+V & \sqrt{2}Q \mp i\sqrt{3/2}\Sigma \\
\sqrt{2}R \mp iS^\dagger/\sqrt{2} & \sqrt{2}Q \pm i\sqrt{3/2}\Sigma^\dagger & P+\Delta_{\mathrm{SO}} \pm iC+V
\end{bmatrix}
\quad (14.31)
$$

where P and Q are the same as above, but R, S, Σ and C now read:

$$R = -\sqrt{3} \left(\frac{\hbar^2}{2m_0} \right) \gamma_\phi k_{||}^2, \qquad S = 2\sqrt{3} \left(\frac{\hbar^2}{2m_0} \right) k_{||} \left[(\sigma - \delta) k_z + k_z \pi \right]$$

$$\Sigma = 2\sqrt{3} \left(\frac{\hbar^2}{2m_0} \right) k_{||} \left\{ \left[\frac{1}{3}(\sigma - \delta) + \frac{2}{3}\pi \right] k_z + k_z \left[\frac{2}{3}(\sigma - \delta) + \frac{1}{3}\pi \right] \right\}$$

$$C = 2 \left(\frac{\hbar^2}{2m_0} \right) k_{||} \left[k_z (\sigma - \delta - \pi) - (\sigma - \delta - \pi) k_z \right]$$

$$\gamma_\phi = \sqrt{\bar{\gamma}^2 + \mu^2 - 2\bar{\gamma}\mu \cos\phi}, \qquad k_{||}^2 = k_x^2 + k_y^2 \tag{14.32}$$

It would now appear unclear what values of γ_2 and γ_3 should be used in γ_ϕ; however, it is a good approximation to use their average values across the structure.

For a layer-type structure (i.e. with a constant composition inside any of the layers) use equation (14.29), or equation (14.31), to find the boundary conditions for the wave function at interfaces. This is achieved by formal integration across the interface, in the same way as with the effective mass Schrödinger equation (resulting in the conclusion that ψ, as well as $(1/m^*)\partial\psi/\partial z$, is conserved across the interface). Such integration of equations (14.29) or (14.31) shows that the amplitudes F_i in the wave function vector are individually conserved, and also that there are particular linear combinations of both the derivatives and the amplitudes of all F_i components that are conserved across the interface (see later).

Methods of calculating the eigenstates of heterostructures may be divided into two groups. One of them, which is practical only for one-dimensional heterostructure potentials (like quantum wells), uses a 'layer approach' and first finds the relevant properties of each single layer in the structure, before proceeding to find its 'global' properties—in particular, the eigenstates.

The other approach considers directly the heterostructure as a whole. It is computationally more demanding than the layer approach, but is equally applicable to quantum wells, wires and dots. In either case, the final ingredient we need for the calculation is the position-dependent potential $V(z)$ (or $V(\mathbf{r})$ in multidimensional structures) that is to be used in the Hamiltonian equations (14.29) or (14.31).

14.8 VALENCE BAND OFFSET

In the case of an unstrained system the valence band offset has a meaning analogous to that in the conduction band: V here shows the valence band edge (of both the HH and LH branches) in any layer. This is precisely how it enters the Hamiltonian. The valence band offsets at heterointerfaces are generally available from the literature, but there is a considerable amount of scatter in this data. In Table 14.1 values are given of the valence band offset (VBO), with respect to vacuum, for a few common semiconductors. For alloys, it is usual to use linear interpolation to approximate. Having found the VBO values for two materials of interest, we subtract the two in order to get the relative offset.

The VBO data in Table 14.1 are given in the real energy scale, hence the material with higher VBO will be the quantum well—in other words, the relative band offset should be multiplied by -1 before being used in either of the Hamiltonians in equation (14.29) or

(14.31), which are written in the inverted energy picture. It should be noted, however, that the valence band offsets determined in this way are only approximate (the exception being GaAs/AlAs), and for more accurate values the literature should be consulted.

The case of strained structures is more complicated, and requires some care in using the available data on valence band offsets. The potential V, which should be inserted into the Hamiltonian, is *not* any of the band edges defined by equation (14.28), because these expressions already include the effects of strain and cannot be filtered out of the Hamiltonian. Instead, V in the Hamiltonian is the potential before the addition of strain.

In many cases data can be found on the so-called average valence band energy E_{av}, which is the weighted mean of the three valence bands. In an unstrained material, where HH and LH band edges are degenerate, the weighted mean (in the inverted energy picture!) is $\Delta_{SO}/3$ above the valence band edge. Alternatively, the interface of two semiconductors may be characterized by the average valence band offset, ΔE_{av}, which is the difference of E_{av}s in the two materials. Since Δ_{SO} is material dependent, the ΔE_{av} is not the same as the valence band offset, even in an unstrained material. It has been established that ΔE_{av} is roughly constant with strain, and can therefore be used as a single parameter to describe the interface.

If a strained semiconductor layer is grown on an unstrained substrate (made of a different material), and the values of ΔE_{av} and Δ_{SO} in the epitaxial layer material are known, then the difference $\Delta E_{av} - \Delta_{SO}/3$ clearly gives the energy of the HH/LH valence band edge in this material *without* strain, measured from the E_{av} value in the substrate. In the same manner the valence band edge can be obtained in all layers of a multilayer structure, measured from the same reference point, regardless of whether a particular layer is in direct contact with the substrate or not. This is the potential that should be used in the Hamiltonian. It can be subtracted from any desired reference energy, if it is preferred to have output energies measured from that point. Since the valence band edge in a layer before strain has no physical significance in the strained system, one reasonable choice for a reference point might be the lowest valence band edge in the quantum well—either HH or LH, whichever came out to be lower (if there are different quantum wells, the deepest one could be chosen). Another choice might be the valence band edge in the substrate, which is unstrained, and still has its HH and LH band edges degenerate.

In some cases, for a particular material interface and particular strain conditions, data may be quoted like 'AB on CD has the valence band offset of ΔE_v (eV)'. This means that strained material AB grown on unstrained substrate CD has such an offset between its valence band edge (either HH or LH, whichever is the lower in this case) and the valence band edge in the substrate. To use this data for a calculation within the same material system, but under different strain conditions, it will be necessary to find whether the HH or LH band in the strained AB material makes its valence band edge (which depends on the sign of the strain), then find P_ϵ and Q_ϵ and then use the appropriate expression from equation (14.28) to get the AB valence band edge without strain.

14.9   THE LAYER (TRANSFER MATRIX) METHOD

In order to describe the procedure of finding the quantized state energies and wave functions, the block-diagonal form of the Hamiltonian will be used, i.e. attention will be focused on just one of the 3×3 blocks. However, the method is straightforwardly applicable to any other size of the Hamiltonian, e.g. the 2×2, or one of the non-block-diagonalized 6×6 or 4×4 forms, or in fact any other form, see [346].

Consider a structure modulated along one dimension (i.e. a quantum well) with arbitrarily varying material composition and potential (see Figs 1.9–1.13 for a few examples). For the purpose of finding its bound states, or the tunnelling probability, the heterostructure of interest is subdivided into a number (N_z, number of coordinate points) of thin layers, and within each layer the potential (which includes any self-consistent potential, if such a calculation was performed) is taken to be constant, as are the values of the Luttinger parameters (note that, under these conditions, $S = \Sigma$ and $C = 0$ in equations (14.31–14.32)). If the structure is step-graded, comprising some number of layers of different material composition and width (perhaps like the structure in Fig. 3.15), and provided any continuously varying potentials (e.g. self-consistent potential) are absent, the computational layers in this calculation coincide with actual material layers, and need not be very thin.

Similarly to the case of quantized states of electrons, the aim is to describe the wave function of a quantized state in terms of its form within each layer. These are then joined at the interfaces using appropriate boundary conditions. Within such an approach the need is not to generate all the energy eigenstates of the bulk Hamiltonian for a specified wave vector, but rather all the solutions that correspond with a definite energy, i.e. the complex band structure has to be found.

In order to find the complex band structure in the valence band, consider a structure grown in the [001] direction, which it is convention to define as the z-axis. Since the structure composition and the potential are modulated (varied) along z, but are constant along the x- and y-axes, the wave function in the x–y plane must behave like a plane wave, hence k_x and k_y must be real, while k_z is arbitrary. The Hamiltonian is written as (to make the writing shorter, the factor $\hbar^2/2m_0$ is taken to be absorbed into the γ parameters):

$$[\mathbf{H}] = [\mathbf{H}_2 k_z^2 + \mathbf{H}_1 k_z + \mathbf{H}_0] \tag{14.33}$$

where $\mathbf{H}_0$, $\mathbf{H}_1$, and $\mathbf{H}_2$ are the 3×3 matrices that are associated with the corresponding powers of the wave vector k_z, that is:

$$\mathbf{H}_2 = \begin{bmatrix} (\gamma_1 - 2\gamma_2) & 0 & 0 \\ 0 & (\gamma_1 + 2\gamma_2) & -2\sqrt{2}\gamma_2 \\ 0 & -2\sqrt{2}\gamma_2 & \gamma_1 \end{bmatrix}$$

$$\mathbf{H}_1 = \begin{bmatrix} 0 & \mp i2\sqrt{3}\gamma_3 k_{||} & \pm i\sqrt{6}\gamma_3 k_{||} \\ \pm i2\sqrt{3}\gamma_3 k_{||} & 0 & \mp i3\sqrt{2}\gamma_3 k_{||} \\ \mp i\sqrt{6}\gamma_3 k_{||} & \pm i3\sqrt{2}\gamma_3 k_{||} & 0 \end{bmatrix}$$

$$\mathbf{H}_0 = \begin{bmatrix} (\gamma_1 + \gamma_2)k_{||}^2 + P_\epsilon + Q_\epsilon + V & -\sqrt{3}\gamma_\phi k_{||}^2 & -\sqrt{6}\gamma_\phi k_{||}^2 \\ -\sqrt{3}\gamma_\phi k_{||}^2 & (\gamma_1 - \gamma_2)k_{||}^2 + P_\epsilon - Q_\epsilon + V & \sqrt{2}(\gamma_2 k_{||}^2 + Q_\epsilon) \\ -\sqrt{6}\gamma_\phi k_{||}^2 & \sqrt{2}(\gamma_2 k_{||}^2 + Q_\epsilon) & \gamma_1 k_{||}^2 + P_\epsilon + \Delta_{\mathrm{SO}} + V \end{bmatrix} \tag{14.34}$$

At any specified value of the energy E, the values of the complex wave vector k_z may be viewed as eigenvalues of the 3×3 non-linear eigenvalue problem $[\mathbf{H}(k_z) - E][F] = 0$, where $[F]$ is the eigenfunction vector of length 3. This eigenproblem is non-linear because eigenvalues (k_z, not E!) appear in powers of both 1 and 2, in contrast to the standard linear eigenproblem where the energy appears only linearly on the diagonal of the matrix. Non-linear eigenproblems of this type are solved by a trick that converts them to a doubled-in-size, 6×6 *linear* eigenvalue problem, for which well-developed techniques exist (the method is readily generalized to handle any polynomial-type non-linear eigenproblem). The linear problem to be solved thus reads:

$$
\begin{bmatrix} \mathbf{0} & \mathbf{1} \\ -\mathbf{H}_2^{-1}(\mathbf{H}_0 - E) & -\mathbf{H}_2^{-1}\mathbf{H}_1 \end{bmatrix} \begin{bmatrix} u \\ k_z u \end{bmatrix} = k_z \begin{bmatrix} u \\ k_z u \end{bmatrix} \tag{14.35}
$$

where $\mathbf{0}$ and $\mathbf{1}$ are the 3×3 null and unity matrices. Note that the first row in equation (14.35) is just an identity. The solution of this non-Hermitian but linear matrix eigenproblem, by standard diagonalization routines [347], delivers the six, generally complex-valued wave vectors k_z, and the corresponding six eigenstates. These states, denoted as $[u]$, are expressed in the basis F_1–F_3 for the upper block. The $[u]$-states are thus particular linear combinations of F_1–F_3 that, at energy E, behave as plane waves, i.e. $\exp(ik_z z)$, within a layer. It should be noted that the first three components of an eigenvector of equation (14.35) are the amplitudes of the basis states, and the other three, *when multiplied by* i, will be their derivatives. In further considerations the notation $[u]$ will denote the vector of length 6, with the amplitudes and *derivatives*.

For convenience, these states may then be divided into two groups, as follows: a state which has purely real k_z may be classified according to the sign of k_z[†], if it is positive the state is 'forward', otherwise it is 'backward'; if k_z is complex, including purely imaginary, it is classified as forward if the wave function decays to the right, otherwise it is backward. There are three forward and three backward states in any one layer and any wave function in the structure may, in this layer, be written as a linear combination of these six states. The vector containing the three coefficients of forward states will be denoted as $[\mathbf{a}_j]$, and that for backward states as $[\mathbf{b}_j]$, where the subscript j denotes that these correspond to layer j in the structure. In a vector of length 6, which gives the amplitudes of these states in a wave function at a particular energy, the choice is to write $[\mathbf{a}_j]$ first and $[\mathbf{b}_j]$ below it.

Having the wave function at the beginning (left-hand side) of the jth layer, written in the basis of eigen-k_z states u, it may be propagated to the beginning of the next layer, by first multiplying the state vector by the corresponding exponentials, i.e. by left-multiplying the state vector with the diagonal matrix $\mathbf{D}_j = \mathrm{diag}(\ldots, \exp(ik_{z\,j}d_j), \ldots)$, where d_j is the width of the jth layer, and then using the boundary conditions (the interface matching matrix) to propagate the wave function just across the interface. However, the interface matching matrix is written in the F-basis, equation (14.1), as noted above, and not in the eigen-k_z states basis. It relates the wave function amplitudes and derivatives, written in the F-basis, and is obtained from equation (14.31) in our case (or, if not using the block-diagonal form, the corresponding form would be obtained from

---

[†]Strictly speaking, it should be tested for the current it carries, where the current density operator is $(2\mathbf{H}_2 k_z + \mathbf{H}_1)/\hbar$, but for the purpose here, the sign test suffices.

equation (14.29)). It demands that:

$$\begin{bmatrix} 1 & 0 \\ \mathbf{B}_1 & \mathbf{B}_2 \end{bmatrix} \cdot \begin{bmatrix} F \\ F' \end{bmatrix} = \text{const} \qquad (14.36)$$

across the interface, where:

$$\mathbf{B}_1 = \begin{bmatrix} 0 & \pm 2\sqrt{3}\pi k_{||} & \mp\sqrt{6}\pi k_{||} \\ \mp 2\sqrt{3}(\sigma - \delta)k_{||} & \pm 2(\sigma - \delta - \pi)k_{||} & \pm\sqrt{2}(2\sigma - 2\delta + \pi)k_{||} \\ \pm\sqrt{6}(\sigma - \delta)k_{||} & \mp\sqrt{2}(\sigma - \delta + 2\pi)k_{||} & \mp 2(\sigma - \delta - \pi)k_{||} \end{bmatrix}$$

$$\mathbf{B}_2 = \begin{bmatrix} \gamma_1 - 2\gamma_2 & 0 & 0 \\ 0 & \gamma_1 + 2\gamma_2 & -2\sqrt{2}\gamma_2 \\ 0 & -2\sqrt{2}\gamma_2 & \gamma_1 \end{bmatrix} \qquad (14.37)$$

and:

$$\begin{bmatrix} F \\ F' \end{bmatrix} = \begin{bmatrix} F_1 \\ F_2 \\ F_3 \\ F_1' \\ F_2' \\ F_3' \end{bmatrix} \qquad (14.38)$$

where the '+' sign applies to the upper block, and the '−' would apply to the lower block (with F_4–F_6 functions) when required. The matrix in equation (14.36) will be denoted as $\mathbf{I}_j$ (the interface matrix, corresponding to the layer j). From $\mathbf{I}_{j+1}[F\ F']^T_{j+1} = \mathbf{I}_j[F\ F']^T_j$ it follows that the matrix $\mathbf{I}_{j+1}^{-1}\mathbf{I}_j$ transfers a state written in the F-basis from the left-hand side into the right-hand side of the interface of layers j and $j+1$.

A wave function written in the u basis, can be re-expressed in the F-basis (amplitudes and derivatives) by multiplying it with the matrix $\mathbf{U}$, made by stacking side-by-side all the eigen-k_z states u_1–u_6 (i.e. eigenvectors of equation (14.35) in which the lower components have been turned into derivatives, as noted above):

$$\mathbf{U} = [u_1|u_2|\ldots|u_6], \qquad [F] = \mathbf{U}[u], \qquad [u] = \mathbf{U}^{-1}[F] \qquad (14.39)$$

Therefore, the transfer matrix $\mathbf{T}^{j,j+1}$, relating the wave functions at the beginning of layer j and the beginning of layer $j+1$ (with the wave function written in the u-basis) is given by:

$$\mathbf{T}^{j,j+1} = \mathbf{U}_{j+1}^{-1}\mathbf{I}_{j+1}^{-1}\mathbf{I}_j\mathbf{U}_j\mathbf{D}_j, \qquad \begin{bmatrix} \mathbf{a}_{j+1} \\ \mathbf{b}_{j+1} \end{bmatrix} = \mathbf{T}^{j,j+1}\begin{bmatrix} \mathbf{a}_j \\ \mathbf{b}_j \end{bmatrix} \qquad (14.40)$$

To give a more verbose description of equation (14.40): having a state at the beginning of layer j written in the u-basis, first propagate it to the end of that layer, but still inside it ($\mathbf{D}_j$), then express it in the F-basis ($\mathbf{U}_j$), then transfer it across the interface into the beginning of layer $j+1$ ($\mathbf{I}_{j+1}^{-1}\mathbf{I}_j$), and finally express it back in the u-basis this time of layer $j+1$ ($\mathbf{U}_{j+1}^{-1}$).

The total transfer matrix of a structure, consisting of N layers, may then clearly be calculated as:

$$\mathbf{T}^{0,N} = \mathbf{T}^{N-1,N}\ldots\mathbf{T}^{1,2}\mathbf{T}^{0,1} \qquad (14.41)$$

How can the bound states of a system be found if the transfer matrix is known? First, note that it can be written in terms of its four 3×3 blocks as:

$$\mathbf{T} = \left[\begin{array}{cc} \mathbf{T}_{11} & \mathbf{T}_{12} \\ \mathbf{T}_{21} & \mathbf{T}_{22} \end{array} \right] \tag{14.42}$$

so that:

$$\left[\begin{array}{c} \mathbf{a}_N \\ \mathbf{b}_N \end{array} \right] = \left[\begin{array}{cc} \mathbf{T}_{11} & \mathbf{T}_{12} \\ \mathbf{T}_{21} & \mathbf{T}_{22} \end{array} \right] \left[\begin{array}{c} \mathbf{a}_0 \\ \mathbf{b}_0 \end{array} \right] \tag{14.43}$$

This is similar to the familiar case of electrons in the conduction band, except that the transfer matrix there is of size 2×2 and its elements are numbers, while here these numbers are replaced by matrix blocks.

Now, assume that, for a particular in-plane wave vector and energy, the complex band structure calculation for the outermost two layers of a quantum well delivers a pair of purely real-valued k_z. In this case (which is analogous to the case of the energy being above the barrier for electrons) there are no bound states. However, if all the k_zs in some energy range turn out to be complex-valued or simply imaginary, there may be bound states because the wave function composed of such states can be made to decay away from the quantum well region. To find these energies and wave functions, it must be noted that, on the left-hand side, a bound state (whose wave function can be normalized to unity) must have $\mathbf{a}_0 = 0$, while $\mathbf{b}_0$ is finite. Therefore, from equation (14.43), on the right-hand side, the wave function has the component $\mathbf{a}_N = \mathbf{T}_{12}\mathbf{b}_0$, which decays further to the right, and the component $\mathbf{b}_N = \mathbf{T}_{22}\mathbf{b}_0$ which grows to the right. However, this latter term cannot exist in a bound state wave function, hence it must be the case that:

$$\mathbf{T}_{22}\mathbf{b}_0 = 0 \tag{14.44}$$

For electronic bound states, where $\mathbf{T}_{22}$ and $\mathbf{b}_0$ are both scalars (numbers), the method would be to search for energies that make $T_{22} = 0$, in which case b_0 is then arbitrary, and eventually determined from the normalization of the wave function. For holes, equation (14.44) is, in fact, a homogeneous system of linear equations in the components of $\mathbf{b}_0$. To have a non-trivial solution, the determinant of the matrix $\mathbf{T}_{22}$ must be equal to zero, and this will happen only for some particular values of the energy. Therefore, when searching for bound states, the energy is varied and the value of $\det|\mathbf{T}_{22}(E)| = 0$ is monitored (note that this determinant is generally a complex number). Upon finding such value(s) of the energy E, the system of equations for $\mathbf{b}_0$ is solved, which can be done to within an arbitrary multiplier (its value is eventually determined from normalization). Finally, starting with this $\mathbf{b}_0$, and $\mathbf{a}_0 = 0$ and using the transfer matrix once again, the wave function for each particular bound state can be generated.

In the energy range where at least one of the wave vectors k_z is real, there are no bound states, but the transmission or reflection coefficient of the structure can be calculated. In this case it would be necessary to specify the incident state (all its components, not just one amplitude) for which the transmission is sought.

Despite a more complicated procedure, which has been described above, the layer method is usually much faster than any direct numerical procedure for finding the bound states. This is because it handles only small matrices, while direct diagonalization uses very large matrices which, in the case of holes, are not tridiagonal. Furthermore, this was a nice illustration of the application of the transfer matrix method to more complex systems, which is why it was considered in detail.

14.10   QUANTUM WELL SUBBANDS

In this section examples of hole subbands will be given, calculated for GaAs quantum wells embedded in the $Al_{0.5}Ga_{0.5}As$ bulk alloy. The valence band edge of GaAs is lower than that of the alloy (in the inverted energy picture), so the GaAs layer forms the quantum well and $Al_{0.5}Ga_{0.5}As$ the barrier. GaAs and AlAs have such similar values of the lattice constant that the structure is taken to be lattice matched, i.e. unstrained. The material parameters were taken from Table 14.1.

Subband dispersion curves (E versus k dependence) in two structures with different quantum well widths are given in Fig. 14.2. The wave vector in both examples was varied along two different directions in the k_x–k_y plane: the $< 01 >$ direction ($k_x \neq 0$ and $k_y = 0$, or vice versa), and the $< 11 >$ direction ($k_x = k_y$). It is immediately apparent that the dispersion is quite anisotropic.

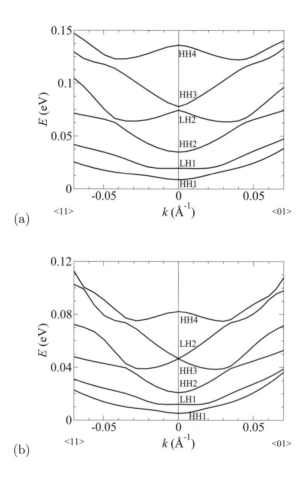

Figure 14.2   The dispersion of the first six hole subbands in (a) an 85 Å wide and (b) a 113 Å wide GaAs quantum well surrounded by $Al_{0.5}Ga_{0.5}As$ barriers. The energy is measured from the valence band edge in the well.

The subbands are denoted as HH or LH according to their composition at $k_x = k_y = 0$. For finite in-plane wave vectors, particularly in the outer parts of Fig. 14.2, the mixing of $|3/2, \pm 3/2\rangle$ and $|3/2, \pm 1/2\rangle$ states becomes so strong that it would be difficult to tell, by looking at the wave function composition alone, whether a subband is of HH or LH type. The label HH, for instance, then just means that this state originated as HH at the Γ-point.

The wave vector dependent mixing of the bulk bands is the reason that the subband dispersion along any particular direction is very non-parabolic, even with occasional reversal ($\partial E/\partial k$ becoming negative, which can be interpreted as a negative mass). The ordering of the first few subbands is the same for both well widths: HH1 is the lowest, then comes LH1, then HH2 (this remains so for any well width), but differences appear for higher subbands. It is also interesting to point out that, in the (relatively low) energy range shown in Fig. 14.2, the content of bulk SO band ($|1/2, \pm 1/2\rangle$ basis states) in the subband wave functions is very small. From <1% at $k_x = k_y = 0$ in the LH subband, it increases but stays typically within the range of <10–15% in either the HH or the LH subbands almost throughout the range shown in Fig. 14.2.

14.11   THE INFLUENCE OF STRAIN

Calculation of the hole subband structure in strained quantum wells proceeds by the same method as in the unstrained case, the only (quantitative) difference being the presence of strain terms in the Hamiltonian. In this section the effects of strain on states in quantum wells will be discussed briefly.

Consider the case of a compressively strained quantum well between two tensile strained barriers. An example of this would be a $Si_{1-x}Ge_x$ layer embedded in Si, grown on a substrate with a composition between the two (this is a common configuration in order to achieve strain balance). The valence band edge in the SiGe alloy is lower than in Si (in the inverted energy picture), so the former is the well and the latter the barrier. The larger lattice constant of Ge and hence the compressive strain in the $Si_{1-x}Ge_x$ layer, imply that the HH band edge in it will decrease from the unstrained value, while the LH band edge would increase (note again the use of the inverted energy picture). The opposite applies in the tensile strained Si barriers. Therefore, the quantum well becomes deeper for heavy-holes and shallower for light-holes. This is a direct consequence of the uniaxial component of strain, which splits the HH and LH band edges apart.

The hydrostatic component also influences the barrier heights, because the hydrostatic deformation potential has different values in the two materials, but this does not affect the bound state energies very much. If the barriers are not very shallow, the bound state energies *measured from the band edge* mostly depend on the well width, and one can perceive that the strain-induced shift of the band edge (HH or LH) will essentially 'drag' with it all of 'its own' bound states (however, due to mixing of the bulk states, it has to be admitted that the meaning of this is somewhat vague for non-zero values of $k_{||}$). In the unstrained case the lowest HH state is below the lowest LH state, therefore compressive strain will displace the two sets of states, leaving HH as the lowest.

In the opposite case of a quantum well under tensile strain, the valence band edges behave in exactly the opposite manner, and this can lead to the unusual situation of the lowest LH state sinking below the lowest HH state. In the SiGe system it is not possible to get the LH subband as the lowest as this would require a tensilely strained well, and in this case the well (the layer with the larger Ge content) is always compressively strained. However, there are other materials (based on III–V alloys) where the well can be put under tensile strain by the appropriate choice of parameters, and the LH1 subband can be made to be the lowest in the system.

14.12 STRAINED QUANTUM WELL SUBBANDS

As an example of a strained quantum well, take a $Si_{0.6}Ge_{0.4}$ layer embedded between Si layers, grown on a $Si_{0.7}Ge_{0.3}$ substrate. The calculated subband dispersion in two structures with different quantum well widths (different $Si_{0.6}Ge_{0.4}$ layer widths), using the material parameters from Table 14.1, and the value $\Delta E_{av} = 0.56$ eV, is shown in Fig. 14.3. Along with the anisotropic dispersion, just as in the case of the unstrained structure in Fig. 14.2, notice that the ordering of subbands now depends on the well width: it is possible to get two (or even more) HH subbands below the lowest LH subband. This occurs because of the strain-induced displacement of the HH and LH band edges. Another interesting feature in Fig. 14.3(b) is the dispersion of the HH2 subband: starting from the zone centre its energy first decreases before acquiring the expected increase with the wave vector, i.e. it shows an inverted-mass feature, which is brought about by mixing with the LH1 subband, lying just above it.

In contrast to the GaAs/AlGaAs system, in this example the SO band contributes significantly to the composition of all subbands. At the Γ-point its contribution to the LH subbands is of the order of $\sim 10\%$ (this is a consequence of strain alone), and then increases significantly for larger in-plane wave vectors for all the subbands.

14.13 DIRECT NUMERICAL METHODS

An alternative method of finding the quantized states is to solve equation (14.29) or (14.31) when written for the whole structure. One possibility is to use the finite difference approximation, which was introduced for approximating derivatives in the effective mass Schrödinger equation of Chapter 3. In the valence band the wave function has more than a single component, hence finite differences never lead to a tridiagonal matrix, even for one-dimensional problems (quantum wells), but rather to a band matrix having non-zero elements along the main and some number of remote diagonals. In the usual implementation the finite difference method implicitly assumes hard wall (box) boundary conditions, as in Fig. 3.6.

Another possibility is to use a plane wave method and expand the wave function components in a Fourier series, hence implicitly assuming the structure to be a superlattice, even though it may be non-periodic—for example, a single quantum well embedded in thick barriers. To prevent interaction between the adjacent wells in implicitly existing neighbouring periods, care has to be taken to ensure that the barriers are thick enough. The plane-wave method is explained in the chapter on empirical pseudopotentials (Chapter 15), and will not be elaborated on here in any more detail.

The advantage of these methods is that they may be used equally well for structures of higher dimensionality (i.e. quantum wires and quantum dots) with no essential complications in the formulation. However, these methods require considerable amounts of computer memory in order to store the necessary matrices, especially in multidimen-

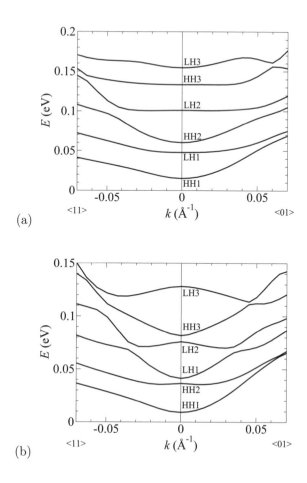

Figure 14.3   The dispersion of the first 6 hole subbands in (a) an 83 Å wide and (b) a 111 Å wide $Si_{0.6}Ge_{0.4}$ quantum well embedded in Si barriers, grown on a $Si_{0.7}Ge_{0.3}$ substrate. The energy is measured from the HH band edge in the strained well material.

sional systems, and such matrices also require large computational time for diagonalization. An alternative is to employ a suitable set of basis functions that are adapted to the problem at hand, in the sense that using a relatively small number of these functions suffices for good convergence.

CHAPTER 15

EMPIRICAL PSEUDO-POTENTIAL THEORY

15.1 PRINCIPLES AND APPROXIMATIONS

The envelope function and effective mass approximations, which as stated earlier can be thought of as an approximation to the band structure of a crystal, rather than to the quantum mechanics, are very successful theories, which have allowed many of the most fundamental properties of semiconductors and their heterostructures to be explained. However, it is clear that it is at least worthwhile considering more complex, perhaps more fundamental, models to see if they can offer more insight. In particular, given the approximations that the 'single band' effective mass and envelope function models made to the crystal potential, an obvious improvement would be to consider the potential microscopically, i.e. including the potentials of the atoms individually.

The complexity of such a procedure—solving an exact solid, is illustrated by the form of the complete Hamiltonian:

$$\mathcal{H} = \mathcal{H}_{\text{electrons}} + \mathcal{H}_{\text{nuclei}} + \mathcal{H}_{\text{electrons–nuclei}} \tag{15.1}$$

where

$$\mathcal{H}_{\text{electrons}} = \sum_{\mu} \left(-\frac{\hbar^2}{2m_0} \nabla_{\mu}^2 + \sum_{\lambda < \mu} \frac{e^2}{|\mathbf{r}_\lambda - \mathbf{r}_\mu|} \right) \tag{15.2}$$

431

Quantum Wells, Wires and Dots, Third Edition. P. Harrison
©2009 John Wiley & Sons, Ltd.

and $\mathbf{r}_\mu$ are the positions of the electrons and m_0 is the mass. In addition:

$$\mathcal{H}_{\text{nuclei}} = \sum_\nu \left(-\frac{\hbar^2}{2M_\nu} \nabla_\nu^2 + \sum_{\lambda<\nu} \frac{Z_\lambda Z_\nu e^2}{|\mathbf{R}_\lambda - \mathbf{R}_\nu|} \right) \tag{15.3}$$

where $\mathbf{R}_\nu$ are the positions, Z_ν the atomic numbers and M_ν the masses of the nuclei. Finally:

$$\mathcal{H}_{\text{electrons–nuclei}} = -\sum_{\mu,\nu} \frac{Z_\nu e^2}{|\mathbf{R}_\nu - \mathbf{r}_\mu|} \tag{15.4}$$

In a typical macroscopic sample of semiconducting crystal, there are a lot of nuclei, and correspondingly a lot of electrons! In fact, for Si, there are about 5×10^{22} atoms per cm^3. The problem, as stated above, is therefore insolvable. There are, however, some approximations [348, 349] that can be made, which have proved to be acceptably accurate and thus lead to a more manageable problem.

(i). By assuming that the electrons below the outer shell are tightly bound to the nucleus simplifies the system slightly—the atoms can then be treated as separate entities and hence the number of particles is reduced. In addition, these inner-shell electrons screen the outer valence electrons from the central nuclear charge, which has the effect of smoothing out the potential of what would otherwise be a rapidly varying term. Thinking ahead slightly, the smoother the potential then the fewer terms may be required if the electron wave functions were to be constructed from some Fourier series.

(ii). The *adiabatic approximation* assumes that a change in the coordinates of a nucleus passes no energy to the electrons, i.e. the electrons respond adiabatically, which then allows the decoupling of the motions of the nuclei and the electrons.

(iii). The *independent electron approximation* removes the complications of the electron–electron interactions and replaces them with a time averaged potential.

These approximations allow the system, i.e. the electron wave functions that represent the complete self-interacting electron cloud within the 'plasma' of the crystal, to be collapsed down to a one-electron problem with the following Hamiltonian:

$$\mathcal{H} = -\frac{\hbar^2}{2m_0} \nabla^2 + V_c \tag{15.5}$$

where V_c is some, as yet undetermined, crystal potential that represents not only the interaction between the electrons and the nuclei of the atoms that constitute the lattice, but also the interaction between the electrons themselves.

15.2   ELEMENTAL BAND STRUCTURE CALCULATION

It is expected *a priori* that the solutions obtained from a model that will rely on the periodicity of the potential are going to be dependent on the electron momentum $\mathbf{k}$, since it is known that electric current flows in semiconductors and from experiences with the

Kronig-Penney model. In addition, as more than one solution may be obtained, then the electron wave function may be most generally labelled as $\psi_{n,\mathbf{k}}$. Given these features, then employment of the time-independent (as the potential is static) Schrödinger equation gives:

$$\mathcal{H}\psi_{n,\mathbf{k}} = E_{n,\mathbf{k}}\psi_{n,\mathbf{k}} \tag{15.6}$$

Being influenced by all the previous success in expanding periodic functions in terms of linear combination of a set of basis functions, where each have the periodicity of the system, and by utilizing the idea of Bloch ([1], p. 133), for example, it is tempting to expand the wave function $\psi_{n,\mathbf{k}}$ in terms of the complete orthonormal set of plane waves:

$$u_{\mathbf{G},\mathbf{k}} = \frac{1}{\sqrt{\Omega}}e^{i(\mathbf{G}+\mathbf{k})\bullet\mathbf{r}} \tag{15.7}$$

and then:

$$\psi_{n,\mathbf{k}}(\mathbf{r}) = \sum_{\mathbf{G}} a_{n,\mathbf{k}}(\mathbf{G})u_{\mathbf{G},\mathbf{k}} \tag{15.8}$$

$$\therefore \psi_{n,\mathbf{k}}(\mathbf{r}) = \frac{1}{\sqrt{\Omega}}\sum_{\mathbf{G}} a_{n,\mathbf{k}}(\mathbf{G})e^{i(\mathbf{G}+\mathbf{k})\bullet\mathbf{r}} \tag{15.9}$$

If $\mathbf{R}$ is a Bravais lattice vector, then:

$$\psi_{n,\mathbf{k}}(\mathbf{r}+\mathbf{R}) = \frac{1}{\sqrt{\Omega}}\sum_{\mathbf{G}} a_{n,\mathbf{k}}(\mathbf{G})e^{i(\mathbf{G}+\mathbf{k})\bullet(\mathbf{r}+\mathbf{R})} \tag{15.10}$$

which gives:

$$\psi_{n,\mathbf{k}}(\mathbf{r}+\mathbf{R}) = \frac{1}{\sqrt{\Omega}}\sum_{\mathbf{G}} a_{n,\mathbf{k}}(\mathbf{G})e^{i(\mathbf{G}+\mathbf{k})\bullet\mathbf{R}}e^{i(\mathbf{G}+\mathbf{k})\bullet\mathbf{r}} \tag{15.11}$$

$$\therefore \psi_{n,\mathbf{k}}(\mathbf{r}+\mathbf{R}) = e^{i\mathbf{k}\bullet\mathbf{R}}\frac{1}{\sqrt{\Omega}}\sum_{\mathbf{G}} a_{n,\mathbf{k}}(\mathbf{G})e^{i\mathbf{G}\bullet\mathbf{R}}e^{i(\mathbf{G}+\mathbf{k})\bullet\mathbf{r}} \tag{15.12}$$

which yields:

$$\psi_{n,\mathbf{k}}(\mathbf{r}+\mathbf{R}) = e^{i\mathbf{k}\bullet\mathbf{R}}\psi_{n,\mathbf{k}}(\mathbf{r}) \tag{15.13}$$

i.e. Bloch's theorem, provided that a set of wave vectors $\mathbf{G}$ can be found for the original expansion, which satisfy:

$$\mathbf{G}\bullet\mathbf{R} = 2\pi n, \quad \text{where} \quad n \in \mathcal{Z} \tag{15.14}$$

Of course, such a set of vectors do exist—they are merely the reciprocal lattice vectors describing the periodicity of the lattice (see Chapter 1).

Therefore, the proposed expansion of the wave function in terms of a linear combination of plane waves satisfies Bloch's theorem and thus it is worthwhile proceeding further in order to deduce the consequences. Hence, substituting the expansion for $\psi_{n,\mathbf{k}}$ given in equation (15.9) into the Schrödinger equation (equation (15.6)), then:

$$\mathcal{H}\sum_{\mathbf{G}} a_{n,\mathbf{k}}(\mathbf{G})u_{\mathbf{G},\mathbf{k}} = E_{n,\mathbf{k}}\sum_{\mathbf{G}} a_{n,\mathbf{k}}(\mathbf{G})u_{\mathbf{G},\mathbf{k}} \tag{15.15}$$

Utilizing the *linearity* of the Hamiltonian, i.e. it acts on each $u_{\mathbf{G},\mathbf{k}}$ in turn, then the above equation can be written:

$$\sum_{\mathbf{G}} a_{n,\mathbf{k}}(\mathbf{G})\mathcal{H}u_{\mathbf{G},\mathbf{k}} = E_{n,\mathbf{k}}\sum_{\mathbf{G}} a_{n,\mathbf{k}}(\mathbf{G})u_{\mathbf{G},\mathbf{k}} \tag{15.16}$$

Multiplying through by a particular plane wave $u^*_{\mathbf{G}',\mathbf{k}}$ and integrating over all space, then:

$$\sum_{\mathbf{G}} a_{n,\mathbf{k}}(\mathbf{G}) \int u^*_{\mathbf{G}',\mathbf{k}}\mathcal{H}u_{\mathbf{G},\mathbf{k}} \ d\tau = \sum_{\mathbf{G}} a_{n,\mathbf{k}}(\mathbf{G})E_{n,\mathbf{k}} \int u^*_{\mathbf{G}',\mathbf{k}}u_{\mathbf{G},\mathbf{k}} \ d\tau \tag{15.17}$$

Writing:

$$\mathcal{H}_{\mathbf{G}',\mathbf{G}} = \int u^*_{\mathbf{G}',\mathbf{k}}\mathcal{H}u_{\mathbf{G},\mathbf{k}} \ d\tau \tag{15.18}$$

and using the orthonormality property of the basis set, i.e.

$$\int u^*_{\mathbf{G}',\mathbf{k}}u_{\mathbf{G},\mathbf{k}} \ d\tau = \delta_{\mathbf{G}',\mathbf{G}} \tag{15.19}$$

equation (15.17) then becomes:

$$\sum_{\mathbf{G}} a_{n,\mathbf{k}}(\mathbf{G})\mathcal{H}_{\mathbf{G}',\mathbf{G}} = \sum_{\mathbf{G}} a_{n,\mathbf{k}}(\mathbf{G})E_{n,\mathbf{k}}\delta_{\mathbf{G}',\mathbf{G}} \tag{15.20}$$

An equation such as this exists for each value of the electron wave vector $\mathbf{k}$. The solution is therefore reduced to finding the eigenvalues and eigenvectors of the square matrix $\mathcal{H}_{\mathbf{G}',\mathbf{G}}$. In principle, the matrix is of infinite order, but in practice the expansion is limited to a finite set of N plane waves $\mathbf{G}$, and hence the problem can be solved by direct diagonalization using one of the many computer libraries that are available. This yields N eigenvalues $E_{n,\mathbf{k}}$ and the corresponding eigenvectors $a_{n,\mathbf{k}}(\mathbf{G})$.

It remains now to construct the individual matrix elements $\mathcal{H}_{\mathbf{G}',\mathbf{G}}$, given by equation (15.18). The Hamiltonian for an electron of rest mass m_0, moving in the crystal potential V_c, is given by:

$$\mathcal{H} = -\frac{\hbar^2}{2m_0}\nabla^2 + V_c \tag{15.21}$$

Using the form of the plane waves given in equation (15.7), then:

$$\mathcal{H}_{\mathbf{G}',\mathbf{G}} = \frac{1}{\Omega} \int e^{-i(\mathbf{G}'+\mathbf{k})\bullet\mathbf{r}} \left(-\frac{\hbar^2}{2m_0}\nabla^2\right) e^{i(\mathbf{G}+\mathbf{k})\bullet\mathbf{r}} \ d\tau$$

$$+ \frac{1}{\Omega} \int e^{-i(\mathbf{G}'+\mathbf{k})\bullet\mathbf{r}} V_c \ e^{i(\mathbf{G}+\mathbf{k})\bullet\mathbf{r}} \ d\tau \tag{15.22}$$

Therefore:

$$\mathcal{H}_{\mathbf{G}',\mathbf{G}} = \frac{\hbar^2}{2m_0\Omega} \int |\mathbf{G}+\mathbf{k}|^2 e^{i(\mathbf{G}-\mathbf{G}')\bullet\mathbf{r}} \ d\tau$$

$$+ \frac{1}{\Omega} \int e^{-i(\mathbf{G}'+\mathbf{k})\bullet\mathbf{r}} V_c \ e^{i(\mathbf{G}+\mathbf{k})\bullet\mathbf{r}} \ d\tau \tag{15.23}$$

Now the integral over all space in the first term only has a value if $\mathbf{G}=\mathbf{G'}$, in which case it is equal to the normalization volume Ω introduced earlier. This can be summarized by using a Kronecker delta, i.e.

$$\mathcal{H}_{\mathbf{G'},\mathbf{G}} = \frac{\hbar^2}{2m_0}|\mathbf{G}+\mathbf{k}|^2 \delta_{\mathbf{G},\mathbf{G'}} + V \tag{15.24}$$

where V is 'the potential' given by:

$$V = \frac{1}{\Omega} \int e^{-i(\mathbf{G'}+\mathbf{k})\bullet\mathbf{r}} \, V_c \, e^{i(\mathbf{G}+\mathbf{k})\bullet\mathbf{r}} \; \mathrm{d}\tau \tag{15.25}$$

Now, the aim of this exercise is to derive a *microscopic* or *atomistic* model of the crystal, so it is clear that the potential V should be written as a sum of some other, as yet undefined, potential that is situated at every atom site $\mathbf{r}_a$, i.e.

$$V_c = \sum_{\mathbf{r}_a} V_a(\mathbf{r} - \mathbf{r}_a) \tag{15.26}$$

Using this form for V_c in equation (15.25) gives:

$$V = \frac{1}{\Omega} \int e^{-i(\mathbf{G'}+\mathbf{k})\bullet\mathbf{r}} \sum_{\mathbf{r}_a} V_a(\mathbf{r} - \mathbf{r}_a) e^{i(\mathbf{G}+\mathbf{k})\bullet\mathbf{r}} \; \mathrm{d}\tau \tag{15.27}$$

$$\therefore V = \frac{1}{\Omega} \sum_{\mathbf{r}_a} \int V_a(\mathbf{r} - \mathbf{r}_a) e^{i(\mathbf{G}-\mathbf{G'})\bullet\mathbf{r}} \; \mathrm{d}\tau \tag{15.28}$$

As the origin for the summation over the atoms is undefined, then it is possible, and indeed mathematically convenient, to perform the transformation $\mathbf{r} \longrightarrow \mathbf{r} + \mathbf{r}_a$, then obtain:

$$V = \frac{1}{\Omega} \sum_{\mathbf{r}_a} \int V_a(\mathbf{r}) e^{i(\mathbf{G}-\mathbf{G'})\bullet(\mathbf{r}+\mathbf{r}_a)} \; \mathrm{d}\tau \tag{15.29}$$

$$\therefore V = \frac{1}{\Omega} \sum_{\mathbf{r}_a} e^{i(\mathbf{G}-\mathbf{G'})\bullet\mathbf{r}_a} \int V_a(\mathbf{r}) e^{i(\mathbf{G}-\mathbf{G'})\bullet\mathbf{r}} \; \mathrm{d}\tau \tag{15.30}$$

The term $\sum_{\mathbf{r}_a} e^{i(\mathbf{G}-\mathbf{G'})\bullet\mathbf{r}_a}$ is known as the geometrical structure factor S (see, for example, [1], p. 104). Utilizing the property discussed earlier that the atomic sites within the crystal are constructed from the Bravais lattice plus the basis, then the sum over the atom sites can be replaced with two separate summations, i.e. one over the Bravais lattice points and one over the points in each basis:

$$S = \sum_{\mathbf{R}} \sum_{\mathbf{t}} e^{i(\mathbf{G}-\mathbf{G'})\bullet(\mathbf{R}+\mathbf{t})} \tag{15.31}$$

where $\mathbf{R}$ is a Bravais lattice vector and $\mathbf{t}$ is a basis vector. Hence:

$$S = \sum_{\mathbf{R}} e^{i(\mathbf{G}-\mathbf{G'})\bullet\mathbf{R}} \sum_{\mathbf{t}} e^{i(\mathbf{G}-\mathbf{G'})\bullet\mathbf{t}} \tag{15.32}$$

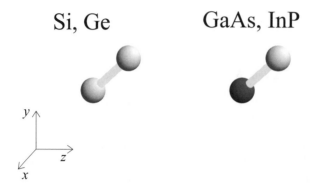

Figure 15.1   The two-atom basis of the diamond structure (left), e.g. Si, Ge, and the zinc blende structure (right), e.g. GaAs, InP

Now the difference between two reciprocal lattice vectors, i.e. $\mathbf{G} - \mathbf{G}'$, must be another reciprocal lattice vector, $\mathbf{G}''$. In addition, the scalar product of a Bravais lattice vector with a reciprocal lattice vector is just an integral number of 2π, i.e.

$$\mathbf{R} \cdot \mathbf{G}'' = 2\pi n, \quad \text{where} \quad n \in \mathbb{Z} \tag{15.33}$$

Therefore, $e^{i(\mathbf{G} - \mathbf{G}') \cdot \mathbf{R}} = 1$, and hence for N Bravais lattice points in the crystal:

$$S = N \sum_{\mathbf{t}} e^{i(\mathbf{G} - \mathbf{G}') \cdot \mathbf{t}} \tag{15.34}$$

As discussed earlier in Chapter 1, *most* of the important semiconductors, both elemental and compound, have a face-centred cubic Bravais lattice with a two-atom basis, as shown in Fig. 15.1. The atomic positions within the basis are described by the *basis vector*, $\mathbf{t} = \pm \mathbf{T}$ (say), where:

$$\mathbf{T} = \frac{A_0}{8} (\hat{\mathbf{i}} + \hat{\mathbf{j}} + \hat{\mathbf{k}}) \tag{15.35}$$

Hence, the summation for the structure factor in equation (15.34) contains just two terms, thus giving:

$$S = N \left[e^{i(\mathbf{G} - \mathbf{G}') \cdot \mathbf{T}} + e^{-i(\mathbf{G} - \mathbf{G}') \cdot \mathbf{T}} \right] \tag{15.36}$$

and therefore:

$$S = N \left[\cos(\mathbf{G} - \mathbf{G}') \cdot \mathbf{T} + i \sin(\mathbf{G} - \mathbf{G}') \cdot \mathbf{T} \right.$$

$$\left. + \cos(\mathbf{G} - \mathbf{G}') \cdot \mathbf{T} - i \sin(\mathbf{G} - \mathbf{G}') \cdot \mathbf{T} \right] \tag{15.37}$$

$$\therefore S = 2N \cos(\mathbf{G} - \mathbf{G}') \cdot \mathbf{T} \tag{15.38}$$

By using this form for the structure factor S, the potential term V, as presented above in equation (15.30), then becomes:

$$V = \frac{2N}{\Omega} \cos(\mathbf{G} - \mathbf{G}') \cdot \mathbf{T} \int V_a(\mathbf{r}) e^{i(\mathbf{G} - \mathbf{G}') \cdot \mathbf{r}} \, d\tau \tag{15.39}$$

Now the total volume, Ω, divided by the number of Bravais lattice points in the crystal N, is equal to the volume occupied by a single Bravais lattice point, Ω_c (say). Note for the face-centred cubic crystal, the primitive cube of side A_0 (the lattice constant) contains four Bravais lattice points (i.e. eight atoms), and hence:

$$\Omega_c = \frac{A_0^3}{4} \tag{15.40}$$

Furthermore, relabelling $\mathbf{G'} - \mathbf{G}$ as the vector $\mathbf{q}$, then:

$$V = \frac{2}{\Omega_c} \cos \mathbf{q} \cdot \mathbf{T} \int V_a(\mathbf{r}) e^{-i\mathbf{q} \cdot \mathbf{r}} \, d\tau \tag{15.41}$$

The factor:

$$\frac{1}{\Omega_c} \int V_a(\mathbf{r}) e^{-i\mathbf{q} \cdot \mathbf{r}} \, d\tau = V_f(q) \tag{15.42}$$

is the Fourier transform of the, as yet, undefined atomic potential V_a, and can be labelled conveniently as $V_f(q)$ where $q = |\mathbf{q}|$. This term is known as the *pseudo-potential form factor* [350].

Table 15.1   The first 65 reciprocal lattice vectors of a face-centred cubic crystal (in units of $2\pi/A_0$)

Direction	Permutations						Magnitude
$\langle 000 \rangle$	$[000]$						0
$\langle 111 \rangle$	$[111]$	$[\bar{1}11]$	$[1\bar{1}1]$	$[11\bar{1}]$	$[\bar{1}\bar{1}1]$	$[1\bar{1}\bar{1}]$	$\sqrt{3}$
	$[\bar{1}1\bar{1}]$	$[\bar{1}\bar{1}\bar{1}]$					
$\langle 200 \rangle$	$[200]$	$[020]$	$[002]$	$[\bar{2}00]$	$[0\bar{2}0]$	$[00\bar{2}]$	2
$\langle 220 \rangle$	$[220]$	$[202]$	$[022]$	$[\bar{2}20]$	$[2 0 \bar{2}]$	$[0 2 \bar{2}]$	$\sqrt{8}$
	$[2\bar{2}0]$	$[\bar{2}02]$	$[0\bar{2}2]$	$[\bar{2}\bar{2}0]$	$[\bar{2}0\bar{2}]$	$[0\bar{2}\bar{2}]$	
$\langle 311 \rangle$	$[311]$	$[131]$	$[113]$	$[\bar{3}11]$	$[\bar{1}31]$	$[\bar{1}13]$	$\sqrt{11}$
	$[3\bar{1}1]$	$[1\bar{3}1]$	$[1\bar{1}3]$	$[11\bar{3}]$	$[31\bar{1}]$	$[13\bar{1}]$	
	$[\bar{3}\bar{1}1]$	$[\bar{1}\bar{3}1]$	$[\bar{1}\bar{1}3]$	$[\bar{3}1\bar{1}]$	$[\bar{1}3\bar{1}]$	$[\bar{1}13]$	
	$[3\bar{1}\bar{1}]$	$[1\bar{3}\bar{1}]$	$[1\bar{1}\bar{1}]$	$[3\bar{1}\bar{1}]$	$[\bar{1}3\bar{1}]$	$[11\bar{3}]$	
$\langle 222 \rangle$	$[222]$	$[\bar{2}22]$	$[2\bar{2}2]$	$[22\bar{2}]$	$[\bar{2}\bar{2}2]$	$[2\bar{2}\bar{2}]$	$\sqrt{12}$
	$[\bar{2}2\bar{2}]$	$[\bar{2}\bar{2}\bar{2}]$					
$\langle 400 \rangle$	$[400]$	$[040]$	$[004]$	$[\bar{4}00]$	$[0\bar{4}0]$	$[00\bar{4}]$	4

It should be noted that, as q (the magnitude of $\mathbf{q}$) is a scalar, then the form factors are spherically symmetric. Physically, they represent the central nuclear potential, plus the potential of the inner electron shells, and so for Si, for example, this would include the $1s^2 2s^2 2p^6$ electrons. The remaining four valence electrons, which in an isolated Si atom are found in the 3s and 3p orbitals, are the subject of the investigation; it is their energy levels and charge distributions that determine the electronic properties of the crystal and thus provide the motivation for this theoretical derivation.

The *empirical* nature of the pseudo-potential method is incorporated by adjusting the values of $V_f(q)$ in order to achieve the closest agreement of the calculated energy levels with those measured by experimental methods. Note, therefore, that $V_f(q)$ summarizes many of the microscopic electrostatic properties of the crystal. For example, it accounts for the nuclear charge, the inner-shell electrons, the screening provided by these electrons, and, under the auspices of the independent electron approximation described earlier, it also accounts for the electron–electron interaction experienced between the valence electrons.

As mentioned above, $\mathbf{q}$ (the difference between two reciprocal vectors) is also a reciprocal lattice vector, and in a bulk crystal, such as a face-centred cubic crystal, it takes discrete values as deduced in Chapter 1 (see Table 15.1). The pseudo-potential form factor $V_f(q)$ is, therefore, also a discrete function, only having non-zero values for particular q. Cohen and Bergstresser [350] found that the experimentally determined band structure features of Si could be reproduced by using the values of $V_f(q)$ given in Table 15.2.

Table 15.2   The form factors (in eV) of the common group IV semiconductor elements, converted from the original values of Cohen and Bergstresser [350]

Material	A_0 (Å)	$V_f(\sqrt{3})$	$V_f(\sqrt{8})$	$V_f(\sqrt{11})$
Si	5.43	−1.43	+0.27	+0.54
Ge	5.66	−1.57	+0.07	+0.41

It should be noted that, as $V_f(q)$ is truncated for $q > \sqrt{11}$ and as $V_f(0)$ only has the effect of shifting the energies up or down, then $V_f(q)$ has only three non-zero values, which occur for $q = \sqrt{3}$, $\sqrt{8}$, and $\sqrt{11}$. Note also that, for $q = 2$, the structure factor for the face-centred cubic crystals of interest here, which is given by equation (15.38) becomes:

$$S = 2N \cos \left(\mathbf{q}_{\bullet} \left(\frac{A_0}{8} \hat{\mathbf{i}} + \frac{A_0}{8} \hat{\mathbf{j}} + \frac{A_0}{8} \hat{\mathbf{k}} \right) \right) \tag{15.43}$$

With $\mathbf{q} = (2\pi/A_0)2\hat{\mathbf{i}}$, for example, this then gives:

$$S = 2N \cos \left(\frac{2\pi}{A_0} \frac{A_0}{8} 2\hat{\mathbf{i}}_{\bullet} \hat{\mathbf{i}} \right) = 2N \cos \frac{\pi}{2} \tag{15.44}$$

which is zero, and hence the value of $V_f(q)$ is irrelevant.

Reverting back to the mathematics, then:

$$V = 2V_f(q) \cos (\mathbf{G} - \mathbf{G}')_{\bullet} \mathbf{T} \tag{15.45}$$

which when substituted back into equation (15.24) finally gives the complete form for the matrix elements as:

$$\mathcal{H}_{\mathbf{G}',\mathbf{G}} = \frac{\hbar^2}{2m_0} |\mathbf{G} + \mathbf{k}|^2 \delta_{\mathbf{G},\mathbf{G}'} + 2V_f(q) \cos (\mathbf{G} - \mathbf{G}')_{\bullet} \mathbf{T} \tag{15.46}$$

Fig. 15.2 displays the results of calculations of the bulk band structure of Si by using the form factors of Table 15.2, together with the 65-element plane wave basis set

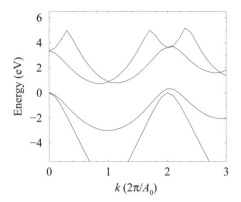

Figure 15.2   The band structure of bulk Si along the $\langle 100 \rangle$ direction, obtained by using the pseudo-potential form factors of Cohen and Bergstresser [350], adjusted to set the top of the valence band to zero

in Table 15.1. The graph shows the three highest-energy valence-band levels, which (in this simple approach, ignoring spin–orbit coupling) are all degenerate at $k = 0$, and two of which remain degenerate across all k. In addition, the two lowest-energy conduction-band states can be clearly seen. In this first calculation, the energy levels are plotted along one of the $\langle 100 \rangle$ directions. The continuous energy *band* nature of the solutions is visible, as expected, which is in contrast to the solutions of the Schrödinger equation in heterostructures, the focus of the majority of this work so far. Furthermore, Fig. 15.2 demonstrates the periodic nature of the band structure, with equal energy levels separated by an electron wave vector equal to [200], i.e. the smallest reciprocal lattice vector along the direction of interest (see Table 15.1). Such periodicity is, of course, merely reflecting the Brillouin zone symmetry structure of the crystal [1] and the edge of the *first* Brillouin is at exactly half of the reciprocal lattice vector in question, i.e. $k = 2\pi/A_0$. Note that there is a slight discrepancy in the energy levels, e.g. exact degeneracy is not reproduced at the top of the valence band at $k = 2 \times 2\pi/A_0$; this is simply a computational deficiency, and as the vast majority of calculations are confined to the first Brillouin zone, is not an issue here.

Of course, there is a periodicity in the energy bands along other crystal directions, e.g. the edge of the Brillouin zone along the $\langle 111 \rangle$ direction is at $k = \sqrt{3}/2 \times 2\pi/A_0$. The data from these two principal axes of symmetry are often gathered together to produce a standard form for displaying the band structure; Fig. 15.3 illustrates this for the two most common group IV elemental semiconductors, i.e. Si and Ge. Again, the form factors given in Table 15.2 were employed, together with the 65-element basis set of Table 15.1. The indirect nature of both materials can be seen from the band-gap minimum for Si occurring towards the X point within the Brillouin zone, while for Ge it occurs towards L. The choice of the basis set, i.e. plane waves, has been discussed, and the *number* of 65 can be justified by the data presented in Table 15.3. This table displays the results of calculations of the energy of the lowest conduction band in Si at two electron wave vectors $\mathbf{k}$. The first is for the Γ point ($\mathbf{k} = 0$), and the second is near the X minima ($\mathbf{k} = 0, 0, 0.8$). In addition, the table shows the discretization of the face-centred reciprocal lattice vectors, plus the largest vector included in the set.

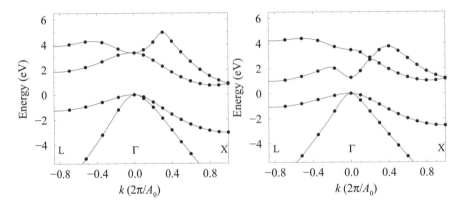

Figure 15.3   The band structures of bulk Si (left) and Ge (right) obtained by using the pseudo-potential form factors of Cohen and Bergstresser [350], adjusted to set the top of the valence band to zero

Table 15.3   The energies of the lowest conduction band of Si at Γ and $[0,0,0.8]$ (near the X minima) as a function of the number of plane waves in the basis set

Number of plane waves	Largest $\mathbf{G}$ $(2\pi/A_0)$	E_Γ (eV)	E_X (eV)
15	$\langle 200 \rangle$	3.212	0.926
27	$\langle 220 \rangle$	3.453	0.982
51	$\langle 311 \rangle$	3.552	1.085
59	$\langle 222 \rangle$	3.389	0.827
65	$\langle 400 \rangle$	3.344	0.741
89	$\langle 331 \rangle$	3.382	0.781
113	$\langle 420 \rangle$	3.412	0.825
137	$\langle 422 \rangle$	3.409	0.821

The convergence of the band structure is perhaps most visible through a plot of the above data, as in Fig. 15.4. The data corresponding to a 65-element basis set is just at the bottom of the shoulder, i.e. at the beginning of the plateau. It is this accuracy that is generally chosen for simple bulk calculations, such as in Fig. 15.3.

In summary, the *empirical* pseudo-potential method is a technique for inputting the experimentally determined features into a theoretical model of the electronic structure of the crystal. At this level, the features that are reproduced are the fundamental band-structure properties, such as the band gap at the centre of the Brillouin Zone, Γ, and at indirect gaps such as the X-valley in Si.

15.3 SPIN–ORBIT COUPLING

Speaking simplistically, the four valence electrons of a Si (or Ge) atom, three of which occupy p-states and one an s-state, produce four equivalent covalent bonds (sp³ hy-

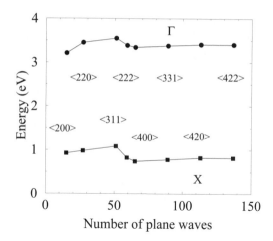

Figure 15.4   The energies of the lowest conduction band of Si at Γ and [0,0,0.8] (near the X minima) as a function of the number of plane waves in the basis set

bridization) directed towards the corners of a tetrahedron, upon crystallization. Four neighbouring atoms each supply a single electron, and hence any one atom completes its outer shell of eight electrons by forming four doubly occupied bonds. The same is also true for a compound semiconductor, but in this case, however, one species, the cation (e.g. Ga) supplies three electrons and the anion (e.g. As) supplies five.

Within an infinite crystal, as has already been shown with the pseudo-potential calculations in the previous section, these bonds form *bands*. Fig. 15.5 illustrates the three uppermost of these *valence bands* for Si and Ge. The lowest-energy valence band is far off scale to the bottom of the figure and represents a well-localized state around the atomic cores, and has an s-type nature.

The other three bands, in the calculations performed so far, are degenerate for stationary electrons (holes). These bands resemble the p-type atomic orbitals and have an orbital angular momentum, $L = 1$ (see, for example [18], p. 41). This couples with the spin angular momentum, $S = \frac{1}{2}$ to give $J = L + S = \frac{3}{2}$ and $J = L - S = \frac{1}{2}$, thus lifting the degeneracy. This is analogous to LS coupling in a low-atomic-number atom (see any atomic physics book for an introduction, for example [3] p. 428 and 441).

In III–V compounds, it has been found that the quadruplet $J = \frac{3}{2}$ is higher in energy than the $J = \frac{1}{2}$ doublet. Note that the degeneracy, whether two-fold or four-fold, arises from the number of possible values that the projection J_z of J can take. As the total energy remains the same, the change in the energy of the quadruplet $J = \frac{3}{2}$ must be half that of the $J = \frac{1}{2}$ doublet.

The quadruplet is degenerate in bulk systems at the zone centre, but for an increasing electron wave vector it splits into two separate spin degenerate bands. The lower-energy band corresponds to the $m_J = \frac{3}{2}$ projection and because of the effective mass interpretation of the E–$\mathbf{k}$ curves, it has a higher effective mass than the higher energy $m_J = \frac{1}{2}$ band. Hence, the two $J = \frac{3}{2}$ bands are labelled as the heavy-hole (HH) ($m_J = \frac{3}{2}$) and the light-hole (LH) ($m_J = \frac{1}{2}$) bands, as shown in Fig. 15.5.

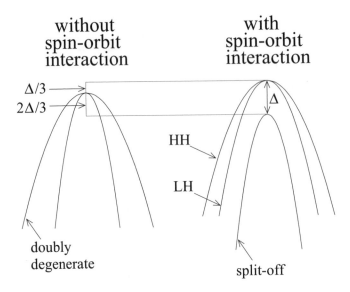

Figure 15.5   The effect of the spin–orbital interaction on the valence band structure

The spin–orbit interaction can be accounted for within the empirical pseudo-potential approach [351, 352] and amounts to an addition term in the potential V as originally defined in equation (15.25), i.e. the crystal potential V_c is replaced with $V_c + V_{so}$.

Inclusion of the spin–orbit interaction is essential if detailed calculations of valence band structure are to be performed, and this will be returned to after the elemental calculation has been generalized to compound semiconductors.

15.4 COMPOUND SEMICONDUCTORS

The generalization of the calculation to compound semiconductors manifests itself merely as a change in the atomic basis.

Recalling equation (15.30):

$$V = \frac{1}{\Omega} \sum_{\mathbf{r}_a} e^{i(\mathbf{G}-\mathbf{G}')\cdot\mathbf{r}_a} \int V_a(\mathbf{r}) e^{i(\mathbf{G}-\mathbf{G}')\cdot\mathbf{r}} \; \mathrm{d}\tau \qquad (15.47)$$

then again, the structure factor can be written as:

$$S = \sum_{\mathbf{r}_a} e^{i(\mathbf{G}-\mathbf{G}')\cdot\mathbf{r}_a} = \sum_{\mathbf{R}}\sum_{\mathbf{t}} e^{i(\mathbf{G}-\mathbf{G}')\cdot(\mathbf{R}+\mathbf{t})} \qquad (15.48)$$

giving as before:

$$S = N \sum_{\mathbf{t}} e^{i(\mathbf{G}-\mathbf{G}')\cdot\mathbf{t}} \qquad (15.49)$$

However, for a compound semiconductor, the sum over $\mathbf{t}$ is over two dissimilar atoms (the cation is at $-\mathbf{T}$ and the anion is at $+\mathbf{T}$, where $\mathbf{T} = \frac{A_0}{8}(\hat{\mathbf{i}} + \hat{\mathbf{j}} + \hat{\mathbf{k}})$) and therefore it cannot be made independent of the form of the atomic potential $V_a(\mathbf{r})$. Hence, the structure factor S must be substituted back into equation (15.47) at this point, thus giving:

$$V = \frac{N}{\Omega} \sum_{\mathbf{t}} e^{i(\mathbf{G}-\mathbf{G}')\cdot\mathbf{t}} \int V_a(\mathbf{r}) e^{i(\mathbf{G}-\mathbf{G}')\cdot\mathbf{r}} \, d\tau \tag{15.50}$$

Performing the sum over the two basis positions, $\mathbf{T}^{\text{cat}} = -\mathbf{T}$, and $\mathbf{T}^{\text{an}} = +\mathbf{T}$, then:

$$V = \frac{N}{\Omega} e^{-i(\mathbf{G}-\mathbf{G}')\cdot\mathbf{T}} \int V_a^{\text{cat}}(\mathbf{r}) e^{i(\mathbf{G}-\mathbf{G}')\cdot\mathbf{r}} \, d\tau$$

$$+ \frac{N}{\Omega} e^{i(\mathbf{G}-\mathbf{G}')\cdot\mathbf{T}} \int V_a^{\text{an}}(\mathbf{r}) e^{i(\mathbf{G}-\mathbf{G}')\cdot\mathbf{r}} \, d\tau \tag{15.51}$$

Recalling that $N/\Omega = 1/\Omega_c$, and again writing $\mathbf{q} = \mathbf{G}' - \mathbf{G}$, then obtain:

$$V = e^{-i(\mathbf{G}-\mathbf{G}')\cdot\mathbf{T}} \frac{1}{\Omega_c} \int V_a^{\text{cat}}(\mathbf{r}) e^{-i\mathbf{q}\cdot\mathbf{r}} \, d\tau + e^{i(\mathbf{G}-\mathbf{G}')\cdot\mathbf{T}} \frac{1}{\Omega_c} \int V_a^{\text{an}}(\mathbf{r}) e^{-i\mathbf{q}\cdot\mathbf{r}} \, d\tau \tag{15.52}$$

Now the definition of the pseudo-potential form factor (from equation (15.42)) is as follows:

$$V_f(q) = \frac{1}{\Omega_c} \int V_a(\mathbf{r}) e^{-i\mathbf{q}\cdot\mathbf{r}} \, d\tau \tag{15.53}$$

$$\therefore V = V_f^{\text{cat}}(q) e^{-i(\mathbf{G}-\mathbf{G}')\cdot\mathbf{T}} + V_f^{\text{an}}(q) e^{i(\mathbf{G}-\mathbf{G}')\cdot\mathbf{T}} \tag{15.54}$$

where the form factors associated with the cation and anion atomic potentials, $V_a^{\text{cat}}(\mathbf{r})$ and $V_a^{\text{an}}(\mathbf{r})$, have been labelled as $V_f^{\text{cat}}(q)$ and $V_f^{\text{an}}(q)$, respectively. This equation can be manipulated further, to give:

$$V = \left[V_f^{\text{an}}(q) + V_f^{\text{cat}}(q) \right] \cos\left(\mathbf{G} - \mathbf{G}'\right)\cdot\mathbf{T}$$

$$+ i \left[V_f^{\text{an}}(q) - V_f^{\text{cat}}(q) \right] \sin\left(\mathbf{G} - \mathbf{G}'\right)\cdot\mathbf{T} \tag{15.55}$$

A quick check on the validity of this extension is to put the cation potential $V_f^{\text{cat}}(q)$ and the anion potential $V_f^{\text{an}}(q)$ equal to the same potential $V_f(q)$, thus reproducing the potential term of the elemental calculation, as in equation (15.45).

Often the 'sum' $\left[V_f^{\text{an}}(q) + V_f^{\text{cat}}(q) \right]$ and 'difference' $\left[V_f^{\text{an}}(q) - V_f^{\text{cat}}(q) \right]$ potentials are relabelled as the symmetric and anti-symmetric form factors, respectively, i.e. $V_f^S(q)$ and $V_f^A(q)$, this gives:

$$V = V_f^S(q) \cos\left(\mathbf{G} - \mathbf{G}'\right)\cdot\mathbf{T} + i V_f^A(q) \sin\left(\mathbf{G} - \mathbf{G}'\right)\cdot\mathbf{T} \tag{15.56}$$

Substituting for the potential V into equation (15.24), finally gives all of the elements of the Hamiltonian matrix as:

$$\mathcal{H}_{\mathbf{G}',\mathbf{G}} = \frac{\hbar^2}{2m_0} |\mathbf{G} + \mathbf{k}|^2 \delta_{\mathbf{G},\mathbf{G}'} + V_f^S(q) \cos\left(\mathbf{G} - \mathbf{G}'\right)\cdot\mathbf{T}$$

$$+ i V_f^A(q) \sin\left(\mathbf{G} - \mathbf{G}'\right)\cdot\mathbf{T} \tag{15.57}$$

Table 15.4 gives examples of the symmetric ($V_f^S(q)$) and anti-symmetric ($V_f^A(q)$) form factors as deduced *empirically* by Cohen and Bergstresser [350]. Note that the value of

Table 15.4   The form factors of a selection of III–V compound semiconductor in eV, converted from the original values of Cohen and Bergstresser [350]

| Material | A_0 (Å) | $V_f^S(q)$ | | | $V_f^A(q)$ | | |
		$\sqrt{3}$	$\sqrt{8}$	$\sqrt{11}$	$\sqrt{3}$	2	$\sqrt{11}$
GaAs	5.64	−3.13	+0.14	+0.82	+0.95	+0.68	+0.14
InAs	6.04	−2.99	0.00	+0.68	+1.09	+0.68	+0.41
GaP	5.44	−2.99	+0.41	+0.95	+1.63	+0.95	+0.27
InP	5.86	−3.13	+0.14	+0.82	+0.95	+0.68	+0.14

the symmetric form factor of $q = 2$, i.e. $V_f^S(2)$, is again unnecessary because of a zero structure factor, as in the elemental (group IV) calculation. In the anti-symmetric case, however, the structure factor corresponding to $q = \sqrt{8}$ is zero, thus allowing this form factor to remain undefined.

Fig. 15.6 displays the results of calculations of the band structures of bulk GaAs and InAs. In contrast to Si and Ge, the minimum band gaps are direct, and thus an electron and hole can recombine without recourse to a momentum change resulting in efficient light emission. Hence, the use of both of these materials as part of the quaternary compound, $In_{1-x}Ga_xAs_yP_{1-y}$ [353] in light-emitting diodes [354].

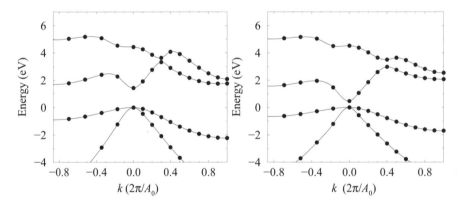

Figure 15.6   The band structures of bulk GaAs (left) and InAs (right) obtained by using the pseudo-potential form factors of Cohen and Bergstresser [350], shifted to set the top of the valence band to zero

In recent years, a new material, namely GaN (and its alloys with In), has come to prominence with the development of room temperature blue-light-emitting diodes [355] and continuous-wave lasers [356,357]. GaN is particularly interesting from the viewpoint of electronic band structure calculations in that it does not take the face-centred cubic zinc blende or diamond structures employed so far in this work, rather the natural crystal structure of GaN, being the hexagonal close-packed wurtzite structure (see, for example [2], p. 47).

Such a change in crystal symmetry does not represent a problem for the analysis derived above, as it is only necessary to change the basis vectors, the reciprocal lattice vectors and, of course, employ the appropriate pseudo-potential form factors (for such a treatise see, for example [358]). Many examples will appear in the following, which illustrate pseudo-potential calculations of systems of symmetry other than zinc blende, e.g. those involving isolated impurities, superlattices and quantum wires.

15.5 CHARGE DENSITIES

In addition to yielding the eigenvalues (energy levels), the diagonalization of equations (15.46) and (15.57) also gives the corresponding eigenvectors, i.e. the expansion coefficients $a_{n,\mathbf{k}}$ for each one of the basis vectors $\mathbf{G}$. It is straightforward to generate the wave function from these eigenvectors by using equation (15.9), i.e.

$$\psi_{n,\mathbf{k}}(\mathbf{r}) = \sum_{\mathbf{G}} a_{n,\mathbf{k}}(\mathbf{G})e^{i(\mathbf{G}+\mathbf{k})\cdot\mathbf{r}} \qquad (15.58)$$

Utilizing the probability interpretation of the wave function, i.e. the probability of finding a particle at a point is proportional to $\psi^*\psi$, then the *charge density* is given by:

$$\rho = \psi_{n,\mathbf{k}}^*(\mathbf{r})\psi_{n,\mathbf{k}}(\mathbf{r}) \qquad (15.59)$$

and can be easily evaluated.

Fig. 15.7 illustrates the results of summing the charge density over all of the four bulk valence band states, across a series of x–y planes through a single face-centred cube of Si. Given the coordinate system employed thus far, the Bravais lattice points are, of course, at the corners and on the faces of the cube of side A_0, with a pair of atoms at $(-\frac{1}{8}, -\frac{1}{8}, -\frac{1}{8})$ and $(+\frac{1}{8}, +\frac{1}{8}, +\frac{1}{8})$, respectively. Therefore, the first ($z = 0$), third ($z = 0.250A_0$) and fifth ($z = 0.5A_0$) planes are cross-sections through the bond centres. The symmetry of the sp^3-hybridized bonds can be seen most clearly across the planes intersecting the centre of atoms, namely the second ($z = 0.125A_0$) and fourth ($z = 0.375A_0$) planes. In addition, this figure illustrates the equality of the atoms in this, the diamond structure, where the symmetry between the planes containing the atoms is merely translational.

In contrast to this, Fig. 15.8 displays the corresponding charge density plot, again summed over all four valence-band states, but this time for the zinc blende structure of GaAs. Again, there is a symmetry between the planes intersecting the bond centres, i.e. the first ($z = 0$), third ($z = 0.250A_0$) and fifth ($z = 0.5A_0$). However, the two planes intersecting the centres of the atoms are quite dissimilar. In particular, the second ($z = 0.125A_0$) plane contains the anions, in this case As, and is quite different from the fourth ($z = 0.375A_0$) plane which contains the cations, i.e. Ga. The fourth plane displays an increase in the charge density (represented by the lighter colour) around the Ga atoms, while the second plane displays a decrease in the charge density around the As atoms, in comparison to the Si structure of Fig. 15.7. This is a reflection of the increased *ionicity* in the Ga–As bond, in comparison with the completely covalent bond in bulk Si. Instead of having an equal share of the two valence electrons that sit between the atomic cores and form the bond, as occurs in the diamond structure, the Ga atom takes a larger share, thus becoming a partially negatively charged centre (i.e. a cation);

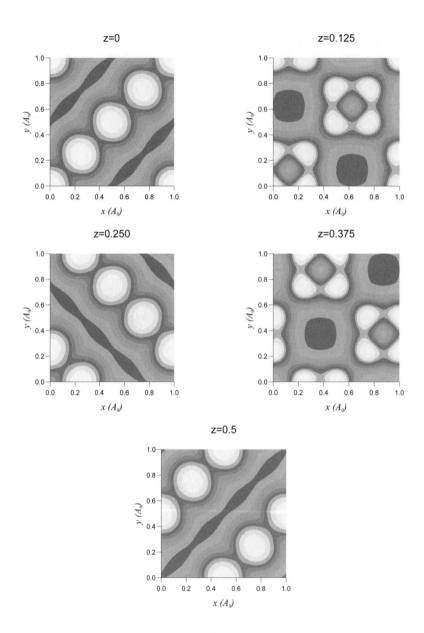

Figure 15.7 Total valence-band charge density for Si across the x–y plane of a single face-centred cubic unit cell, for a variety of z values (given in A_0); note that the lighter the colour, then the higher the charge density. The Bravais lattice points are at the corners and centre of the $z = 0$ plane and on the edges of the $z = 0.5A_0$ plane

the As atom therefore has a smaller share, thus becoming a partially positively charged centre (i.e. an anion).

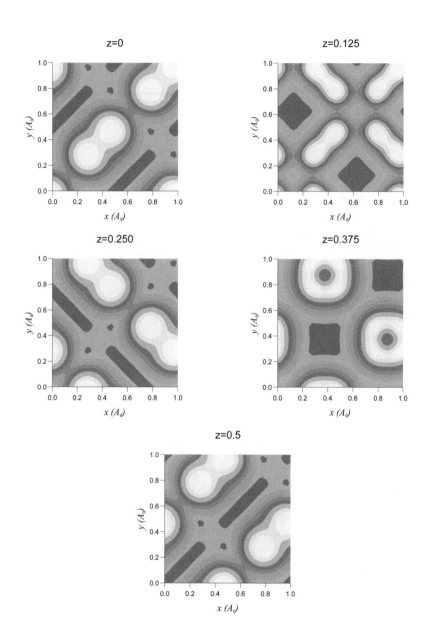

Figure 15.8   Total valence-band charge density for GaAs across the x–y plane of a single face-centred cubic unit cell, for a variety of z values (given in A_0); note that the lighter the colour, then the higher the charge density. The Bravais lattice points are at the corners and centre of the $z = 0$ plane and on the edges of the $z = 0.5A_0$ plane

15.6 CALCULATING THE EFFECTIVE MASS

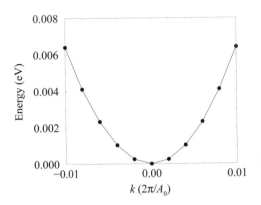

Figure 15.9 Empirical pseudo-potential calculations of the band structure of bulk GaAs around the centre (Γ) of the Brillouin zone (solid circles), together with a parabolic fit to the data (line)

Empirical pseudo-potential calculations can be used to calculate material parameters other than just the main energy gaps. Fig. 15.9 displays the results of calculations of the lowest conduction band of GaAs at points along one of the $\langle 100 \rangle$ directions around the centre of the Brillouin zone (the minimum has been adjusted to zero). In addition to the data points, a parabolic fit $E = Ak^2$ has also been included, suggested by the definition of the effective mass, i.e.

$$E = \frac{\hbar^2 k^2}{2m^*} \tag{15.60}$$

From this, the effective mass m^* of the electrons around the Γ conduction band minimum can be calculated, in this case giving $m^* = 0.074 m_0$. Given the crudity of the form factors employed, this compares well with the accepted value of $0.067 m_0$ [14]. This implies that form factors that have been deduced empirically by comparing calculated energy gaps with those measured experimentally can also give good information as to the effective masses. This approach has been used successfully in other materials, and in one case has produced addition insight into the behaviour of excitons in quantum wells [102]. More sophisticated potentials will be introduced later, which in addition to allowing for more accurate determination of effective masses will also allow for the calculation of deformation potentials in strained crystals.

15.7 ALLOYS

Mixing two binary (or indeed elemental) semiconductors together to form an alloy is a very common and often used technique for producing a whole new range of materials whose fundamental properties, e.g. the bandgap and the Γ valley effective mass, can be tuned by adjusting the proportions of the constituents [14]. The simplest semiconductor alloy would be that formed from the two common elemental semiconductors Si and Ge [359], but perhaps the most common are the alloys of GaAs, such as $Ga_{1-x}Al_xAs$, a ternary alloy formed from GaAs and AlAs.

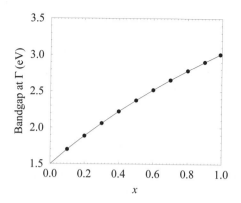

Figure 15.10   The zone-centre bandgap of a $Ga_{1-x}Al_xAs$ alloy calculated by linearly interpreting the pseudo-potential form factors

Generally, such alloys are assumed to have properties that vary linearly between the two constituents, a procedure that is known as the *virtual crystal approximation*, thus giving material parameters that are simply taken in proportion from their components. For example, the bandgap of $Ga_{1-x}Al_xAs$ is given by:

$$E_g^{Ga_{1-x}Al_xAs} = (1-x)E_g^{GaAs} + xE_g^{AlAs} \qquad (15.61)$$

However, in some instances such a linear interpolation is only approximate and the actual dependence is more complicated, e.g. it is quite surprising that there is still some disagreement as to the particular form of the bandgap of $Ga_{1-x}Al_xAs$, as highlighted by the range of measured values quoted by Adachi ([14], p. 146). A non-linear dependence is often written as:

$$E_g^{Ga_{1-x}Al_xAs} = (1-x)E_g^{GaAs} + xE_g^{AlAs} + C_{GaAl}x(x-1) \qquad (15.62)$$

where $C_{GaAl}x(x-1)$ is known as the *bowing parameter*.

Such detailed reviews of the properties of alloys lay beyond the focus of this present work, although it is sufficient to say here that the particulars are well documented (see, for example [14, 359, 360]) and that in this work it is sufficient to know that equation (15.61) can be generalized to represent any material parameter, e.g. the lattice constant:

$$A_0^{Ga_{1-x}Al_xAs} = (1-x)A_0^{GaAs} + xA_0^{AlAs} \qquad (15.63)$$

or the pseudo-potential form factors:

$$V_f(q)^{Ga_{1-x}Al_xAs} = (1-x)V_f(q)^{GaAs} + xV_f(q)^{AlAs} \qquad (15.64)$$

Fig. 15.10 displays the results of a simple demonstration calculation showing the bandgap at the centre of the Brillouin zone as a function of the proportion of Al in the $Ga_{1-x}Al_xAs$ alloy; note that a slight degree of bowing *is* evident here.

The reader may be interested to know that the microscopic properties of alloys are themselves a thriving research topic, one aspect of which is whether the constituent atoms, for example, Si and Ge or Ga and Al, are arranged randomly or otherwise (see, for example [361–363]).

15.8 ATOMIC FORM FACTORS

It is interesting to decompose the symmetric and anti-symmetric form factors to yield the individual *atomic* form factors, which are given simply by:

$$V_f^{\text{an}}(q) = \frac{1}{2} \left[V_f^S(q) + V_f^A(q) \right] \tag{15.65}$$

$$V_f^{\text{cat}}(q) = \frac{1}{2} \left[V_f^S(q) - V_f^A(q) \right] \tag{15.66}$$

The term 'atomic' is introduced here to specify unambiguously that these are pseudo-potential form factors that can be identified with an individual atom, unlike the symmetric, $V_f^S(q)$, and anti-symmetric, $V_f^A(q)$, form factors which are identified with a cation–anion basis.

Table 15.5   The atomic form factors (in eV) of a selection of III–V compound semiconductors, decomposed from their original symmetric and anti-symmetric values as in Table 15.4

Cation	Anion	$V_f^{\text{cat}}(q)$		$V_f^{\text{an}}(q)$	
		$\sqrt{3}$	$\sqrt{11}$	$\sqrt{3}$	$\sqrt{11}$
Ga	As	−2.04	+0.34	−1.09	+0.48
In	As	−2.04	+0.14	−0.95	+0.54
Ga	P	−2.31	+0.34	−0.68	+0.61
In	P	−2.04	+0.34	−1.09	+0.48

Table 15.5 gives the atomic form factors for the cations (Ga, In) and anions (As, P) calculated in this manner. Note at this point that it is not possible to deduce the atomic form factors $V_f^{\text{an}}(q)$ and $V_f^{\text{cat}}(q)$ at $q = 2$ and $\sqrt{8}$ as this would rely on *both* the symmetric $(V_f^S(q))$ and anti-symmetric $(V_f^A(q))$ form factors being known at these points. There are quite some differences between the potentials of the same species in different materials, e.g. the As associated with GaAs, compared with the As associated with InAs. This appears to be a general result, with the physical interpretation being that the *core* potential i.e. nucleus plus inner electron shells, of the As atom is indeed different in both materials, with this difference arising from the crystal environment. Such a result is disappointing, as a single well-defined atomic potential suitable for all compounds would be a very powerful parameter indeed, allowing the properties of new materials to be predicted without recourse to empirical deductions of the form factors.

15.9 GENERALIZATION TO A LARGE BASIS

Thus far, the background to the empirical pseudo-potential method has been explored and applied to the calculation of the electronic band structure of bulk elemental and compound semiconductors. It is convenient now to generalize this method to allow for the calculation of the electronic properties of systems of different symmetries, in particular quantum wells, wires and dots. The application of this theory will be covered in the following two chapters.

Consider now the first, and hence most general, expression for the crystal potential (equation (15.30)), i.e.

$$V = \frac{1}{\Omega} \sum_{r_a} e^{i(\mathbf{G}-\mathbf{G}')\cdot r_a} \int V_a(\mathbf{r}) e^{i(\mathbf{G}-\mathbf{G}')\cdot \mathbf{r}} \, d\tau \tag{15.67}$$

Again, writing $\mathbf{q} = \mathbf{G}' - \mathbf{G}$, then:

$$V = \frac{\Omega_c}{\Omega} \sum_{r_a} e^{-i\mathbf{q}\cdot r_a} \frac{1}{\Omega_c} \int V_a(\mathbf{r}) e^{-i\mathbf{q}\cdot \mathbf{r}} \, d\tau \tag{15.68}$$

where Ω_c is again the volume of the primitive cell, which in the case of a face-centred cubic crystal is $A_0^3/4$. Now, clearly the normalized integral of the atomic potential $V_a(\mathbf{r})$ is still the pseudo-potential form factor, although now it is acknowledged that the generalization may allow for many atom types at, as yet, unspecified positions. Hence, it is important to write:

$$\frac{1}{\Omega_c} \int V_a(\mathbf{r}) e^{-i\mathbf{q}\cdot \mathbf{r}} \, d\tau \quad \text{as} \quad V_f^{r_a}(q) \tag{15.69}$$

implying the empirical pseudo-potential form factor of the atom at position r_a. Therefore:

$$V = \frac{\Omega_c}{\Omega} \sum_{r_a} e^{-i\mathbf{q}\cdot r_a} V_f^{r_a}(q) \tag{15.70}$$

As with bulk semiconductors, the atomic positions can always be written as a sum of a Bravais lattice vector and a basis vector, i.e.

$$r_a = \mathbf{R} + \mathbf{t} \tag{15.71}$$

which, as before, replaces the summation over the atomic positions r_a with two summations, i.e. one over $\mathbf{R}$ and one over $\mathbf{t}$, thus giving the following:

$$V = \frac{\Omega_c}{\Omega} \sum_{\mathbf{R}} e^{-i\mathbf{q}\cdot\mathbf{R}} \sum_{\mathbf{t}} e^{-i\mathbf{q}\cdot\mathbf{t}} V_f^{r_a}(q) \tag{15.72}$$

Clearly, whatever the symmetry, the expansion set of plane waves will always have the periodicity of the system, so just as before the scalar product of a reciprocal lattice vector, $\mathbf{q} = \mathbf{G}' - \mathbf{G}$, with a Bravais lattice vector, $\mathbf{R}$, will be equal to an integral multiple of 2π, and hence $e^{-i\mathbf{q}\cdot\mathbf{R}} = 1$. If there are N of these new generalized bases in the total volume of the crystal, then:

$$V = \frac{\Omega_c N}{\Omega} \sum_{\mathbf{t}} e^{-i\mathbf{q}\cdot\mathbf{t}} V_f^{r_a}(q) \tag{15.73}$$

Now the total volume of the crystal, Ω, divided by the number of the new general Bravais lattice points, N, is equal to the volume occupied by one of the bases, Ω_b (say), and therefore:

$$V = \frac{\Omega_c}{\Omega_b} \sum_{\mathbf{t}} e^{-i\mathbf{q}\cdot\mathbf{t}} V_f^{r_a}(q) \tag{15.74}$$

The diamond, zinc blende and wurtzite crystals discussed so far all have a two-atom basis, and hence Ω_c (the volume of the primitive cell) has been the volume occupied by two atoms. In this more general basis with N_a atoms (say), the volume occupied by its primitive cell would be:

$$\Omega_b = \frac{N_a}{2}\Omega_c \tag{15.75}$$

Therefore, the final expression for the crystal potential follows as:

$$V = \frac{2}{N_a}\sum_t e^{-i\mathbf{q}\bullet\mathbf{t}}\; V_f^{\mathbf{r}_a}(q) \tag{15.76}$$

and the full expression for the Hamiltonian matrix elements is:

$$\mathcal{H}_{\mathbf{G}',\mathbf{G}} = \frac{\hbar^2}{2m_0}|\mathbf{G}+\mathbf{k}|^2\delta_{\mathbf{G},\mathbf{G}'} + \frac{2}{N_a}\sum_t e^{-i\mathbf{q}\bullet\mathbf{t}}\; V_f^{\mathbf{r}_a}(q) \tag{15.77}$$

At this point, the final equation appears to be merely a re-expression of the Hamiltonian matrix elements of the elemental and compound bulk semiconductors, and it can indeed be shown to reproduce those expressions. For example, consider a ($N_a=$) 2-atom basis, with a cation at $-\mathbf{T} = -\frac{A_0}{8}(\hat{\mathbf{i}}+\hat{\mathbf{j}}+\hat{\mathbf{k}})$ and an anion at $+\mathbf{T}$; then the potential term in equation (15.77) gives:

$$V = \frac{2}{2}\left[e^{i\mathbf{q}\bullet\mathbf{T}}\; V_f^{\mathrm{cat}}(q) + e^{-i\mathbf{q}\bullet\mathbf{T}}\; V_f^{\mathrm{an}}(q)\right] \tag{15.78}$$

which, recalling that $\mathbf{q} = \mathbf{G}' - \mathbf{G}$, is equivalent to equation (15.54).

However, the above is much more than just a generalized form for bulk semiconductor calculations; by thoughtful choice of the atomic basis and the primitive cell, equation (15.77) *can* be used to calculate the electronic structure of heterostructures of all dimensions, i.e. quantum wells, wires and dots. Such calculations are often referred to as *large-cell* calculations; however, the computational method of summing over atoms in a more extensive basis suggests the term *'large-basis'* calculations to be more appropriate. The promise of this generalization will be explored fully in subsequent chapters; however, for now, it is worthwhile pursuing these ideas for bulk systems.

Table 15.6   The form factors (in eV) of a selection of III–V compound semiconductors, converted from the original values of Cohen and Bergstresser [350], where the symmetric form factor at $q = 2$ and the anti-symmetric at $q = \sqrt{8}$ have been deduced by linear interpolation from the two adjacent values

	$V_f^S(q)$				$V_f^A(q)$			
	$\sqrt{3}$	2	$\sqrt{8}$	$\sqrt{11}$	$\sqrt{3}$	2	$\sqrt{8}$	$\sqrt{11}$
GaAs	−3.13	−2.33	+0.14	+0.82	+0.95	+0.68	+0.34	+0.14
InAs	−2.99	−2.26	0.00	+0.68	+1.09	+0.68	+0.51	+0.41
GaP	−2.99	−2.16	+0.41	+0.95	+1.63	+0.95	+0.52	+0.27
InP	−3.13	−2.33	+0.14	+0.82	+0.95	+0.68	+0.34	+0.14

In particular, the brief digress into *atomic form factors* was in anticipation of this formalism—this large-basis method relies on pseudo-potential form factors, which are associated with individual atoms, and hence, in order to perform calculations of bulk compound semiconductors, the symmetric and anti-symmetric form factors listed in Table 15.4 need to be decomposed (as in Section 15.8), into their atomic components. In addition, however, the 'structure factor' associated with these atomic potentials is now of the form $e^{-i\mathbf{q}\cdot\mathbf{t}}$, which is never zero, and hence the *atomic* form factors in this method need to be specified at all values of $\mathbf{q}$ that can arise from the basis set of reciprocal lattice vectors $\mathbf{G}$.

For a compound semiconductor, this implies that it is necessary to know the atomic form factors at $q = 2$ and $\sqrt{8}$. For the example III–V materials given earlier in Table 15.4, this can be achieved as a zeroth-order approximation simply by linearly interpolating between the existing symmetric and anti-symmetric potentials (as in Table 15.6), and then decomposing the resulting form factors to give the data presented in Table 15.7.

Table 15.7   The atomic form factors (in eV) of a selection of III–V compound semiconductors, decomposed from their original symmetric and anti-symmetric values and now including the interpolated potentials for $q = 2$ and $\sqrt{8}$ (as indicated by italics)

| | $V_f^{\text{cat}}(q)$ | | | | $V_f^{\text{an}}(q)$ | | | |
	$\sqrt{3}$	2	$\sqrt{8}$	$\sqrt{11}$	$\sqrt{3}$	2	$\sqrt{8}$	$\sqrt{11}$
GaAs	-2.04	*-1.51*	*-0.10*	$+0.34$	-1.09	*-0.83*	*$+0.24$*	$+0.48$
InAs	-2.04	*-1.47*	*-0.26*	$+0.14$	-0.95	*-0.79*	*$+0.26$*	$+0.55$
GaP	-2.31	*-1.56*	*-0.06*	$+0.34$	-0.68	*-0.61*	*$+0.47$*	$+0.61$
InP	-2.04	*-1.51*	*-0.10*	$+0.34$	-1.09	*-0.83*	*$+0.24$*	$+0.48$

Fig. 15.11 shows the band structures of GaP and InP, calculated by using this large-basis method with a 65-element plane-wave basis set and the atomic form factors of Table 15.7. The validity of this approach, which turns out to be a small generalization for the bulk case, is substantiated by the close agreement obtained with the original calculations of Cohen and Bergstresser [350].

15.10   SPIN–ORBIT COUPLING WITHIN THE LARGE BASIS APPROACH

In constructing spin-dependent solutions the basis set must contain spin-dependent (s or s') terms, hence the matrix elements given in equation (15.18) i.e.

$$\mathcal{H}_{\mathbf{G}',\mathbf{G}} = \int u^*_{\mathbf{G}',\mathbf{k}}\mathcal{H}u_{\mathbf{G},\mathbf{k}}\ d\tau \tag{15.79}$$

become:

$$\mathcal{H}_{(\mathbf{G}',s'),(\mathbf{G},s)} = \int u^*_{\mathbf{G}',\mathbf{k},s'}\mathcal{H}u_{\mathbf{G},\mathbf{k},s}\ d\tau \tag{15.80}$$

The additional potential V_{so} representing the spin–orbit interaction in the total Hamiltonian $\mathcal{H}$, leads to an additional term within each matrix element [351, 364, 365]. Under

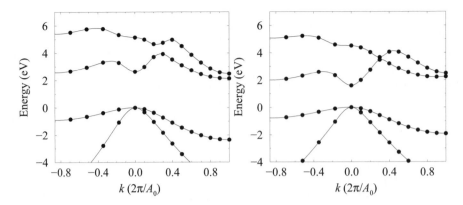

Figure 15.11   The band structures of bulk GaP (left) and InP (right) obtained by using the interpolated atomic form factors given in Table 15.7, shifted to set the top of the valence band to zero

the auspices of the large basis approach, this term is again a sum over the contributions from all the N_a atoms within the basis, and by analogy to the crystal potential of equation (15.76), is given by:

$$V_{\text{so}} = \frac{1}{N_a} \sum_t e^{-i\mathbf{q}\bullet\mathbf{t}} \left[-i\lambda^{\mathbf{r}_a} \left(\mathbf{G}' + \mathbf{k} \right) \times \left(\mathbf{G} + \mathbf{k} \right)_{\bullet} \sigma_{s',s} \right] \qquad (15.81)$$

where $\lambda^{\mathbf{r}_a}$ is a parameter quantifying the magnitude of the interaction and $\mathbf{q} = \mathbf{G}' - \mathbf{G}$. The superscript $\mathbf{r}_a$ indicates that λ is dependent upon the atomic species at all the basis sites within the unit cell. The entity $\sigma_{s',s} = \langle s'|\sigma|s \rangle$ where:

$$\sigma = \sigma_1 \hat{\mathbf{i}} + \sigma_2 \hat{\mathbf{j}} + \sigma_3 \hat{\mathbf{k}} \qquad (15.82)$$

and σ_1, σ_2 and σ_3 are the Pauli spin matrices:

$$\sigma_1 = \begin{pmatrix} 0 & 1 \\ 1 & 0 \end{pmatrix} \quad \sigma_2 = \begin{pmatrix} 0 & -i \\ i & 0 \end{pmatrix} \quad \sigma_3 = \begin{pmatrix} 1 & 0 \\ 0 & -1 \end{pmatrix} \qquad (15.83)$$

Thus:

$$\sigma = \begin{pmatrix} \hat{\mathbf{k}} & \hat{\mathbf{i}} - i\hat{\mathbf{j}} \\ \hat{\mathbf{i}} + i\hat{\mathbf{j}} & -\hat{\mathbf{k}} \end{pmatrix} \qquad (15.84)$$

As partial analytical verification for this representation presented here, consider the case of a zinc-blende basis consisting of two atoms, one cation and one anion. In this example, N_a would be 2 and $\mathbf{t} = \pm\mathbf{T}$, where, again, $\mathbf{T} = \frac{A_0}{8}(\hat{\mathbf{i}} + \hat{\mathbf{j}} + \hat{\mathbf{k}})$, i.e.

$$V_{\text{so}} = \frac{1}{2} \left[(-i\lambda^{\text{cat}})e^{i\mathbf{q}\bullet\mathbf{T}} + (-i\lambda^{\text{an}})e^{-i\mathbf{q}\bullet\mathbf{T}} \right] \left(\mathbf{G}' + \mathbf{k} \right) \times \left(\mathbf{G} + \mathbf{k} \right)_{\bullet} \sigma_{s',s} \qquad (15.85)$$

which gives:

$$V_{\text{so}} = \frac{1}{2} \left[-i\lambda^{\text{cat}} \left(\cos\mathbf{q}\bullet\mathbf{T} + i\sin\mathbf{q}\bullet\mathbf{T} \right) - i\lambda^{\text{an}} \left(\cos\mathbf{q}\bullet\mathbf{T} - i\sin\mathbf{q}\bullet\mathbf{T} \right) \right]$$

$$\times \left(\mathbf{G'} + \mathbf{k}\right) \times \left(\mathbf{G} + \mathbf{k}\right)_\bullet \sigma_{s',s} \tag{15.86}$$

Gathering terms in cos and sin, then:

$$V_{so} = \frac{1}{2}\left[-i(\lambda^{cat} + \lambda^{an})\cos\mathbf{q}_\bullet\mathbf{T} - i(\lambda^{cat} - \lambda^{an})i\sin\mathbf{q}_\bullet\mathbf{T}\right]$$

$$\times \left(\mathbf{G'} + \mathbf{k}\right) \times \left(\mathbf{G} + \mathbf{k}\right)_\bullet \sigma_{s',s} \tag{15.87}$$

which gives:

$$V_{so} = \left[-i\left(\frac{\lambda^{cat} + \lambda^{an}}{2}\right)\cos\mathbf{q}_\bullet\mathbf{T} + \left(\frac{\lambda^{cat} - \lambda^{an}}{2}\right)\sin\mathbf{q}_\bullet\mathbf{T}\right]$$

$$\times \left(\mathbf{G'} + \mathbf{k}\right) \times \left(\mathbf{G} + \mathbf{k}\right)_\bullet \sigma_{s',s} \tag{15.88}$$

The half sum and half difference of the λ parameters are often labelled as symmetric λ^S and anti-symmetric λ^A contributions to the spin–orbit interaction, thus giving:

$$V_{so} = \left(-i\lambda^S\cos\mathbf{q}_\bullet\mathbf{T} + \lambda^A\sin\mathbf{q}_\bullet\mathbf{T}\right)\left(\mathbf{G'} + \mathbf{k}\right) \times \left(\mathbf{G} + \mathbf{k}\right)_\bullet\sigma_{s',s} \tag{15.89}$$

which is the same result as Chelikowsky and Cohen [365].

15.11 COMPUTATIONAL IMPLEMENTATION

With the aim of a computational implementation of this extension to the pseudo-potential method*, first consider the new extended basis set. The addition of spin information means that the basis set of plane waves has to be doubled (one set for spin 'up' and one set for spin 'down'). The procedure is to split the new matrix vertically and horizontally into four blocks, each one of which is the same size as the original matrix:

$$\begin{pmatrix} \begin{array}{c|c} \text{Block 1} & \text{Block 2} \\ s = +\frac{1}{2}, s' = +\frac{1}{2} & s = +\frac{1}{2}, s' = -\frac{1}{2} \\ \hline \text{Block 3} & \text{Block 4} \\ s = -\frac{1}{2}, s' = +\frac{1}{2} & s = -\frac{1}{2}, s' = -\frac{1}{2} \end{array} \end{pmatrix} \tag{15.90}$$

Each block has exactly the same crystal potential components as the original matrix, but with the additional spin-dependent component given in equation (15.81). A Kronecker delta $\delta_{s',s}$ acts upon the original components of each matrix element just to ensure that the only spin-dependent part is the new term V_{so}. Thus the complete matrix elements, including the spin–orbit interaction within the large basis approach, are given by:

$$\mathcal{H}_{(\mathbf{G'},s'),(\mathbf{G},s)} = \mathcal{H}_{\mathbf{G'},\mathbf{G}}\delta_{s',s} + V_{so} \tag{15.91}$$

where, for the large basis approach, $\mathcal{H}_{\mathbf{G'},\mathbf{G}}$ is given by equation (15.77). In full, this would be:

$$\mathcal{H}_{(\mathbf{G'},s'),(\mathbf{G},s)} = \frac{\hbar^2}{2m_0}|\mathbf{G} + \mathbf{k}|^2\delta_{\mathbf{G'},\mathbf{G}}\delta_{s',s} + \frac{2}{N_a}\sum_t e^{-i\mathbf{q}_\bullet\mathbf{t}}V_f^{r_a}(q)\delta_{s',s} +$$

*The author would like to thank Fei Long for his assistance

$$\frac{1}{N_a} \sum_t e^{-i\mathbf{q} \cdot \mathbf{t}} \left[-i\lambda^{r_a} \left(\mathbf{G}' + \mathbf{k} \right) \times \left(\mathbf{G} + \mathbf{k} \right)_{\bullet} \sigma_{s',s} \right] \tag{15.92}$$

Thus computationally the quickest way to generate this new extended matrix is to calculate the original crystal potential terms for just the lower triangle of the first 'block' and then copy it into the lower triangle of the fourth block. The spin–orbit term can then be added, with the only other **k**-dependent part, the kinetic energy term, added last along the leading diagonal.

For computational reasons it may be worthwhile considering the form of the spin–orbit contribution to the matrix elements, as this can be simplified for more efficient evaluation. With this in mind, consider the forms of $\sigma_{s',s}$. The spin states are represented by the spinors $(1,0)$ and $(0,1)$, hence:

$$\sigma_{\frac{1}{2},\frac{1}{2}} = (1,0) \begin{pmatrix} \hat{\mathbf{k}} & \hat{\mathbf{i}} - i\hat{\mathbf{j}} \\ \hat{\mathbf{i}} + i\hat{\mathbf{j}} & -\hat{\mathbf{k}} \end{pmatrix} \begin{pmatrix} 1 \\ 0 \end{pmatrix} = (1,0) \begin{pmatrix} \hat{\mathbf{k}} \\ \hat{\mathbf{i}} + i\hat{\mathbf{j}} \end{pmatrix} = \hat{\mathbf{k}} \tag{15.93}$$

$$\sigma_{\frac{1}{2},-\frac{1}{2}} = (1,0) \begin{pmatrix} \hat{\mathbf{k}} & \hat{\mathbf{i}} - i\hat{\mathbf{j}} \\ \hat{\mathbf{i}} + i\hat{\mathbf{j}} & -\hat{\mathbf{k}} \end{pmatrix} \begin{pmatrix} 0 \\ 1 \end{pmatrix} = (1,0) \begin{pmatrix} \hat{\mathbf{i}} - i\hat{\mathbf{j}} \\ -\hat{\mathbf{k}} \end{pmatrix} = \hat{\mathbf{i}} - i\hat{\mathbf{j}} \tag{15.94}$$

$$\sigma_{-\frac{1}{2},\frac{1}{2}} = (0,1) \begin{pmatrix} \hat{\mathbf{k}} & \hat{\mathbf{i}} - i\hat{\mathbf{j}} \\ \hat{\mathbf{i}} + i\hat{\mathbf{j}} & -\hat{\mathbf{k}} \end{pmatrix} \begin{pmatrix} 1 \\ 0 \end{pmatrix} = (0,1) \begin{pmatrix} \hat{\mathbf{k}} \\ \hat{\mathbf{i}} + i\hat{\mathbf{j}} \end{pmatrix} = \hat{\mathbf{i}} + i\hat{\mathbf{j}} \tag{15.95}$$

$$\sigma_{-\frac{1}{2},-\frac{1}{2}} = (0,1) \begin{pmatrix} \hat{\mathbf{k}} & \hat{\mathbf{i}} - i\hat{\mathbf{j}} \\ \hat{\mathbf{i}} + i\hat{\mathbf{j}} & -\hat{\mathbf{k}} \end{pmatrix} \begin{pmatrix} 0 \\ 1 \end{pmatrix} = (0,1) \begin{pmatrix} \hat{\mathbf{i}} - i\hat{\mathbf{j}} \\ -\hat{\mathbf{k}} \end{pmatrix} = -\hat{\mathbf{k}} \tag{15.96}$$

The vector product $(\mathbf{G}' + \mathbf{k}) \times (\mathbf{G} + \mathbf{k})$ is clearly equal to some other vector $A_x\hat{\mathbf{i}} + A_y\hat{\mathbf{j}} + A_z\hat{\mathbf{k}}$, say. Then in Block 1 of the Hamiltonian matrix:

$$(\mathbf{G}' + \mathbf{k}) \times (\mathbf{G} + \mathbf{k})_{\bullet} \sigma_{s',s} = A_z \tag{15.97}$$

In Block 2:
$$(\mathbf{G}' + \mathbf{k}) \times (\mathbf{G} + \mathbf{k})_{\bullet} \sigma_{s',s} = A_x - iA_y \tag{15.98}$$

In Block 3:
$$(\mathbf{G}' + \mathbf{k}) \times (\mathbf{G} + \mathbf{k})_{\bullet} \sigma_{s',s} = A_x + iA_y \tag{15.99}$$

In Block 4:
$$(\mathbf{G}' + \mathbf{k}) \times (\mathbf{G} + \mathbf{k})_{\bullet} \sigma_{s',s} = -A_z \tag{15.100}$$

15.12   DEDUCING THE PARAMETERS AND APPLICATION

The primary aim of the book is to demystify the theory of semiconductor heterostructures, and make such calculations more *accessible*. This current topic of spin–orbit coupling within the empirical pseudo-potential approach is one example where this is paramount, as the literature can be quite confusing. The spin–orbit interaction V_{so} given in equation (15.81) is an energy, and λ is usually quoted in Rydbergs (a unit of energy), thus the vectors $\mathbf{G}$ and $\mathbf{k}$ must be dimensionless. The question is what dimensions? In this work the result of the vector and scalar product, i.e. $(\mathbf{G}' + \mathbf{k}) \times (\mathbf{G} + \mathbf{k})_{\bullet} \sigma_{s',s}$, is made dimensionless by dividing by $(2\pi/A_0)^2$. As the structure factor $e^{-i\mathbf{q}\cdot\mathbf{t}}$ is dimensionless, then the contribution V_{so} to the matrix element takes the units of the parameter λ.

Not all authors use this convention and this is reflected in the literature by the usual choice to quote the spin–orbit splitting Δ_0 between the doubly degenerate light- and heavy-hole bands and the split-off band. In some instances the symmetric λ^S and antisymmetric λ^A parameters are quoted for binary compounds and it is a simple task to obtain the spin–orbit parameters for the individual atoms as required here:

$$\lambda^{\text{cat}} = \lambda^S - \lambda^A \qquad \lambda^{\text{an}} = \lambda^S + \lambda^A \qquad (15.101)$$

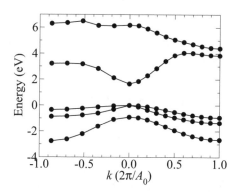

Figure 15.12   The band structure of bulk CdTe calculated with 169 plane waves and adjusted to set the top of the valence band to zero

Fig. 15.12 shows the results of a calculation of the band structure of CdTe using the original form factors of Cohen and Bergstresser [350] with the spin–orbit parameters λ^S and λ^A quoted by Bloom and Bergstresser [366]. It was found that the λ values had to be multiplied by a factor of 0.343 in order to obtain a splitting Δ_0 of 905 meV in agreement with the values of 910 and 900 meV quoted by Bloom and Bergstresser. This is a relatively quick process because, to *first-order*, the valence-band splitting is proportional to the λ parameters, so if the result for Δ_0 is half the experimental value, scaling the lambda parameters by a factor of 2 will 'get you close'. The splitting of the light- and heavy-hole at the X ($k = (0, 0, 2\pi/A_0)$) is 415 meV and at the L point ($k = (\pi/A_0, \pi/A_0, \pi/A_0)$) is 525 meV, which are both in reasonable agreement with the values quoted by Chelikowsky and Cohen [365] of 380 and 530 meV, respectively. These results serve to substantiate the validity of the spin–orbit extension. Note though that there is some discrepancy in the calculations here and disagreement between the results quoted [365, 366]. This serves as an illustration of the uncertainty surrounding the knowledge of remote areas of the band structure (away from high symmetry points and band edges), but also it must be noted that the spin–orbit parameters should be chosen self-consistently with the atomic form factors—a tedious process that is not going to be attempted here.

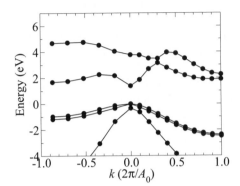

Figure 15.13   The band structure of bulk GaAs using the spin-free form factors of Mader and Zunger [367] together with the spin–orbit parameters described in the text

Taking μ' as some adjustable parameter (which can again absorb any differences in units employed), Chelikowsky and Cohen [365] give the spin–orbit splittings for a binary compound AB as:

$$\lambda^A = \mu' \qquad \lambda^B = \alpha\mu' \qquad (15.102)$$

where α is the ratio of the spin–orbit splittings of the free atoms, where the latter can be looked up in a book [368]. Unfortunately, the book only gives data for atoms found in II–VI compounds, so it is difficult to know what α should be for any III–V material. Thus any calculation for a III–V material, such as GaAs, would be dependent upon two parameters, which means a lot of work may be required to limit the values they take and produce good agreement with experiment.

To 'cut a corner', consider just *interpolating* the spin–orbit splittings of the free Ga and As atoms from their neighbouring entries in the tabulated data of Herman and Skillman (p. 2–6), using their notation:

$$E(\text{S-O})(\text{GA}) = \frac{E(\text{S-O})(\text{Zn}) + E(\text{S-O})(\text{Ge})}{2} \qquad (15.103)$$

$$E(\text{S-O})(\text{AS}) = \frac{E(\text{S-O})(\text{Ge}) + E(\text{S-O})(\text{Se})}{2} \qquad (15.104)$$

i.e.

$$E(\text{S-O})(\text{GA}) = \frac{-0.00596 - 0.00515}{2} = -0.00556 \qquad (15.105)$$

$$E(\text{S-O})(\text{AS}) = \frac{-0.00515 - 0.01026}{2} = -0.00771 \qquad (15.106)$$

Giving:

$$\alpha = \frac{E(\text{S-O})(\text{AS})}{E(\text{S-O})(\text{GA})} = 1.38 \qquad (15.107)$$

Using this, then the value of μ' was adjusted (and found to be 0.000402) to give a valence-band splitting of 350 meV. This in itself is not enough to substantiate the calculation;

however, at the same time the light-hole–heavy-hole splitting at X is 97 meV, which is incredibly close to the 100 meV quoted by Chelikowsky and Cohen, see Fig. 15.13.

This shows the robustness of the method—while this GaAs calculation has been forced through and is not necessarily scientifically rigorous, this calculation does demonstrate that the technique can be used to obtain the valence band structure of III–V materials to some degree of acceptability and, should further accuracy be required, then more time could justifiably be invested in determining a better set of parameters.

It is a much simpler task to deduce the spin–orbit parameters of the elemental group IV semiconductors because, with only one atom type, there is only one parameter. Using the methods outlined above λ was found to be 0.000106 and 0.00058 Rydberg in order to give valence band splittings of 44 and 296 meV for Si and Ge, respectively. This gave rise to the band structures plotted in Fig. 15.14.

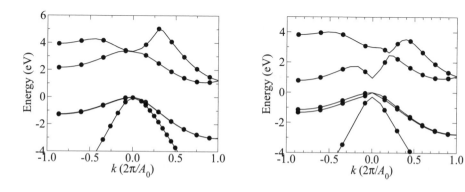

Figure 15.14   The band structures of bulk Si (left) and Ge (right) including spin–orbit coupling along the usual L$\rightarrow \Gamma \rightarrow$X line

The calculations are based on the spin-free atomic form factors for Si and Ge heterostructures given by Friedel *et al.* [369]. They are in excellent agreement with the data presented by Chelikowsky and Cohen and, in particular for Ge, the light-hole–heavy-hole splitting at the L-point is 186 meV in comparison with the quoted value of 200 meV.

15.13   ISOELECTRONIC IMPURITIES IN BULK

A generalization of the standard two-atom basis within the empirical pseudo-potential method was demonstrated in Section 15.9, although hitherto the calculations presented have merely seemed more complex than previously and the investment seems to have produced no visible benefit. In fact, the generalization was motivated by the desire to study more complex distributions of atoms within a crystal, where primarily this implies quantum well heterostructures, wires and dots (the latter having their own chapters). However, in addition, some bulk systems of interest still remain and the techniques are now at hand for their solutions.

The first case in question is that of a single impurity atom within a host semiconductor. This topic is itself extensive, and has already been given a substantial chapter in this text within the envelope function and effective mass approximations. The interest here lies now with the ability to explore the microscopic electronic properties from a more fundamental point of view. Indeed, such a detailed study has already been undertaken by Jaros [349]—here, a very brief demonstration will be given as to how the large basis calculation *may* be applied to such instances.

The first point to be addressed is: 'What is the nature of the atomic basis?'. Clearly, a host crystal needs to be specified and a single atom, representing the impurity, needs to be placed substitutionally on one of the lattice sites. It is perhaps convenient in this scenario to consider a cube of Si, and in this case the impurity atom chosen will be Ge, which would, of course, be *isoelectronic*.

Figure 15.15   The non-primitive eight atom basis of a group-IV face-centred cubic crystal (diamond structure)

Ashcroft and Mermin ([1], p. 75), give the non-primitive face-centred cubic four-point basis vectors as:

$$\mathbf{T_0} = \mathbf{0}; \quad \mathbf{T_1} = \frac{A_0}{2}(\hat{\mathbf{i}}+\hat{\mathbf{j}}); \quad \mathbf{T_2} = \frac{A_0}{2}(\hat{\mathbf{j}}+\hat{\mathbf{k}}); \quad \mathbf{T_3} = \frac{A_0}{2}(\hat{\mathbf{k}}+\hat{\mathbf{i}}) \qquad (15.108)$$

These are chosen to represent the cubic symmetry of the face-centred cubic lattice more clearly. When applied to the diamond or zinc blende structures, such a set of vectors would then imply an eight-atom basis, with atoms at $\pm\mathbf{T}=\pm\frac{A_0}{8}(\hat{\mathbf{i}}+\hat{\mathbf{j}}+\hat{\mathbf{k}})$ for each of the above, i.e.

$$\mathbf{r}_a = \mathbf{T}_i \pm \frac{A_0}{8}(\hat{\mathbf{i}}+\hat{\mathbf{j}}+\hat{\mathbf{k}}), \quad \text{where} \quad i = 0, 1, 2, 3 \qquad (15.109)$$

as displayed in Fig. 15.15. The Bravais lattice vectors then simply follow as:

$$\mathbf{a}_1 = A_0\hat{\mathbf{i}}; \quad \mathbf{a}_2 = A_0\hat{\mathbf{j}}; \quad \mathbf{a}_3 = A_0\hat{\mathbf{k}} \qquad (15.110)$$

Such an alternative way of constructing the crystal is useful for defining large cells within a host semiconductor. In this case, where a description of a single isolated impurity is the aim, it is necessary to generalize this to produce a larger cube. For example, if n_x, n_y and n_z of these cubes were stacked together along the x-, y-, and z- axes, respectively, then the atomic positions within this larger basis would be given by:

$$\mathbf{r}_a = m_x A_0\hat{\mathbf{i}} + m_y A_0\hat{\mathbf{j}} + m_z A_0\hat{\mathbf{k}} + \mathbf{T}_i \pm \frac{A_0}{8}(\hat{\mathbf{i}}+\hat{\mathbf{j}}+\hat{\mathbf{k}}) \qquad (15.111)$$

where again the index $i = 0$, 1, 2 and 3 and the additional indices $m_{x,y,z} = 0$, 1, 2,...,$n_{x,y,z}$. Fig. 15.16 illustrates a large basis produced by stacking eight face-centred cubic cells together, i.e. $n_x = n_y = n_z = 2$, thus giving 64 atoms in total.

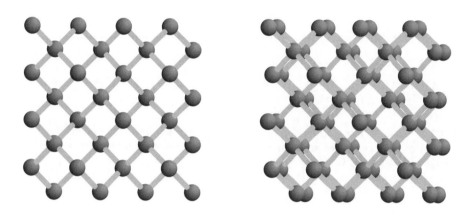

Figure 15.16   A 64-atom basis for the diamond structure, shown looking along one of the $\langle 100 \rangle$ directions (left) and slightly off-axis (right)

For this large $(n_x \times n_y \times n_z)$ cuboid, the primitive reciprocal lattice vectors would simply generalize to:

$$\mathbf{b}_1 = \frac{1}{n_x} \frac{2\pi}{A_0} \hat{\mathbf{i}}; \quad \mathbf{b}_2 = \frac{1}{n_y} \frac{2\pi}{A_0} \hat{\mathbf{j}}; \quad \mathbf{b}_3 = \frac{1}{n_z} \frac{2\pi}{A_0} \hat{\mathbf{k}} \qquad (15.112)$$

Hence, the basis–$\mathbf{G}$ set for expansion is given by all linear combinations of the above, i.e.

$$\mathbf{G} = \beta_1 \mathbf{b}_1 + \beta_2 \mathbf{b}_2 + \beta_3 \mathbf{b}_3, \quad \text{where} \quad \beta_i \in \mathcal{Z} \qquad (15.113)$$

Note that this set of reciprocal lattice vectors implies that the magnitude of the difference vector, $\mathbf{q} = \mathbf{G}' - \mathbf{G}$, can take many more values than the discrete set of the face-centred cubic structure (given in Table 15.1). For this reason (and many others), more recently deduced empirical pseudo-potential form factors are often represented by a continuous curve rather than by a discrete set of points. Fig. 15.17 illustrates this for the Si and Ge form factors of Friedel *et al.* [369], and compares them with the original discrete form factors of Cohen and Bergstresser [350], employed earlier in this chapter.

It can be seen that the agreement between the two representations of the pseudo-potential is close and both have been truncated at $q = 4$, although at the higher end of the domain there is a slight difference. The continuous potentials of Friedel *et al.* [369] have been deduced by comparison with a variety of experimental results and have been shown to reproduce a large amount of experimental data, including bulk-band structure, deformation potentials, and heterostructure band offsets.   Truncating the reciprocal lattice vectors of equation (15.112) at the same point as the potential ($q = 4$) gives a basis set of 2109 plane waves. Hence, the Hamiltonian matrix $\mathcal{H}_{\mathbf{G}',\mathbf{G}}$ has $(2109)^2$

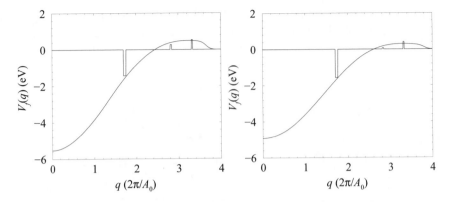

Figure 15.17 The discrete (δ-function like) pseudo-potential form factors of the face-centred cubic lattice [350] compared with the continuous form necessary for systems of reduced dimensionality, for Si (left) and Ge (right) [369]

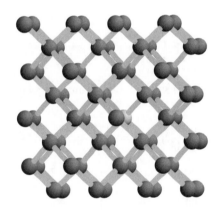

Figure 15.18 A 64-atom basis for the diamond structure, with a single substitutional Ge impurity (lighter colour, near centre)

elements, each of which is a complex number. Using double-precision arithmetic (8 bytes per number), this then implies the Hamiltonian matrix will occupy $(2109^2 \times 2 \times 8)$ bytes ≈ 71 Mbytes. Even when allowing for the operating system and the code to diagonalize it, the memory requirement for this size of calculation is now well within the reach of a standard desktop computer. However, it can be seen that the memory demands for this type of calculation can grow very quickly and this is one of the detrimental points of the large-basis method. While it is very simple, convenient and flexible, simply employing direct diagonalization of the matrix as the computational solution is quite inefficient. However, a calculation *of this size* is manageable and all is now in place to proceed with it.

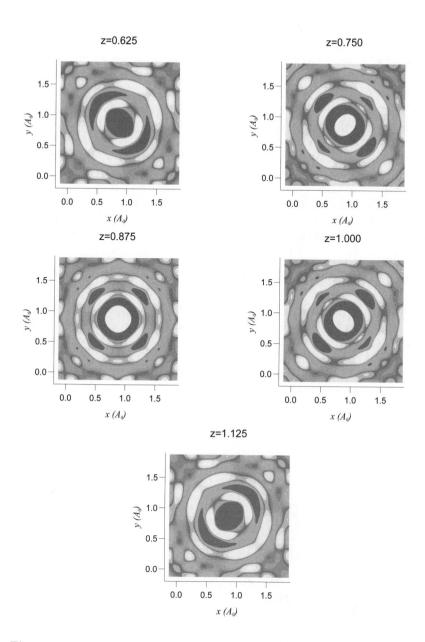

Figure 15.19 The change in the total valence-band charge density (ignoring spin–orbit coupling) around a single Ge atom within a Si host across the x–y plane of four face-centred cubic unit cells, for a variety of z values (given in A_0); note the lighter the colour, then the higher the increase in the charge density

In this analysis one of the 64 Si atoms in the basis is substituted by Ge, as shown in Fig. 15.18. The choice of which atom is not important, as the periodic nature of the pseudo-potential calculations ensures that they are all equivalent. Just as the two-atom basis of the simple elemental Si calculation produces four valence-band states, this more complex 64-atom basis gives $64 \times 2 \ (= 128)$ valence-band states. Fig. 15.19 shows the change in the *total* valence-band charge density on substitution of the single Ge isoelectronic impurity at an electron wave vector $\mathbf{k} = 0$; note the lighter the colour, then the larger the increase.

In this particular example, the Si atom with coordinates $(0.875, 0.875, 0.875)A_0$ was substituted, and hence the third plane, with $z = 0.875 \ A_0$, intersects the centre of the Ge atom. Focusing on this plot, the charge density on the Ge atomic site is higher than that on the surrounding crystal and is fairly well localized, which (just about) justifies the initial choice of a 64-atom basis. The remaining four figures display the change in the charge density above and below the plane of this isolated Ge atom, where the symmetry is apparent equidistant from the centre of the impurity, and it can be seen that a quarter of a lattice constant away $(1.4 \ \text{Å})$, as shown on the first and fifth planes, the charge density has returned to its background (Si) value.

15.14   THE ELECTRONIC STRUCTURE AROUND POINT DEFECTS

Such an approach can also be used to model point defects. For example, instead of substituting a Si atom for an isoelectronic Ge atom, the atom can be removed completely to represent a vacancy. Now this is a simplified picture, in that for this example it is assumed that all of the other atoms remain on their diamond lattice sites. This, of course, will not be the case, as there will be some localized relaxation of the crystal around the vacancy, i.e. some movement of the Si atoms.

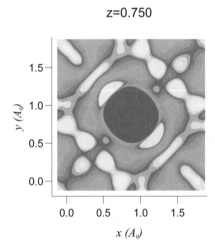

Figure 15.20   The change in the total valence-band charge density (ignoring spin–orbit coupling) around a vacancy within a Si host across the x–y plane of four face-centred cubic unit cells; note that the lighter the colour, then the higher the increase in the charge density

Again employing the 64-atom cubic basis of above, Fig. 15.20 displays the change in the electronic structure around a single vacancy when compared to the unperturbed Si host crystal. The plot centres the vacancy site and represents a cross-section of four (2×2) face-centred cubic unit cells. The absence of charge is clearly illustrated by the darker shading, and interestingly suggests a 'halo' of increased charge density around the vacancy. Such a movement of charge will aid shielding of the 'hole' from the remainder of the crystal.

CHAPTER 16

MICROSCOPIC ELECTRONIC PROPERTIES OF HETEROSTRUCTURES

16.1 THE SUPERLATTICE UNIT CELL

The large-basis approach to a pseudo-potential calculation *can* be applied to semiconductor heterostructures, where in particular, the short-period nature of the repeating unit cell in a superlattice is an ideal example. In order to proceed with such a calculation, it is necessary only to deduce the Bravais lattice vectors describing the crystal symmetry of the superlattice and from these derive the reciprocal lattice vectors.

Fig. 16.1 displays the now familiar zinc blende crystal lattice along one of the $\langle 100 \rangle$ directions. The crystal *can* be thought of as being composed of a series of 'one-dimensional' spirals that project into the plane of the paper. Such a spiral begins on an atom (in this case a cation) labelled '1' and then proceeds anticlockwise through atoms '2', '3' and '4', with the fifth atom in the spiral being exactly one lattice constant behind the first. This is illustrated more clearly in Fig. 16.2. Choosing the z-axis along the length of the

Quantum Wells, Wires and Dots, Third Edition. P. Harrison
©2009 John Wiley & Sons, Ltd.

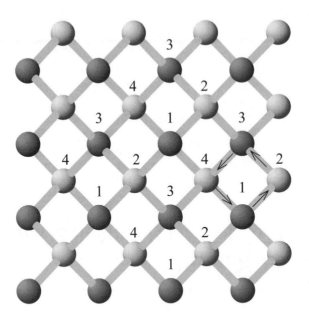

Figure 16.1   Looking along one of the $\langle 100 \rangle$ axes of a zinc blende crystal lattice

spiral, then the atomic positions, in units of A_0, follow as:

$$\mathbf{r}_1 = \left(-\frac{1}{8}, -\frac{1}{8}, -\frac{1}{8}\right) \tag{16.1}$$

$$\mathbf{r}_2 = \left(\frac{1}{8}, \frac{1}{8}, \frac{1}{8}\right) \tag{16.2}$$

$$\mathbf{r}_3 = \left(-\frac{1}{8}, \frac{3}{8}, \frac{3}{8}\right) \tag{16.3}$$

$$\mathbf{r}_4 = \left(-\frac{3}{8}, \frac{1}{8}, \frac{5}{8}\right) \tag{16.4}$$

The spiral then continues with a period (in this bulk case) of $A_0 \hat{\mathbf{k}}$, and thus:

$$\mathbf{r}_5 = \mathbf{r}_1 + (0, 0, 1) \tag{16.5}$$

$$\mathbf{r}_6 = \mathbf{r}_2 + (0, 0, 1) \tag{16.6}$$

This is easily generalized to a superlattice by simply defining the atomic positions with the appropriate choice of species. Fig. 16.3 illustrates this for a diamond (just to keep the illustration simple) heterostructure. Again, the atomic positions within the unit cell are repetitions of the four primitive vectors $\mathbf{r}_1$ to $\mathbf{r}_4$ given in equations (16.1) to

Figure 16.2 A single spiral of zinc blende: (left) Looking along one of the $\langle 100 \rangle$ directions; (centre) slightly off axis; (right) side view

(16.4) above. Thus for a superlattice of period $n_z A_0$, the position vectors of the atoms within this larger basis are given by:

$$\mathbf{t} = \mathbf{r}_i + i_{n_z} A_0 \hat{\mathbf{k}}, \quad i = 1, 2, 3, 4 \text{ and } i_{n_z} = 0, 1, 2, \ldots, n_z - 1 \qquad (16.7)$$

Figure 16.3 Off-axis view along the single spiral unit cell of a diamond heterostructure

The Bravais lattice vector describing the periodicity of the superlattice along the axis of symmetry (usually defined as the z-axis) is therefore clearly just $n_z A_0 \hat{\mathbf{k}}$. Referring back to Fig. 16.1, then the Bravais lattice vectors describing the in-plane $(x\text{--}y)$ periodicity are those linking the atoms labelled '1', i.e. in units of A_0, $(\frac{1}{2}, \frac{1}{2}, 0)$ and $(\frac{1}{2}, -\frac{1}{2}, 0)$.

In summary, the Bravais lattice vectors for a superlattice are:

$$\mathbf{a_1} = A_0 \left(\frac{1}{2}\hat{\mathbf{i}} + \frac{1}{2}\hat{\mathbf{j}}\right) \tag{16.8}$$

$$\mathbf{a_2} = A_0 \left(\frac{1}{2}\hat{\mathbf{i}} - \frac{1}{2}\hat{\mathbf{j}}\right) \tag{16.9}$$

$$\mathbf{a_3} = n_z A_0 \hat{\mathbf{k}} \tag{16.10}$$

The primitive reciprocal lattice vectors then follow from the usual vector product relationships (see Chapter 1), as:

$$\mathbf{b_1} = \frac{2\pi}{A_0} \left(\hat{\mathbf{i}} + \hat{\mathbf{j}}\right) \tag{16.11}$$

$$\mathbf{b_2} = \frac{2\pi}{A_0} \left(\hat{\mathbf{i}} - \hat{\mathbf{j}}\right) \tag{16.12}$$

$$\mathbf{b_3} = \frac{2\pi}{A_0} \frac{1}{n_z} \hat{\mathbf{j}} \tag{16.13}$$

The complete set of reciprocal lattice vectors required for the pseudo-potential calculation is created by taking linear combinations of the above primitive vectors. Although this could be worked out analytically, it is quicker to take the short cut of generating this set numerically with a simple computer program. In this manner, Fig. 16.4 plots the number of plane waves in the expansion set (when limiting the maximum magnitude to $4 \times (2\pi/A_0)$) as a function of the superlattice period.

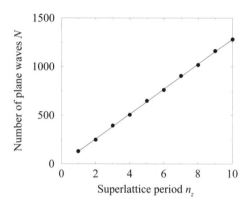

Figure 16.4   The number of plane waves required in the expansion set for the large-basis approach to superlattices

A linear fit to the data shown in Fig. 16.4 gives the number of plane waves as:

$$N \approx 128 n_z \tag{16.14}$$

which, as can be seen, can become quite large for relatively short periods. In particular, a superlattice of period $n_z = 10 A_0$ corresponds to 56.5 Å in the GaAs/Ga$_{1-x}$Al$_x$As system. This simplistic approach generates 1281 plane waves thus giving a Hamiltonian matrix $\mathcal{H}_{\mathbf{G'},\mathbf{G}}$ of order 1281. With each element being a complex ($2\times$) double (8 bytes), then this matrix requires $2 \times 8 \times 1281^2 \sim 26$ Mbytes of storage, which is now well within the reach of a modern desktop computer.

16.2   APPLICATION OF LARGE BASIS METHOD TO SUPERLATTICES

While not perhaps being the most computationally 'slick' method, the large-basis approach to the empirical pseudo-potential calculation of an extended crystallographic unit cell is straightforward and offers an excellent introduction to such calculations for researchers new to the field. In the last section, the application of this technique has been introduced for a particular classification of semiconductor heterostructures, namely superlattices with a relatively short period. In these first example calculations, this period will be considered to be constructed from a single layer of GaAs coupled to a single layer of AlAs, as illustrated schematically in Fig. 16.5.

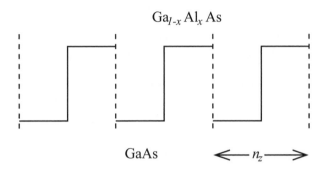

Figure 16.5   Schematic illustration of the repeating unit cell, of length $n_z A_0$, within a superlattice

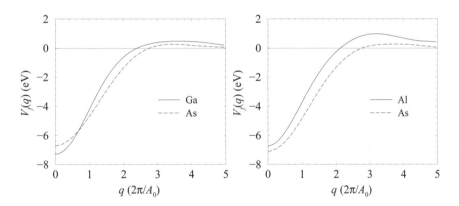

Figure 16.6   The continuous atomic pseudo-potentials of GaAs (left) and AlAs (right) of Mäder and Zunger [367]

As mentioned often before, a knowledge of the pseudo-potentials for superlattices is required at many more points than the small discrete set akin to bulk materials. With this aim, the pseudo-potentials of Mäder and Zunger [367], which have been shown to reproduce many experimentally measured parameters as well as the results of *ab initio* calculations, will be employed. The required continuous nature of these potentials is illustrated in Fig. 16.6 for both atomic species in GaAs and AlAs.

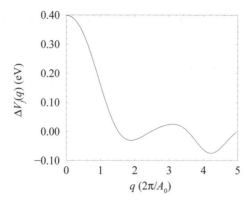

Figure 16.7   The difference in the atomic potentials of As in GaAs and AlAs

Again, notice how the atomic potential of As in GaAs is different from that in AlAs. This difference, $\Delta V_f(q) = V_f^{\text{As in GaAs}}(q) - V_f^{\text{As in AlAs}}(q)$, is plotted in Fig. 16.7, just to re-emphasize the point made in Section 15.8, namely that, unfortunately, universal atomic potentials do not actually exist.

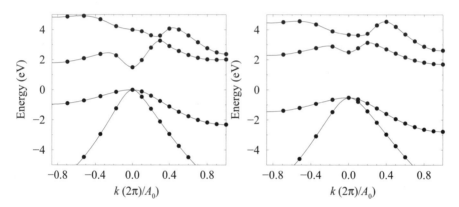

Figure 16.8   The bulk band structures of GaAs (left) and AlAs (right), illustrating the valence band offset built into the potentials of Mäder and Zunger [367]

Fig. 16.8 shows the band structures of bulk GaAs and AlAs, *ignoring spin–orbit coupling*, as produced from the continuous potentials. Both band structures have been translated upwards in energy such that the top of the valence band of GaAs is at zero. At the same time, the top of the valence band for AlAs is at -0.51 eV, which gives the offset in the valence band as 33% of the total, in line with experimentally determined values (see, for example [14], p. 179), and recent calculations (see, for example [370]). This illustrates an important point, namely that the correct choice of potentials can not only just be used to reproduce experimental data for bulk material, but in addition they can retain information on the alignment (or band offset) of two dissimilar materials.

Therefore, all is in place for the calculation to proceed: the superlattice unit cell has been defined, the reciprocal lattice vectors calculated, and the pseudo-potentials have been shown to reproduce the bulk band structure *and* the band alignment between

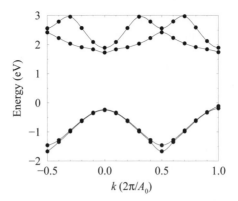

Figure 16.9   The band structure of a 1 ml GaAs/1 ml AlAs superlattice around the fundamental gap

the materials of interest. Fig. 16.9 displays the results of calculations of the energy level structure around the fundamental gap of a 1 ml (monolayer) GaAs/1 ml AlAs superlattice (sometimes written as $(GaAs)_1(AlAs)_1$), i.e. $n_z=1$, for a range of electron wave vectors k along the z-axis (parallel to the growth direction). The cyclic nature of the dispersion curves is evident with a period of $2\pi/A_0$, which represents the primitive reciprocal lattice vector along that direction, as given above in equation (16.13). Thus, the edges of the superlattice Brillouin zone are at $k = \pm\pi/(n_zA_0)$, which is again the commonly found result.

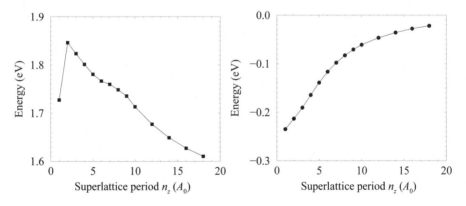

Figure 16.10   The energies of the lowest conduction-band state (left) and highest valence-band state (right) as functions of the superlattice period

Following on from this, Fig. 16.10 displays the results of calculations of the energies at the centre of the superlattice Brillouin zone ($k = 0$) for the uppermost valence band and the lower-most conduction band, as a function of the superlattice period (expressed in units of A_0). In both cases, the quantum-confinement energy decreases as the superlattice period increases, an effect which was commonly observed in the discussions in Chapters 2 and 3. Note that the energies are given in relation to the top of the valence band of bulk GaAs, which is set to zero, with the bottom of the conduction band coming out as 1.499 eV in these calculations. One interesting feature is the sudden reduction in the quantum-confinement energy of the conduction-band state when the period decreases to just A_0. In fact, it would be expected *a priori* that the conduction band minimum would tend towards the band edge of the bulk alloy $Ga_{0.5}Al_{0.5}As$ in the limit of short-period GaAs/AlAs superlattices. By taking a linear combination of the bulk band structures as suggested by the virtual crystal approximation, this would imply an energy of around 2 eV. This behaviour will be discussed in more detail in the next section.

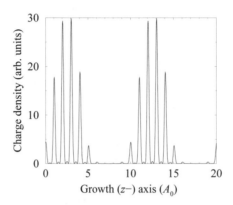

Figure 16.11   The charge density of the uppermost valence-band state along the $x = y = 0$ (z-) axis for two complete superlattice periods

Finally, for this series of calculations, Fig. 16.11 plots the charge density along the centre ($x = y = 0$) of the n_z=10 superlattice spiral, for the uppermost valence-band state. The confinement in the GaAs layers is clear, and in addition to this, the lowest conduction-band state is also confined in these layers, thus illustrating the Type-I nature of the band alignment. It can be seen from this figure that the wave function does consist of two components, where one is rapidly varying and the second is an envelope. This reflects well the idea behind the envelope function approximation, which was discussed (and then used extensively) earlier. The rapidly varying component has a period that is the same as the atomic spacing, and indeed along the axis used in the plot, i.e. the $x = y = 0$ axis, the peaks in the charge density correspond to bond centres. The envelope function is obtained by joining together the peaks in the charge density plot by using the eye, with its form being familiar from Chapter 2.

16.3   COMPARISON WITH ENVELOPE FUNCTION APPROXIMATION

Fig. 16.12 compares the results of a Kronig–Penney superlattice calculation under the envelope function/effective mass approximations (see Section 2.13), with the empirical pseudo-potential calculations of the lowest-energy conduction band state as a function of the superlattice period, as previously displayed in Fig. 16.10. It is this comparison that draws attention to the complexity of the data. For guidance, it is necessary to refer to the specialist treatise on empirical pseudo-potential calculations of GaAs/AlAs superlattices from which the pseudo-potentials employed in these calculations are taken (see [367], p. 17396).

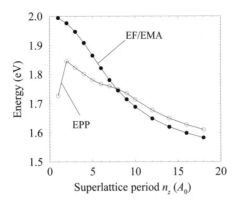

Figure 16.12   Comparison of the lowest-energy conduction band state given by the empirical pseudo-potential (EPP) calculation in the previous section with that given by the Kronig–Penney model of a superlattice under the envelope function/effective mass approximations (EF/EMA)

Mäder and Zunger indicate that these simple bulk-like potentials are not sufficient to describe the microscopic structure of such a short-period superlattice and hence the result for $n_z=1$ could be prone to error. In particular, no account has been made for the change in the coordination of the interface anions (see Section 16.5). For $2 \leq n_z \leq 8$, the conduction-band minimum for the superlattice originates from the X valleys of the bulk and not from the Γ valley—this is possible because of the indirect nature of AlAs. As the X valley in bulk AlAs lies below that of the Γ valley, then narrow quantum wells can produce such high confinement energies that the eigenstate is influenced by these outlying valleys (in essence!). For $n_z > 8$, the simpler Kronig–Penney model describes the behaviour of the energy state with superlattice period quite well *qualitatively*.

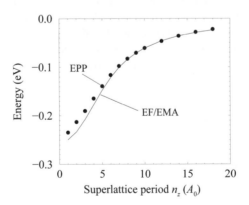

Figure 16.13   Comparison of the highest-energy valence-band state given by the empirical pseudo-potential (EPP) calculation in the previous section with that given by the Kronig–Penney model of a superlattice under the envelope function/effective mass approximations (EF/EMA)

Fig. 16.13 displays the results of an identical series of calculations for the uppermost (heavy-hole) valence-band state. The effective mass in the AlAs barriers was taken (and fixed) as the bulk value of 0.51 m_0 (Adachi [14], p. 254). In this case, however, the hole effective mass in the GaAs well regions was used as a parameter and varied in order to produce the best fit to the empirical pseudo-potential (EPP) results. It can be seen that the Kronig–Penney (EF/EMA) results for an effective mass of $0.45m_0$ fit the pseudo-potential data very well at the larger well and barrier widths, but the match is poorer at narrower widths. This is the result found by Long *et al.* [21], and indeed should be expected, as the envelope function approximation (see Chapter 1) hinges on the point that the wave function can be considered as a product of two components, with one being a rapidly varying Bloch function (which is factorized out) and the second a slowly varying envelope. It stands to reason that, when the period of the envelope approaches that of the Bloch function, as happens here in short-period superlattices, the approximation becomes poorer.

In essence, the result here can be summarized by saying that the effective mass of a particle in a quantum well is a function of the well width. The fact that a constant effective mass predicts too high a confinement energy allows the further deduction that the effective mass must be increased as the well width decreases in order to produce agreement with the empirical pseudo-potential calculations.

In conclusion, the microscopic nature of the pseudo-potential calculation gives more detail and allows for more complexity in the eigenstates of heterostructures than methods based on the envelope function/effective mass approximation. In particular, use of the constant effective mass approximation has been shown to breakdown for short-period superlattices.

16.4   IN-PLANE DISPERSION

In addition to computing the dispersion along the line of symmetry, i.e. the growth (z-) axis, the large-basis method, when applied to superlattices as in Section 16.2, can be used to calculate the in-plane (x–y) dispersion. Such knowledge is fundamental for describing the transport properties of electronic devices, which exploit in-plane transport for their operation, such as High Electron Mobility Transistors (HEMTs), as well as optical devices that are influenced by the carriers populating the in-plane momentum states, such as intersubband lasers (see Section 11.2).

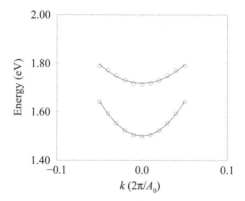

Figure 16.14   The in-plane [100] dispersion curve around the centre of the Brillouin zone for a $(GaAs)_{10}(AlAs)_{10}$ superlattice (top) in comparison with that for bulk (bottom)

Fig. 16.14 displays the results of calculations of the in-plane dispersion (top curve), in this case along the [100] direction, for the $(GaAs)_{10}(AlAs)_{10}$ superlattice of the previous section. On the same axes, the bulk dispersion curve (bottom) is also shown for comparison. The subband minima are both measured from the top of the valence band and hence the minimum of the bottom curve represents the band gap of the bulk material, while the difference in the minima represents the quantum confinement energy of this, i.e. the lowest confined state in the superlattice.

Remembering that, for low electron momenta the E–k dispersion curves are parabolic, then as before the effective mass is given by:

$$m^* = \hbar^2 \left(\frac{\partial^2 E}{\partial k^2} \right)^{-1} \tag{16.15}$$

The fitting of parabolas (the solid curves) to the data points in Fig. 16.14 thus allows the effective masses to be calculated (as before in Chapter 15). Using the potentials of Mäder and Zunger [367], this procedure gives the effective mass along any of the $\langle 100 \rangle$ directions in the bulk crystal as $0.082m_0$. In contrast to this, the in-plane electron effective mass for the superlattice is $0.15m_0$, which is obviously quite different.

This would seem to be a general result, at least for relatively short period superlattices; note the period for this example is $10A_0 = 56.5$ Å. *A priori*, it might be expected that as there is no confinement in the x–y plane (i.e. parallel to the layers), that the dispersion curves would resemble that of the bulk. In fact, as there is confinement along the z-direction, which leads to a shift in the band minimum, the band structure around the minima is clearly different from that of the bulk, the immediate consequence of which is that the effective mass increases.

16.5   INTERFACE COORDINATION

In bulk zinc blende material, each anion, e.g. As in GaAs, is bonded to four Ga cations. It has been shown that the atomic pseudo-potential of this As is different from the As″ (say) in AlAs, which is merely a reflection of the different chemical nature of the bonding between Ga atoms and Al atoms arising from their different electronegativities. Although disappointing in that universal atomic potentials cannot be deduced, the calculations thus far have shown how to deal with such a problem.

Consider now a heterojunction between two compound semiconductors that share a common species, e.g. GaAs/AlAs, as illustrated in Fig. 16.15. The As anion at the interface, as indicated, is bonded to two Ga atoms and two Al atoms, and hence they have neither the character of a fully tetrahedrally coordinated As or As″. In fact, they have a character that is intermediate between the two, which can be described by the mean:

$$As' = \frac{As + As''}{2} \tag{16.16}$$

and hence the interface properties *can* be described better by the mean in the atomic pseudo-potentials, i.e.

$$V_f^{As'}(q) = \frac{V_f^{As}(q) + V_f^{As''}(q)}{2} \tag{16.17}$$

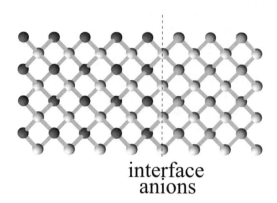

interface
anions

Figure 16.15 The atoms near the heterojunction between two compound semiconductors

16.6 STRAIN-LAYERED SUPERLATTICES

As already mentioned, the more recently deduced pseudo-potentials generally take account of a much greater diversity of experimental information than just the main energy gaps. In particular, one of these additional points is the reproducibility of the deformation potentials from the pseudo-potential calculations (see, for example [367,370] for III–V compounds and [369] for SiGe). One consequence of this development is that the pseudo-potentials can be used to accurately describe the microscopic electronic structure in *strain-layered* systems.

Strain arises in heterosystems that are formed from materials with difference lattice constants (see Kelly [7], p. 317, or Adachi [14], p. 271, for an introduction). For example, the lattice constant of GaAs is often quoted as 5.653 Å and that of AlAs as 5.660 Å, i.e. a difference of just 0.1 %. Hence, heterosystems formed from these materials (and all of the intermediate $Ga_{1-x}Al_xAs$ alloys) have a quite constant lattice spacing, and for this reason strain in this material system is usually ignored.

The lattice constant of Si is 5.43 Å and that of Ge 5.66 Å, which represents a lattice mismatch of $(5.66 - 5.43)/5.43 = 4$ %. While this may appears to be a low value, this is not the case for a semiconductor structure (for an introduction to strain in SiGe, see Meyerson [371]). If a layer of Ge is grown on top of a Si substrate, then the first few layers of Ge will assume the in-plane lattice constant of Si, i.e. the Ge atoms will be squeezed beyond their normal equilibrium separation by the crystal potential of Si. To compensate for this, the lattice constant along the growth (z-) axis increases, thus distorting the usual cubic cell into a cuboid. Given the pseudo-potentials now available, it would suffice to calculate the atomic positions, in order to calculate the effect of strain on the electronic structure. One very simplistic model, perhaps suitable for thin strained layers, would be to assume that the material attempts to keep its unit cell at constant volume. Hence, if A_z is the lattice constant along the growth axis then:

$$A_x A_y A_z = A_0^3 \qquad (16.18)$$

Therefore, considering Ge on Si, then the in-plane lattice constants of the Ge layer will equate to Si, i.e. 5.43 Å, and hence:

$$A_z = \frac{5.66^3}{5.43^2} \text{ Å} = 6.15 \text{ Å} \tag{16.19}$$

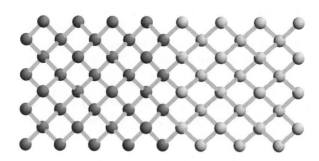

Figure 16.16   The atomic positions at a Si–Ge heterojunction

The application of this model is illustrated for just such a Si–Ge heterojunction in Fig. 16.16—there is an increase in the atomic spacing along the growth (z-) axis as the heterojunction is crossed from the left-hand region of Si atoms to the Ge atoms on the right-hand side. More sophisticated models for the atomic positions in strain-layered semiconductors are available. In particular, Morrison *et al.* [372] compared the results of pseudo-potential calculations of SiGe superlattices, keeping the lattice constant equal to that of a SiGe alloy buffer layer, with those produced with the atomic spacing calculated by minimizing the elastic energy [373].

If the strained Ge layer is allowed to grow thicker than the *critical thickness*, then dislocations form that relieve the strain, thus leading to a reduction in the total structural energy of the crystal. Empirical pseudo-potential calculations of the electronic structure around dislocations, and indeed any in-depth study of the particulars of strain-layered superlattices, lie beyond the scope of this present work. The interested reader is therefore referred to the literature (see, for example [374, 375] for SiGe, and [376] for $In_x Ga_{1-x} As/GaAs$).

The latter of these two material systems is exploited in a relatively common electronic device, namely the pseudomorphic-High Electron Mobility Transistor, or p-HEMT [377].

16.7   THE SUPERLATTICE AS A PERTURBATION

The direct approach to the solution of superlattices by the large-basis method works well, as shown so far; however, this is not always the case. The large number of plane waves required in the expansion generate a very large Hamiltonian matrix, $\mathcal{H}_{G',G}$, which in turn requires a computer with a large amount of memory for solution. Such machines are not always available and thus it may be necessary to resort to alternative methods. In the case of simple quantum wells and superlattices, one such method is to treat the superlattice as a perturbation on a bulk semiconductor. For the case of quantum dots and wires, an even more powerful computational technique may be required, which will be discussed later in Chapter 17.

Concentrating upon one-dimensional periodic potentials as generated by a pseudo-potential description of a quantum well, then as already mentioned, the superlattice can be thought of as a perturbation on a bulk 'host' semiconductor [378]. For example, for a GaAs/Ga$_{1-x}$Al$_x$As superlattice, the original host semiconductor may be chosen as GaAs, with the superlattice being formed by exchanging some of the Ga cations for Al. As the word *perturbation* implies, the superlattice wave function is then constructed from a linear combination of the orthogonal bulk eigenfunctions.

Recalling the bulk Schrödinger equation, i.e.

$$\mathcal{H}\psi_{n,\mathbf{k}} = E_{n,\mathbf{k}}\psi_{n,\mathbf{k}} \tag{16.20}$$

then introduction of a perturbation, V_{sl}, which is acknowledged to represent the additional potential introduced on formation of the superlattice (sl), yields a new Schrödinger equation with a new set of eigenfunctions, i.e.

$$(\mathcal{H} + V_{sl})\Psi_{N,\xi} = E_{N,\xi}\Psi_{N,\xi} \tag{16.21}$$

where the indices N and ξ represent the principal quantum number and the wave vector, respectively, of the superlattice eigenfunctions $\Psi_{N,\xi}$. As this is a perturbative approach, it is expected that these eigenfunctions will be expanded in terms of the bulk eigenfunctions, $\psi_{n,\mathbf{k}_\xi}$, i.e.

$$\Psi_{N,\xi} = \sum_{n,\mathbf{k}_\xi} A^N_{n,\mathbf{k}_\xi} \psi_{n,\mathbf{k}_\xi} \tag{16.22}$$

The additional qualification ξ on the wave vectors of the bulk eigenstates indicates that the superlattice eigenstate $\Psi_{N,\xi}$ is summed, not just over a particular set of eigenfunctions at a fixed $\mathbf{k}$, but also over the set of bulk wave vectors, which are *zone-folded* into the superlattice Brillouin Zone ([5], p. 95). This is illustrated in Fig. 16.17.

The smallest reciprocal lattice vector directed along one of the $\langle 100 \rangle$ directions, as highlighted by Table 15.1, is the [200] vector. Hence, the von Laue condition, i.e.

$$\mathbf{k}\cdot\hat{\mathbf{G}} = \frac{1}{2}|\mathbf{G}| \tag{16.23}$$

(see [1], p. 98) is fulfilled for an electron wave vector $\mathbf{k}$ equal to [100], and therefore the edge of the Brillouin zone along the z-axis is at [001], as indicated in the figure. A superlattice, of period $A_{sl} = n_z A_0$ along the z-axis, has its symmetry described by a new set of reciprocal lattice vectors, and indeed the principal primitive is given by:

$$\mathbf{b}_3 = \frac{2\pi}{A_{sl}}\hat{\mathbf{k}} = \frac{1}{n_z}\frac{2\pi}{A_0}\hat{\mathbf{k}} \tag{16.24}$$

which is the same result as that obtained for the simple cubic lattice (see equation (15.112)). Hence, the edge of the superlattice Brillouin zone, which is often referred to as the *mini-zone* is at:

$$\frac{1}{2}|\mathbf{b}_3| = \frac{2\pi}{2A_{sl}} = \frac{1}{2n_z}\frac{2\pi}{A_0} \tag{16.25}$$

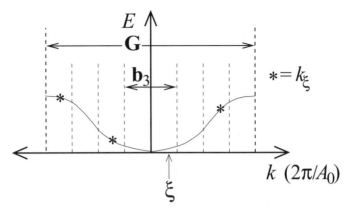

Figure 16.17   Schematic representation of the effect of zone-folding of the bulk Brillouin zone to produce the reduced superlattice mini-zone

Fig. 16.17 illustrates the mini-zone for a superlattice with period $A_{sl} = 2A_0$, i.e. $n_z = 2$. Thus, the edge of the mini-zone is a quarter of the wave vector that it is for bulk, along this direction. The set of wave vectors $\mathbf{k}_\xi$ are those points within the bulk Brillouin zone that are separated by a superlattice reciprocal lattice vector $\mathbf{b}_3$ from the electron wave vector within the mini-zone. The former are illustrated by the asterisks (*), while the latter is indicated by ξ in the figure.

Therefore, a pseudo-potential calculation of a superlattice of period $A_{sl} = n_z A_0$ requires $N_{\mathbf{k}_\xi} = 2n_z$ wave vector points from the bulk Brillouin zone. The points themselves are given by:

$$\mathbf{k}_\xi = \xi + (-n_z + i_{\mathbf{k}_\xi})\mathbf{b}_3, \quad \text{where} \quad i_{\mathbf{k}_\xi} = 0, 1, 2, \ldots, 2n_z - 1 \tag{16.26}$$

and $\mathbf{b}_3$ is given by equation (16.24).

Following standard perturbative theory expansion procedures, substitute $\Psi_{N,\mathbf{k}_\xi}$ from equation (16.22) into the perturbed Schrödinger equation (equation 16.21), to give the following:

$$\sum_{n,\mathbf{k}_\xi} A^N_{n,\mathbf{k}_\xi}(\mathcal{H} + V_{sl})\psi_{n,\mathbf{k}_\xi} = E_{N,\xi} \sum_{n,\mathbf{k}_\xi} A^N_{n,\mathbf{k}_\xi}\psi_{n,\mathbf{k}_\xi} \tag{16.27}$$

Using equation (16.20), then:

$$\sum_{n,\mathbf{k}_\xi} A^N_{n,\mathbf{k}_\xi}(E_{n,\mathbf{k}_\xi} + V_{sl})\psi_{n,\mathbf{k}_\xi} = E_{N,\xi} \sum_{n,\mathbf{k}_\xi} A^N_{n,\mathbf{k}_\xi}\psi_{n,\mathbf{k}_\xi} \tag{16.28}$$

Multiplying by $\psi^*_{n',\mathbf{k}'_\xi}$, and integrating over all space:

$$\sum_{n,\mathbf{k}_\xi} A^N_{n,\mathbf{k}_\xi}\int \psi^*_{n',\mathbf{k}'_\xi}(E_{n,\mathbf{k}_\xi} + V_{sl})\psi_{n,\mathbf{k}_\xi} \; \mathrm{d}\tau = E_{N,\xi} \sum_{n,\mathbf{k}_\xi} A^N_{n,\mathbf{k}_\xi}\int \psi^*_{n',\mathbf{k}'_\xi}\psi_{n,\mathbf{k}_\xi} \; \mathrm{d}\tau \tag{16.29}$$

Using the orthonormality of the bulk eigenvectors, i.e. the integral over all space of the product $\psi^*_{n',\mathbf{k}'_\xi}\psi_{n,\mathbf{k}_\xi}$ is non-zero only when $n = n'$ and $\mathbf{k}_\xi = \mathbf{k}'_\xi$, which in mathematical language would be written as:

$$\int \psi^*_{n',\mathbf{k}'_\xi}\psi_{n,\mathbf{k}_\xi} \; \mathrm{d}\tau = \delta_{n'n}\delta_{\mathbf{k}'_\xi,\mathbf{k}_\xi} \tag{16.30}$$

then:

$$\sum_{n,\mathbf{k}_\xi} A^N_{n,\mathbf{k}_\xi} \left(E_{n,\mathbf{k}_\xi} \delta_{n'n} \delta_{\mathbf{k}'_\xi,\mathbf{k}_\xi} + \int \psi^*_{n',\mathbf{k}'_\xi} V_{sl} \psi_{n,\mathbf{k}_\xi} \ \mathrm{d}\tau \right) =$$

$$E_{N,\xi} \sum_{n,\mathbf{k}_\xi} A^N_{n,\mathbf{k}_\xi} \delta_{n'n} \delta_{\mathbf{k}'_\xi,\mathbf{k}_\xi} \tag{16.31}$$

Write this as:

$$\sum_{n,\mathbf{k}_\xi} A^N_{n,\mathbf{k}_\xi} \mathcal{H}'_{(n',\mathbf{k}'_\xi),(n,\mathbf{k}_\xi)} = E_{N,\xi} \sum_{n,\mathbf{k}_\xi} A^N_{n,\mathbf{k}_\xi} \delta_{n'n} \delta_{\mathbf{k}'_\xi,\mathbf{k}_\xi} \tag{16.32}$$

where the perturbed Hamiltonian matrix elements are given by:

$$\mathcal{H}'_{(n',\mathbf{k}'_\xi),(n,\mathbf{k}_\xi)} = E_{n,\mathbf{k}_\xi} \delta_{n'n} \delta_{\mathbf{k}'_\xi,\mathbf{k}_\xi} + \int \psi^*_{n',\mathbf{k}'_\xi} V_{sl} \psi_{n,\mathbf{k}_\xi} \ \mathrm{d}\tau \tag{16.33}$$

or

$$\mathcal{H}'_{(n',\mathbf{k}'_\xi),(n,\mathbf{k}_\xi)} = E_{n,\mathbf{k}_\xi} \delta_{n'n} \delta_{\mathbf{k}'_\xi,\mathbf{k}_\xi} + V' \tag{16.34}$$

where the potential term has been relabelled as follows:

$$V' = \int \psi^*_{n',\mathbf{k}'_\xi} V_{sl} \psi_{n,\mathbf{k}_\xi} \ \mathrm{d}\tau \tag{16.35}$$

The problem is therefore reduced to finding the eigenvalues, $E_{N,\xi}$, and eigenvectors, $A^N_{n,\mathbf{k}_\xi}$, of the square matrix, $\mathcal{H}'_{(n',\mathbf{k}'_\xi),(n,\mathbf{k}_\xi)}$. This is a direct analogy of the original equation (equation (15.20)), for bulk, and can again be solved by direct diagonalization. Furthermore, if N_n bulk energy levels at any one of the $N_{\mathbf{k}_\xi}$ bulk wave vectors are included, then $\mathcal{H}'_{(n',\mathbf{k}'_\xi),(n,\mathbf{k}_\xi)}$ is of the order $N_n N_{\mathbf{k}_\xi}$.

It remains then to calculate the matrix elements of $\mathcal{H}'$ (for short), and given that the bulk has already been solved, this manifests itself merely as deducing the integral constituting the second term of equation (16.34). Consider substituting the bulk eigenvectors for their plane wave summations, as detailed in the previous chapter, i.e.

$$\psi_{n,\mathbf{k}} = \frac{1}{\sqrt{\Omega}} \sum_{\mathbf{G}} a_{n,\mathbf{k}}(\mathbf{G}) e^{i(\mathbf{G}+\mathbf{k})\cdot\mathbf{r}} \tag{16.36}$$

then:

$$V' = \frac{1}{\Omega} \sum_{\mathbf{G}',\mathbf{G}} a^*_{n',\mathbf{k}\xi'}(\mathbf{G}') a_{n,\mathbf{k}\xi}(\mathbf{G}) \int e^{-i(\mathbf{G}'+\mathbf{k}\xi')\cdot\mathbf{r}} V_{sl} e^{i(\mathbf{G}+\mathbf{k}\xi)\cdot\mathbf{r}} \ \mathrm{d}\tau \tag{16.37}$$

$$\therefore V' = \frac{1}{\Omega} \sum_{\mathbf{G}',\mathbf{G}} a^*_{n',\mathbf{k}'_\xi}(\mathbf{G}') a_{n,\mathbf{k}_\xi}(\mathbf{G}) \int e^{-i(\mathbf{G}'-\mathbf{G}+\mathbf{k}'_\xi-\mathbf{k}_\xi)\cdot\mathbf{r}} V_{sl} \ \mathrm{d}\tau \tag{16.38}$$

The perturbation V_{sl} can be written as a sum of the atomic potentials over all of the sites where atoms have been exchanged. If $V_{a'}$ is the atomic potential of the newly introduced species at a position $\mathbf{r}'_a$, and V_a is the atomic potential of the original species, then the potential due to the creation of the superlattice is just the difference between the two summed over all of the sites, i.e.

$$V_{sl} = \sum_{\mathbf{r}'_a} [V_{a'}(\mathbf{r}-\mathbf{r}'_a) - V_a(\mathbf{r}-\mathbf{r}'_a)] \tag{16.39}$$

Again, specifying *all* of the atomic positions $\mathbf{r}'_a$ as a combination of a Bravais lattice vector $\mathbf{R}_{sl}$ and a basis $\mathbf{t}_{sl}$ (note that the set of basis vectors $\mathbf{t}_{sl}$ are the position vectors of the exchanged atom positions within a superlattice unit cell), i.e.

$$\mathbf{r}'_a = \mathbf{R}_{sl} + \mathbf{t}_{sl} \tag{16.40}$$

then:

$$V_{sl} = \sum_{\mathbf{R}_{sl}} \sum_{\mathbf{t}_{sl}} [V_{a'}(\mathbf{r} - \mathbf{R}_{sl} - \mathbf{t}_{sl}) - V_a(\mathbf{r} - \mathbf{R}_{sl} - \mathbf{t}_{sl})] \tag{16.41}$$

Consider the vector $\mathbf{G}' - \mathbf{G} + \mathbf{k}'_\xi - \mathbf{k}_\xi = \mathbf{g}$ (say), in equation (16.38). From Fig. 16.17, $\mathbf{k}'_\xi - \mathbf{k}_\xi$ is clearly a reciprocal lattice vector of the superlattice, and in addition, as both $\mathbf{G}'$ and $\mathbf{G}$ are bulk reciprocal lattice vectors, then they are also reciprocal lattice vectors of the reduced symmetry system of the superlattice, albeit some large multiple of the primitive superlattice reciprocal lattice vectors. Hence, $\mathbf{g}$ is also a reciprocal lattice vector of the superlattice.

Substituting for $\mathbf{g}$ and V_{sl} from equation (16.41) into equation (16.38) then gives:

$$V' = \frac{1}{\Omega} \sum_{\mathbf{G}',\mathbf{G}} a^*_{n',\mathbf{k}'_\xi}(\mathbf{G}') a_{n,\mathbf{k}_\xi}(\mathbf{G})$$

$$\times \int e^{-i\mathbf{g} \cdot \mathbf{r}} \sum_{\mathbf{R}_{sl}} \sum_{\mathbf{t}_{sl}} [V_{a'}(\mathbf{r} - \mathbf{R}_{sl} - \mathbf{t}_{sl}) - V_a(\mathbf{r} - \mathbf{R}_{sl} - \mathbf{t}_{sl})] \ d\tau \tag{16.42}$$

Making the transformation $\mathbf{r} - \mathbf{R}_{sl} - \mathbf{t}_{sl} \longrightarrow \mathbf{r}$ then:

$$V' = \sum_{\mathbf{R}_{sl}} e^{-i\mathbf{g} \cdot \mathbf{R}_{sl}} \sum_{\mathbf{t}_{sl}} e^{-i\mathbf{g} \cdot \mathbf{t}_{sl}}$$

$$\times \frac{1}{\Omega} \sum_{\mathbf{G}',\mathbf{G}} a^*_{n',\mathbf{k}'_\xi}(\mathbf{G}') a_{n,\mathbf{k}_\xi}(\mathbf{G}) \int e^{-i\mathbf{g} \cdot \mathbf{r}} [V_{a'}(\mathbf{r}) - V_a(\mathbf{r})] \ d\tau \tag{16.43}$$

Now the product of any Bravais lattice vector of the superlattice $\mathbf{R}_{sl}$ and a reciprocal lattice vector $\mathbf{g}$ is an integral number of 2π, and hence for N_{sl} superlattice unit cells within the total volume Ω of the crystal:

$$\sum_{\mathbf{R}_{sl}} e^{-i2\pi n} = N_{sl}, \quad \text{where} \quad n \in \mathbb{Z} \tag{16.44}$$

and:

$$\frac{N_{sl}}{\Omega} = \frac{1}{\Omega_{sl}} \tag{16.45}$$

and therefore:

$$V' = \frac{1}{\Omega_{sl}} \sum_{\mathbf{G}',\mathbf{G}} a^*_{n',\mathbf{k}'_\xi}(\mathbf{G}') a_{n,\mathbf{k}_\xi}(\mathbf{G}) \sum_{\mathbf{t}_{sl}} e^{-i\mathbf{g} \cdot \mathbf{t}_{sl}} \int e^{-i\mathbf{g} \cdot \mathbf{r}} [V_{a'}(\mathbf{r}) - V_a(\mathbf{r})] \ d\tau \tag{16.46}$$

Note that, as in the bulk calculation, the integral over all space of the atomic potential multiplied by the exponential factor is assigned as the empirical pseudo-potential form factor, i.e.

$$\frac{1}{\Omega_c} \int e^{-i\mathbf{g} \cdot \mathbf{r}} V_a(\mathbf{r}) \ d\tau = V_f^{\mathbf{r}'_a}(g) \tag{16.47}$$

and:

$$\frac{1}{\Omega_c} \int e^{-i\mathbf{g} \cdot \mathbf{r}} V_{a'}(\mathbf{r}) \; d\tau = V_{f'}^{\mathbf{r}'_a}(g) \tag{16.48}$$

where $g = |\mathbf{g}|$.

As already mentioned, $\mathbf{g}$ is a superlattice reciprocal lattice vector, and as the period of the superlattice is much greater than the period of the bulk crystal, then $\mathbf{g}$ is much smaller than the bulk analogy $\mathbf{q}$. Thus, the pseudo-potential form factor $V_f(g)$ needs to be known at intermediate points between the discrete bulk reciprocal lattice vectors— just as in the case of the large-basis calculation.

Therefore, the potential term becomes:

$$V' = \frac{\Omega_c}{\Omega_{sl}} \sum_{\mathbf{G}',\mathbf{G}} a^*_{n',\mathbf{k}'_\xi}(\mathbf{G}') a_{n,\mathbf{k}_\xi}(\mathbf{G}) \sum_{\mathbf{t}_{sl}} e^{-i\mathbf{g} \cdot \mathbf{t}_{sl}} \left[V_{f'}^{\mathbf{r}'_a}(g) - V_f^{\mathbf{r}'_a}(g) \right] \tag{16.49}$$

Hence, the final form of the Hamiltonian matrix elements $\mathcal{H}'$ is obtained by substituting the potential matrix elements V' given in equation (16.49) into equation (16.34):

$$\mathcal{H}'_{(n',\mathbf{k}'_\xi),(n,\mathbf{k}_\xi)} = E_{n,\mathbf{k}_\xi} \delta_{n'n} \delta_{\mathbf{k}'_\xi,\mathbf{k}_\xi}$$

$$+ \frac{\Omega_c}{\Omega_{sl}} \sum_{\mathbf{G}',\mathbf{G}} a^*_{n',\mathbf{k}'_\xi}(\mathbf{G}') a_{n,\mathbf{k}_\xi}(\mathbf{G}) \sum_{\mathbf{t}_{sl}} e^{-i\mathbf{g} \cdot \mathbf{t}_{sl}} \left[V_{f'}^{\mathbf{r}'_a}(g) - V_f^{\mathbf{r}'_a}(g) \right] \tag{16.50}$$

Recalling the discussion in Section 15.8 that there is no single As potential for all semiconductors, it is important to realize that, when creating a $Ga_{1-x}Al_xAs$ /GaAs superlattice for solution by this perturbative approach, e.g. by substituting atoms onto some of the bulk sites, it is necessary to substitute both Al cations and the As anions associated with AlAs, as these are different from those associated with GaAs.

Equation (16.50) represents the elements of the square matrix $\mathcal{H}'_{(n',\mathbf{k}'_\xi),(n,\mathbf{k}_\xi)}$, the eigenvalues and eigenvectors of which represent, respectively, the energy levels and wave functions of the superlattice, where they are found by direct diagonalization.

16.8   APPLICATION TO GAAS/ALAS SUPERLATTICES

In this section, the series of calculations given in Section 16.2 will be revisited and the lowest conduction-band energy calculated for the same range of GaAs/AlAs superlattices, again as a function of period, but this time by using the perturbative approach in order to implement a empirical pseudo-potential calculation rather than the more direct large-basis method.

While the superlattice period controls the number of bulk electron wave vectors, given by the set $\mathbf{k}_\xi$, there is freedom as to the number of *bands* included in the perturbative expansion set. Such a choice is usually made by considering the system of interest and its relation to the bulk band structure. In particular, calculations of the lowest conduction-band state in a superlattice *could* be thought of as being adequately represented by considering just the lowest bulk-conduction-band state, whereas the triply degenerate (when excluding spin–orbit coupling) valence band suggests that all three may be needed for the calculation of the uppermost superlattice valence states.

Fig. 16.18 compares the results of calculations obtained by using both the perturbative method and the large-basis (LB) method, for the lowest-conduction-band state

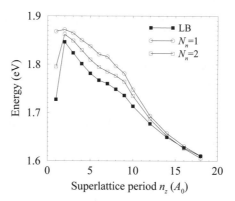

Figure 16.18 Comparison of the perturbative approach to the pseudo-potential calculation of a superlattice, with N_n=1 and 2 conduction band(s), with that of the large-basis (LB) method of earlier

in a GaAs/AlAs superlattice as a function of the period $(n_z(A_0))$. As expected for a perturbative technique, the energy levels obtained are higher than those given by the direct large-basis solution.

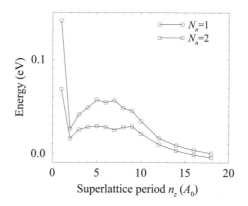

Figure 16.19 The energy differences between the perturbative and large basis approaches of Fig. 16.18

It can be seen that some of the details of the data, previously commented upon in Section 16.2, are not reproduced when just one bulk conduction band (N_n=1) is used to provide the expansion set. In particular, the low-energy state of the $(GaAs)_1(AlAs)_1$ short-period superlattice is not reproduced, although there is some evidence of a change in the nature of the states, as evidenced by the functional form of the E–n_z data, at around $n_z = 8$ or 9, as discussed earlier.

When the expansion set is increased to include a second bulk conduction band, the results are much closer to those of the large-basis calculation, as illustrated by the $N_n = 2$ data in Fig. 16.18. Thus, the obvious conclusion can be drawn that the more bulk bands included in the expansion set, then the better the approximation produced by the perturbative approach. These data are summarized in Fig. 16.19, which displays the energy difference between the calculations. Analysis of this figure allows a second conclusion to be drawn, i.e. that the larger the superlattice period, then the better the approximation given by the perturbative approach.

Table 16.1   The lowest conduction-band energy level of a $(GaAs)_{10}(AlAs)_{10}$ superlattice as a function of the number of bulk conduction bands N_n included in the basis set

N_n	E_1 (eV)
1	1.747636
2	1.734434
3	1.734429
4	1.734421
5	1.734421

Table 16.1 takes this line of enquiry further, displaying the results of calculations of the lowest conduction-band state of the $(GaAs)_{10}(AlAs)_{10}$ superlattice (i.e. $n_z = 10$), as a function of the number of bulk-conduction bands (N_n) included in the basis set. The improvement, evidenced by the lowering in the energy level, as implied by the variational principle, is clear when moving from just one bulk conduction band $N_n=1$ to $N_n=2$. The data show that further improvements are marginal, i.e. less than 1 meV, and are really not worth the additional computational time. The 'best' value offered by this approach, i.e. 1.734 eV, compares well with the 1.713 eV given by the large basis calculation. The perturbative approach attempts to replace the rather crude set of plane waves with a more sophisticated basis set. However, a discrepancy still exists because, no matter how many bulk bands are included in the superlattice expansion set, each one is still only constructed from the same set of N (in this case, 65) plane waves. This difference between the two approaches diminishes as the superlattice period increases.

The calculations in this section lead to the conclusion that the large-basis approach gives lower, and therefore more accurate (from the variational principle) energy levels. However, good approximations can be made by using the perturbative approach, particularly for the longer superlattice periods. The further consideration of computational expense is discussed below in Section 16.11.

16.9   INCLUSION OF REMOTE BANDS

It has been concluded in the previous section that, for these particular GaAs/ AlAs superlattices, if the basis set for the perturbative approach for the superlattice calculation is *limited* to bulk-conduction bands, then two bands are sufficient.

However, the question remains, *Can an extended basis including more remote bands, give improved energies?*. This implies, for the present series of calculations, the necessary inclusion of one or more of the bulk-valence bands in the basis set.

Table 16.2 compares the results of just such calculations, including the uppermost bulk-valence-band state, for the series of $(GaAs)_{n_z}(AlAs)_{n_z}$ superlattices considered earlier. It can be seen that the difference in the lowest conduction-band energy level, calculated including the remote valence band (2CB+1VB), is always less than 1 meV than the original (2CB) data, across the entire range of superlattice periods being con-

Table 16.2   The effect of the inclusion of the uppermost bulk-valence band in the basis set of a $(GaAs)_{n_z}(AlAs)_{n_z}$ superlattice

n_z	E_1^{2CB} (eV)	$E_1^{2CB+1VB}$ (eV)
1	1.795507	1.796218
2	1.861993	1.862025
3	1.849019	1.849090
4	1.829507	1.829511
5	1.809960	1.809997
6	1.794834	1.794835
7	1.784877	1.784882
8	1.776331	1.776331
9	1.764226	1.764241
10	1.734433	1.734456
12	1.686980	1.687005
14	1.652077	1.652094

sidered. Therefore, the inclusion of remote bands for these particular superlattices is unnecessary.

16.10   THE VALENCE BAND

As mentioned earlier, the triple degeneracy of the valence band at the centre of the bulk Brillouin zone, produced in this series of calculations without spin–orbit coupling, suggests that all three bands may be needed in any valence-band calculation. Fig. 16.20 displays the highest-valence-band state of the GaAs/AlAs superlattices, calculated by using the three uppermost bulk-valence-bands, compared with the results of earlier produced by using the large-basis (LB) method.

Again, the same conclusions can be drawn as for the conduction-band calculations, namely that the perturbative approach does not give energies as low as the large basis method, while the discrepancy between the two decreases as the superlattice period increases. This is highlighted in Fig. 16.21.

16.11   COMPUTATIONAL EFFORT

The main computational difficulty with the large-basis approach is the demands on computer memory necessary to store the large Hamiltonian matrix. In this respect, the perturbative method can be quite computationally efficient. As mentioned above, the order of the Hamiltonian matrix requiring diagonalization in this latter method is given by:

$$O(\mathcal{H}'_{(n',\mathbf{k}'_\xi),(n,\mathbf{k}_\xi)}) = N_n N_{\mathbf{k}_\xi} = 2N_n n_z \tag{16.51}$$

where n_z is the period of the superlattice in lattice constants. In comparison, the number of plane waves required in the expansion set for the large-basis calculation was deduced

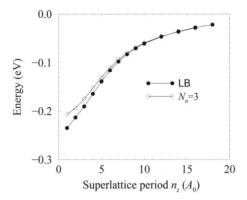

Figure 16.20 Comparison of the perturbative approach to the pseudo-potential calculation of a superlattice, with $N_n = 3$ valence bands, with that of the large-basis (LB) method considered earlier

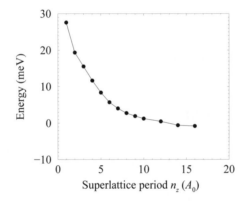

Figure 16.21 The difference between the energies of Fig. 16.20

empirically in Section 16.2 as approximately $128n_z$; hence, for the large basis approach:

$$O(\mathcal{H}_{\mathbf{G'},\mathbf{G}}) \approx 128n_z \tag{16.52}$$

Thus, it can be seen that the order of both matrices scales linearly with the superlattice period n_z. However, it is clear, that if the number of bands N_n required in the perturbative calculation is less than 64, then the Hamiltonian matrix $\mathcal{H}'_{(n',\mathbf{k}'_\xi),(n,\mathbf{k}_\xi)}$ will be smaller than $\mathcal{H}_{\mathbf{G'},\mathbf{G}}$.

The computational effort associated with the actual process of diagonalization of these matrices scales as the square of the order, and hence this part of the perturbative approach will require less computer time, as well as less memory, than the large basis approach.

16.12   SUPERLATTICE DISPERSION AND THE INTERMINIBAND LASER

As yet, the dispersion curves for electron movement along the growth (z-) axis, i.e., perpendicular transport, have not been calculated by using the perturbative approach to the superlattice.

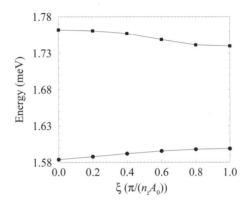

Figure 16.22   The two lowest energy dispersion curves (minibands) along the growth (z-) axis of a $(GaAs)_{10}(Ga_{0.8}Al_{0.2}As)_{10}$ superlattice

Using the virtual crystal approximation to obtain the form factors for the alloy, Fig. 16.22 plots the lowest two conduction bands of the superlattice, for a range of electron wave vectors ξ across the Brillouin zone of a $(GaAs)_{10}(Ga_{0.8}Al_{0.2}As)_{10}$ superlattice. Just as the superlattice Brillouin zone is often referred to as the *mini-zone*, these superlattice dispersion curves are called *minibands* (see Section 2.13). As discovered in Section 2.13, when using the Kronig–Penney model, the lowest-energy miniband has its minimum at the zone centre, while the second miniband has its minimum at the zone edges.

Thus, if carriers are injected into the upper of these two minibands, then they will rapidly lower their energy, via the fast non-radiative *intraminiband* scattering processes of phonon emission and carrier scattering. These have times similar to those characterizing intrasubband processes, perhaps of the order of 100 fs (see Chapter 10). Given the functional form of the E–$\mathbf{k}$ curves within the superlattice Brillouin zone, this energy lowering implies an increasing momentum and the carriers thus move towards the zone edge.

At this point, the carriers have reached the bottom of the miniband and can no longer lower their energy through intraminiband relaxation processes, and they must wait the relatively long time for an *interminiband* scattering event to occur. Following this, any carrier scattering to the states near the zone edge of the lowest miniband will scatter rapidly to lower momentum states. Such a scenario of a long lifetime in the upper level, compared to a short lifetime in the lower level, for the selection of electron momenta near the superlattice zone edge, is reminiscent of the requirements of the intersubband laser of Chapter 10. Indeed, the combination of these scattering processes within the two lowest minibands of a superlattice forms the basis of a new form of semiconductor laser known as the *interminiband laser* [41, 42]. One exciting feature of this device is

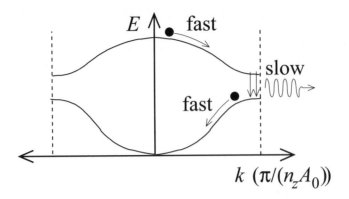

Figure 16.23 The carrier dynamics within an interminiband laser

that it lases without an overall population inversion: it is only necessary to obtain a localized inversion in the states near the zone edge for stimulated emission to occur.

16.13 ADDITION OF ELECTRIC FIELD

An electric field can be considered merely as a perturbation on the semiconductor system, just as the superlattice itself was. However, given the inherent periodic nature of the pseudo-potential method, the electric field itself has to be periodic and repeated within each unit cell, as shown in Fig. 16.24.

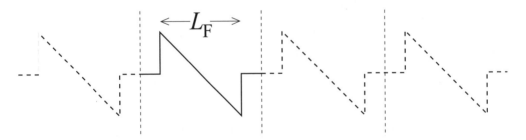

Figure 16.24 The necessary periodic nature of any electric field accounted for with pseudo-potential theory

Taking the first and last atomic positions within the unit cell of interest as the extent of the electric field, and defining the zero-field point as the centre of the cell, i.e. $(z_{N_a} + z_1)/2$, then the potential energy due to the electric field would be:

$$V_F = 0, \qquad\qquad z < z_1 \qquad\qquad (16.53)$$

$$V_F = -qF\left(z - \frac{z_{N_a} + z_1}{2}\right), \qquad z_1 < z < z_{N_a} \qquad (16.54)$$

$$V_F = 0, \qquad\qquad z_{N_a} < z \qquad\qquad (16.55)$$

where z_1 and z_{N_a} are the z-coordinates of the first (1) and last (N_a) atoms, respectively. With this definition, the field does look like that in Fig. 16.24, with small zero-field regions of width $A_0/2$ in-between the regions of linear sloping potential.

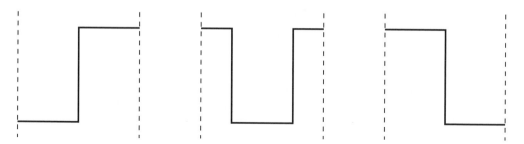

Figure 16.25 The possible unit cells for a pseudo-potential study of a superlattice

The pseudo-potential calculations so far have all centred around true superlattices, i.e. systems of quantum wells with significant overlap between the wave functions of adjacent wells. Single isolated quantum wells (SQWs) can be considered by using pseudo-potential theory, just by making the barriers within each period thick in size, thus producing a large distance between the wells. Hitherto, it was not relevant where the atoms were exchanged within the unit cell to produce a superlattice; for example, as shown in Fig. 16.25, the well could be formed at the beginning of the unit cell, the middle or the end. However, when incorporating an electric field as well, it is important that the field extends either side of the 'region of interest'—this is achieved simply by ensuring the quantum wells are centred in the unit cell.

The effect of an electric field can be calculated for a single quantum well, or for a system of several quantum wells, by the appropriate choice of unit cell (see, for example Fig. 16.26).

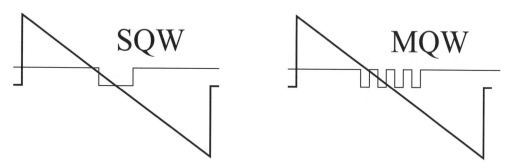

Figure 16.26 The unit cells required to study the effect of an electric field on a single quantum well (SQW) or a multiple quantum well (MQW)

Therefore, an electric field can be considered as an additional perturbation with the same periodicity as the superlattice unit cell, whether that cell contains one or more quantum wells. Thus, the original Schrödinger equation for the superlattice (equation 16.21) i.e.

$$(\mathcal{H} + V_{sl})\Psi_{N,\xi} = E_{N,\xi}\Psi_{N,\xi} \tag{16.56}$$

would have an additional term representing the perturbation due to the electric field, i.e.

$$(\mathcal{H} + V_{sl} + V_F)\Psi_{N,\xi} = E_{N,\xi}\Psi_{N,\xi} \tag{16.57}$$

the consequence of which is that the potential term in equation (16.38) becomes:

$$V'' = \frac{1}{\Omega} \sum_{G',G} a^*_{n',k'_\xi}(G') a_{n,k_\xi}(G) \int e^{-i(G'-G+k'_\xi-k_\xi)\bullet r}(V_{sl} + V_F) \ d\tau \tag{16.58}$$

The manipulation of the first term (V_{sl}) clearly proceeds as before, thus giving the original perturbing potential V' due to the superlattice potential, as defined in equation (16.49), so therefore:

$$V'' = V' + \frac{1}{\Omega} \sum_{G',G} a^*_{n',k'_\xi}(G') a_{n,k_\xi}(G) \int e^{-i(G'-G+k'_\xi-k_\xi)\bullet r} V_F \ d\tau \tag{16.59}$$

Consider just the integral component, and again writing $\mathbf{g} = \mathbf{G}' - \mathbf{G} + \mathbf{k}'_\xi - \mathbf{k}_\xi$, then:

$$\frac{1}{\Omega} \int e^{-i(G'-G+k'_\xi-k_\xi)\bullet r} V_F \ d\tau = \frac{1}{\Omega} \int e^{-ig\bullet r} V_F \ d\tau \tag{16.60}$$

Given the form of V_F in equation (16.55) and writing the origin of the electric field potential as $z_0 = (z_{N_a} + z_1)/2$, then obtain:

$$\frac{1}{\Omega} \int e^{-ig\bullet r} V_F \ d\tau = -\frac{qF}{\Omega} \int_{-\infty}^{+\infty} \int_{-\infty}^{+\infty} \int_{-\infty}^{+\infty} (z - z_0) e^{-ig\bullet r} \ dz \ dx \ dy \tag{16.61}$$

The in-plane $(x–y)$ integrals only have value when the x- and y-components of $\mathbf{g}$ are zero, and are then equal to the length of the crystal in that dimension, i.e.

$$\frac{1}{\Omega} \int e^{-ig\bullet r} V_F \ d\tau = -\frac{qFL_xL_y}{\Omega} \int_{-\infty}^{+\infty} (z - z_0) e^{-ig_z z} \ dz \ \delta_{0,g_x} \delta_{0,g_y} \tag{16.62}$$

where the the vector coefficients g_x and g_y are defined by $\mathbf{g} = g_x\hat{\mathbf{i}} + g_y\hat{\mathbf{j}} + g_z\hat{\mathbf{k}}$, and thus the Dirac δ-functions ensure that the integral is non-zero for $\mathbf{g}$ vectors along the axis of the field only. If there are N_{sl_z} unit cells along the z-direction, then:

$$\frac{1}{\Omega} \int e^{-ig\bullet r} V_F \ d\tau = -\frac{qFL_xL_yN_{sl_z}}{\Omega} \int_{z_1}^{z_{N_a}} (z - z_0) e^{-ig_z z} \ dz \ \delta_{0,g_x} \delta_{0,g_y} \tag{16.63}$$

However, the total volume of the crystal $\Omega = L_xL_yN_{sl_z}n_zA_0$, where, of course, n_zA_0 is the superlattice period. Hence:

$$\frac{1}{\Omega} \int e^{-ig\bullet r} V_F \ d\tau = -\frac{qF}{n_zA_0} \int_{z_1}^{z_{N_a}} (z - z_0) e^{-ig_z z} \ dz \ \delta_{0,g_x} \delta_{0,g_y} \tag{16.64}$$

Integrating by parts, then obtain:

$$\frac{1}{\Omega} \int e^{-ig\bullet r} V_F \ d\tau = -\frac{qF}{n_zA_0} \left\{ \left[(z - z_0) \frac{e^{-ig_z z}}{-ig_z} \right]_{z_1}^{z_{N_a}} \right.$$

$$-\frac{1}{-ig_z}\int_{z_1}^{z_{Na}}e^{-ig_z z}\ dz\Bigg\}\delta_{0,g_x}\delta_{0,g_y} \tag{16.65}$$

$$\therefore \frac{1}{\Omega}\int e^{-i\mathbf{g}\bullet\mathbf{r}}V_F\ d\tau = -\frac{qF}{n_z A_0}\left[(z-z_0)\frac{ie^{-ig_z z}}{g_z}+\frac{e^{-ig_z z}}{g_z^2}\right]_{z_1}^{z_{Na}}\delta_{0,g_x}\delta_{0,g_y} \tag{16.66}$$

and:

$$\frac{1}{\Omega}\int e^{-i\mathbf{g}\bullet\mathbf{r}}V_F\ d\tau = -\frac{qF}{n_z A_0}\left[\left(\frac{i(z-z_0)}{g_z}+\frac{1}{g_z^2}\right)e^{-ig_z z}\right]_{z_1}^{z_{Na}}\delta_{0,g_x}\delta_{0,g_y} \tag{16.67}$$

which upon evaluation, gives:

$$\frac{1}{\Omega}\int e^{-i\mathbf{g}\bullet\mathbf{r}}V_F\ d\tau = -\frac{qF}{n_z A_0}\Bigg\{\left(\frac{i(z_{Na}-z_0)}{g_z}+\frac{1}{g_z^2}\right)e^{-ig_z z_{Na}}$$

$$-\left(\frac{i(z_1-z_0)}{g_z}+\frac{1}{g_z^2}\right)e^{-ig_z z_1}\Bigg\}\delta_{0,g_x}\delta_{0,g_y} \tag{16.68}$$

The total potential term for both the superlattice perturbation and the electric field perturbation is obtained by substituting equation (16.68) into equation (16.59), thus giving:

$$V'' = V' + \sum_{\mathbf{G}',\mathbf{G}}a^*_{n',\mathbf{k}'_\xi}(\mathbf{G}')a_{n,\mathbf{k}_\xi}(\mathbf{G})\left(-\frac{qF}{n_z A_0}\right)$$

$$\times\Bigg\{\left(\frac{i(z_{Na}-z_0)}{g_z}+\frac{1}{g_z^2}\right)e^{-ig_z z_{Na}}-\left(\frac{i(z_1-z_0)}{g_z}+\frac{1}{g_z^2}\right)e^{-ig_z z_1}\Bigg\}\delta_{0,g_x}\delta_{0,g_y} \tag{16.69}$$

Using the definition for the superlattice perturbing potential V' in equation (16.49), then the final form for the Hamiltonian matrix elements *including* an electric field is:

$$\mathcal{H}'_{(n',\mathbf{k}'_\xi),(n,\mathbf{k}_\xi)} = E_{n,\mathbf{k}_\xi}\delta_{n'n}\delta_{\mathbf{k}'_\xi,\mathbf{k}_\xi} + \sum_{\mathbf{G}',\mathbf{G}}a^*_{n',\mathbf{k}'_\xi}(\mathbf{G}')a_{n,\mathbf{k}_\xi}(\mathbf{G})$$

$$\times\Bigg\{\frac{\Omega_c}{\Omega_{sl}}\sum_{\mathbf{t}_{sl}}e^{-i\mathbf{g}\bullet\mathbf{t}_{sl}}\left[V^{\mathbf{r}'_a}_{f'}(g)-V^{\mathbf{r}'_a}_f(g)\right]-\frac{qF}{n_z A_0}\delta_{0,g_x}\delta_{0,g_y}$$

$$\times\Bigg\{\left(\frac{i(z_{Na}-z_0)}{g_z}+\frac{1}{g_z^2}\right)e^{-ig_z z_{Na}}-\left(\frac{i(z_1-z_0)}{g_z}+\frac{1}{g_z^2}\right)e^{-ig_z z_1}\Bigg\}\Bigg\} \tag{16.70}$$

which is an extension of the earlier form in equation (16.50). Inspection of equation (16.70) does raise a small problem, namely divergence of the electric field perturbation when $g_z = 0$. For this particular instance, it is necessary to revisit equation (16.64) and put $g_z = 0$, i.e.

$$\frac{1}{\Omega}\int e^{-i\mathbf{g}\bullet\mathbf{r}}V_F\ d\tau = -\frac{qF}{n_z A_0}\int_{z_1}^{z_{Na}}(z-z_0)\ dz\ \delta_{0,g_x}\delta_{0,g_y} \tag{16.71}$$

and then:

$$\frac{1}{\Omega}\int e^{-i\mathbf{g}\bullet\mathbf{r}}V_F\ d\tau = -\frac{qF}{n_z A_0}\left[\frac{z^2}{2}-zz_0\right]_{z_1}^{z_{Na}}\delta_{0,g_x}\delta_{0,g_y} \tag{16.72}$$

Recalling that $z_0 = (z_{N_a} + z_1)/2$, then this becomes:

$$\frac{1}{\Omega} \int e^{-i\mathbf{g} \bullet \mathbf{r}} V_F \; d\tau = 0 \qquad (16.73)$$

Therefore, it has been shown that an electric field can be included in the pseudo-potential formalism, with the result being an additional potential term in the Hamiltonian matrix.

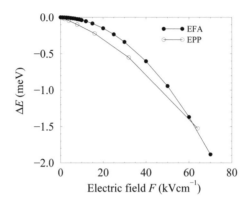

Figure 16.27   The change in the energy of the lowest eigenstate of a 28.25 Å GaAs single quantum well surrounded by $Ga_{0.8}Al_{0.2}As$ barriers, as a function of the applied electric field

Consider the application of an electric field to a single 10 ml (28.25 Å) thick GaAs quantum well surrounded by 20 ml (56.5 Å) thick $Ga_{0.8}Al_{0.2}As$ barriers. Thus, in the periodic formalism characteristic of pseudo-potentials, this would imply a separation of 2×20 ml (113 Å), which should be enough for them to act as independent quantum wells.

Fig. 16.27 displays the results of calculations of the change in the energy (ΔE) of the lowest conduction-band energy level as a function of the electric field, as deduced by the empirical pseudo-potential (EPP) method. For comparison, the figure also shows data obtained by the envelope function approximation (EFA) for the same system. The parabolic nature of this, i.e. the quantum-confined Stark effect, is clearly evidenced in the empirical pseudo-potential calculation and, in fact, the energy changes are remarkably similar over the range of electric fields employed. Thus Fig. 16.27 seems to imply at first sight, that the computationally 'long-winded' approach of the empirical pseudo-potential method is no better than much quicker methods based on the envelope function approximation. Indeed, if Stark effect energy level changes are all that are of interest, then this would be the case. However, the pseudo-potential method gives much more, e.g. the same calculation can also be used to yield the in-plane dispersion curve, and the pseudo-potential method can account much more fundamentally for valence band mixing and strain, *at the same time* as modelling the electric field.

CHAPTER 17

APPLICATION TO QUANTUM WIRES AND DOTS

17.1 RECENT PROGRESS

The difficulties associated with applying the techniques outlined in this work to lower-dimensional quantum wires and dots, have been hinted at already. In the work on point defects in bulk material (Section 15.13), in which a single impurity atom was placed within a larger cube of bulk material, more than 2000 plane waves were needed in the expansion set. In effect, such a system represents the ultimate quantum dot, with the confining potential existing on account of the single atom; after all, quantum dots are often referred to as *artificial atoms* (see, for example [379]). Single-atom quantum dots are also *single-electron* quantum dots [380, 381]. To deal with larger quantum dots, remembering that real dots often have a base length in the range 100–400 Å requires a very large number of plane waves, and indeed the problem becomes too large to handle with present-day computers. The perturbative approach, as derived for the superlatice in Chapter 16, can be extended to deal with quantum dots [382], and this appears to be a promising improvement for modest-sized unit cells.

Recently, Zunger and co-workers have demonstrated a new computational technique, which deals with this large-matrix problem. Instead of seeking solutions of the usual Schrödinger equation, i.e.

$$\mathcal{H}\psi_{n,\mathbf{k}} = E_{n,\mathbf{k}}\psi_{n,\mathbf{k}} \qquad (17.1)$$

Quantum Wells, Wires and Dots, Third Edition. P. Harrison
©2009 John Wiley & Sons, Ltd.

solutions are sought to the alternative expression:

$$\left(\mathcal{H} - E_{\text{ref}}\right)^2 \psi_{n,\mathbf{k}} = \left(E_{n,\mathbf{k}} - E_{\text{ref}}\right)^2 \psi_{n,\mathbf{k}} \tag{17.2}$$

where the reference energy, E_{ref}, can be chosen to lie within the fundamental gap, and hence the valence- and conduction- band-edge states, which are often those of primary interest, are transformed from being arbitrarily high-energy states to being the lowest states. The technique is to minimize the expectation value:

$$\left\langle \psi \left| \left(\mathcal{H} - E_{\text{ref}}\right)^2 \right| \psi \right\rangle \tag{17.3}$$

where the standard empirical pseudo-potential operator of equation (15.5) can be employed. The approach has become known as the *folded spectrum method*; for more details of the theory, see Wang and Zunger [383], while for an example of its application, see Wang *et al.* [384].

At the present time, such a technique lies beyond the scope of this introductory text, and although limited, the large-basis method coupled with direct diagonalization of the Hamiltonian matrix, will be explored to its limits with application to the reduced-dimensional systems of quantum wires and dots.

17.2 THE QUANTUM-WIRE UNIT CELL

The unit cell that has to be employed for a quantum wire is just a special case of the extended cube utilized earlier in Section 15.13 for impurities in bulk material. The periodic nature of the pseudo-potential method implies that, in fact, the crystal potential will be an infinite sequence of parallel quantum wires, as shown in Fig. 17.1. However, making use of the symmetry properties of the wire, the latter need only be one lattice constant in length, i.e. assuming the wire to lie along the z-axis, then the number of lattice constants in this direction is given by $n_z = 1$ (see Fig. 17.2).

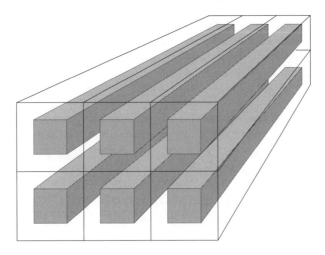

Figure 17.1   The periodic nature of the quantum wire unit cell

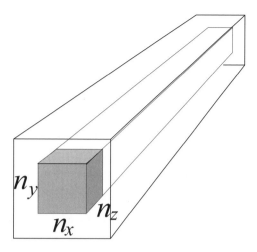

Figure 17.2   The quantum-wire unit cell; note the depth of one lattice constant, with sufficient barrier to encompass the wire, and if modelling a single wire, to localize the charge

A barrier or 'cladding' material surrounds the quantum wire; in this method, it is important that this is of sufficient thickness to isolate the wires from their hypothetical neighbours—just as in the case of modelling single quantum wells using pseudo-potential theory. This thickness depends on the confinement of the carriers within the wire. For example, for wide wires, the confinement is high and hence the overlap of the wave function with that in an adjacent wire is low. Therefore, only a thin barrier layer may be needed.

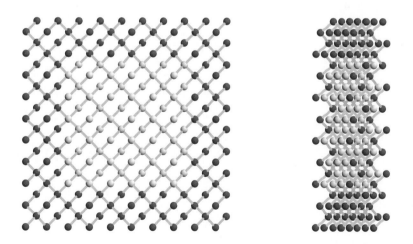

Figure 17.3   The quantum-wire unit cell; note the depth of 1 lattice constant

For the purpose of this illustrative example, consider a square cross-sectional Ge quantum wire surrounded by Si barriers. The atomic positions within the unit cell are illustrated in Fig. 17.3.

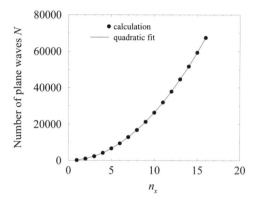

Figure 17.4   The number of plane waves in the expansion set versus the length of the wire side n_x

The unit cell is just five lattice constants square, i.e. n_x=5, n_y=5, and as mentioned above, just one lattice constant deep, i.e. n_z=1. About the only useful wire that can be accommodated in this box would be three lattice constants square with a single lattice constant barrier, thus giving a total of two lattice constants between the wires; the reason for these limitations on the wire geometry being that even for this small cross-section wire, the number of plane waves required in the expansion would normally be 6625, which would give a Hamiltonian matrix $\mathcal{H}_{\mathbf{G},\mathbf{G}'}$ occupying $6625^2 \times 8 = 351$ Mbytes of computer memory. This is only just in reach of high-end desktop computers at the present time, so for the purpose of these illustrative calculations, the expansion set will be reduced by truncating at a maximum reciprocal lattice vector of $3 \times (2\pi/A_0)$, rather than the usual 4. This reduces the expansion set to 2751, which still represents 60 Mbytes.

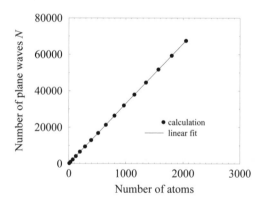

Figure 17.5   The number of plane waves in the expansion set versus the number of atoms in the unit cell

Fig. 17.4 illustrates how the untruncated expansion set increases with the number of lattice constants along the side of the wire unit cell. The accuracy of the fit indicates that the expansion set increases as the square of n_x. Given that the number of atoms within the unit cell is also proportional to the area, which in this case of a *square* unit cell is proportional to n_x^2, then the expansion set may also be expected to be proportional to the number of atoms in the unit cell. This is confirmed by the linear fit shown in Fig. 17.5.

As machine specifications increase, larger wire unit cells will be tractable, thus making this straightforward direct diagonalization method more useful. For example, at the time of writing, desktop machines with 1 Gbyte of RAM are becoming more common, which would allow the full expansion set to be used, or a quantum-wire unit cell of twice the side (four times the area) to be tackled.

17.3 CONFINED STATES

The results of a direct diagonalization of the Hamiltonian matrix for the quantum-wire unit cell in Fig. 17.3 are shown in Fig. 17.6. The latter illustrates the charge density of the lowest conduction-band state for an area the size of the unit cell and across the $z = 0$ plane. The origin of the plot has been shifted slightly in order to centralize the wire within the unit cell.

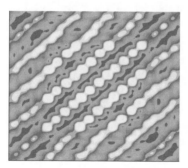

Figure 17.6   The charge density of the lowest conduction-band state over the cross-section of a Ge quantum wire embedded in a Si host

17.4 V-GROOVED QUANTUM WIRES

Due to the limitations on unit-cell size, some 'artistic licence' has to be invoked so as to produce a rough approximation to a cross-section of a 'V-groove' quantum wire (illustrated in Fig. 17.7).

The same method of calculation as used in the previous section can be employed to generate the corresponding charge density, again for the lowest conduction-band state, as now given in Fig. 17.8. The change in the charge distribution in comparison with a square cross-section wire can be seen by referring back to Fig. 17.6, i.e. a relatively small change in the number of atoms in the wire unit cell, which changes the symmetry of the wire, leads to a quite different charge distribution.

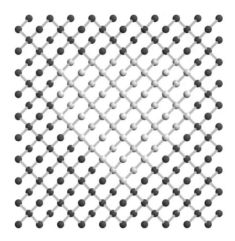

Figure 17.7   A 'V-grooved' quantum-wire unit cell

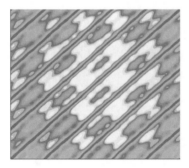

Figure 17.8   The charge density of the lowest conduction-band state over the cross-section of a 'V-grooved' Ge quantum wire embedded in a Si host

17.5   ALONG-AXIS DISPERSION

The shortest reciprocal lattice vector along any of the mutually perpendicular Cartesian axes in bulk is of the form $(0,0,2)$ when expressed in units of $2\pi/A_0$ (see, for example, Table 15.1). Therefore, the edge of the Brillouin zone in this direction is $(0,0,1)$, i.e. the point usually referred to as 'X'. However, for quantum wires, the shortest 'along-axis' reciprocal lattice vector is actually $(0,0,1)$, and the edge of the Brillouin zone is therefore $(0,0,\frac{1}{2})$. This is a little counter-intuitive, since along the axis of the wire, the material might be considered as being just infinitely extended bulk, and therefore be expected to exhibit bulk-material properties. In fact, as the calculations of the near-band-edge dispersion curves displayed in Fig. 17.9 show, this is not the case, and indeed zone-folding does occur.

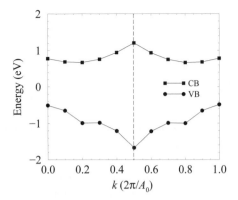

Figure 17.9   The along-axis dispersion of a Ge quantum wire: CB, conduction band; VB, valence band

17.6 TINY QUANTUM DOTS

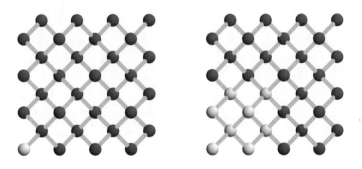

Figure 17.10   Two of the units cells for the 'tiny' quantum dot calculations: containing 1 Ge atom (left); and 8 Ge atoms (right)

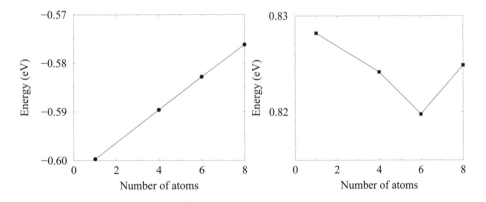

Figure 17.11   The energy of the upper-most valence state (left) and lower most conduction state (right) as a function of the number of Ge atoms embedded in a Si host

The problem of the very large basis set is even more acute for a quantum dot, which requires a three-dimensional cubic (or cuboid) unit cell to encase it. Again, with the present computational methods, only small calculations can be attempted, which do, however, illustrate how such calculations are set up.

Consider a unit cell of $2 \times 2 \times 2$ lattice constants; this contains 64 atoms. The (almost ridiculously small) simplest quantum dot that could be placed within this unit cell would be just one atom. Taking silicon as the host material and germanium as the dot, then this particular case is just the isoelectronic impurity dealt with earlier (see Chapter 15). However, in this case, the single Ge atom will be placed at the extreme corner of the unit cell (see Fig. 17.10), which will show the equivalence with the previous example, and will also be the approach used for subsequent calculations. This method of choosing the unit cell is somewhat easier than that used in the quantum-wire calculations described earlier in this chapter.

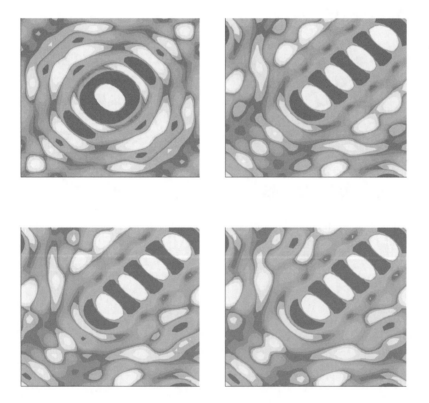

Figure 17.12 The difference in the total valence-band charge density between the 'tiny' quantum dots and bulk Si for different numbers of Ge atoms: (top left) 1; (top right) 4; (bottom left) 6; and (bottom right) 8

After choosing the unit cell, which in turn determines the reciprocal lattice vectors in the expansion set, then the empirical pseudo-potential method can yield information such as the energy of the uppermost valence-band state or the lowest conduction-band state. These are illustrated for this simple example in Fig. 17.11. In addition, it is possible to plot the charge density for any state or number of states (see Fig. 17.12). Note that the plots are centred on the point occupied by the single Ge atom in Fig. 17.10 (left).

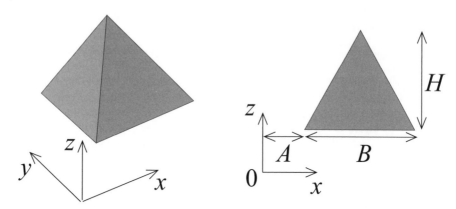

Figure 17.13   Schematic representations of a pyramidal quantum dot

17.7   PYRAMIDAL QUANTUM DOTS

As mentioned in Chapter 8, the deposition of a thin layer of one material on top of a substrate where there is a large difference in lattice constants, *can* lead to the formation of pyramidal-shaped quantum dots. These quantum dots can be quite large, perhaps a few hundred Angstroms across, and are totally beyond the range of direct diagonalization of the empirical pseudopotential Hamiltonian matrix. However, solution has been achieved by using the folded-spectrum method [385].

The setting up of such a calculation requires a knowledge of the atomic positions, so with this aim consider the schematic representation of a pyramidal dot shown in Fig. 17.13. Allowing the dot of base length B and height H to be encased within a cubic unit cell of side $A + B + A$, then the equations of the planes represented by the diagonally sloping sides in the right-hand diagram of Fig. 17.13 are as follows:

$$z = \frac{2H}{B}x - \frac{2H}{B} + A \quad \text{and} \quad z = -\frac{2H}{B}x + \frac{2HA}{B} + 2H + A \qquad (17.4)$$

Similar equations follow, but this time in terms of y for the remaining two planes. Fig. 17.14 illustrates the resulting atomic positions for an elemental quantum dot, such as Ge, as might occur when deposited on a Si substrate (the Si atoms have been removed for clarity). Note that, at certain ratios of the height and base, horizontal ridges form around the circumference of the pyramid.

17.8   TRANSPORT THROUGH DOT ARRAYS

Thus far, only the zone-centre, i.e. $\mathbf{k} = 0$ states have been looked at, but clearly it is possible to calculate transport or other properties that relate to larger momentum states.

Figure 17.14   The atomic positions for self-assembled pyramidal quantum dots formed from face-centred-cubic materials

For example, an earlier figure, i.e. Fig. 17.9, showed the along-axis dispersion curve of a quantum wire. Transport through the dots is possible, and could be studied by using the empirical pseudo-potential method, although given the necessary periodicity of the technique, solvable problems would be limited to *arrays* of dots.

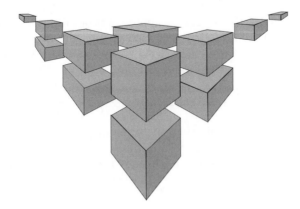

Figure 17.15   A schematic representation of how an array of cubic quantum dots embedded in a host crystal might look

Fig. 17.15 illustrates a periodic array of cubic quantum dots, while Fig. 17.16 indicates the nature of possible solutions for such a system.

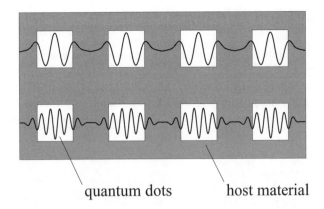

quantum dots   host material

Figure 17.16   Possible transport modes through arrays of quantum dots

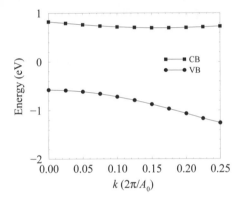

Figure 17.17   Lowest conduction-band (CB) and highest valence band (VB) dispersion curves through a periodic array of Ge quantum dots embedded in a Si host crystal

Fig. 17.17 shows the calculated dispersion curves for the uppermost valence-band state and the lower most conduction-band state for a periodic array of cubic eight-atom Ge quantum dots embedded in a Si host crystal. The dispersion curves along this [001] axis resemble those of a SiGe superlattice, with a zone-folded conduction-band minimum that is nearer the zone-centre than in the bulk material; the difference here though is that the [010] and [100] directions, i.e. the in-plane directions of the superlattices also have this same dispersion curve. Thus, the outlying X-valleys in the conduction band are brought nearer the zone centre *in all directions*. This contrasts with those superlattices that retain valleys near the zone boundary for electron motion in the conduction band. It may be expected, therefore, that introducing periodic potentials in all directions, as occurs with arrays of quantum dots, might lead to a more direct bandgap in SiGe, and hence a better quantum efficiency for light emission.

17.9   ANTI-WIRES AND ANTI-DOTS

The 'reversing' of the materials from which the quantum dots were grown, e.g. depositing a thin wetting layer of Si on to a Ge substrate, could lead to the formation of Si quantum dots. However, as Si has a wider bandgap than Ge, these would repel carriers rather than attracting them, thus leading to use of the term *anti-dots*. *Anti-wires* represent a similar concept in one-dimensional systems. Fig. 17.18 shows possible transport trajectories for the former case.

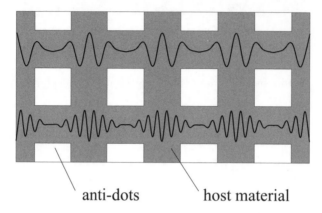

Figure 17.18   Possible transport modes through arrays of anti-dots

The development of dot arrays is really still in its infancy, although work has been reported of transport through both dot [386] and anti-dot [387, 388] structures. These systems of dots may have applications in future generations of logic [389, 390] and *Coulomb blockade*-based [391] devices.

CONCLUDING REMARKS

A detailed study of this book will have revealed that, besides containing quite a considerable amount of information, there is also a fair bit missing, and indeed there remain one or two question marks over some of the content that has been presented. The reason for this is that the whole area is still a very active field. It is not 'all done and dusted', and there still remain some major pieces of theory that need to be developed. It is hoped that this present work will provide a stimulus to other workers in the field to fill the gaps—gaps that may well have only become apparent upon reading this summary.

MATERIALS PARAMETERS

The results of the calculations in the main text depend on the assumed material parameters, a brief summary of which are presented here.

GaAs/Ga$_{1-x}$Al$_x$As

- Bandgap, $E_g = (1.426 + 1.247x)$ eV

- Band alignment: 33% of total discontinuity in valence band, i.e. $\Delta V_{\mathrm{VB}} = 0.33$; $\Delta V_{\mathrm{CB}} = 0.67$

- Electron effective mass, $m^* = (0.067 + 0.083x)\ m_0$

- Heavy-hole effective mass, $m^* = (0.62 + 0.14x)\ m_0$

- Lattice constant $A_0 = 5.65$ Å

- Low frequency (static) dielectric constant $\epsilon_s = 13.18\epsilon_0$

- High frequency dielectric constant $\epsilon_\infty = 10.89\epsilon_0$

- Material density $\rho = 5317.5$ kgm^{-3}

- Longitudinal Optical (LO) phonon energy $E_{\mathrm{LO}} = 36$ meV

509

Quantum Wells, Wires and Dots, Third Edition. P. Harrison
©2009 John Wiley & Sons, Ltd.

- Deformation potential $D_A = 7.0$ eV

- Velocity of sound $v_s = 5117.0$ m^{-1}

CdTe/Cd$_{1-x}$Mn$_x$Te

- Bandgap $E_g = (1.606 + 1.587x)$ eV

- Band alignment: 30% of total discontinuity in valence band, i.e. $\Delta V_{\text{VB}} = 0.30$; $\Delta V_{\text{CB}} = 0.70$

- Electron effective mass, $m^* = (0.11 + 0.067x)\, m_0$

- Heavy-hole effective mass, $m^* = (0.60 + 0.21x + 0.15x^2)\, m_0$

In$_{1-x-y}$Al$_x$Ga$_y$As/AlAs

- Total band discontinuity, $\Delta V = [2.093x + 0.629y + 0.577x^2 + 0.436y^2 + 1.013xy - 2.0x^2(1 - x - y)]$ eV

- Band alignment: 47% of total discontinuity in valence band, i.e. $\Delta V_{\text{VB}} = 0.47$; $\Delta V_{\text{CB}} = 0.53$

- Electron effective mass, $m^* = (0.0427 + 0.0685x)\, m_0$

REFERENCES

1. N. W. Ashcroft and N. D. Mermin, *Solid State Physics*, Saunders College Publishing, Philadelphia, 1976.

2. J. S. Blakemore, *Solid State Physics*, University Press, Cambridge, Second edition, 1985.

3. R. M. Eisberg, *Fundamentals of Modern Physics*, John Wiley & Sons, Ltd, New York, 1961.

4. R. T. Weidner and R. L. Sells, *Elementary Modern Physics*, Allyn and Bacon, Boston, Third edition, 1980.

5. M. Jaros, *Physics and Applications of Semiconductor Microstructures*, Clarendon Press, Oxford, 1989.

6. J. H. Davies and A. R. Long, Eds., *Physics of Nanostructures*, IOP Publishing, Bristol, 1992.

7. M. J. Kelly, *Low Dimensional Semiconductors: Materials, Physics, Technology, Devices*, Clarendon Press, Oxford, 1995.

8. R. Turton, *The Quantum Dot: A Journey into the Future of Microelectronics*, W. H. Freeman Spectrum, Oxford, 1995.

9. E. L. Ivchenko and G. Pikus, *Superlattices and other Heterostructures: Symmetry and Optical Phenomena*, Springer-Verlag, Berlin, 1995.

10. A. Shik, *Quantum Wells: Physics and Electronics of Two-Dimensional Systems*, World Scientific, London, 1997.

11. P. K. Basu, *Theory of Optical Processes in Semiconductors*, Clarendon Press, Oxford, 1997.

Quantum Wells, Wires and Dots, Third Edition. P. Harrison
©2009 John Wiley & Sons, Ltd.

12. P. A. M. Dirac, *The Principles of Quantum Mechanics*, Clarendon Press, Oxford, Fourth edition, 1967.

13. S. Nakamura and G. Fasol, *The Blue Laser Diode*, Springer, Berlin, 1997.

14. S. Adachi, *GaAs and Related Materials*, World Scientific, Singapore, 1994.

15. Landolt and Bornstein, Eds., *Numerical Data and Functional Relationships in Science and Technology*, vol. 22a of *Series III*, Springer-Verlag, Berlin, 1987.

16. A. Tredicucci, C. Gmachl, F. Capasso, D. L. Sivco, and A. L. Hutchinson, 'Long wavelength superlattice quantum cascade lasers at $\lambda \approx 17$ μm', *Appl. Phys. Lett.*, **74**:638, 1999.

17. G. Bastard, 'Superlattice band structure in the envelope function approximation', *Phys. Rev. B*, **24**:5693, 1981.

18. G. A. Bastard, *Wave Mechanics Applied to Semiconductor Heterostructures*, Les Editions de Physique, Paris, 1988.

19. M. G. Burt, 'The justification for applying the effective-mass approximation to microstructures', *J. Phys.: Condens. Matter*, **4**:6651, 1992.

20. M. G. Burt, 'Fundamentals of envelope function theory for electronic states and photonic modes in nanostructures', *J. Phys.:Condensed Matter*, **9**:R53, 1999.

21. Fei Long, W. E. Hagston, and P. Harrison, 'Breakdown of the envelope function/effective mass approximation in narrow quantum wells', in *The Proceedings of the 23rd International Conference on the Physics of Semiconductors*, Singapore, 1996, pp. 1819–1822, World Scientific.

22. J. W. Leech, *Classical Mechanics*, Chapman and Hall, London, Second edition, 1965.

23. I. S. Gradshteyn and I. M. Ryzhik, *Table of Integrals, Series, and Products*, Academic Press, London, Fifth edition, 1994.

24. G. T. Einevoll and L. J. Sham, 'Boundary conditions for envelope functions at interfaces between dissimilar materials', *Phys. Rev. B*, **49**:10533, 1994.

25. I. Galbraith and G. Duggan, 'Envelope-function matching conditions for GaAs/(Al,Ga)As heterojunctions', *Phys. Rev.*, **38**:10057, 1988.

26. J. W. Conley, C. B. Duke, G. D. Mahan, and J. J. Tiemann, 'Electron tunneling in metal-semiconductor barriers', *Phys. Rev.*, **150**:466, 1966.

27. D. J. BenDaniel and C. B. Duke, 'Space-charge effects on electron tunneling', *Phys. Rev.*, **152**:683, 1966.

28. W. E. Hagston, P. Harrison, T. Piorek, and T. Stirner, 'Boundary conditionson current carrying states and the implications for observation of Bloch oscillations', *Superlatt. Microstruct.*, **15**:199–202, 1994.

29. L. I. Schiff, *Quantum Mechanics*, McGraw-Hill, London, 1968.

30. O. von Roos, 'Position dependent effective masses in semiconductor theory', *Phys. Rev. B*, **27**:7547, 1983.

31. R. A. Morrow and K. R. Brownstein, 'Model effective mass Hamiltonians for abrupt heterojunctions and the associated wave function matching conditions', *Phys. Rev. B*, **30**:678, 1984.

32. R. A. Morrow, 'Establishment of an effective-mass Hamiltonian for abrupt heterojunctions', *Phys. Rev. B*, **35**:8074, 1987.

33. Ch. Schnittler and M. Kirilov, 'Hamiltonian and boundary conditions for electrons in semiconductor heterostructures', *phys. stat. sol. (b)*, **176**:143, 1993.

34. J. Khurgin, 'Novel configuration of self-electro-optic effect device based on asymmetric quantum wells', *Appl. Phys. Lett.*, **53**:779, 1988.

35. Masahiko Morita, Katsuyuki Goro, and Takeo Suzuki, 'Quantum-confined Stark effect in stepped-potential wells', *Japanese J. Appl. Phys.*, **29**:L1663, 1990.

36. N. Susa and T. Nakahara, 'Large blue shifts induced by Stark effect in asymmetric coupled quantum well', *Electr. Lett.*, **28**:941, 1992.

37. D. A. B. Miller, D. S. Chemla, T. C. Damen, A. C. Gossard, W. Wiegmann, T. H. Wood, and C. A. Burrus, 'Band-edge electroabsorption in quantum well structures - the quantum-confined stark-effect', *Phys. Rev. Lett.*, **53**:2173–2176, 1984.

38. Milton Abramowitz and Irene A. Stegun, *Handbook of Mathematical Functions*, Dover Publications Inc., New York, 1965.

39. S. Vatannia and G. Gildenblat, 'Airy's function implementation of the transfer-matrix method for resonant tunneling in variably spaced finite superlattices', *IEEE J. Quant. Electr.*, **32**:1093, 1996.

40. M. P. Halsall, J. E. Nicholls, J. J. Davies, B. Cockayne, and P. J. Wright, 'CdS/CdSe intrinsic stark superlattices', *J. Appl. Phys.*, **71**:907–915, 1992.

41. L. Friedman, R. A. Soref, and G. Sun, 'Quantum parallel laser: A unipolar superlattice interminiband laser', *IEEE Photonics Technology Letters*, **9**:593–595, 1997.

42. G. Scamarcio, F. Capasso, J. Faist, C. Sirtori, D. L. Sivco, A. L. Hutchinson, and A. Y. Cho, 'Tunable interminiband infrared emission in superlattice electron transport', *Appl. Phys. Lett.*, **70**:1796–1798, 1997.

43. G. Scamarcio, F. Capasso, C. Sirtori, J. Faist, A. L. Hutchinson, D. L. Sivco, and A. Y. Cho, 'High-power infrared (8-micrometer wavelength) superlattice lasers', *Science*, **276**:773–776, 1997.

44. S. M. Sze, *Physics of Semiconductor Devices*, John Wiley & Sons, Ltd, New York, Second edition, 1981.

45. D. K. Ferry, *Quantum Mechanics: An Introduction for Device Physicists and Electrical Engineers*, IOP Publishing, London, 1995.

46. S. Datta, *Quantum Phenomena*, vol. Volume VIII of *Modular Series on Solid State Devices*, Addison-Wesley, New York, 1989.

47. S. Luyri, *High Speed Semiconductor Devices*, p. 399, Wiley-Interscience, New York, 1990.

48. J. P. Sun, G. I. Haddad, P. Mazumder, and J. N. Schulman, 'Resonant tunnelling diodes:Models and properties', *Proc. of the IEEE*, **86**:639, 1998.

49. Hiroshi Mizuta and Tomonori Tanoue, *The Physics and Applications of Resonant Tunnelling Diodes*, Cambridge University Press, Cambridge, 1995.

50. J. K. Furdyna, 'Diluted magnetic semiconductors', *J. Appl. Phys.*, **64**:R29, 1988.

51. J. A. Gaj, *Diluted Magnetic Semiconductors*, vol. 25, chapter Chapter 7, Academic, Boston, 1988.

52. N. Malkova and U. Ekenberg, 'Spin properties of quantum wells with magnetic barriers. I. A k.p analysis for structures with normal band ordering', *Phys. Rev.*, **66**:155324, 2002.

53. T. Wenckebach, *Essentials of Semiconductor Physics*, John Wiley & Sons, Ltd., Chichester, 1999.

54. I. Savić, V. Milanović, Z. Ikonić, D. Indjin, V. Jovanović, and P. Harrison, 'Dilute magnetic semiconductor quantum-well structures for magnetic field tunable far-infrared/Terahertz absorption', *IEEE J. Quant. Electr.*, **40**:1614–1621, 2004.

55. R. L. Liboff, *Introductory Quantum Mechanics*, Addison-Wesley, San Francisco, Fourth edition edition, 2003.

56. R. Turton, *The Physics of Solids*, Oxford University Press, Oxford, 2000.

57. A. R. Sugg and J-P. C. Leburton, 'Modeling of modulation-doped multiple-quantum-well structures in applied electric fields using the transfer-matrix technique', *IEEE J. Quant. Electr.*, **27**:224, 1991.

58. J. P. Killingbeck, *Microcomputer Algorithms*, Hilger, Bristol, 1992.

59. E. H. Li, 'Interdiffusion as a means of fabricating parabolic quantum wells for the enhancement of the nonlinear third-order susceptibility by triple resonance', *Appl. Phys. Lett.*, **69**:460–462, 1996.

60. S. Flügge, *Practical Quantum Mechanics*, Springer-Verlag, Berlin, 1970.

61. T. Stirner, 'Notes on the Pöschl Teller potential hole', unpublished.

62. S. R. Jackson, J. E. Nicholls, W. E. Hagston, P. Harrison, T. Stirner, J. H. C. Hogg, B. Lunn, and D. E. Ashenford, 'Magneto-optical study of exciton binding energies band offsets and the role of interface potentials in $CdTe/Cd_{1-x}Mn_xTe$ multiple quantum wells', *Phys. Rev. B*, **50**:5392–5403, 1994.

63. F. Schwabl, *Quantenmechanik*, Springer, Berlin, 1990.

64. Y. Hirayama, J. H. Smet, L. H. Peng, C. G. Fonstad, and E. P. Ippen, 'Feasibility of 1.55 mu-m intersubband photonic devices using ingaas/alas pseudomorphic quantum-well structures', *Japanese J. Appl. Phys. Part 1-Regular Papers Short Notes & Review Papers*, **33**:890–895, 1994.

65. H. Asai and Y. Kawamura, 'Intersubband absorption in $In_{0.53}Ga_{0.47}As/In_{0.52}Al_{0.48}As$ multiple quantum-wells', *Phys. Rev. B-Condensed Matter*, **43**:4748–4759, 1991.

66. A. Raymond, J. L. Robert, and C. Bernard, 'The electron effective mass in heavily doped GaAs', *J. Phys. C: Solid State Phys.*, **12**:2289, 1979.

67. P. Harrison and R. W. Kelsall, '$1.55\mu m$ intersubband pumping of an $In_{0.53}Ga_{0.47}As/$ AlAs:InP symmetric double quantum well terahertz laser', *Physica E*, **2**:468–472, 1998.

68. J. H. Smet, L. H. Peng, Y. Hirayama, and C. G. Fonstad, 'Electron intersubband transitions to 0.8 eV (1.55 μm) in InGaAs/AlAs single quantum wells', *Appl. Phys. Lett.*, **64**:986, 1994.

69. W. J. Duffin, *Electricity and Magnetism*, McGraw-Hill, Third edition, 1980.

70. C. M. Snowden, Ed., *Semiconductor Device Modelling*, Springer-Verlag, 1988.

71. C. M. Snowden and R. E. Miles, Eds., *Compound Semiconductor Device Modelling*, Springer-Verlag, 1993.

72. P. H. Ladbrooke, *MMIC design: GaAs FETs and HEMTs*, Artech House, 1989.

73. J. M. Golio, *Microwave MESFETs and HEMTs*, Artech House, 1991.

74. E. H. Li, Ed., *Selected Papers on Quantum Well Intermixing for Photonics*, SPIE Optical Engineering Press, Bellingham, 1998.

75. A. Fick, 'Generalized Fick law for anomalous diffusion', *Ann. Phys.*, **170**:59, 1855.

76. W. D. Callister Jr., *Materials Science and Engineering*, John Wiley & Sons, Ltd., New York, 1985.

77. P. Harrison, W. E. Hagston, and T. Stirner, 'Excitons in diffused quantum wells', *Phys. Rev. B*, **47**:16404–16409, 1993.

78. Nguyen The Khoi P. Kossacki, J. A. Gaj, G. Karczewski, T. Wojtowicz, E. Janik, A. Zakrzewski, M.Kutrowski, and J. Kossut, 'Rapid thermal processing of semimagnetic superstructures studied by magnetoreflectivity', *Superlatt. Microstruct.*, **16**:63, 1994.

79. M. T. Furtado and M. S. S. Loural, 'Direct evaluation of interdiffusion coefficients in quantum well heterostructures usinfg photoluminescence', *Superlatt. Microstruct.*, **14**:21, 1993.

80. D. Tönnies, G. Bacher, A. Forchel, A. Waag, and G. Landwehr, 'Photoluminescence study of strong interdiffusion in CdTe/CdMnTe quantum wells induced by rapid thermal annealing', *Appl. Phys. Lett.*, **64**:766, 1994.

81. J. Crank, *The Mathematics of Diffusion*, Oxford University Press, London, 1956.

82. P. Shewmon, *Diffusion in Solids*, McGraw-Hill, New York, 1963.

83. B. Tuck, 'Some explicit solutions to the non-linear diffusion equation', *J. Phys. D: Appl. Phys.*, **9**:123, 1976.

84. D. Shaw, 'Diffusion mechanisms in II-VI materials', *J. Crystal Growth*, **86**:778, 1988.

85. K. Binder, 'Atomistic modeling of materials properties by Monte-Carlo simulation', *Adv. Mater.*, **4**:540, 1992.

86. P. Harrison, 'Differentiating between constant and concentration-dependent diffusion coefficients via the optical spectroscopy of excitons in quantum wells', *Semicond. Sci. Technol.*, **11**:1022–1025, 1996.

87. J. M. Fatah, I. Karla, P. Harrison, T. Stirner, W. E. Hagston, and J. H. C. Hogg, 'Defect induced diffusion mechanisms in ion implanted quantum well structures', in *Proceedings of the 22nd International Conference on the Physics of Semiconductors*, Singapore, 1994, pp. 2275–2278, World Scientific.

88. I. Karla, D. Shaw, W. E. Hagston, J. H. C. Hogg, S. Chalk, J. E. Nicholls, and C. Peili, 'Measurement of interdiffusion in II-VI quantum-well structures using optical methods', *J. Appl. Phys.*, **79**:1895, 1996.

89. I. Karla, J. H. C. Hogg, W. E. Hagston, J. Fatah, and D. Shaw, 'Monitoring of intermixing and interdiffusion by x-ray diffraction of ion-implanted quantum-well structures', *J. Appl. Phys.*, **79**:1898, 1996.

90. B. Elman, E. S. Koteles, P. Melman, and C. A. Armiento, 'GaAs/AlGaAs quantum-well intermixing using shallow ion implantation and rapid thermal annealing', *J. Appl. Phys.*, **66**:2104, 1989.

91. M. K. Chai, S. F. Wee, K. P. Homewood, W. P. Gillin, T. Cloitre, and R. L. Aulombard, 'An optical study of interdiffusion in ZnSe/ZnCdSe', *Appl. Phys. Lett.*, **69**:1579, 1996.

92. I. V. Bradley, W. P. Gillin, K. P. Homewood, and R. P. Webb, 'The effects of ion implantation on the interdiffusion coefficients in $In_{1-x}Ga_xAs$/GaAs quantum well structures', *J. Appl. Phys.*, **73**:1686, 1993.

93. M. A. Litovskii and R. SH. Malkovich, 'Method of determining the diffusion profile for a concentration dependent diffusion coefficient', *phys. stat. sol. (a)*, **36**:K145, 1976.

94. A. D. Pelton and T. H. Etsell, 'Analytical solution of Fick's second law when the diffusion coefficient varies directly as concentration', *Acta Metallurgica*, **20**:1269, 1972.

95. F. C. Frank and D. Turnbull, 'Mechanisms of diffusion of copper in germanium', *Phys. Rev.*, **104**:617, 1956.

96. T. Taskin, S. Gardelis, J. H. Evans, B. Hamilton, and A. R. Peaker, 'Sharp 1.54 μm luminescence from porous erbium implanted silicon', *Electr. Lett.*, **31**:2132, 1995.

97. J. F. Ziegler, J. P. Bierrack, and U. Littmark, *The Stopping and Range of Ions in Matter*, vol. 1, Pergamon Press, 1985.

98. P. Harrison, 'Numerical solution to the general one-dimensional diffusion equation in semiconductor heterostructures', *phys. stat. sol. (b)*, **197**:81–90, 1996.

99. J. M. Fatah, P. Harrison, T. Stirner, J. H. C. Hogg, and W. E. Hagston, 'Double crystal X-ray diffraction simulation of diffusion in semiconductor microstructures', *J. Appl. Phys.*, **83**:4037–4041, 1998.

100. E. F. Schubert, Ed., *Delta-doping of Semiconductors*, Cambridge University Press, Cambridge, 1996.

101. S. M. Sze, *Semiconductor Devices: Physics and Technology*, Wiley, New York, 1985.

102. P. Harrison, Fei Long, and W. E. Hagston, 'Empirical pseudo-potential calculation of the in-plane effective masses of electron and holes of two-dimensional excitons in CdTe quantum wells', *Superlatt. Microstruct.*, **19**:123–130, 1996.

103. C. Mailhiot, Yia-Chung Chang, and T. C. McGill, 'Energy spectra of donors in GaAs/Ga$_{1-x}$Al$_x$As quantum well structures in the effective mass approximation', *Phys. Rev. B*, **26**:4449, 1982.

104. R. L. Greene and K. K. Bajaj, 'Energy levels of hydrogenic impurity states in GaAs-Ga$_{1-x}$As$_x$As quantum well structures', *Solid State Commun.*, **45**:825, 1983.

105. G. N. Carneiro, G. Weber, and L. E. Oliveira, 'Binding energies and intra-donor absorption spectra in GaAs-GaAlAs quantum wells', *Semicond. Sci. Technol.*, **10**:41, 1995.

106. S. Chaudhuri and K. K. Bajaj, 'Effect of non-parabolicity on the energy levels of hydrogenic donors in GaAs-Ga$_{1-x}$Al$_x$As quantum-well structures', *Phys. Rev. B*, **29**:1803, 1984.

107. Chong ru Huo, Ben-Yuan Gu, and Lei Gu, 'General variational expressions for the calculation of the binding eneries of anisotropic donor states in stepped quantum wells', *J. Appl. Phys.*, **70**:4357, 1991.

108. R. G. Roberts, P. Harrison, and W. E. Hagston, 'The symmetry of donor bound electron wavefunctions in quantum wells', *Superlatt. Microstruct.*, **23**:289–296, 1998.

109. W. E. Hagston, P. Harrison, and T. Stirner, 'Neutral donors and spin-flip Raman spectra in dilute magnetic semiconductor microstructures', *Phys. Rev. B*, **49**:8242–8248, 1994.

110. W. T. Masselink, Yia-Chung Chang, and H. Morkoc, 'Binding-energies of acceptors in GaAs-Al$_x$Ga$_{1-x}$As quantum-wells', *Phys. Rev. B*, **28**:7373, 1983.

111. R. C. Miller, A. C. Gossard, W. T. Tsang, and O. Munteanu, 'Extrinsic photoluminescence from GaAs quantum wells', *Phys. Rev. B*, **25**:3871, 1982.

112. S. Fraizzoli, F. Bassani, and R. Buczko, 'Shallow donor impurities in GaAs-Ga$_{1-x}$Al$_x$As quantum well structures: Role of the dielectric constant mismatch', *Phys. Rev. B*, **41**:5096, 1990.

113. U. Ekenberg, 'Non-parabolicity effects in a quantum well—sublevel shift, parallel mass, and Landau levels', *Phys. Rev. B*, **40**:7714, 1989.

114. S. R. Parihar and S. A. Lyon, *Quantum Well Intersubband Transitions Physics and Devices*, p. 403, Kluwer, Netherlands, 1994.

115. J. K. Furdyna and J. Kossut Volume Editors, 'Diluted magnetic semiconductors,' in *Semiconductors and Semimetals*, R. K. Willardson and Treatise Editors A. C. Beer, Eds., vol. 25. Academic Press, Boston, 1988.

116. A. Twardowski, *Diluted Magnetic Semiconductors*, World Scientific, Singapore, 1996.

117. J. M. Fatah, T. Piorek, P. Harrison, T. Stirner, and W. E. Hagston, 'Numerical simulation of anti-ferromagnetic spin-pairing effects in diluted magnetic semiconductors and enhanced paramagnetism at interfaces', *Phys. Rev. B*, **49**:10341–10344, 1994.

118. J. A. Gaj, C. Bodin-Deshayes, P. Peyla, G. Feuillet J. Cibert, Y. Merle d'Aubigne, R. Romestain, and A. Wasiela, 'Magneto-optical study of interface mixing in CdTe-(Cd,Mn)Te system', in *Proceedings of the 21st International Conference on the Physics of Semiconductors*, Singapore, 1992, p. 1936, World Scientific.

119. A. K. Ramdas and S. Rodriguez, 'Effect of strain on vibrational modes in strained layer superlattices', in *Semiconductors and Semimetals*, R. K. Willardson and A. C. Beer, Eds., Boston, 1988, vol. 25, p. 345, Academic Press.

120. M. P. Halsall, S. V. Railson, D. Wolverson, J. J. Davies, B. Lunn, and D. E. Ashenford, 'Spin-flip Raman scattering in $CdTe/Cd_{1-x}Mn_xTe$ multiple quantum wells: A model system for the study of electron-donor binding in semiconductor heterostructures', *Phys. Rev. B*, **50**:11755, 1994.

121. D. R. Yakovlev, 'Two dimensional magnetic polarons in semimagnetic quantum well structures', in *Festkörperprobleme/Advances in Solid State Physics*, U. Roessler, Ed., Braunschweig, 1992, vol. 32, p. 251, Vieweg.

122. W. E. Hagston, P. Harrison, J. H. C. Hogg, S. R. Jackson, J. E. Nicholls, T. Stirner, B. Lunn, and D. E. Ashenford, 'An MBE investigation of interface disorder effects in magnetic II-VI quantum wells', *J. Vac. Sci. Technol. B*, **11**:881, 1993.

123. T.Stirner, P.Harrison, W.E.Hagston, and J.P.Goodwin, 'Bandgap renormalisation and observation of the type I-type II transition in quantum well systems', *J. Appl. Phys.*, **73**:5081–5087, 1993.

124. S. O. Kasap, *Principles of Electrical Engineering Materials and Devices*, Irwin McGraw-Hill, Boston, 1997.

125. G. Xiao, J. Lee, J. J. Liou, and A. Ortiz-Conde, 'Incomplete ionization in a semiconductor and its implications to device modeling', *Microelectronics Reliability*, **39**:1299, 1999.

126. G. L. Pearson and J. Bardeen, 'Electrical properties of pure silicon and silicon alloys containing Boron and Phosphorous', *Phys. Rev.*, **75**:865, 1949.

127. M. Avon and J. S. Prener, Eds., *Physics and Chemistry of II-VI compounds*, North-Holland, Amsterdam, 1967.

128. C. P. Hilton, W. E. Hagston, and J. E. Nicholls, 'Variational-methods for calculating the exciton binding-energies in quantum-well structures', *J. Phys. A*, **25**:2395, 1992.

129. P. Hilton, J. P. Goodwin, P. Harrison, and W. E. Hagston, 'Theory of exciton energy levels in multiply periodic systems', *J. Phys. A: Math. Gen.*, **25**:5365–5372, 1992.

130. P. Harrison, J. P. Goodwin, and W. E. Hagston, 'Exciton energy levels and band-offset determination in magnetic superlattices', *Phys. Rev. B*, **46**:12377–12383, 1992.

131. P. Harrison, T. Piorek, W. E. Hagston, and T. Stirner, 'The symmetry of the relative motion of excitons in semiconductor heterostructures', *Superlatt. Microstruct.*, **20**:45–57, 1996.

132. R. G. Roberts, P. Harrison, T. Stirner, and W. E. Hagston, 'Stark ladders in strongly coupled finite superlattices', *J. de Physique IV*, **3-C5**:203–206, 1993.

133. P. Harrison and W. E. Hagston, 'The effect of linear and non-linear diffusion on exciton energies in quantum wells', *J. Appl. Phys.*, **79**:8451–4855, 1996.

134. G. Bastard, E. E. Mendez, L. L. Chang, and L. Esaki, 'Exciton binding energy in quantum wells', *Phys. Rev. B*, **26**:1974, 1982.

135. S. K. Chang, A. V. Nurmikko, Wu J. W, L. A. Kolodziejski, and R. L. Gunshor, 'Band offsets and excitons in CdTe (Cd,Mn)Te quantum wells', *Phys. Rev. B*, **37**:1191, 1988.

136. M. M. Dignam and J. E. Sipe, 'Exciton state in Type I and Type II GaAs-Ga$_{1-x}$Al$_x$As superlattices', *Phys. Rev. B*, **41**:2865, 1990.

137. U. Ekenberg and M. Altarelli, 'Exciton binding-energy in a quantum-well with inclusion of valence band coupling and nonparabolicity', *Phys. Rev. B*, **35**:7585, 1987.

138. E. L. Ivchenko, A. V. Kavokin, G. R. Posina V. P. Kochereshko, I. N. Uraltsev, D. R. Yakovlev, R. N. Bicknell-Tassius, A. Waag, and G. Landwehr, 'Exciton oscillator strength in magnetic-field-induced spin superlattices CdTe (Cd,Mn)Te', *Phys. Rev. B*, **46**:7713, 1992.

139. Abdsadek Bellabchara, Pierre Lefebvre, Philippe Christol, and Henry Mathieu, 'Improved modeling of excitons in type-II semiconductor heterostructures by use of a three-dimensional variational function', *Phys. Rev. B*, **50**:11840, 1994.

140. S. V. Branis J. Cen and K. K. Bajaj, 'Exciton binding energies in finite-barrier Type II quantum-well structures in a magnetic field', *Phys. Rev. B*, **44**:12848, 1991.

141. Y. Shinozuka and M. Matsuura, 'Wannier excitons in quantum wells', *Phys. Rev. B*, **28**:4878, 1983.

142. Spiros V. Branis, J. Cen, and K. K. Bajaj, 'Effect of magnetic-fields on exciton binding energies in Type II GaAs-AlAs quantum-well structures', *Phys. Rev. B*, **44**:11196, 1991.

143. Yuan ping Feng, Hiap Sing Tan, and Harold N. Spector, 'Quantum well excitons in an electric field: Two versus three dimensional behaviour', *Superlatt. Microstruct.*, **17**:267, 1995.

144. J. W. Wu and A. V. Nurmikko, 'Wannier excitons in semiconductor quantum wells with small valence band-offsets—a generalised variational approach', *Phys. Rev. B*, **38**:1504, 1988.

145. M. J. L. S. Haines, N. Ahmed, S. J. A. Adams, K. Mitchell, I. R. Agool, C. R. Pidgeon, B. C. Cavenett, E. P. O'Reilly, A. Ghiti, and M. T. Emeny, 'Exciton-binding-energy maximum in Ga$_{1-x}$In$_x$As/GaAs quantum wells', *Phys. Rev. B*, **43**:11944, 1991.

146. E. E. Mendez, F. Agulló-Reuda, and J. M. Hong, 'Stark localisation in GaAs-GaAlAs superlattices under an electric field', *Phys. Rev. Lett.*, **60**:2426, 1988.

147. A. V. Kavokin, V. P. Kochereshko, G. R. Posina, I. N. Uraltsev, D. R. Yakovlev, G. Landwehr, R. N. Bicknell-Tassius, and A. Waag, 'Effect of the electron Coulomb potential on hole confinement in II-VI quantum wells', *Phys. Rev. B*, **46**:9788, 1992.

148. J. Warnock, B. T. Jonker, A. Petrou, W. C. Chou, and X. Liu, 'Exciton energies in shallow quantum wells and spin superlattices', *Phys. Rev. B*, **48**:17321–17330, 1993.

149. T. Piorek, P. Harrison, and W. E. Hagston, 'The relative importance of self-consistency and variable symmetry in the calculation of exciton energies in type-I and type-II semiconductor heterostructures', *Phys. Rev. B*, **52**:14111–14117, 1995.

150. T. Piorek, W. E. Hagston, and P. Harrison, 'Spontaneous symmetry breaking of excitons in multiple quantum wells', *Solid State Commun.*, **99**:601–605, 1996.

151. C. Benoit á la Guillaume, 'Non-existence of spontaneous symmetry breaking of excitons in multiple-quantum-wells', *Solid State Commun.*, **101**:847, 1997.

152. J. F. Nye, *Physical Properties of Crystals*, Clarendon Press, Oxford, 1957.

153. J. W. Matthews and A. E. Blakeslee, 'Defects in epitaxial multilayers: I. Misfit dislocations', *J. Cryst. Growth*, **27**:118, 1974.

154. V. D. Jovanonić, Z. Ikonić, D. Indjin, P. Harrison, V. Milanović, and R. A. Soref, 'Designing strain-balanced GaN/AlGaN quantum well structures: Application to intersubband devices at 1.3 and 1.55 μm', *J. Appl. Phys.*, **93**:3194, 2003.

155. J Faist, F Capasso, Dl Sivco, Al Hutchinson, Sng Chu, and Ay Cho, 'Short wavelength ($\lambda \sim 3.4 \mu$m) quantum cascade laser based on strained compensated InGaAs/AlInAs', *Appl. Phys. Lett.*, **72**:680–682, 1998.

156. O. Ambacher, J. Majewski, C. Miskys, A. Link, M. Hermann, M. Eickhoff, M. Stutzmann, F. Bernardini, V. Fiorentini, V. Tilak, B. Schaff, and L. F. Eastman, 'Pyroelectric properties of Al(In)GaN/GaN hetero- and quantum well structures', *J. Phys.: Condens. Matter*, **14**:3399, 2002.

157. D. L. Smith and C. Mailhiot, 'Theory of semiconductor superlattice electronic structure', *Rev. Mod. Phys.*, **62**:173, 1990.

158. C. Gmachl, H. M. Ng, and A. Y. Cho, 'Intersubband absorption in GaN/AlGaN multiple quantum wells in the wavelength range of $\lambda \sim$1.75–4.2 μm', *Appl. Phys. Lett.*, **77**:334, 2000.

159. J. S. Im, H. Kollmer, J. Off, A. Sohmer, F. Scholz, and A. Hangleiter, 'Reduction of oscillator strength due to piezoelectric fields in GaN/Al$_x$Ga$_{1-x}$N quantum wells', *Phys. Rev. B*, p. 9435, 1998.

160. S. Gangopadhyay and B. R. Nag, 'Energy levels in finite barrier triangular and arrowhead-shaped quantum wires', *J. Appl. Phys.*, **81**:7885, 1997.

161. I. Kamiya, I. Tanaka, and H. Sakaki, 'Optical properties of near surface-InAs quantum dots and their formation processes', *Physica E*, **2**:637, 1998.

162. E. Palange, G. Capellini, L. Di Gaspare, and F. Evangelisti, 'Atomic force microscopy and photoluminescence study of Ge layers and self-organised Ge quantum dots on Si(100)', *Appl. Phys. Lett.*, **68**:2982, 1996.

163. D. Gershoni, H. Temkin, G. J. Dolan, J. Dunsmuir, S. N. G. Chu, and M. B. Panish, 'Effects of two-dimensional confinement on the optical properties of InGaAs/InP quantum wire structures', *Appl. Phys. Lett.*, **53**:995, 1988.

164. M. Califano and P. Harrison, 'Approximate methods for the solution of quantum wires and dots: connection rules between pyramidal, cuboid and cubic dots', *J. Appl. Phys.*, **86**:5054–5059, 1999.

165. S. Gangopadhyay and B. R. Nag, 'Energy levels in three-dimensional quantum-confinement structures', *Nanotechnology*, **8**:14, 1997.

166. I. N. Stranski and L. Von Krastanov, '', *Akad. Wiss. Lit. Mainz Math.Natur. Kl. IIb*, **146**:797, 1939.

167. M. A. Cusack, P. R. Briddon, and M. Jaros, 'Electronic structure of InAs/GaAs self-assembled quantum dots', *Phys. Rev. B*, **54**:R2300, 1996.

168. M. Califano and P. Harrison, 'Composition, volume and aspect ratio dependence of the strain distribution, band lineups and electron effective masses in self-assembled pyramidal In$_{1-x}$Ga$_x$As/GaAs and Si$_x$Ge$_{1-x}$/Si quantum dots', *J. Appl. Phys.*, **91**:389–398, 2002.

169. M. Califano and P. Harrison, 'Presentation and experimental validation of a single-band, constant-potential model for self-assembled InGa/GaAs quantum dots', *Phys. Rev. B*, **61**:10959–10965, 2000.

170. M. Califano and P. Harrison, 'Quantum box energies as a route to the ground state levels of self-assembled quantum dots', *J. Appl. Phys.*, **88**:5870–5874, 2000.

171. P. Harrison, *Computational Methods in Physics, Chemistry and Mathematical Biology: An Introduction*, John Wiley & Sons, Ltd, Chichester, U. K., 2001.

172. D. El-Moghraby, R. G. Johnson, and P. Harrison, 'Calculating modes of quantum wire and dot systems using a finite differencing technique', *Comp. Phys. Commun.*, **150**:235–246, 2003.

173. D. El-Moghraby, R. G. Johnson, and P. Harrison, 'The effect of inter-dot separation on the finite difference solution of vertically aligned coupled quantum dots', *Comp. Phys. Commun.*, **155**:236–243, 2003.

174. Arvind Kumar, Steven E. Laux, and Frank Stern, 'Electron states in a GaAs quantum dot in a magnetic field', *Phys. Rev. B*, **42**:5166–5175, 1990.

175. D. J. Eaglesham and M. Cerullo, 'Dislocation-free Stranski–Krastanow growth of Ge on Si(100)', *Phys. Rev. Lett.*, **64**:1943–1946, 1990.

176. H. Sunamura, N. Usami, Y. Shiraki, and S. Fukatsu, 'Island formation during growth of Ge on Si(100): a study using photoluminescence spectroscopy', *Appl. Phys. Lett.*, **66**:3024–3026, 1995.

177. M. Grundmann, J. Christen, N. N. Ledentsov, J. Böhrer, D. Bimberg, S. S. Ruvimov, P. Werner, U. Richter, U. Gösele, J. Heydenreich, V. M. Ustinov, A. Yu. Egorov, A. E. Zhukov, P. S. Kop'ev, and Zh. I. Alferov, 'Ultranarrow luminescence lines from single quantum dots', *Phys. Rev. Lett.*, **74**:4043–4046, 1995.

178. J. M. Moison, F. Houzay, F. Barthe, L. Leprince, E. André, and O. Vatel, 'Self-organized growth of regular nanometer-scale InAs dots on GaAs', *Appl. Phys. Lett.*, **64**:196–198, 1994.

179. D. Leonard, K. Pond, and P. M. Petroff, 'Critical layer thickness for self-assembled InAs islands on GaAs', *Phys. Rev. B*, **50**:11687–11692, 1994.

180. M. Fricke, A. Lorke, J.P. Kotthaus, G. Medeiros-Ribeiro, and P.M. Petroff, 'Shell structure and electron–electron interaction in self-assembled InAs quantum dots', *Europhys. Lett.*, **36**:197–202, 1996.

181. S. Sauvage, P. Boucaud, F. H. Julien, J.-M. Gérard, and J.-Y. Marzin, 'Infrared spectroscopy of intraband transitions in self-organized InAs/GaAs quantum dots', *J. Appl. Phys.*, **82**:3396–3401, 1997.

182. N. Liu, J. Tersoff, O. Baklenov, A. L. Holmes, and C. K. Shih, 'Nonuniform composition profile in $In_{0.5}Ga_{0.5}As$ alloy quantum dots', *Phys. Rev. Lett.*, **84**:334–337, 2000.

183. D. Bimberg, M. Grundmann, and N. N. Ledentsov, *Quantum Dot Heterostructures*, Wiley, New York, 1998.

184. J. A. Barker and E. P. O'Reilly, 'Theoretical analysis of electron-hole alignment in InAs-GaAs quantum dots', *Phys. Rev. B*, **61**:13840–13851, 2000.

185. L. R. C. Fonseca, J. L. Jimenez, J. P. Leburton, and Richard M. Martin, 'Self-consistent calculation of the electronic structure and electron–electron interaction in self-assembled InAs-GaAs quantum dot structures', *Phys. Rev. B*, **57**:4017–4026, 1998.

186. M. A. Cusack, P. R. Briddon, and M. Jaros, 'Electronic structure of InAs/GaAs self-assembled quantum dots', *Phys. Rev. B*, **54**:R2300–R2303, 1996.

187. D. Gershoni, H. Temkin, G. J. Dolan, J. Dunsmuir, S. N. G. Chu, and M. B. Panish, 'Effects of two-dimensional confinement on the optical properties of InGaAs/InP quantum wire structures', *Appl. Phys. Lett.*, **53**:995–997, 1988.

188. S. Gangopadhyay and B. R. Nag, 'Energy levels in three-dimensional quantum-confinement structures', *Nanotechnology*, **8**:14–17, 1997.

189. M. Califano and P. Harrison, 'Presentation and experimental validation of a single-band, constant-potential model for self-assembled InAs/GaAs quantum dots', *Phys. Rev. B*, **61**:10959–10965, 2000.

190. I. S. Gradshteyn and I. M. Ryzhik, *Table of Integrals, Series, and Products*, Academic Press, San Diego, London, Fifth edition, 1980.

191. E. Anderson, Z. Bai, C. Bischof, J. Demmel, J. Dongarra, J. Du Croz, A. Greenbaum, S. Hammarling, A. McKenney, S. Ostruchov, and D. Sorensen, *LAPACK Users' Guide*, Society for Industrial and Applied Mathematics, Philadelphia, 1995.

192. M. Grundmann, O. Stier, and D. Bimberg, 'InAs/GaAs pyramidal quantum dots: Strain distribution, optical phonons, and electronic structure', *Phys. Rev. B*, **52**:11969–11981, 1995.

193. L. R. C. Fonseca, J. L. Jimenez, and J. P. Leburton, 'Electronic coupling in InAs/GaAs self-assembled stacked double-quantum-dot systems', *Phys. Rev. B*, **58**:9955–9960, 1998.

194. Al.L. Efros and A.L. Efros, 'Interband absorption of light in a semiconductor sphere', *Sov. Phys. Semicond.*, **16**:772–775, 1982.

195. D. Ninno, M.A. Gell, and M. Jaros, 'Electronic-structure and optical-transitions in GaAs-$Ga_{1-x}Al_xAs(001)$ superlattices', *J. Phys. C: Solid State Physics*, **19**:3845–3853, 1986.

196. O. Stier, M. Grundmann, and D. Bimberg, 'Electronic and optical properties of strained quantum dots modeled by 8-band k·p theory', *Phys. Rev. B*, **59**:5688–5701, 1999.

197. K. H. Schmidt, G. Medeiros-Ribeiro, M. Oestreich, P. M. Petroff, and G. H. Döhler, 'Carrier relaxation and electronic structure in InAs self-assembled quantum dots', *Phys. Rev. B*, **54**:11346–11353, 1996.

198. M. Grundmann, N. N. Ledentsov, O. Stier, D. Bimberg, V. M. Ustinov, P. S. Kop'ev, and Zh. I. Alferov, 'Excited states in self-organized InAs/GaAs quantum dots: theory and experiment', *Appl. Phys. Lett.*, **68**:979–981, 1996.

199. Susumu Noda, Tomoki Abe, and Masatoshi Tamura, 'Mode assignment of excited states in self-assembled InAs/GaAs quantum dots', *Phys. Rev. B*, **58**:7181–7187, 1998.

200. Y. Toda, S. Shinomori, K. Suzuki, and Y. Arakawa, 'Near-field magneto-optical spectroscopy of single self-assembled InAs quantum dots', *Appl. Phys. Lett.*, **73**:517–519, 1998.

201. J.Y. Marzin and G. Bastard, 'Calculation of the energy-levels in InAs/GaAs quantum dots', *Solid State Commun.*, **92**:437–442, 1994.

202. M. Califano, *Development of Computational Models for the Electronic Structure of Self-Assembled Quantum Dots*, Ph.D. thesis, University of Leeds, 2002.

203. M. A. Cusack, P. R. Briddon, and M. Jaros, 'Absorption spectra and optical transitions in InAs/GaAs self-assembled quantum dots', *Phys. Rev. B*, **56**:4047–4050, 1997.

204. N. Nishiguchi and K. Yoh, 'Energy-dependent effective mass approximation in one-dimensional quantum dots', *Jpn. J. Appl. Phys.*, **36**:3928–3931, 1997.

205. A. D. Andreev, J. R. Downes, D. A. Faux, and E. P. O'Reilly, 'Strain distributions in quantum dots of arbitrary shape', *J. Appl. Phys.*, **86**:297–305, 1999.

206. A. J. Williamson, L. W. Wang, and Alex Zunger, 'Theoretical interpretation of the experimental electronic structure of lens-shaped self-assembled InAs/GaAs quantum dots', *Phys. Rev. B*, **62**:12963–12977, 2000.

207. A. Franceschetti, H. Fu, L. W. Wang, and A. Zunger, 'Many-body pseudopotential theory of excitons in InP and CdSe quantum dots', *Phys. Rev. B*, **60**:1819–1829, 1999.

208. R. Loudon, *The Quantum Theory of Light*, Oxford University Press, Oxford, Second edition, 1983.

209. F. W. Sears and G. L. Salinger, *Thermodynamics, Kinetic Theory and Statistical Thermodynamics*, Addison-Wesley, Reading, Massachusetts, Third edition, 1975.

210. B. K. Ridley, *Quantum Processes in Semiconductors*, Clarendon Press, Oxford, Second edition, 1988.

211. M. Lundstrom, *Fundamentals of Carrier Transport*, Modular series on solid state devices. Addison-Wesley, Reading, Wokingham, 1990.

212. B. K. Ridley, *Electrons and Phonons in Semiconductor Multilayers*, Cambridge University Press, Cambridge, 1997.

213. P. Kinsler, 'Private communication', .

214. T. Piorek, *Aspects of low-dimensional diluted semimagnetic structures*, Ph.D. thesis, University of Hull, 1996.

215. S.-H. Park, D. Ahn, and Y.-T. Lee, 'Screening effects on electron-longitudinal optical-phonon intesubband scattering in wide quantum well and comparison with experiment', *Jpn. J. Appl. Phys.*, **39**:6601, 2000.

216. B. K. Ridley, 'Electron-hybridon interaction in a quantum well', *Phys. Rev. B*, **47**:4592, 1993.

217. E. Molinari, C. Bungaro, M. Gulia, P. Lugli, and H. Rücker, 'Electron–phonon interactions in two-dimensional systems: a microscopic approach', *Semicond. Sci. Technol.*, **7**:B67, 1992.

218. P. Kinsler, R. W. Kelsall, and P. Harrison, 'Interface phonons in asymmetric quantum well structures', *Superlatt. Microstruct.*, **25**:163–166, 1999.

219. R. L. Liboff, *Introductory Quantum Mechanics*, Holden-Day, San Francisco, 1980.

220. P. Kinsler, P. Harrison, and R. W. Kelsall, 'Intersubband electron–electron scattering in asymmetric quantum wells designed for far-infrared emission', *Phys. Rev. B*, **58**:4771–4778, 1998.

221. S. M. Goodnick and P. Lugli, 'Effect of electron–electron scattering on non-equilibrium transport in quantum well systems', *Phys. Rev. B*, **37**:2578, 1988.

222. J. H. Smet, C. G. Fonstad, and Q. Hu, 'Intrawell and interwell intersubband transitions in multiple quantum wells for far-infrared sources', *J. Appl. Phys.*, **79**:9305, 1996.

223. J. M. Ziman, *Electrons and Phonons*, Oxford University Press, Oxford, 1960.

224. N. Takenaka, M. Inoue, and Y. Inuishi, 'Influence of inter-carrier scattering on hot electron distribution function in GaAs', *J. Phys. Soc. Japan*, **47**:861, 1979.

225. M. Moško, A. Moškova, and V. Cambel, 'Carrier-carrier scattering in photoexcited intrinsic GaAs quantum wells and its effect on femtosecond plasma thermalisation', *Phys. Rev. B*, **51**:16860, 1995.

226. P. Harrison and R. W. Kelsall, 'The relative importance of electron–electron and electron–phonon scattering in terahertz quantum cascade lasers', *Solid State Electr.*, **42**:1449–1451, 1998.

227. C. Sirtori, J. Faist, F. Capasso, D. L. Sivco, A. L. Hutchinson, and A. Y. Cho, 'Long wavelength infrared ($\lambda \approx 11$ μm) quantum cascade lasers', *Appl. Phys. Lett.*, **69**:2810, 1996.

228. T. Ando, A. B. Fowler, and F. Stern, 'Electronic properties of two-dimensional systems', *Rev. Mod. Phys.*, **54**:437, 1982.

229. P. F. Maldague, 'Many-body corrections to the polarizability of the two-dimensional electron gas', *Surf. Sci.*, **73**:296, 1978.

230. P. Kinsler, P. Harrison, and R. W. Kelsall, 'Intersubband terahertz lasers using four-level asymmetric quantum wells', *J. Appl. Phys.*, **85**:23–28, 1999.

231. K. Donovan, P. Harrison, and R. W. Kelsall, 'Stark ladders as tunable far-infrared emitters', *J. Appl. Phys.*, **84**:5175–5179, 1998.

232. P. S. Zory, *Quantum Well Lasers*, Academic Press, Boston, 1993.

233. D. D. Coon and R. P. G. Karunasiri, 'New mode of infrared detection using quantum wells', *Appl. Phys. Lett.*, **45**:649, 1984.

234. W. T. Tsang, Ed., *Lightwave Communications Technology*, vol. 22 of *Semiconductors and Semimetals*, Academic Press, Orlando, 1985.

235. P. N. J. Dennis, *Photodetectors: An Introduction to Current Technology*, Updates in applied physics and electrical technology. Plenum Press, New York, 1986.

236. M. A. Trishenkov, *Detection of Low-Level Optical Signals : Photodetectors, Focal Plane Arrays and Systems*, vol. 4 of *Solid-state Science and Technology Library*, Kluwer Academic Publisher, Boston, 1997.

237. A. N. Baranov, V. V. Sherstnev, C. Alibert, and A. Krier, 'New III-V semiconductor lasers emitting near 2.6 μm', *J. Appl. Phys.*, **79**:3354, 1996.

238. R. L. Gunshor and A. V. Nurmikko, Eds., *II-VI Blue/Green Light Emitters: Device Physics and Epitaxial Growth*, Academic Press, San Diego, 1997.

239. E. L. Ivchenko and G. Pikus, *Superlattices and other Heterostructures: Symmetry and Optical Phenomena*, Springer-Verlag, Berlin, Second edition, 1995.

240. B. K. Ridley, *Quantum Processes in Semiconductors*, Clarendon, Oxford, Third edition, 1993.

241. L. C. West and S. J. Eglash, 'First observation of an extremely large-dipole infrared transition within the conduction band of a GaAs quantum well', *Appl. Phys. Lett.*, **46**:1156, 1985.

242. D. Kaufman, A. Sa'ar, and N. Kuze, 'Anisotropy, birefringence, and optical-phase retardation related to intersubband transitions in multiple-quantum-well structures', *Appl. Phys. Lett.*, **64**:2543, 1994.

243. V. Berger, 'Three-level laser based on intersubband transitions in asymmetric quantum wells: a theoretical study', *Semicond. Sci. Technol.*, **9**:1493, 1994.

244. F. H. Julien, Z. Moussa, P. Boucaud, Y. Lavon, A. Sa'ar, J. Wang, J. P. Leburton, V. Berger, J. Nagle, and R. Planel, 'Intersubband mid-infrared emission in optically pumped quantum wells', *Superlatt. Microstruct.*, **19**:69, 1996.

245. J. Katz, Y. Zhang, and W. I. Wang, 'Normal incidence infra-red absorption from intersubband transitions in *p*-type GaInAs/AlInAs quantum wells', *Electr. Lett.*, **28**:932, 1992.

246. W. Batty and K. A. Shore, 'Normal-incidence TE inter-subband transitions', *IEE Proc. Optoelectron.*, **145**:21, 1998.

247. M. J. Burt, 'The evaluation of the matrix element for interband optical-transitions in quantum-wells using envelope functions', *J. Phys.:Cond. Matter*, **5**:4091, 1993.

248. D. T. F. Marple, 'Refractive index of ZnSe, ZnTe, and CdTe', *J. Appl. Phys.*, **35**:539, 1964.

249. B. O. Seraphin and H. E. Bennett, *Semiconductors and Semimetals*, vol. 3, p. 499, Academic, New York, 1967.

250. P. Harrison, R. W. Kelsall, P. Kinsler, and K. Donovan, 'Quantum well intersubband transitions as a source of terahertz radiation', in *1998 IEEE Sixth International Conference on Terahertz Electronics Proceedings*, P. Harrison, Ed., 1998, pp. 74–78.

251. U. Bockelmann and G. Bastard, 'Phonon scattering and energy relaxation in two, one and zero-dimensional electron gases', *Phys. Rev. B*, **42**:8947, 1990.

252. L. Zheng and S. DasSarma, 'Inelastic lifetimes of confined two-component electron systems in semiconductor quantum-wire and quantum-well structures', *Phys. Rev. B*, **54**:13908, 1996.

253. E. H. Hwang and S. DasSarma, 'Electron–phonon and electron–electron interactions in one-dimensional GaAs quantum wire nanostructures', *Superlatt. Microstruct.*, **21**:1, 1997.

254. C. R. Bennett and B. Tanatar, 'Energy relaxation via confined and interface phonons in quantum-wire systems', *Phys. Rev. B*, **55**:7165, 1997.

255. M. Brasken, M. Lindberg, and J. Tulkki, 'Carrier dynamics in strain-induced quantum dots', *Physica Status Solidi A*, **164**:427, 1997.

256. K. Kral and Z. Khas, 'Electron self-energy in quantum dots', *Phys. Rev. B*, **57**:R2061, 1998.

257. J. Faist, F. Capasso, D. L. Sivco, C. Sirtori, A. L. Hutchinson, and A. Y. Cho, 'Quantum cascade laser', *Science*, **264**:553, 1994.

258. C. Sirtori, J. Faist, F. Capasso, D. L. Sivco, A. L. Hutchinson, and A. Y. Cho, 'Mid-infrared (8.5 μm) semiconductor lasers operating at room temperature', *IEEE Photon. Tech. Lett.*, **9**:294–296, 1997.

259. J. Devenson, O. Cathabard, R. Teissier, and AN Baranov, 'InAs/ AlSb quantum cascade lasers emitting at 2.75–2.97 μm', *Appl. Phys. Lett.*, **91**:251102, 2007.

260. S.Y. Zhang, D.G. Revin, J.W. Cockburn, K. Kennedy, A.B. Krysa, and M. Hopkinson, 'λ 3.1 μm room temperature InGaAs/AlAsSb/InP quantum cascade lasers', *Appl. Phys. Lett.*, **94**:031106, 2009.

261. H. Page, C. Backer, A. Roberston, G. Glastre, V. Ortiz, and C. Sirtori, '300 K operation of a GaAs-based quantum-cascade laser at λ approximate to 9 μm', *Appl. Phys. Lett.*, **78**:3529, 2001.

262. C. Pflügl, W. Schrenk, S. Andres, G. Strasser, C. Becker, C. Sirtori, Y. Bonetti, and A. Muller, ', *Appl. Phys. Lett.*, **83**:4698, 2003.

263. J. Heinrich, R. Langhans, M. S. Vitiello, G. Scamarcio, D. Indjin, C. A. Evans, Z. Ikonič, P. Harrison, S. Höfling, and A. Forchel, 'Wide wavelength tuning of GaAs/Al$_x$Ga$_{1-x}$As bound-to-continuum quantum cascade lasers by aluminum content control', *Appl. Phys. Lett.*, **92**:141111, 2008.

264. S. Andres, W. Schrenk, E. Gornik, and G. Strasser, 'A physical model of quantum cascade lasers', *Appl. Phys. Lett.*, **80**:1864, 2002.

265. H. Page, S. Dhillon, M. Calligaro, C. Becker, V. Ortiz, and C. Sirtori, 'Proton-implanted shallow-ridge quantum-cascade laser', *IEEE J. Quant. Electr.*, **40**:665, 2004.

266. R. Köhler, A. Tredicucci, F. Beltram, H. E. Beere, E. H. Linfield, A. G. Davies, D. A. Ritchie, R. C. Iotti, and F. Rossi, 'Terahertz semiconductor heterostructure laser', *Nature*, **417**:156, 2002.

267. S. Kumar, B. S. Williams, Q. Hu, and J. L. Reno, '1.9 THz quantum cascade laser with one-well injector', *Appl. Phys. Lett.*, **88**:121123, 2006.

268. C. Worrall, J. Alton, M. Houghton, S. Barbieri, H. E. Beere, D. Ritchie, and C. Sirtori, 'Continuous wave operation of a superlattice quantum cascade laser emitting at 2 THz', *Optics Express*, **14**:171, 2006.

269. A. Wade, G. Fedorov, D. Smirnov, S. Kumar, BS Williams, Q. Hu, and JL Reno, 'Magnetic-field-assisted terahertz quantum cascade laser operating up to 225 K', *Nature Photonics*, **3**:41, 2008.

270. P. Harrison, 'The nature of the electron distribution functions in quantum cascade lasers', *Appl. Phys. Lett.*, **75**:2800–2802, 1999.

271. C. Sirtori, P. Kruck, S. Barbieri, P. Collot, J. Nagle, M. Beck, J. Faist, and U. Oesterle, 'GaAs/Al$_x$Ga$_{1-x}$As quantum cascade lasers', *Appl. Phys. Lett.*, **73**:3486, 1998.

272. V. D. Jovanović, S. Höfling, D. Indjin, N. Vukmirović, Z. Ikonić, P. Harrison, J. P. Reithmaier, and A. Forchel, 'Influence of doping density on electron dynamics in GaAs/AlGaAs quantum cascade lasers', *J. Appl. Phys.*, **99**:103106, 2006.

273. A. Valavanis, L. Lever, C. A. Evans, Z. Ikonić, and R. W. Kelsall, 'Theory and design of quantum cascade lasers in (111) n-type Si/SiGe', *Phys. Rev. B*, **78**(3):035420, 2008.

274. D. Indjin, P. Harrison, R. W. Kelsall, and Z. Ikonic, 'Influence of leakage current on temperature performance of GaAs/AlGaAs quantum cascade lasers', *Appl. Phys. Lett.*, **81**:400–402, 2002.

275. D. Indjin, S. Tomić, Z. Ikonić, P. Harrison, R. W. Kelsall, V. Milanović, and S. Kóćinac, 'Gain-maximised GaAs/AlGaAs quantum cascade laser with digitally graded active region', *Appl. Phys. Lett.*, **81**:2163–2165, 2002.

276. D. Indjin, P. Harrison, R. W. Kelsall, and Z. Ikonic, 'Self-consistent scattering theory of transport and output characteristics of quantum cascade lasers', *J. Appl. Phys.*, **91**:9019–9026, 2002.

277. D. Indjin, P. Harrison, R. W. Kelsall, and Z. Ikonić, 'Mechanisms of temperature performance degradation in terahertz quantum-cascade lasers', *Appl. Phys. Lett.*, **82**:1347–1349, 2003.

278. D. Indjin, P. Harrison, R. W. Kelsall, and Z. Ikonić, 'Self-consistent scattering model of carrier dynamics in GaAs-AlGaAs Terahertz quantum cascade lasers', *IEEE Photonics Tech. Lett.*, **15**:15–17, 2003.

279. D. Indjin, Z. Ikonić, V. D. Jovanović, P. Harrison, and R. W. Kelsall, 'Mechanisms of carrier transport and temperature performance evaluation in terahertz quantum cascade lasers', *Semicond. Sci. Technol.*, **19**:S104–S106, 2004.

280. CA Evans, VD Jovanovic, D. Indjin, Z. Ikonic, and P. Harrison, 'Thermal effects in InGaAs/AlAsSb quantum-cascade lasers', *Optoelectr., IEE Proc.*, **153**(6):287–292, 2006.

281. C. A. Evans, D. Indjin, Z. Ikonić, P. Harrison, M. S. Vitiello, V. Spagnolo, and G. Scamarcio, 'Thermal modeling terahertz quantum-cascade lasers: Comparison of optical waveguides', *IEEE J. Quant. Electr.*, **44**:680, 2008.

282. V. Spagnolo, G. Scamarcio, H. Page, and C. Sirtori, 'Simultaneous measurement of the electronic and lattice temperatures in GaAs/AlGaAs quantum-cascade lasers: Influence on the optical performance', *Appl. Phys. Lett.*, **84**:3690, 2004.

283. V. Spagnolo, G. Scamarcio, H. Page, and C. Sirtori, 'Influence of the band-offset on the electronic temperature of GaAs/Al(Ga)As superlattice quantum cascade lasers', *Semicond. Sci. Technol*, **19**:S110, 2004.

284. V. D. Jovanović, D. indjin, N. Vukmirović, Z. Ikonić, P. Harrison, E. H. Linfield, H. Page, X. Marcadet, C. Sirtori, C. Worrall, H. E. Beere, and D. A. Ritchie, 'Mechanisms of dynamic range limitations in GaAs/AlGaAs quantum-cascade lasers: Influence of injector doping', *Appl. Phys. Lett.*, **86**:211117, 2005.

285. S. Höfling, VD Jovanović, D. Indjin, JP Reithmaier, A. Forchel, Z. Ikonić, N. Vukmirović, P. Harrison, A. Mirčetić, and V. Milanović, 'Dependence of saturation effects on electron confinement and injector doping in GaAs/ AlGaAs quantum-cascade lasers', *Appl. Phys. Lett.*, **88**:251109, 2006.

286. K. Ohtani, Y. Moriyasu, H. Ohnishi, and H. Ohno, 'Above room-temperature operation of InAs/ AlGaSb superlattice quantum cascade lasers emitting at 12 μm', *Appl. Phys. Lett.*, **90**:261112, 2007.

287. E. Mujagić, M. Austerer, S. Schartner, M. Nobile, L.K. Hoffmann, W. Schrenk, G. Strasser, M.P. Semtsiv, I. Bayrakli, M. Wienold, et al., 'Impact of doping on the performance of short-wavelength InP-based quantum-cascade lasers', *J. Appl. Phys.*, **103**:033104, 2008.

288. D. Indjin, P. Harrison, and R. W. Kelsall, 'Self-consistent scattering theory evaluation of thermionic emission and leakage currents in III-V quantum cascade lasers', *Russ. Acad. Sci. J. Phys.*, **67**:259–261, 2003.

289. V. D. Jovanović, P. Harrison, Z. Ikonić, and D. Indjin, 'Physical model of quantum-well infrared photodetectors', *J. Appl. Phys.*, **96**:269–272, 2004.

290. N. Vukmirović, Z. Ikonić, I. Savić, D. Indjin, and P. Harrison, 'A microscopic model of elcetron transport in quantum dot infrared photodetectors', *J. Appl. Phys.*, **100**:074502, 2006.

291. N. Vukmirović, D. Indjin, Z. Ikonić, and P. Harrison, 'Origin of detection wavelength tuning in quantum dots-in-a-well infrared photodetectors', *Appl. Phys. Lett.*, **88**:251107, 2006.

292. S. Barik, H. H. Tan, C. Jagadish, N. Vukmirović, and P. Harrison, 'Selective wavelength tuning of self-assembled InAs quantum dots grown on InP', *Appl. Phys. Lett.*, **88**:193112, 2006.

293. L. Fu, H. H. Tan, I. McKerracher, J. Wong-Leung, C. Jagadish, N. Vukmirović, and P. Harrison, 'Effects of rapid thermal annealing on device characteristics of InGaAs/GaAs quantum dot infrared photodetectors', *J. Appl. Phys.*, **99**:114517, 2006.

294. L. Fu, I. McKerracher, HH Tan, C. Jagadish, N. Vukmirović, and P. Harrison, 'Effect of GaP strain compensation layers on rapid thermally annealed InGaAs/ GaAs quantum dot infrared photodetectors grown by metal-organic chemical-vapor deposition', *Appl. Phys. Lett.*, **91**:073515, 2007.

295. P. Harrison, Z. Ikonić, N. Vukmirović, D. Indjin, R. W. Kelsall, and V. D. Jovanović, 'On the incoherence of quantum transport in semiconductor heterostructure optoelectronic devices', in *Proceedings of the 10th Biennial Baltic Electronics Conference*, 2006, p. 11.

296. P. Harrison, D. Indjin, I. Savić, Z. Ikonić, C. A. Evans, N. Vukmirović, R. W. Kelsall, J. McTavish, V. D. Jovanović, and V. Milanović, 'On the coherence/incoherence of electron transport in semiconductor heterostructure optoelectronic devices', *Proc. SPIE*, **6909**:690912, 2008.

297. R. C. Iotti and F. Rossi, 'Carrier thermalisation versus phonon-assisted relaxation in quantum cascade lasers: a Monte Carlo approach', *Appl. Phys. Lett.*, **78**:2902, 2001.

298. I. Savić, N. Vukmirović, Z. Ikonić, D. Indjin, R.W. Kelsall, P. Harrison, and V. Milanović, 'Density matrix theory of transport and gain in quantum cascade lasers in a magnetic field', *Phys Rev B*, **76**:165310, 2007.

299. H.C. Liu and F. Capasso, Eds., *Intersubband Transitions in Quantum Wells: Physics and Device Application 1; Semiconductors and semimetals*, vol. 62, Academic Press, 2000.

300. R. Paiella, *Intersubband Transitions in Quantum Structures*, McGraw-Hill, New York, 2006.

301. Gerald Bastard, *Wave Mechanics Applied to Semiconductor Heterostructures*, John Wiley & Sons, Ltd, New York, 1990.

302. Shun Lien Chuang, *Physics of Optoelectronic Devices*, John Wiley & Sons, Ltd, New York, 1995.

303. Z. Ikonić, V. Milanović, and D. Tjapkin, 'Bound-free intraband absorption in GaAs-$Al_xGa_{1-x}As$ semiconductor quantum wells', *Appl. Phys. Lett.*, **54**(3):247–249, 1989.

304. Y. Fu and M. Willander, *Physical Models of Semiconductor Quantum Devices*, Kluwer Academic Publisher, Norwell, Massachusetts, 1999.

305. H. C. Liu, 'Dependence of absorption spectrum and responsivity on the upper state position in quantum well intersubband photodetectors', *J. Appl. Phys.*, **73**(6):3062–3067, 1993.

306. M. K. Gurnick and T. A. DeTemple, 'Synthetic Non-Linear Semiconductors', *IEEE J. Quant. Electr.*, **19**(5):791–794, 1983.

307. P. Boucaud, F. H. Julien, D. D. Yang, J-M. Lourtioz, E. Rosencher, P. Bois, and J. Nagle, 'Detailed analysis of second-harmonic generation near 10.6 μm in GaAs/AlGaAs asymmetric quantum wells', *Appl. Phys. Lett.*, **57**(3):215–217, 1990.

308. Carlo Sirtori, Federico Capasso, Deborah L. Sivco, A. L. Hutchinson, and Alfred Y. Cho, 'Resonant Stark tuning of second-order susceptibility in coupled quantum wells', *Appl. Phys. Lett.*, **60**(2):151–153, 1992.

309. Chenhsin Lien, Yimin Huang, and Tan-Fu Lei, 'The double resonant enhancement of optical second harmonic susceptibility in the compositionally asymmetric coupled quantum well', *J. Appl. Phys.*, **75**(4):2177–2183, 1994.

310. Dongxia Qu, Feng Xie, Gary Shu, Safiyy Momen, Evgenii Narimanov, Claire F. Gmachl, and Deborah L. Sivco, 'Second-harmonic generation in quantum cascade lasers with electric field and current dependent nonlinear susceptibility', *Appl. Phys. Lett.*, **90**(3):031105, 2007.

311. E. Rosencher and Ph. Bois, 'Model system for optical nonlinearities: asymmetric quantum wells', *Phys. Rev. B*, **44**(20):11315–11327, 1991.

312. D. Indjin, V. Milanović, and Z. Ikonić, 'Application of Bragg-confined semiconductor structures for higher-energy resonant intersubband second-harmonic generation', *Phys. Rev. B*, **55**(15):9722–9730, 1997.

313. J. Radovanović, G. Todorović, V. Milanović, Z. Ikonić, and D. Indjin, 'Two methods of quantum well profile optimization for maximal nonlinear optical susceptibilities', *Phys. Rev. B*, **63**(11):115327, 2001.

314. F. Capasso, C. Sirtori, and A.Y. Cho, 'Coupled-quantum-well semiconductors with giant electric-field tunable nonlinear-optical properties in the infrared', *IEEE J. Quant. Electr.*, **30**(5):1313–1326, 1994.

315. D. Indjin, Z. Ikonić, V. Milanović, and J. Radovanović, 'Optimization of resonant second- and third-order nonlinearities in step and continuously graded semiconductor quantum wells', *IEEE J. Quant. Electr.*, **34**(5):795–802, 1998.

316. Robert W. Boyd, *Nonlinear Optics*, Academic Press, San Diego, 2003.

317. C. Pflügl, W. Schrenk, S. Anders, G. Strasser, C. Becker, C. Sirtori, Y. Bonetti, and A. Muller, 'High-temperature performance of GaAs-based bound-to-continuum quantum-cascade lasers', *Appl. Phys. Lett.*, **83**(23):4698–4700, 2003.

318. J. Heinrich, R. Langhans, M. S. Vitiello, G. Scamarcio, D. Indjin, C. A. Evans, Z. Ikonić, P. Harrison, S. Höfling, and A. Forchel, 'Wide wavelength tuning of GaAs/$Al_xGa_{1-x}As$ bound-to-continuum quantum cascade lasers by aluminum content control', *Appl. Phys. Lett.*, **92**(14):141111, 2008.

319. B. E. A. Saleh and M. C. Teich, *Fundamentals of Photonics*, John Wiley & Sons, Ltd., 1991.

320. G. Lifante, *Integrated Photonics Fundamentals*, John Wiley & Sons, Ltd., 2003.

321. F. Graham Smith, T. A. King, and D. Wilkins, *Optics and Photonics: An Introduction*, John Wiley & Sons, Ltd., 2nd edition, 2007.

322. E. Anemogiannis and E. N. Glytsis, 'Multilayer waveguides: efficient numerical analysis of general structures', *IEEE J. Light. Technol.*, **10**(10):1344–1351, October 1992.

323. J. Stiens, R. Vounckx, and I. Veretennicoff, 'Slab plasmon polaritons and waveguide modes in four-layer resonant semiconductor waveguides', *J. Appl. Phys.*, **81**(1):1–10, January 1997.

324. E. Anemogiannis, E. N. Glytsis, and T. K. Gaylord, 'Determination of guided and leaky modes in lossless and lossy planar multilayer optical waveguides: reflection pole method and wavevector density method', *IEEE J. Light. Technol.*, **17**(5):929–941, May 1999.

325. E. D. Palik, Ed., *Handbook of Optical Constant of Solids*, vol. 2, Elsevier, 1998.

326. S. Adachi, *GaAs and Related Materials: Bulk Semiconducting and Superlattice Properties*, World Scientific, 1994.

327. M. A. Ordal, R. J. Bell, R. W. Alexander Jr., L. L. Long, and M. R. Querry, 'Optical properties of fourteen metals in the infrared and far infrared: Al, Co, Cu, Au, Fe, Pb, Mo, Ni, Pd, Pt, Ag, Ti, V, and W', *Appl. Opt.*, **24**(24):4493, December 1985.

328. D. Caughey and R. Thomas, 'Carrier mobilities in silicon empirically related to doping and field', *Proc. IEEE*, **52**:2192, 1967.

329. V. Palankovski, *Simulation of Heterojunction Bipolar Transistors*, Ph.D. thesis, Technischen Universität Wien, 2000.

330. S. M. Sze, *Physics of Semiconductor Devices*, John Wiley & Sons, Ltd., 2nd edition, 1981.

331. E. D. Palik, Ed., *Handbook of Optical Constants of Solids*, vol. 2, chapter 24, p. 513, Elsevier, 1998.

332. V. M. Ustinov, A. E. Zhukov, A. Y. Egorov, and N. A. Maleev, *Quantum Dot Lasers*, Oxford University Press, 2003.

333. J. Faist, F. Capasso, D. L. Sivco, C. Sirtori, A. L. Hutchinson, and A. Y. Cho, 'Quantum cascade laser', *Science*, **264**:553, 1994.

334. C. Sirtori, P. Kruck, S. Barbieri, P. Collot, J. Nagle, M. Beck, J. Faist, and U. Oesterle, 'GaAs/Al$_x$Ga$_{1-x}$As quantum cascade lasers', *Appl. Phys. Lett.*, **73**:3486, 1998.

335. R. Köhler, A. Tredicucci, F. Beltram, H. E. Beere, E. H. Linfield, A. G. Davies, D. A. Ritchie, R. C. Iotti, and F. Rossi, 'Terahertz semiconductor heterostructure laser', *Nature*, **417**(156):156–159, 2002.

336. C. Sirtori, P. Kruck, S. Barbieri, H. Page, and J. Nagle, 'Low-loss Al-free waveguides for unipolar semiconductor lasers', *Appl. Phys. Lett.*, **75**(25):3911–3913, 1999.

337. T. D. Visser, B. Demeulenaere, J. Haes, D. Lenstra, R. Baets, and H. Blok, 'Confinement and modal gain in dielectric waveguides', *J. Lightwave Technol.*, **14**:885, 1996.

338. C. Sirtori, C. Gmachl, F. Capasso, J. Faist, D. L. Sivco, A. L. Hutchinson, and A. Y. Cho, 'Long-wavelength ($\lambda \approx 8$–11.5 μm) semiconductor lasers with waveguides based on surface plasmons', *Opt. Lett.*, **23**(17):1366–1368, 1998.

339. R. Colombelli, F. Capasso, C. Gmachl, A. L. Hutchinson, D. L. Sivco, A. Tredicucci, M. C. Wanke, A. M. Sergent, and A. Y. Cho, 'Far-infrared surface-plasmon quantum-cascade lasers at 21.5 μm and 24 μm wavelengths', *Appl. Phys. Lett.*, **78**(18):2620–2622, 2001.

340. J. M. Luttinger and W. Kohn, 'Motion of electrons and holes in perturbed periodic fields', *Phys. Rev.*, **97**:869, 1955.

341. C. R. Pidgeon and R. N. Brown, 'Interband magneto-absorption and Faraday rotation in InSb', *Phys. Rev.*, **146**:575, 1966.

342. B. A. Foreman, 'Effective-mass Hamiltonian and boundary conditions for the valence bands of semiconductor microstructures', *Phys. Rev. B*, **48**:4964, 1993.

343. I. Vurgaftman, J. R. Meyer, and L. R. Ram-Mohan, 'Band parameters for III-V compound semiconductors and their alloys', *J. Appl. Phys.*, **89**:5815, 2001.

344. A. Kahan, M. Chi, and L. Friedman, ', *J. Appl. Phys.*, **75**:8012, 1994.

345. C. Y. P. Chao and S. L. Chuang, 'Spin-orbit-coupling effects on the valence-band structure of strained semiconductor quantum wells', *Phys. Rev. B*, **46**:4110, 1992.

346. B. Chen, M. Lazzouni, and L. R. Ram-Mohan, 'Diagonal representation for the transfer matrix method for obtaining electronic energy levels in layered semiconductor heterostructures', *Phys. Rev. B*, **45**:1204, 1992.

347. E. Anderson, Z. Bai, C. Bischof, J. Demmel, J. Dongarra, J. Du Croz, A. Greenbaum, S. Hammarling, A. McKenney, S. Ostrouchov, and D. Sorensen, *LAPACK Users' Guide*, Society for Industrial and Applied Mathematics, Philadelphia, Second edition, 1995.

348. W. A. Harrison, *Pseudopotentials in the Theory of Metals*, W. A. Benjamin, New York, 1966.

349. M. Jaros, *Deep Levels in Semiconductors*, Hilger, Bristol, 1982.

350. M. L.Cohen and T. K. Bergstresser, 'Band structures and pseudopotential form factors for fourteen semiconductors of the diamond and zinc-blende structures', *Phys. Rev.*, **141**:789, 1966.

351. G. Weisz, 'Band structure and Fermi surface of white tin', *Phys. Rev.*, **149**:504, 1966.

352. S. Bloom and T. K. Bergstresser, 'Band structure of α-Sn, InSb and CdTe including spin-orbit effects', *Solid State Commun.*, **6**:465, 1970.

353. S. Adachi, *Physical Properties of III-V Semiconductor Compounds: InP, InAs, GaAs, GaP, InGaAs, and InGaAsP*, John Wiley & Sons, Ltd., New York, 1992.

354. L. A. Coldren and S. W. Corzine, *Diode Laser and Photonic Integrated Circuits*, John Wiley & Sons, Ltd., New York, 1995.

355. S. Nakamura, M. Senoh, N. Iwasa, S. Nagahama, T. Yamada, and T. Mukai, 'Superbright green InGaN single-quantum-well-structure light-emitting-diodes', *Jpn. J. Appl. Phys. Part 2*, **34**:L1332, 1995.

356. S. Nakamura, M. Senoh, S. Nagahama, N. Iwasa, T. Yamada, T. Matsushita, Y. Sugimoto, and H. Kiyoku, 'Continuous-wave operation of InGaN multi-quantum-well-structure laser diodes at 233 K', *Appl. Phys. Lett.*, **69**:3034, 1996.

357. S. Nakamura, 'Characteristics of room temperature-CW operated InGaN multi-quantum-well-structure laser diodes', *Materials Research Society J. of Nitride Semicond. Research*, **2**:5, 1997.

358. Y. C. Yeo, T. C. Chong, and M. F. Li, 'Electronic band structures and effective mass parameters of wurtzite GaN and InN', *J. Appl. Phys.*, **83**:1429, 1998.

359. S. C. Jain, *Germanium-Silicon Strained Layers and Heterostructures*, Advances in electronics and electron physics. Academic Press, Boston, 1994.

360. An-Ban Chen and Arden Sher, *Semiconductor Alloys : Physics and Materials Engineering*, Plenum, New York, 1995.

361. G. P. Strivastava, J. L. Martins, and A. Zunger, 'Atomic structure and ordering in semiconductor alloys', *Phys. Rev. B*, **31**:2561, 1985.

362. S.-H. Wei and A. Zunger, 'Band gap narrowing in ordered and disordered semiconductor alloys', *Appl. Phys. Lett.*, **56**:662, 1990.

363. S.-H. Wei and A. Zunger, 'Optical properties of zinc blende alloys: effects of epitaxial strain and atomic ordering', *Phys. Rev. B*, **49**:14337, 1994.

364. J. P. Walter and M. L. Cohen, 'Calculated and Measured Reflectivity of ZnTe and ZnSe', *Phys. Rev. B*, **1**:2661, 1970.

365. J. R. Chelikowsky and M. L. Cohen, 'Nonlocal pseudopotential calculations for the electronic structure of eleven zinc-blende semiconductors', *Phys. Rev. B*, **14**:556, 1976.

366. S. Bloom and T. K. Bergstresser, 'Band structure of α-Sn, InSb and CdTe including spin-orbit effects', *Solid State Commun.*, **6**:465, 1968.

367. K. A. Mäder and A. Zunger, 'Empirical atomic pseudopotentials for AlAs/GaAs superlattices, alloys, and nanostructures', *Phs. Rev. B*, **50**:17393, 1994.

368. F. Herman and S. Skillman, *Atomic Structure Calculations*, Prentice-Hall, Englewood Cliffs, N.J., 1963.

369. P. Friedel, M. S. Hybertsen, and M. Schlüter, 'Local empirical pseudopotential approach to the optical properties of Si/Ge superlattices', *Phys. Rev. B*, **39**:7974, 1989.

370. T. Mattila, L.-W. Wang, and A. Zunger, 'Electronic consequences of lateral modulation in semiconductor alloys', *Phys. Rev. B*, 1999, BY6515 accepted.

371. B. S. Meyerson, 'High-speed Silicon-Germanium electronics', *Sci. Am.*, **3**:42, 1994.

372. I. Morrison, M. Jaros, and K. B. Wong, 'Stain-induced confinement in $Si_{0.75}Ge_{0.25}$ $(Si/Si_{0.5}Ge_{0.5})$ (001) superlattice systems', *Phys. Rev. B*, **35**:9693, 1987.

373. G. C. Osbourn, 'Strained-layer superlattices from lattice mismatched materials', *J. Appl. Phys.*, **53**:1586, 1982.

374. J. C. Bean, L. C. Feldman, A. T. Fiory, S. Nakahara, and I. K. Robinson, '$Ge_x Si_{1-x}$/Si strained layer superlattices grown by molecular beam epitaxy', *J. Vac. Sci. Technol. A*, **2**:436, 1984.

375. E. Kasper, H. J. Herzog, and F. Schaffler, 'Si/Ge multilayer structures', in *Physics, Fabrication and Applications of Multilayer Structures*, P. Dhez and C. Weisbuch, Eds., New York, 1988, NATO ASI, vol. 182 of *B, Physics*, p. 229, Plenum.

376. I. J. Fritz, S. T. Picraux, L. R. Dawson, T. J. Drummond, W. D. Laidig, and N. G. Anderson, 'Dependence of critical layer thickness on strain for $In_x Ga_{1-x}$As/GaAs strained layer superlattices', *Appl. Phys. Lett.*, **46**:967, 1985.

377. R. L. Ross, S. P. Svensson, and P. Lugli, Eds., *Pseudomorphic HEMT Technology and Applications*, vol. 309 of *NATO ASI, E*, Kluwer Academic, Dordrecht ; London, 1996.

378. M. A. Gell, D. Ninno, M. Jaros, and D. C. Herbert, 'Zone folding, morphogenesis of charge densities, and the role of periodicity in GaAs-$Al_x Ga_{1-x}$As (001) superlattices', *Phys. Rev. B*, **34**:2416, 1986.

379. S. Tarucha, D. G. Austing, T. Honda, R. van der Hage, and L. P. Kouwenhoven, 'Atomic-like properties of semiconductor quantum dots', *Jpn. J. Appl. Phys.*, **36**:3917, 1997.

380. P. Harrison, 'Proposal for neutral donors in quantum wells to act as charge storage centres for room temperature single electron memories', in *Proceedings of the 6th International Symposium on Nanostructures: Physics and Technology*, 1998, pp. 237–240.

381. M. P. Halsall, P. Harrison, H. Pellemans, and C. R. Pidgeon, 'Free-electron laser studies of intra-acceptor transitions in GaAs: a potential far-infrared emission system', in *Terahertz Spectroscopy and Applications*, J. M. Chamberlain, Ed. SPIE, 1999, vol. 3617, pp. 171–176.

382. R. J. Turton and M. Jaros, 'Effects of interfacial ordering on the optical properties of Si-Ge superlattices', *Semicond. Sci. Technol.*, **8**:2003, 1993.

383. L.-W. Wang and A. Zunger, 'Solving Schrödinger's equation around a desired energy: Application to silicon quantum dots', *J. Chem. Phys.*, **100**:2394, 1994.

384. L.-W. Wang, A. Franceschetti, and A. Zunger, 'Million-atom pseudopotential calculation of Γ-X mixing in GaAs/AlAs superlattices and quantum dots', *Phys. Rev. Lett.*, **78**:2819, 1997.

385. L. W. Wang, J. N. Kim, and A. Zunger, 'Electronic structures of [110]-faceted self-assembled pyramidal InAs/GaAs quantum dots', *Phys. Rev. B*, **59**, 1999.

386. H. Ueno, K. Moriyasu, Y. Wada, S. Osako, H. Kubo, N. Mori, and C. Hamaguchi, 'Conductance through laterally coupled quantum dots', *Jpn. J. Appl. Phys. Part 1*, **38**:332, 1999.

387. T. Ando, S. Uryu, S. Ishizaka, and T. Nakanishi, 'Quantum transport in anti-dot lattices', *Chaos, Solitons and Fractals*, **8**:1057, 1997.

388. V. Y. Demikhovskii and A. A. Perov, 'Electron states in quantum-dot and antidot arrays placed in a strong magnetic field', *Physics of the Solid State*, **6**:1035, 1998.

389. P. D. Tougaw and C. S. Lent, 'Dynamic behavior of quantum cellular automata', *J. Appl. Phys.*, **80**:4722, 1996.

390. C. K. Wang, I.I. Yakimenko, I. V. Zozoulenko, and K. F. Berggren, 'Dynamical response in an array of quantum-dot cells', *J. Appl. Phys.*, **84**:2684, 1998.

391. M. Hirasawa, S. Katsumoto, A. Endo, and Y. Iye, 'Coulomb blockade in arrays of quantum dots', *Physica B*, **251**:252, 1998.

INDEX